T0265149

CHAPMAN & HALL/CRC
Monographs and Surveys in
Pure and Applied Mathematics   127

CANONICAL PROBLEMS

IN SCATTERING AND

POTENTIAL THEORY

PART II:

Acoustic and Electromagnetic Diffraction by Canonical Structures

CHAPMAN & HALL/CRC
Monographs and Surveys in Pure and Applied Mathematics

Main Editors

H. Brezis, *Université de Paris*
R.G. Douglas, *Texas A&M University*
A. Jeffrey, *University of Newcastle upon Tyne (Founding Editor)*

Editorial Board

R. Aris, *University of Minnesota*
G.I. Barenblatt, *University of California at Berkeley*
H. Begehr, *Freie Universität Berlin*
P. Bullen, *University of British Columbia*
R.J. Elliott, *University of Alberta*
R.P. Gilbert, *University of Delaware*
R. Glowinski, *University of Houston*
D. Jerison, *Massachusetts Institute of Technology*
K. Kirchgässner, *Universität Stuttgart*
B. Lawson, *State University of New York*
B. Moodie, *University of Alberta*
L.E. Payne, *Cornell University*
D.B. Pearson, *University of Hull*
G.F. Roach, *University of Strathclyde*
I. Stakgold, *University of Delaware*
W.A. Strauss, *Brown University*
J. van der Hoek, *University of Adelaide*

CHAPMAN & HALL/CRC
Monographs and Surveys in
Pure and Applied Mathematics   127

CANONICAL PROBLEMS

IN SCATTERING AND

POTENTIAL THEORY

PART II:

Acoustic and Electromagnetic Diffraction by Canonical Structures

S. S. VINOGRADOV

P. D. SMITH

E. D. VINOGRADOVA

CRC Press
Taylor & Francis Group
Boca Raton London New York

CRC Press is an imprint of the
Taylor & Francis Group, an **informa** business

A CHAPMAN & HALL BOOK

First published 2002 by CRC Chapman & Hall

Published 2019 by CRC Press
Taylor & Francis Group
6000 Broken Sound Parkway NW, Suite 300
Boca Raton, FL 33487-2742

© 2002 by Taylor & Francis Group, LLC
CRC Press is an imprint of Taylor & Francis Group, an Informa business

First issued in paperback 2019

No claim to original U.S. Government works

ISBN-13: 978-0-367-45494-4 (pbk)
ISBN-13: 978-1-58488-163-6 (hbk)

This book contains information obtained from authentic and highly regarded sources. Reasonable efforts have been made to publish reliable data and information, but the author and publisher cannot assume responsibility for the validity of all materials or the consequences of their use. The authors and publishers have attempted to trace the copyright holders of all material reproduced in this publication and apologize to copyright holders if permission to publish in this form has not been obtained. If any copyright material has not been acknowledged please write and let us know so we may rectify in any future reprint.

Except as permitted under U.S. Copyright Law, no part of this book may be reprinted, reproduced, transmitted, or utilized in any form by any electronic, mechanical, or other means, now known or hereafter invented, including photocopying, microfilming, and recording, or in any information storage or retrieval system, without written permission from the publishers.

For permission to photocopy or use material electronically from this work, please access www.copyright.com (http://www.copyright.com/) or contact the Copyright Clearance Center, Inc. (CCC), 222 Rosewood Drive, Danvers, MA 01923, 978-750-8400. CCC is a not-for-profit organiza-tion that provides licenses and registration for a variety of users. For organizations that have been granted a photocopy license by the CCC, a separate system of payment has been arranged.

Trademark Notice: Product or corporate names may be trademarks or registered trademarks, and are used only for identification and explanation without intent to infringe.

**Visit the Taylor & Francis Web site at
http://www.taylorandfrancis.com**

**and the CRC Press Web site at
http://www.crcpress.com**

Library of Congress Card Number 2001028226

Library of Congress Cataloging-in-Publication Data

Vinogradov, Sergey S. (Sergey Sergeyevich)
   Canonical problems in scattering and potential theory / Sergey S. Vinogradov, Paul D. Smith, Elena D. Vinogradova.
     p. cm.— (Monographs and surveys in pure and applied mathematics ; 122)
   Includes bibliographical references and index.
   Contents: pt. 1. Canonical structures in potential theory
   ISBN 1-58488-162-3 (v. 1 : alk. paper)
   1. Potential theory (Mathematics) 2. Scattering (Mathematics) I. Smith, P.D. (Paul Denis), 1955- II. Vinogradova, Elena D. (Elena Dmitrievna) III. Title. IV. Chapman & Hall/CRC monographs and surveys in pure and applied mathematics ; 122.

QA404.7 . V56 2001
515'.9—dc21   2001028226

To our children

Contents

Preface

Scattering of electromagnetic or acoustic waves is of widespread interest, because of the enormous number of technological applications developed during the last century. The advent of powerful computing resources has facilitated numerical modelling and simulation of many concrete diffraction problems and the many methods developed and refined over the last three decades have had a significant impact in providing numerical solutions and insight into the important mechanisms operating in scattering problems.

It is fair to say that the study of diffraction from closed bodies with smooth surfaces is well developed, from an analytical and numerical point of view, and computational algorithms have attained a good degree of accuracy and generality. However, the accuracy of present-day purely numerical methods can be difficult to ascertain, particularly for objects of some complexity incorporating edges, re-entrant structures, and dielectrics. Historically, these structures have been less tractable to analytical methods.

Our objective in the second part of this two-volume text on scattering and potential theory is to describe a class of analytic and semi-analytic techniques for accurately determining the diffraction from structures comprising edges and other complex cavity features. This is a natural development of techniques developed in Part I [1] for the corresponding potential problem. Various classes of canonical scatterers of particular relevance to edge-cavity structures are examined.

There are several reasons for focusing on such canonical objects. The exact solution to a diffraction problem is interesting in its own right as well as of direct technological interest. As Bowman et al. [13] state, most of our understanding of how scattering takes place is obtained by detailed examination of such representative scatterers. Their classic text is a collation of well developed analysis for closed bodies of simple geometric shape; our analysis quantifies the effect of edges, cavities, and inclusions by examining open scatterers of similarly simple geometric shape. Such results are invaluable for assessing the relative importance of these effects in other, more general structures. Some solutions developed in the text are analytic, others are semi-analytic, providing a linear system of equations for which the solution accuracy can be rigorously determined. Canonical scattering problems highlight the generic difficulties that numerical methods must successfully tackle for more general structures. Reliable benchmarks, against which a solution obtained by such general-purpose numerical methods can be verified, are needed to establish confidence in the validity of these computational methods where analysis becomes impossible. This is important in wider contexts such as inverse scattering.

Mathematically, we solve a class of mixed boundary value problems and develop numerical formulations that are computationally stable, rapidly convergent and of guaranteed accuracy. The diffraction problems are formulated as dual (or multiple) series equations, or dual (or multiple) integral equations. These were intensively studied in Part I with analytical regularisation techniques that transform the part of the series equations to a well behaved set of equations (technically, second-kind Fredholm equations).

All the problems considered in this book are analysed from a common perspective: a mathematically correct formulation of the scattering problem is subjected to rigorous analysis followed by solution and extensive physical interpretation. Chapters 2–6 examine a variety of acoustic and electromagnetic scattering problems for spheres with apertures and slots, with some consideration of dielectric or metallic inclusions. In Chapter 7 we examine the analogous problems for open spheroidal structures using a specially adapted version of the method of regularisation. Although the main thrust is three-dimensional, some canonical two-dimensional structures, such as slotted cylinders and strips, are considered in Chapter 8. The adaptation of regularisation methods to noncanonical structures is illustrated by the study of the singly-slotted cylinder of arbitrary cross-section. Discs of circular and elliptic shape are analysed. Illustrative applications of these methods to periodic structures and waveguides are provided.

We hope this book will be useful to both new researchers and experienced specialists in mathematics, physics and electrical engineering. In common with Part I the analysis is presented in a concrete, rather than an abstract or formal style. It is suitable for postgraduate courses in diffraction and potential theory and related mathematical methods. It is also suitable for advanced-level undergraduates, particularly for project material.

1

Mathematical Aspects of Wave Scattering.

The scattering and diffraction of waves is universally and ubiquitously observable. It involves the transfer of energy and information without bulk motion; light, sound and elastic waves as well as less obvious varieties are important to us in natural and technological senses. Whilst human perception of the natural world depends heavily on interpreting light and sound waves, the last century has seen the birth of many new technologies exploiting electromagnetic waves, such as telecommunications, radar (including remote sensing and imaging radars) and lasers; in the acoustic domain, ultrasonic imaging finds application in non-destructive testing and medicine.

Further development and exploitation depends crucially on mathematical modelling and analysis of wave phenomena – radiation, propagation, interaction and diffraction by obstacles and detection of the scattered field by sensors. Such models allow us to interpret the intrinsic information contained in waves that are otherwise invisible (or inaudible), to infer the presence or absence of certain types of scattering features in the environment and ideally, to identify the shape and location of the objects responsible for the wave-scattering.

At a fundamental level Maxwell's equations or the scalar wave equation completely describe all electromagnetic or acoustic phenomena, when supplemented with appropriate boundary, radiation and edge conditions. A variety of analytical and numerical techniques have been devised to solve these equations, but certain classes of direct and inverse scattering problems still challenge us to devise reliable, robust and efficient methods for their solution.

Our principal interest is in the scattering and diffraction from canonical structures having edges and cavity regions that possibly enclose other scatterers. Apart from their intrinsic interest, such results provide insight into the relative importance of the diffraction mechanisms operating in *complex* scatterers. Because these canonical objects contain geometrical features that are representative of the larger class of such scatterers, their intensive study offers insight into the important mechanisms (such as edge scattering, aperture coupling mechanisms, and internal cavity oscillations) that may dominate the scattering response of other scatterers with similar features. Moreover the analytical and semi-analytical solutions developed in this book provide accurate benchmarks for the development and testing of approximate scattering methods and general purpose numerical algorithms designed for three-dimensional finite open scatterers with cavities.

The purpose of this chapter is to survey mathematical aspects of wave scattering. Thus we shall formulate the appropriate wave equations and discuss the conditions under which existence and uniqueness of solutions can be assured. There are several classic texts that provide a greater depth of our necessarily very brief treatment, for example [41], [42], [17], [74] and [38]. The method of separation of variables is surveyed as are Green's functions, integral equation formulations of scattering and dual series equation formulations of scattering that are shown to be equivalent. Methods based upon dual series equations and dual integral equations are central to our studies in subsequent chapters on diffraction from spherical and spheroidal cavities and other structures with edges. These methods were introduced in Part I [1] in the context of potential theory, and this volume extends their application to the wave diffraction context. The chapter concludes with a brief survey of various numerical methods for scattering.

1.1   The Equations of Acoustic and Electromagnetic Waves.

In this section we describe the equations governing the propagation of acoustic and electromagnetic waves, namely Maxwell's equations and the Helmholtz equation. Our treatment is brief and the reader is referred to [41] for a more substantive development.

Acoustics may be described as the theory of the propagation of small disturbances in fluids, liquid or gaseous. The propagation arises from the rarefaction and compression of the fluid that causes a change in the density. In a medium in which viscosity and thermal conductivity can be neglected, and the flow is isentropic, i.e., is constant throughout the medium, the pressure p is a function of density ρ only. The equations of motion connecting the density $\rho = \rho(\vec{r}, t)$ and the fluid velocity $\vec{v} = \vec{v}(\vec{r}, t)$ as functions of position $\vec{r}$ and time t are given in [41]. The first is the conservation of mass law,

$$\frac{\partial \rho}{\partial t} + \text{div}(\rho \vec{v}) = 0. \tag{1.1}$$

Defining the total derivative by

$$\frac{Df}{Dt} = \frac{\partial f}{\partial t} + \vec{v} \cdot \text{grad } f$$

for each scalar function $f = f(\vec{r}, t)$, Equation (1.1) is equivalent to

$$\frac{D\rho}{Dt} + \rho \, \text{div}(\vec{v}) = 0. \tag{1.2}$$

In the absence of external stimulus, the second equation of motion is then

$$\frac{D\vec{v}}{Dt} = -\frac{1}{\rho} \text{grad } p = -\frac{a^2}{\rho} \text{grad } \rho, \tag{1.3}$$

where

$$a^2 = \frac{dp}{d\rho}$$

is the square of the speed of sound; it may vary throughout the medium.

Suppose that at rest the fluid is uniform and has constant density ρ_0. Now consider disturbances in which the velocity stays small, so that the corresponding perturbation in density $\rho_1 = \rho - \rho_0$ is small. Neglecting second order terms of magnitude $\rho |\vec{v}|$, the governing equations become

$$\frac{\partial \rho_1}{\partial t} + \rho_0 \, \text{div}(\vec{v}) = 0, \tag{1.4}$$

$$\frac{\partial \vec{v}}{\partial t} = -\frac{a_0^2}{\rho_0} \text{grad } \rho_1, \tag{1.5}$$

where

$$a_0^2 = \left(\frac{dp}{d\rho}\right)_{\rho = \rho_0}.$$

The sound speed a_0 is uniform throughout the medium. Eliminating $\vec{v}$ from (1.4) and (1.5) produces the *wave equation*

$$\nabla^2 \rho_1 = \frac{1}{a_0^2} \frac{\partial^2 \rho_1}{\partial t^2}. \tag{1.6}$$

Introducing the *velocity potential* $U = U(\vec{r}, t)$ via

$$\vec{v} = -\operatorname{grad} U,$$

the perturbation in sound pressure about its mean value $p_0 = p(\rho_0)$ equals

$$p_1 = \rho_0 \frac{\partial U}{\partial t}.$$

The velocity potential also satisfies the wave equation

$$\nabla^2 U = \frac{1}{a_0^2} \frac{\partial^2 U}{\partial t^2}. \tag{1.7}$$

When the density perturbation ρ_1 varies harmonically in time it may be represented by

$$\rho_1 = \rho_1(\vec{r}) e^{-i\omega t}, \tag{1.8}$$

and equation (1.6) reduces to the *Helmholtz equation*

$$\nabla^2 \rho_1(\vec{r}) + k^2 \rho_1(\vec{r}) = 0 \tag{1.9}$$

where $k = \omega/a_0$ is the wave number; the velocity potential also satisfies the Helmholtz equation, and

$$p_1 = -i\omega \rho_0 U. \tag{1.10}$$

We shall be interested in the scattering or diffraction of acoustic energy by obstacles that are hard (or rigid) or are soft. The appropriate boundary condition to be applied at each point P on the surface S of a hard scatterer is

$$\vec{n} \cdot \vec{v} = 0, \tag{1.11}$$

where $\vec{n}$ is a unit normal vector at P, i.e., the normal component of the fluid velocity vanishes on the surface. Equivalently

$$\frac{\partial U}{\partial n} = \vec{n} \cdot \operatorname{grad} U = 0 \tag{1.12}$$

on S. The soft boundary condition asserts that the pressure perturbation p_1 vanishes on S, or equivalently,

$$U = 0 \tag{1.13}$$

on S. For a fixed interface where there can be motion on both sides, we require that the normal component of velocity be continuous across the interface,

$$\overrightarrow{n} \cdot \overrightarrow{v_1} = \overrightarrow{n} \cdot \overrightarrow{v_2} \tag{1.14}$$

where $\overrightarrow{v_1}$ and $\overrightarrow{v_2}$ denote the velocity on either side of the interface. This is equivalent to continuity of $\frac{\partial U}{\partial n}$ across the interface. Also continuity of pressure will be enforced so that U is continuous across the interface.

The *acoustic (or sound) energy density* is

$$\frac{1}{2}\rho_0 \left|\overrightarrow{v}\right|^2 + \frac{1}{2}a_0^2 \left|\rho_1\right|^2 / \rho_0 \tag{1.15}$$

and may be recognised as a sum of densities of kinetic acoustic energy and potential acoustic energy arising from compression of the fluid. In terms of the velocity potential (in the time harmonic case) it equals

$$W_s = \frac{1}{2}\rho_0 \left\{ |\operatorname{grad} U|^2 + k^2 |U|^2 \right\}. \tag{1.16}$$

Related to the energy density is the *acoustic intensity* that is the rate of energy flux per unit area across a surface and is defined by

$$\overrightarrow{I} = a_0^2 \rho_1 \overrightarrow{v} = -\rho_0 \frac{\partial U}{\partial t} \operatorname{grad} U. \tag{1.17}$$

In the time harmonic case the *complex acoustic intensity* is employed and the change in energy flux per unit area per oscillation cycle equals

$$\frac{1}{2}a_0^2 \operatorname{Re}\left(\rho_1 \overrightarrow{v^*}\right) = -\frac{1}{2}\operatorname{Re}\left(i\omega\rho_0 U \operatorname{grad} U^*\right) = -\frac{1}{2}\omega\rho_0 \operatorname{Im}\left(U \operatorname{grad} U^*\right), \tag{1.18}$$

where the star $(*)$ denotes complex conjugate. The energy crossing a closed surface S with outward unit normal $\overrightarrow{n}$ is thus

$$\int_S \overrightarrow{I} \cdot \overrightarrow{n}\, dS = -\frac{1}{2}\omega\rho_0 \operatorname{Im} \int_S U \frac{\partial U^*}{\partial n}\, dS. \tag{1.19}$$

If U represents the total velocity potential in a source free region bounded by S, this integral vanishes; this statement represents the conservation of energy law.

The electromagnetic field is described by Maxwell's equations (see [41])

$$\operatorname{curl} \vec{E} \;=\; -\frac{\partial \vec{B}}{\partial t} \tag{1.20}$$

$$\operatorname{curl} \vec{H} \;=\; \frac{\partial \vec{D}}{\partial t} + \vec{J} \tag{1.21}$$

$$\operatorname{div} \vec{D} \;=\; \rho \tag{1.22}$$

$$\operatorname{div} \vec{B} \;=\; 0 \tag{1.23}$$

where the current density $\vec{J}$ and charge density ρ are regarded as sources of the electromagnetic field. The vectors $\vec{E}$ and $\vec{H}$ are known as the *electric* and *magnetic intensity*, respectively; the vectors $\vec{B}$ and $\vec{D}$ are known as the *magnetic* and *electric flux density*, respectively. The current and charge densities are connected by the *equation of continuity* or conservation of charge law

$$\operatorname{div} \vec{J} + \frac{\partial \rho}{\partial t} = 0. \tag{1.24}$$

For electromagnetic fields varying with time harmonic dependence $e^{-i\omega t}$, Maxwell's equations take the form

$$\begin{aligned}
\operatorname{curl} \vec{E} &= i\omega \vec{B} \\
\operatorname{curl} \vec{H} &= -i\omega \vec{D} + \vec{J} \\
\operatorname{div} \vec{D} &= \rho \\
\operatorname{div} \vec{B} &= 0
\end{aligned} \tag{1.25}$$

where all field quantities are vector functions of position (but not time).

The macroscopic electromagnetic equations must be supplemented by the *constitutive equations* connecting field intensities with flux densities. In *free space*

$$\vec{D} = \varepsilon_0 \vec{E}, \qquad \vec{B} = \mu_0 \vec{H} \tag{1.26}$$

where ε_0 and μ_0 are the *vacuum permittivity* and *permeability*, respectively. They are connected by the relation

$$c = (\varepsilon_0 \mu_0)^{-\frac{1}{2}},$$

where c is the speed of light in free space ($\approx 3 \cdot 10^8$ m/sec); μ_0 has value $4\pi \cdot 10^{-7}$ henry/m and so $\varepsilon_0 = 1/\mu_0 c^2 = (1/36\pi) \cdot 10^{-9}$ farad/m. The quantity

$$Z_0 = (\mu_0/\varepsilon_0)^{\frac{1}{2}}$$

is known as the impedance of free space; it has approximate value 120π ohms.

In isotropic bodies the constitutive relations are

$$\vec{D} = \varepsilon\vec{E}, \qquad \vec{B} = \mu\vec{H} \qquad (1.27)$$

where ε and μ are the medium (possibly position dependent) permittivity and permeability, respectively. In isotropic metals the current density $\vec{J}$ and applied field $\vec{E}$ are connected by the medium conductivity σ via

$$\vec{J} = \sigma\vec{E}. \qquad (1.28)$$

For most metals σ is so large that it is reasonable in theoretical investigations to replace the metal by a fictitious *perfect conductor* in which σ is taken to be infinite.

Thus in a homogeneous isotropic medium Maxwell's equations become

$$
\begin{aligned}
\operatorname{curl}\vec{E} &= i\omega\mu\vec{H} \\
\operatorname{curl}\vec{H} &= -i\omega\varepsilon\vec{E} + \vec{J} \\
\operatorname{div}\vec{E} &= \rho/\varepsilon \\
\operatorname{div}\vec{H} &= 0.
\end{aligned}
\qquad (1.29)
$$

When the medium is free from charges and currents $\left(\rho = 0,\ \vec{J} = 0\right)$ Maxwell's equations can be reduced to the pair of vector wave equations relatively $\vec{E}$ and $\vec{H}$

$$\operatorname{curl}\operatorname{curl}\vec{E} - k^2\vec{E} = 0 \qquad (1.30)$$

$$\operatorname{curl}\operatorname{curl}\vec{H} - k^2\vec{H} = 0 \qquad (1.31)$$

where

$$k^2 = \omega^2\varepsilon\mu = \frac{\omega^2}{c^2}\varepsilon_r\mu_r \qquad (1.32)$$

is the square of the relative wavenumber, and $\varepsilon_r = \varepsilon/\varepsilon_0$ and $\mu_r = \mu/\mu_0$ are the relative permittivity and permeability of the medium, respectively. Throughout the text we will denote ε_r and μ_r by ε and μ, respectively.

For any vector $\vec{A}$ with Cartesian form $\vec{A} = A_1\vec{i_1} + A_2\vec{i_2} + A_3\vec{i_3}$, the identity

$$\operatorname{curl}\operatorname{curl}\vec{A} = \operatorname{grad}\operatorname{div}\vec{A} - \nabla^2\vec{A} \qquad (1.33)$$

holds, where the interpretation of the final term on the right hand side of (1.33) is

$$\nabla^2\vec{A} = \nabla^2 A_1\vec{i_1} + \nabla^2 A_2\vec{i_2} + \nabla^2 A_3\vec{i_3}.$$

suitable direction we conclude that any electromagnetic field, in a homogeneous isotropic medium free of charges and currents, can be expressed in terms of an electric Hertz vector and a magnetic Hertz vector both aligned in a direction that can be chosen at will. Thus, in the absence of charges and currents, any electromagnetic field has a representation in terms of the two scalars Π and $\Pi^{(m)}$.

There is a useful classification of electromagnetic waves that we shall employ extensively in this book. If the direction of wave propagation is defined by the unit vector $\vec{\xi}$ and the structure of the electromagnetic field is such that $E_\xi \neq 0$ and $H_\xi = 0$, the waves are termed *transverse-magnetic* (TM waves) or *waves of E-type*, or TM to $\vec{\xi}$ (if the direction $\vec{\xi}$ needs specification). If the structure of the electromagnetic field is such that $E_\xi = 0$ and $H_\xi \neq 0$ the waves are termed *transverse-electric* (TE waves) or *waves of H-type*, or TE to $\vec{\xi}$. Formulae (1.49)–(1.50) show that any electromagnetic field may be expressed as the sum of a TM field and a TE field generated by scalar functions Π and $\Pi^{(m)}$, respectively.

In Cartesian coordinates (x, y, z) the TM waves are obtained from a scalar Π_z via

$$E_x = \frac{\partial^2 \Pi_z}{\partial x \partial z}, \quad E_y = \frac{\partial^2 \Pi_z}{\partial y \partial z}, \quad E_z = \left(\frac{\partial^2}{\partial z^2} + k^2 \right) \Pi_z, \quad (1.51)$$

$$H_x = -ik \frac{\partial \Pi_z}{\partial y}, \quad H_y = ik \frac{\partial \Pi_z}{\partial x}, \quad H_z = 0; \quad (1.52)$$

and the TE waves are obtained from a scalar $\Pi_z^{(m)}$ via

$$E_x = ik \frac{\partial \Pi_z^{(m)}}{\partial y}, \quad E_y = -ik \frac{\partial \Pi_z^{(m)}}{\partial x}, \quad E_z = 0, \quad (1.53)$$

$$H_x = \frac{\partial^2 \Pi_z^{(m)}}{\partial x \partial z}, \quad H_y = \frac{\partial^2 \Pi_z^{(m)}}{\partial y \partial z}, \quad H_z = \left(\frac{\partial^2}{\partial z^2} + k^2 \right) \Pi_z^{(m)}. \quad (1.54)$$

In polar cylindrical coordinates (ρ, ϕ, z) it is usually preferable to operate with two scalar functions that are the z-components of electric Π_z and magnetic $\Pi_z^{(m)}$. The TM waves are governed by

$$E_\rho = \frac{\partial^2 \Pi_z}{\partial \rho \partial z}, \quad E_\phi = \frac{1}{\rho} \frac{\partial^2 \Pi_z}{\partial \phi \partial z}, \quad E_z = \left(\frac{\partial^2}{\partial z^2} + k^2 \right) \Pi_z,$$

$$H_\rho = -\frac{ik}{\rho} \frac{\partial \Pi_z}{\partial \phi}, \quad H_\phi = ik \frac{\partial \Pi_z}{\partial \rho}, \quad H_z = 0,$$

and the TE waves are governed by

$$H_\rho = \frac{\partial^2 \Pi_z^{(m)}}{\partial \rho \partial z}, \quad H_\phi = \frac{1}{\rho}\frac{\partial^2 \Pi_z^{(m)}}{\partial \phi \partial z}, \quad H_z = \left(\frac{\partial^2}{\partial z^2} + k^2\right)\Pi_z^{(m)},$$

$$E_\rho = \frac{ik}{\rho}\frac{\partial \Pi_z^{(m)}}{\partial \phi}, \quad E_\phi = -ik\frac{\partial \Pi_z^{(m)}}{\partial \rho}, \quad E_z = 0. \tag{1.55}$$

The reduction of the Maxwellian vector equations to scalar Helmholtz equations that are themselves separable is termed *separation* of Maxwell's equations. This procedure that is distinctive of Cartesian coordinates and other orthogonal curvilinear coordinate systems can be effected in only very few coordinate systems. Let us determine those systems where such separation is possible.

First consider the homogeneous form Maxwell's equations that follow from (1.29) by setting $\overrightarrow{J} = 0, \rho = 0$

$$\operatorname{curl} \overrightarrow{E} = iw\mu \overrightarrow{H} \tag{1.56}$$

$$\operatorname{curl} \overrightarrow{H} = -iw\varepsilon \overrightarrow{E} \tag{1.57}$$

In theoretical studies it is common to use a more symmetrical form of these equations. The quantities $Z = (\mu/\varepsilon)^{\frac{1}{2}}$ and $v = (\varepsilon\mu)^{-\frac{1}{2}}$ are the impedance and velocity of propagation of electromagnetic waves, respectively, in medium with the parameters ε and μ. Note that

$$iw\mu = ikZ, iw\varepsilon = ik/Z$$

where $k = w/v$ is the relative wavenumber. If we replace $\overrightarrow{H}$ by $Z\overrightarrow{H}$, the equations (1.56)–(1.57) take the symmetrical form

$$\operatorname{curl} \overrightarrow{E} = ik \overrightarrow{H} \tag{1.58}$$

$$\operatorname{curl} \overrightarrow{H} = -ik \overrightarrow{E} \tag{1.59}$$

It should be noted that this form of Maxwell's equations (1.58)–(1.59) will mostly be used throughout the text; the impedance Z usually refers to the impedance Z_0 of free space.

In an orthogonal curvilinear coordinate system with coordinates u_1, u_2, u_3 and Lamé coefficients h_1, h_2, h_3 (see Section 1.1 of Volume I) the equations (1.58)–(1.59) take the form

$$\frac{\partial}{\partial u_\alpha}\left(h_{u_\beta}E_{u_\beta}\right) - \frac{\partial}{\partial u_\beta}\left(h_{u_\alpha}E_{u_\alpha}\right) = ikh_{u_\alpha}h_{u_\beta}H_{u_\gamma} \tag{1.60}$$

$$\frac{\partial}{\partial u_\alpha}\left(h_{u_\beta}H_{u_\beta}\right) - \frac{\partial}{\partial u_\beta}\left(h_{u_\alpha}H_{u_\alpha}\right) = -ikh_{u_\alpha}h_{u_\beta}E_{u_\gamma} \qquad (1.61)$$

where the index triple (α, β, γ) signifies any cyclic permutation of the triple $(1, 2, 3)$.

Now suppose that the total electromagnetic field is representable as a sum of two particular types of waves, characterised by

$$E_{u_l} \neq 0, H_{u_l} = 0 \qquad (1.62)$$

and

$$E_{u_l} = 0, H_{u_l} \neq 0, \qquad (1.63)$$

respectively, where index u_l ($l = 1, 2$ or 3) identifies the field component along the coordinate line u_l. Thus, (1.62) describes electromagnetic waves of TM type (E-polarised), and (1.63) describes electromagnetic waves of TE type (H-polarised).

Let us now suppose that (1.62) holds. Then fix the value of γ to be l and obtain

$$\frac{\partial}{\partial u_j}\left(h_{u_k}E_{u_k}\right) = \frac{\partial}{\partial u_k}\left(h_{u_j}E_{u_j}\right) \qquad (1.64)$$

where both remaining indexes j, k are automatically fixed. This relationship (1.64) will be satisfied identically if we represent E_{u_j} and E_{u_k} in the form

$$E_{u_j} = \frac{1}{h_{u_j}}\frac{\partial U^*}{\partial u_j}, \quad E_{u_k} = \frac{1}{h_{u_k}}\frac{\partial U^*}{\partial u_k} \qquad (1.65)$$

where U^* is some function. Now substitute (1.65) into (1.61), with $\gamma = j, k$ and $H_{u_l} = 0$ to obtain

$$\frac{\partial}{\partial u_l}\left(h_{u_k}H_{u_k}\right) = ik\frac{h_{u_k}h_{u_l}}{h_{u_j}}\frac{\partial U^*}{\partial u_j}$$

$$\frac{\partial}{\partial u_l}\left(h_{u_j}H_{u_j}\right) = -ik\frac{h_{u_l}h_{u_l}}{h_{u_k}}\frac{\partial U^*}{\partial u_k} \qquad (1.66)$$

These relationships (1.66) are satisfied identically by setting

$$U^* = \frac{\partial U}{\partial u_l}, H_{u_k} = \frac{ik}{h_{u_j}}\frac{\partial U}{\partial u_j}, H_{u_k} = -\frac{ik}{h_{u_k}}\frac{\partial U}{\partial u_k} \qquad (1.67)$$

where U is some function.

Furthermore, our assumption will hold if and only if the metric coefficients of orthogonal curvilinear coordinate system u_1, u_2, u_3 satisfy the conditions

$$h_{u_l} = 1, \frac{\partial}{\partial u_l} \left(\frac{h_{u_j}}{h_{u_k}} \right) = 0. \tag{1.68}$$

Let us substitute the expressions (1.67) in that equation (1.59) for which $\gamma = l$. We find that

$$E_{u_l} = -\frac{1}{h_{u_j} h_{u_k}} \left[\frac{\partial}{\partial u_j} \left(\frac{h_{u_k}}{h_{u_j}} \frac{\partial U}{\partial u_j} \right) + \frac{\partial}{\partial u_k} \left(\frac{h_{u_j}}{h_{u_k}} \frac{\partial U}{\partial u_k} \right) \right]. \tag{1.69}$$

In addition, starting from (1.67) one finds from (1.65) that

$$E_{u_j} = \frac{1}{h_{u_j}} \frac{\partial^2 U}{\partial u_j \partial u_l}, E_{u_k} = \frac{1}{h_{u_k}} \frac{\partial^2 U}{\partial u_k \partial u_l}. \tag{1.70}$$

Thus, all components of the electromagnetic field $\overrightarrow{E}$ and $\overrightarrow{H}$ are expressible in terms of a single function U. Substitutions of these expressions into the two as yet unused equations from (1.60) shows that the function U must be a solution of the differential equation

$$\frac{1}{h_{u_j} h_{u_k}} \left[\frac{\partial}{\partial u_j} \left(\frac{h_{u_k}}{h_{u_j}} \frac{\partial U}{\partial u_j} \right) + \frac{\partial}{\partial u_k} \left(\frac{h_{u_j}}{h_{u_k}} \frac{\partial U}{\partial u_k} \right) \right] + \frac{\partial^2 U}{\partial u_l^2} + k^2 U = 0. \tag{1.71}$$

We conclude from (1.71) that the expressions for electromagnetic field components are

$$
\begin{aligned}
E_{u_j} &= \frac{1}{h_{u_j}} \frac{\partial^2 U}{\partial u_j \partial u_l}, & H_{u_j} &= -\frac{ik}{h_{u_k}} \frac{\partial U}{\partial u_k}, \\
E_{u_k} &= \frac{1}{h_{u_k}} \frac{\partial^2 U}{\partial u_k \partial u_l}, & H_{u_k} &= \frac{ik}{h_{u_j}} \frac{\partial U}{\partial u_j}, \\
E_{u_l} &= \frac{\partial^2 U}{\partial u_l^2} + k^2 U, & H_{u_l} &= 0
\end{aligned}
\tag{1.72}
$$

In the TE case (1.63) we obtain analogously

$$
\begin{aligned}
E_{u_j} &= \frac{ik}{h_{u_k}} \frac{\partial V}{\partial u_k}, & H_{u_j} &= \frac{1}{h_{u_j}} \frac{\partial^2 V}{\partial u_j \partial u_l}, \\
E_{u_k} &= -\frac{ik}{h_{u_j}} \frac{\partial V}{\partial u_j}, & H_{u_j} &= \frac{1}{h_{u_k}} \frac{\partial^2 V}{\partial u_k \partial u_l}, \\
E_{u_l} &= 0, & H_{u_l} &= \frac{\partial^2 V}{\partial u_l^2} + k^2 V,
\end{aligned}
\tag{1.73}
$$

where V is an auxiliary scalar function that satisfies the equation (1.71).

Thus, in those orthogonal curvilinear coordinates u_1, u_2, u_3, where conditions (1.68) are satisfied, the total field is expressible in terms of two scalar functions U and V as

$$
\begin{aligned}
E_{u_j} &= \frac{1}{h_{u_j}} \frac{\partial^2 U}{\partial u_j \partial u_l} + \frac{ik}{h_{u_k}} \frac{\partial V}{\partial u_k} \\
E_{u_k} &= \frac{1}{h_{u_k}} \frac{\partial^2 U}{\partial u_k \partial u_l} - \frac{ik}{h_{u_j}} \frac{\partial V}{\partial u_j} \\
E_{u_l} &= \frac{\partial^2 U}{\partial u_l^2} + k^2 U \\
H_{u_j} &= \frac{1}{h_{u_j}} \frac{\partial^2 V}{\partial u_j \partial u_l} - \frac{ik}{h_{u_k}} \frac{\partial U}{\partial u_k} \\
H_{u_k} &= \frac{1}{h_{u_k}} \frac{\partial^2 V}{\partial u_k \partial u_l} + \frac{ik}{h_{u_j}} \frac{\partial U}{\partial u_j} \\
H_{u_l} &= \frac{\partial^2 V}{\partial u_l^2} + k^2 V.
\end{aligned}
\tag{1.74}
$$

The conditions (1.68) on the Lamé coefficients are satisfied in generalised cylindrical coordinates (rectangular Cartesian, elliptic, parabolic and other systems constructed from two-dimensional planar orthogonal systems combined with the third Cartesian coordinate z, in a direction normal to the plane) and also in spherical coordinates.

In standard cylindrical polar coordinates, based on the circular cylinder, the Lamé coefficients are $h_\rho = 1, h_\varphi = \rho, h_z = 1$, so conditions (1.68) are satisfied and (1.74) takes the form

$$
\begin{aligned}
E_\rho &= \frac{\partial^2 U}{\partial \rho \partial z} + \frac{ik}{\rho} \frac{\partial V}{\partial \varphi} \\
E_\varphi &= \frac{1}{\rho} \frac{\partial^2 U}{\partial \varphi \partial z} - ik \frac{\partial V}{\partial \rho} \\
E_z &= \frac{\partial^2 U}{\partial z^2} + k^2 U \\
H_\rho &= \frac{\partial^2 V}{\partial \rho \partial z} - \frac{ik}{\rho} \frac{\partial U}{\partial \varphi} \\
H_\varphi &= \frac{1}{\rho} \frac{\partial^2 V}{\partial \rho \partial z} + ik \frac{\partial U}{\partial \rho} \\
H_z &= \frac{\partial^2 V}{\partial z^2} + k^2 V.
\end{aligned}
\tag{1.75}
$$

In this case we may identify the functions U, V with the electric and magnetic scalar Hertz potentials:

$$U \equiv \Pi_z, \quad V \equiv \Pi_z^{(m)}. \tag{1.76}$$

In spherical coordinates the Lamé coefficients are $h_r = 1, h_\theta = r, h_\varphi = r \sin \theta$, and conditions (1.68) are satisfied and (1.74) takes the form

$$
\begin{aligned}
E_r &= \frac{\partial^2 U}{\partial r^2} + k^2 U \\[2mm]
E_\theta &= \frac{1}{r} \frac{\partial^2 U}{\partial \theta \partial r} + \frac{ik}{r \sin \theta} \frac{\partial V}{\partial \varphi} \\[2mm]
E_\varphi &= \frac{1}{r \sin \theta} \frac{\partial^2 U}{\partial \varphi \partial r} - \frac{ik}{r} \frac{\partial V}{\partial \theta} \\[2mm]
H_r &= \frac{\partial^2 V}{\partial r^2} + k^2 V \\[2mm]
H_\theta &= \frac{1}{r} \frac{\partial^2 V}{\partial \theta \partial r} - \frac{ik}{r \sin \theta} \frac{\partial U}{\partial \varphi} \\[2mm]
H_\varphi &= \frac{1}{r \sin \theta} \frac{\partial^2 V}{\partial \varphi \partial r} + \frac{ik}{r} \frac{\partial U}{\partial \theta}.
\end{aligned}
\tag{1.77}
$$

where U and V are solutions Φ of the differential equation

$$\frac{\partial^2 \Phi}{\partial r^2} + \frac{1}{r^2 \sin \theta} \frac{\partial}{\partial \theta} \left(\sin \theta \frac{\partial \Phi}{\partial \theta} \right) + \frac{1}{r^2 \sin^2 \theta} \frac{\partial^2 \Phi}{\partial \varphi^2} + k^2 \Phi = 0. \tag{1.78}$$

With the substitution $\Phi = r\overline{\Phi}$ or $U = r\overline{U}, V = r\overline{V}$, (1.78) is reduced to the Helmholtz equation

$$\frac{\partial^2 \overline{\Phi}}{\partial r^2} + \frac{2}{r} \frac{\partial \overline{\Phi}}{\partial r} + \frac{1}{r^2 \sin \theta} \frac{\partial}{\partial \theta} \left(\sin \theta \frac{\partial \overline{\Phi}}{\partial \theta} \right) + \frac{1}{r^2 \sin^2 \theta} \frac{\partial^2 \overline{\Phi}}{\partial \varphi^2} + k^2 \overline{\Phi} = 0. \tag{1.79}$$

To avoid some further confusion, we will always use the notation U and V for scalar functions $\overline{U}$ and $\overline{V}$ in the spherical coordinate system context, and consequently the expressions for electromagnetic field components take the form

$$E_r = \frac{\partial^2 (rU)}{\partial r^2} + k^2 (rU)$$

$$E_\theta = \frac{1}{r} \frac{\partial^2 (rU)}{\partial \theta \partial r} + \frac{ik}{\sin \theta} \frac{\partial V}{\partial \varphi}$$

$$E_\varphi = \frac{1}{r \sin \theta} \frac{\partial^2 (rU)}{\partial \varphi \partial r} - ik \frac{\partial V}{\partial \theta} \qquad (1.80)$$

$$H_r = \frac{\partial^2 (rV)}{\partial r^2} + k^2 (rV)$$

$$H_\theta = \frac{1}{r} \frac{\partial^2 (rV)}{\partial \theta \partial r} - \frac{ik}{\sin \theta} \frac{\partial U}{\partial \varphi}$$

$$H_\varphi = \frac{1}{r \sin \theta} \frac{\partial^2 (rV)}{\partial \varphi \partial r} + ik \frac{\partial U}{\partial \theta}.$$

The functions U and V will be termed by their historical names as the *electric Debye potential* and the *magnetic Debye potential*.

The representation of electromagnetic field significantly simplifies when the field does not depend upon one of the coordinates and may be done concretely in terms of the electromagnetic field components themselves.

Suppose the electromagnetic field does not vary along the z-axis and thus is essentially two-dimensional ($\partial/\partial z \equiv 0$). Equations (1.58)–(1.59) separate into two independent parts. In Cartesian coordinates (x, y, z) we obtain

$$H_x = \frac{1}{ik} \frac{\partial E_z}{\partial y}, \quad H_y = -\frac{1}{ik} \frac{\partial E_z}{\partial x}, \quad -ik E_z = \frac{\partial H_y}{\partial x} - \frac{\partial H_x}{\partial y} \qquad (1.81)$$

and

$$E_x = -\frac{1}{ik} \frac{\partial H_z}{\partial y}, \quad E_y = \frac{1}{ik} \frac{\partial H_z}{\partial x}, \quad ik H_z = \frac{\partial E_y}{\partial x} - \frac{\partial E_x}{\partial y}, \qquad (1.82)$$

where both E_z and H_z are solutions F of the Helmholtz equation

$$\nabla^2 F + k^2 F = 0. \qquad (1.83)$$

The electromagnetic field for which only the three components H_x, H_y, E_z do not vanish is called *E-polarised*; the electromagnetic field for which only the three components E_x, E_y, H_z do not vanish is called *H-polarised*.

In polar cylindrical coordinates we have two similar representations,

$$H_\rho = \frac{1}{ik\rho} \frac{\partial E_z}{\partial \phi}, \quad H_\phi = -\frac{1}{ik} \frac{\partial E_z}{\partial \rho}, \quad -ik E_z = \frac{1}{\rho} \left[\frac{\partial}{\partial \rho} (\rho H_\phi) - \frac{\partial H_\rho}{\partial \phi} \right]$$

$$(1.84)$$

and

$$E_\rho = -\frac{1}{ik\rho}\frac{\partial H_z}{\partial \phi}, \quad E_\phi = \frac{1}{ik}\frac{\partial H_z}{\partial \rho}, \quad ikH_z = \frac{1}{\rho}\left[\frac{\partial}{\partial \rho}(\rho E_\phi) - \frac{\partial E_\rho}{\partial \phi}\right], \quad (1.85)$$

where again E_z and H_z are solutions of the Helmholtz equation (1.83).

The representation of an axially-symmetric electromagnetic field is also significantly simpler. We now consider Maxwell's equations in polar cylindrical, spherical and prolate spheroidal coordinates. Axial symmetry results from setting $\partial/\partial\phi \equiv 0$; in cylindrical coordinates we obtain two types of electromagnetic field,

$$E_\rho = \frac{1}{ik}\frac{\partial H_\phi}{\partial z}, \quad E_z = -\frac{1}{ik}\frac{1}{\rho}\frac{\partial}{\partial \rho}(\rho H_\phi), \quad ikH_\phi = \frac{\partial E_\rho}{\partial z} - \frac{\partial E_z}{\partial \rho} \quad (1.86)$$

and

$$H_\rho = -\frac{1}{ik}\frac{\partial E_\phi}{\partial z}, \quad H_z = \frac{1}{ik}\frac{1}{\rho}\frac{\partial}{\partial \rho}(\rho E_\phi), \quad -ikE_\phi = \frac{\partial H_\rho}{\partial z} - \frac{\partial H_z}{\partial \rho}, \quad (1.87)$$

where both the azimuthal electric E_ϕ and azimuthal magnetic H_ϕ components of the field are solutions Φ of the partial differential equation

$$\frac{1}{\rho}\frac{\partial}{\partial \rho}\left(\rho\frac{\partial \Phi}{\partial \rho}\right) + \frac{\partial^2 \Phi}{\partial z^2} - \frac{1}{\rho^2}\Phi + k^2\Phi = 0. \quad (1.88)$$

Equation (1.88) differs from the axisymmetric Helmholtz equation in polar cylindrical coordinates ($\frac{\partial}{\partial\phi} \equiv 0$) by an additional term $(-\rho^{-2}\Phi)$, i.e., the Hertz vector components are not solutions of the Helmholtz equation. However it is readily seen that the functions $\Phi(\rho, z)\cos\phi$ and $\Phi(\rho, z)\sin\phi$ are solutions of the Helmholtz equation.

In spherical coordinates we also have two types of electromagnetic field, governed by

$$E_r = -\frac{(ik)^{-1}}{r\sin\theta}\frac{\partial}{\partial\theta}(\sin\theta\, H_\phi), \quad E_\theta = \frac{(ik)^{-1}}{r}\frac{\partial}{\partial r}(rH_\phi),$$
$$ikH_\phi = \frac{1}{r}\left\{\frac{\partial}{\partial r}(rE_\theta) - \frac{\partial E_r}{\partial\theta}\right\} \quad (1.89)$$

and

$$H_r = \frac{(ik)^{-1}}{r\sin\theta}\frac{\partial}{\partial\theta}(\sin\theta\, E_\phi), \quad H_\theta = -\frac{(ik)^{-1}}{r}\frac{\partial}{\partial r}(rE_\phi),$$
$$-ikE_\phi = \frac{1}{r}\left\{\frac{\partial}{\partial r}(rH_\theta) - \frac{\partial}{\partial\theta}H_r\right\} \quad (1.90)$$

where both H_ϕ and E_ϕ are solutions F of the partial differential equation

$$\frac{1}{r^2}\frac{\partial}{\partial r}\left(r^2\frac{\partial F}{\partial r}\right) + \frac{1}{r^2\sin\theta}\frac{\partial}{\partial\theta}\left(\sin\theta\frac{\partial F}{\partial\theta}\right) - \frac{1}{r^2\sin^2\theta}F + k^2 F = 0. \quad (1.91)$$

Equation (1.91) differs from the Helmholtz equation in the presence of the term $-\left(r^2\sin^2\theta\right)^{-1}F$; however the functions $F\left(r,\theta\right)\cos\phi$ and $F\left(r,\theta\right)\sin\phi$ satisfy the Helmholtz equation.

In prolate spheroidal coordinates (ξ,η,ϕ) related to rectangular coordinates by the transformation [29],

$$x = \frac{d}{2}\left[(1-\eta^2)(\xi^2-1)\right]^{\frac{1}{2}}\cos\phi,$$

$$y = \frac{d}{2}\left[(1-\eta^2)(\xi^2-1)\right]^{\frac{1}{2}}\sin\phi,$$

$$z = \frac{d}{2}\eta\xi \quad (1.92)$$

where d is the interfocal distance, and $-1 \le \eta \le 1, 1 \le \xi < \infty, 0 \le \phi \le 2\pi$; the Lamé coefficients are

$$h_\xi = \frac{d}{2}\left(\frac{\xi^2-\eta^2}{\xi^2-1}\right)^{\frac{1}{2}}, h_\eta = \frac{d}{2}\left(\frac{\xi^2-\eta^2}{1-\eta^2}\right)^{\frac{1}{2}}, h_\phi = \frac{d}{2}\left[(1-\eta^2)(\xi^2-1)\right]^{\frac{1}{2}}.$$

$$(1.93)$$

Since the conditions (1.68) are not satisfied in spheroidal coordinates, it is not possible to represent the electric and magnetic field vectors $\overrightarrow{E},\overrightarrow{H}$ in terms of two scalar potentials that satisfy two scalar equations, each of which is separable. However the separation of Maxwell's equations can be achieved if the scattered field does not depend on the azimuthal coordinate ϕ $(\partial/\partial\phi \equiv 0)$. In this case the electromagnetic problem reduces to two scalar problems, because Maxwell's equations (1.58), (1.59) separate into two independent systems of equations:

$$H_\xi = \frac{1}{ic\left(\xi^2-\eta^2\right)^{\frac{1}{2}}}\frac{\partial}{\partial\eta}\left[(1-\eta^2)^{\frac{1}{2}}E_\phi\right] \quad (1.94)$$

$$H_\eta = \frac{-1}{ic\left(\xi^2-\eta^2\right)^{\frac{1}{2}}}\frac{\partial}{\partial\xi}\left[(\xi^2-1)^{\frac{1}{2}}E_\phi\right] \quad (1.95)$$

$$E_\phi = \frac{(1-\eta^2)^{\frac{1}{2}}}{ic\left(\xi^2-\eta^2\right)}\frac{\partial}{\partial\eta}\left[(\xi^2-\eta^2)^{\frac{1}{2}}H_\xi\right]$$

$$-\frac{(\xi^2-1)^{\frac{1}{2}}}{ic\left(\xi^2-\eta^2\right)}\frac{\partial}{\partial\xi}\left[(\xi^2-\eta^2)^{\frac{1}{2}}H_\eta\right], \quad (1.96)$$

and

$$E_\xi = \frac{-1}{ic\left(\xi^2 - \eta^2\right)^{\frac{1}{2}}} \frac{\partial}{\partial \eta}\left[\left(1 - \eta^2\right)^{\frac{1}{2}} H_\phi\right] \tag{1.97}$$

$$E_\eta = \frac{1}{ic\left(\xi^2 - \eta^2\right)^{\frac{1}{2}}} \frac{\partial}{\partial \xi}\left[\left(\xi^2 - 1\right)^{\frac{1}{2}} H_\phi\right] \tag{1.98}$$

$$H_\phi = \frac{-\left(1 - \eta^2\right)^{\frac{1}{2}}}{ic\left(\xi^2 - \eta^2\right)} \frac{\partial}{\partial \eta}\left[\left(\xi^2 - \eta^2\right)^{\frac{1}{2}} E_\xi\right]$$

$$+ \frac{\left(\xi^2 - 1\right)^{\frac{1}{2}}}{ic\left(\xi^2 - \eta^2\right)} \frac{\partial}{\partial \xi}\left[\left(\xi^2 - \eta^2\right)^{\frac{1}{2}} E_\eta\right]. \tag{1.99}$$

Here $c = \frac{d}{2}k$ is the nondimensional frequency parameter.

The system (1.96) describes a field of magnetic type ($H_\xi \neq 0, E_\xi = 0$) whereas the system (1.99) describes a field of electric type ($H_\xi = 0, E_\xi \neq 0$).

From the system (1.99) we deduce the second order partial differential equation satisfied by the component $H_\phi = H_\phi\left(\xi, \eta\right)$,

$$\frac{\partial}{\partial \xi}\left[\left(\xi^2 - 1\right)\frac{\partial}{\partial \xi} H_\phi\right] + \frac{\partial}{\partial \eta}\left[\left(1 - \eta^2\right)\frac{\partial}{\partial \eta} H_\phi\right] +$$

$$\left[c^2\left(\xi^2 - \eta^2\right) - \frac{1}{1 - \eta^2} - \frac{1}{\xi^2 - 1}\right] H_\phi = 0. \tag{1.100}$$

This equation is separable (see the next section). The other field components (E_ξ, E_η) are readily obtained from $H_\phi = H_\phi\left(\xi, \eta\right)$.

After this brief introduction to the representation of acoustic and electromagnetic fields, it is clear that the Helmholtz equation plays a pivotal role in describing wave fields of various physical origins. Thus, in the next section we examine its solution by separation of variables in some different coordinate systems.

1.2 Solution of the Helmholtz Equation: Separation of Variables.

In this section we use the method of separation of variables to construct a general solution of the Helmholtz equation in the basic coordinate systems that we intend to exploit in solving wave-scattering problems for structures with edges. Laying aside its physical interpretation suppose that the function ψ satisfies the Helmholtz equation

$$\nabla^2 \psi + k^2 \psi = 0. \tag{1.101}$$

In Cartesian coordinates (x, y, z)

$$\frac{\partial^2 \psi}{\partial x^2} + \frac{\partial^2 \psi}{\partial y^2} + \frac{\partial^2 \psi}{\partial z^2} + k^2 \psi = 0, \tag{1.102}$$

and in accordance with the method of separation of variables (see Chapter 2 of Volume I) we seek solution in the form

$$\psi(x, y, z) = X(x) Y(y) Z(z).$$

Equation (1.102) is transformed to

$$\frac{1}{X} X'' + \frac{1}{Y} Y'' + \frac{1}{Z} Z'' + k^2 = 0, \tag{1.103}$$

where the primes denote derivatives (with respect to the relevant argument). Thus there are constants (*separation constants*) ν and μ such that

$$\frac{1}{X} X'' = -\nu^2, \tag{1.104}$$

$$\frac{1}{Y} Y'' = -\mu^2, \tag{1.105}$$

and

$$\frac{1}{Z} Z'' = -\left(\nu^2 + \mu^2 - k^2\right) = 0. \tag{1.106}$$

The solutions of equations (1.104)–(1.106) define *partial solutions* or *separated solutions* of equation (1.103),

$$\psi_{\nu\mu}(x, y, z) = \{A(\nu) \cos \nu x + B(\nu) \sin \nu x\} \{C(\mu) \cos \mu y + D(\mu) \sin \mu y\}$$
$$\times \left\{ E(\nu, \mu) e^{\sqrt{\nu^2 + \mu^2 - k^2} z} + F(\nu, \mu) e^{-\sqrt{\nu^2 + \mu^2 - k^2} z} \right\}, \tag{1.107}$$

where A, B, C, D, E and F are functions only of the separation constants. The general solution is given by integration over the spectral parameters ν and μ, lying within the intervals $\nu \in (0, \infty)$ and $\mu \in (0, \infty)$,

$$\psi(x, y, z) = \int_0^\infty \int_0^\infty \psi_{\nu\mu}(x, y, z) d\nu d\mu. \tag{1.108}$$

Both parameters ν and μ in representation (1.107) are continuous. The choice is not unique. In some problems additional conditions may dictate that one or both separation constant may be discrete. For example, suppose that $\psi(x, y, z)$ does not depend on the variable y, and is periodic in the coordinate x with period l, i.e.,

$$\psi(x, z) = \psi(x + nl, z) \tag{1.109}$$

where n is integral. Then nontrivial separated solutions are found only when the separation constant ν takes the values $\nu_n = 2\pi n/l = \gamma_n$. In this case equation (1.107) is replaced by

$$\psi_n(x, z) = \{A_n \cos(\gamma_n x) + B_n \sin(\gamma_n x)\} \left\{ E_n e^{z\sqrt{\gamma_n^2 - k^2}} + F_n e^{-z\sqrt{\gamma_n^2 - k^2}} \right\} \tag{1.110}$$

and the general solution is given by

$$\psi(x, z) = \sum_{n=0}^{\infty} \psi_n(x, z). \tag{1.111}$$

This situation occurs in the examination of scattering by a periodic structure consisting of an array of infinitely long and parallel thin strips.

Now consider the Helmholtz equation in polar cylindrical coordinates (ρ, ϕ, z),

$$\frac{1}{\rho} \frac{\partial}{\partial \rho} \left(\rho \frac{\partial \psi}{\partial \rho} \right) + \frac{1}{\rho^2} \frac{\partial^2 \psi}{\partial \phi^2} + \frac{\partial^2 \psi}{\partial z^2} + k^2 \psi = 0. \tag{1.112}$$

As always, we can obtain three representations that are discontinuous in each coordinate. However as we do not intend to solve wave-scattering problems for wedges, we omit consideration of solutions that are discontinuous in coordinate ϕ. The method of separation of variables seeks solutions in the form

$$\psi(\rho, \phi, z) = R(\rho) \Phi(\phi) Z(z), \tag{1.113}$$

with

$$\frac{1}{\rho R} \frac{d}{d\rho} \left(\rho \frac{dR}{d\rho} \right) + \frac{1}{\rho^2 \Phi} \frac{d^2 \Phi}{d\phi^2} + \frac{1}{Z} \frac{d^2 Z}{dz^2} + k^2 = 0. \tag{1.114}$$

It is evident that the function Φ may be assumed to be periodic, with period 2π; so, the relevant separation constant is discrete,

$$\frac{1}{\Phi} \frac{d^2 \Phi}{d\phi^2} = -m^2 \tag{1.115}$$

where $m = 0, 1, 2, \ldots$; for suitable constants A_m and B_m, the solution is

$$\Phi = A_m \cos m\phi + B_m \sin m\phi. \tag{1.116}$$

As there are no special requirements on the functions $R(\rho)$ and $Z(z)$, the second separation constant ν is continuous and we obtain a pair of ordinary differential equations for these functions,

$$\frac{1}{\rho R}\frac{d}{d\rho}\left(\rho\frac{dR}{d\rho}\right) - \frac{m^2}{\rho^2} = -\nu^2, \tag{1.117}$$

$$\frac{1}{Z}\frac{d^2 Z}{dz^2} - \left(\nu^2 - k^2\right) = 0. \tag{1.118}$$

Rearrange (1.117) in the standard form of Bessel's differential equation

$$x\frac{d}{dx}\left(x\frac{dR}{dx}\right) + \left(x^2 - m^2\right) R = 0, \tag{1.119}$$

where $x = \nu\rho$; the general solution is a linear combination of the Bessel function $J_m(x)$ and the Neumann function $Y_m(x)$ (of order m),

$$R_m(\nu\rho) = C_m(\nu) J_m(\nu\rho) + D_m(\nu) Y_m(\nu\rho) \tag{1.120}$$

where $C_m(\nu)$ and $D_m(\nu)$ are independent of ρ. Equation (1.118) has the obvious solution

$$Z(z) = E_m(\nu) e^{-z\sqrt{\nu^2-k^2}} + D_m(\nu) e^{+z\sqrt{\nu^2-k^2}} \tag{1.121}$$

where $E_m(\nu)$ and $F_m(\nu)$ are independent of z. Thus the general solution that is discontinuous in z is given by

$$\psi(\rho, \phi, z) = \sum_{m=0}^{\infty} (A_m \cos m\phi + B_m \sin m\phi) \times$$

$$\int_0^{\infty} \{C_m(\nu) J_m(\nu\rho) + D_m(\nu) Y_m(\nu\rho)\} \times$$

$$\left\{E_m(\nu) e^{-\sqrt{\nu^2-k^2}z} + D_m(\nu) e^{\sqrt{\nu^2-k^2}z}\right\} d\nu. \tag{1.122}$$

The representation (1.122) is useful for the analysis of wave-scattering by a circular disc.

Let us now construct the solution discontinuous in coordinate ρ. The differential equation (1.115) is unchanged, but for the other differential equations different choices of separation constant will be made, so that

$$\frac{d^2 Z}{dz^2} + \nu^2 Z = 0 \tag{1.123}$$

and

$$\rho^2 \frac{d^2 R}{d\rho^2} + \rho \frac{dR}{d\rho} - \left[(\nu^2 - k^2) \rho^2 + m^2 \right] R = 0. \tag{1.124}$$

Equation (1.123) has solutions in terms of trigonometric functions,

$$Z_\nu = E(\nu) \cos(\nu z) + D(\nu) \sin(\nu z) \tag{1.125}$$

where $D(\nu)$ and $E(\nu)$ are independent of z. Equation (1.124) coincides with the differential equation for modified Bessel functions, having solutions

$$R_{m\nu}(\rho) = C_m(\nu) I_m(\sqrt{\nu^2 - k^2}\rho) + D_m(\nu) K_m(\sqrt{\nu^2 - k^2}\rho) \tag{1.126}$$

where $C_m(\nu)$ and $D_m(\nu)$ are independent of ρ. The general solution is

$$\psi(\rho, \phi, z) = \sum_{m=0}^{\infty} (A_m \cos m\phi + B_m \sin m\phi) \times$$

$$\int_0^\infty \left\{ C_m(\nu) I_m \left(\sqrt{\nu^2 - k^2}\rho \right) + D_m(\nu) K_m \left(\sqrt{\nu^2 - k^2}\rho \right) \right\}$$

$$\times \left\{ E(\nu) \cos(\nu z) + D(\nu) \sin(\nu z) \right\} d\nu. \tag{1.127}$$

If the function $\psi(\rho, \phi, z)$ is periodic in coordinate z with period l,

$$\psi(\rho, \phi, z) = \psi(\rho, \phi, z + nl), \tag{1.128}$$

where n is integral, the continuous parameter ν is replaced by a discrete parameter $\nu_n = 2\pi n/l = \gamma_n$. Periodicity condition (1.128) occurs for example in connection with the periodic structure of a finite hollow cylinder array.

Let us now find separated solution representations of the Helmholtz equation in polar spherical coordinates (r, θ, ϕ)

$$\frac{1}{r^2} \frac{\partial}{\partial r} \left(r^2 \frac{\partial \psi}{\partial r} \right) + \frac{1}{r^2 \sin\theta} \frac{\partial}{\partial \theta} \left(\sin\theta \frac{\partial \psi}{\partial \theta} \right) + \frac{1}{r^2 \sin^2\theta} \frac{\partial^2 \psi}{\partial \phi^2} + k^2 \psi = 0. \tag{1.129}$$

Following the method of separation of variables, we seek a solution in the form

$$\psi(r, \theta, \phi) = R(r) \Theta(\theta) \Phi(\phi) \tag{1.130}$$

and its substitution in (1.129) produces

$$\frac{1}{Rr^2} \frac{d}{dr} \left(r^2 \frac{dR}{dr} \right) + \frac{1}{\Theta r^2 \sin\theta} \frac{d}{d\theta} \left(\sin\theta \frac{d\Theta}{d\theta} \right) + \frac{1}{\Phi r^2 \sin^2\theta} \frac{d^2\Phi}{d\phi^2} + k^2 = 0. \tag{1.131}$$

Amongst possible variants, we select the solutions that are discontinuous in coordinate r. These are especially useful in solving scattering problems for spherical geometry. The solution must be periodic in the coordinate ϕ (with period 2π), and we use a discrete integral separation constant m,

$$\frac{1}{\Phi}\frac{d^2\Phi}{d\phi^2} = -m^2. \tag{1.132}$$

This has solutions

$$\Phi_m(\phi) = A_m \cos m\phi + B_m \sin m\phi \tag{1.133}$$

where A_m and B_m are constants.

The separated equation for the function Θ also employs a discrete and integral separation constant n,

$$\frac{1}{\sin\theta}\frac{d}{d\theta}\left(\sin\theta\frac{d\Theta}{d\theta}\right) + \left[n(n+1) - \frac{m^2}{\sin^2\theta}\right]\Theta = 0. \tag{1.134}$$

With the substitution $z = \cos\theta$ this equation is the well-known differential equation for the associated Legendre functions,

$$\frac{d}{dz}\left[(1-z^2)\frac{d\Theta}{dz}\right] + \left(n(n+1) - \frac{m^2}{1-z^2}\right)\Theta = 0 \tag{1.135}$$

having general solution

$$\Theta(\theta) = C_n^m P_n^m(\cos\theta) + D_n^m Q_n^m(\cos\theta) \tag{1.136}$$

where C_n^m and D_n^m are constants; however because we seek only bounded solutions, we may set $D_n^m \equiv 0$.

The separated equation for the radial function R is (with $x = kr$)

$$\frac{d}{dx}\left(x^2\frac{dR}{dx}\right) + \left[x^2 - n(n+1)\right]R = 0. \tag{1.137}$$

The substitution $R = x^{-\frac{1}{2}}R_1$ transforms (1.137) to Bessel's differential equation,

$$x\frac{d}{dx}\left(x\frac{dR_1}{dx}\right) + \left[x^2 - \left(n+\frac{1}{2}\right)^2\right]R_1 = 0, \tag{1.138}$$

with general solution

$$R_1 = E_n J_{n+\frac{1}{2}}(x) + F_n H^{(1)}_{n+\frac{1}{2}}(x) \tag{1.139}$$

where E_n and F_n are constants. The functions

$$j_n(x) = \left(\frac{\pi}{2x}\right)^{\frac{1}{2}} J_{n+\frac{1}{2}}(x), \quad h_n^{(1)}(x) = \left(\frac{\pi}{2x}\right)^{\frac{1}{2}} H_{n+\frac{1}{2}}^{(1)}(x) \qquad (1.140)$$

are the *spherical Bessel function* and *spherical Hankel function (of first kind)*, respectively. In this notation the radial separated solution is

$$R(r) = E_n j_n(kr) + F_n h_n^{(1)}(kr). \qquad (1.141)$$

In electromagnetic wave-scattering by spherical structures, it is usually convenient to use the so-called *spherical Bessel functions in Debye notation*

$$\psi_n(x) = x j_n(x) = \left(\frac{\pi x}{2}\right)^{\frac{1}{2}} J_{n+\frac{1}{2}}(x),$$

$$\zeta_n(x) = x h_n^{(1)}(x) = \left(\frac{\pi x}{2}\right)^{\frac{1}{2}} H_{n+\frac{1}{2}}^{(1)}(x). \qquad (1.142)$$

Thus, the general solution representation of (1.131) that is discontinuous in coordinate r is

$$\psi(r, \theta, \phi) = \sum_{m=0}^{\infty} (A_m \cos m\phi + B_m \sin m\phi) \times$$

$$\sum_{n=m}^{\infty} \left\{ E_n j_n(kr) + D_n h_n^{(1)}(kr) \right\} P_n^m(\cos\theta). \qquad (1.143)$$

For the prolate spheroidal system (ξ, η, ϕ), introduced in equation (1.92) the scalar wave equation

$$(\Delta + k^2) U = 0 \qquad (1.144)$$

takes the form

$$\frac{\partial}{\partial \xi}\left[(\xi^2 - 1)\frac{\partial U}{\partial \xi}\right] + \frac{\partial}{\partial \eta}\left[(1 - \eta^2)\frac{\partial U}{\partial \eta}\right]$$

$$+ \frac{\xi^2 - \eta^2}{(\xi^2 - 1)(1 - \eta^2)}\frac{\partial^2 U}{\partial \phi^2} + c^2(\xi^2 - \eta^2)U = 0 \qquad (1.145)$$

where c is a dimensionless parameter proportional to the ratio of focal distance d to the wavelength λ: $c = k\frac{d}{2} = \pi\frac{d}{\lambda}$. The method of separation of variables provides solutions of the form

$$U_{mn} = R_{mn}(c, \xi) S_{mn}(c, \eta) \left\{ \begin{array}{c} \cos m\phi \\ \sin m\phi \end{array} \right\}$$

where $m = 0, 1, 2, \ldots$, and $n = m, m + 1, \ldots$; the functions $R_{mn}(c, \xi)$ ($\xi \in (1, \infty)$) and $S_{mn}(c, \eta)$ ($\eta \in (-1, 1)$) are the prolate radial and angle spheroidal functions, respectively. These functions (with m and n zero or positive integers) appear in many physical applications exploiting spheroidal geometry. They are discussed in depth by Flammer [29].

The prolate radial and angle spheroidal functions satisfy the ordinary differential equations,

$$\frac{d}{d\xi}\left[(\xi^2 - 1)\frac{d}{d\xi}R\right] + \left[-\lambda + c^2\xi^2 - \frac{m^2}{\xi^2 - 1}\right]R = 0, \qquad 1 \leq \xi < \infty$$
(1.146)

$$\frac{d}{d\eta}\left[(1 - \eta^2)\frac{d}{d\eta}S\right] + \left[\lambda - c^2\eta^2 - \frac{m^2}{1 - \eta^2}\right]S = 0, \qquad -1 \leq \eta \leq 1,$$
(1.147)

respectively, where the separation constants $\lambda = \lambda_{mn}(c)$ lie in an infinite discrete set.

The solutions of the equation (1.147) associated with the eigenvalues $\lambda_{mn}(c)$, that are defined over the interval $-1 \leq \eta \leq 1$ and are finite at $\eta = \pm 1$, are the *prolate spheroidal angle functions of the first kind, of order m and degree n*, denoted $S_{mn}(c, \eta)$. They possess a Fourier series expansion in associated Legendre functions (see [29], [50], [122]) of the form

$$S_{mn}(c, \eta) = \sum_{r=0,1}^{\infty}{}' d_r^{mn}(c)P_{m+r}^m(\eta)$$
(1.148)

where the prime over the summation sign indicates that the summation is over only even values of r when $n - m$ is even, and over only odd values of r when $n - m$ is odd. The angle functions $S_{mn}(c, \eta)$ reduce to the associated Legendre functions of the first kind of integer order and degree, as c goes to zero. They are normalised by the requirement that the norm N_{mn} of the p.a.s.f. defined by

$$\int_{-1}^{1} S_{mn}(c, \eta) S_{mn}(c, \eta)\, d\eta = \begin{Bmatrix} N_{mn}^2, & n = m \\ 0, & n \neq m \end{Bmatrix}$$
(1.149)

has the value fixed by Flammer [29] to be

$$N_{mn} = \left\{2 \sum_{r=0,1}^{\infty}{}' \frac{(r + 2m)!\, (d_r^{mn}(c))^2}{(2r + 2m + 1)\, r!}\right\}^{\frac{1}{2}},$$
(1.150)

where the Fourier coefficients $d_r^{mn}(c)$ are those that appear in (1.148).

A related constant that appears in calculations with this function is

$$\chi_{mn}(c) = N_{mn}(c) \frac{c^m i^{m-n}}{2^m m!} \lim_{\eta \to 1-0} \frac{\sqrt{1-\eta^2}}{S_{mn}(c,\eta)} \tag{1.151}$$

$$= N_{mn}(c) c^m i^{m-n} \left(\sum_{r=0,1}^{\infty} {}' d_r^{mn}(c) \frac{(r+2m)!}{r!} \right)^{-1} . \tag{1.152}$$

The *radial prolate spheroidal functions* of first, second, third and fourth kind are defined as the solutions of the equation (1.146) at $\lambda = \lambda_{mn}(c)$, over the range $1 \le \xi < \infty$, that possess the following asymptotics

$$R_{mn}^{(1)}(c,\xi) = \frac{1}{c\xi} \cos\left[c\xi - \frac{\pi}{2}(n+1) \right] + O\left((c\xi)^{-2} \right)$$

$$R_{mn}^{(2)}(c,\xi) = \frac{1}{c\xi} \sin\left[c\xi - \frac{\pi}{2}(n+1) \right] + O\left((c\xi)^{-2} \right)$$

$$R_{mn}^{(3)}(c,\xi) = \frac{1}{c\xi} \exp\left[i\left(c\xi - \frac{\pi}{2}(n+1) \right) \right] + O\left((c\xi)^{-2} \right)$$

$$R_{mn}^{(4)}(c,\xi) = \frac{1}{c\xi} \exp\left[-i\left(c\xi - \frac{\pi}{2}(n+1) \right) \right] + O\left((c\xi)^{-2} \right) \tag{1.153}$$

as $c\xi \to \infty$, respectively. It is obvious that $R_{mn}^{(3)}(c,\xi) = R_{mn}^{(1)}(c,\xi) + iR_{mn}^{(2)}(c,\xi)$ and $R_{mn}^{(4)}(c,\xi) = R_{mn}^{(1)}(c,\xi) - iR_{mn}^{(2)}(c,\xi)$. Any two functions chosen from these four are linearly independent.

When $\frac{d}{2}$ goes to zero and ξ goes to infinity, so that $\frac{d}{2}\xi = r$ remains constant, the four functions (1.153) reduce to the spherical Bessel, Neumann and Hankel functions of the first and second kind, respectively. In this book we use the radial functions of first and third kinds; their Wronskian equals

$$W\left[R_{1l}^{(1)}(c,\xi), R_{1l}^{(3)}(c,\xi) \right] =$$

$$R_{1l}^{(1)}(c,\xi) \frac{d}{d\xi} R_{1l}^{(3)}(c,\xi) - R_{1l}^{(3)}(c,\xi) \frac{d}{d\xi} R_{1l}^{(1)}(c,\xi) = \frac{i}{c(\xi^2-1)}. \tag{1.154}$$

The radial functions of the first and second kinds have the following series representations in spherical Bessel functions $j_{m+r}(c\xi)$ and spherical Neumann functions $n_{m+r}(c\xi)$,

$$\left. \begin{array}{c} R_{mn}^{(1)}(c,\xi) \\ R_{mn}^{(2)}(c,\xi) \end{array} \right\} = \left(\sum_{r=0,1}^{\infty} {}' \frac{(2m+r)!}{r!} d_r^{mn}(c) \right)^{-1} \left(\frac{\xi^2-1}{\xi^2} \right)^{m/2} \times$$

$$\sum_{r=0,1}^{\infty} {}' \left(i^{r+m-n} \right) \frac{(2m+r)!}{r!} d_r^{mn}(c) \left\{ \begin{array}{c} j_{m+r}(c\xi) \\ n_{m+r}(c\xi) \end{array} \right. \tag{1.155}$$

where the Fourier coefficients $d_r^{mn}(c)$ are those that appear in (1.148).

In the axisymmetric form of Maxwell's equations, the method of separation of variables is applicable. If solutions of the form $H_\phi(\xi, \eta) = R(\xi)\, S(\eta)$ are sought to Equation (1.100), the pair of ordinary differential equations

$$\frac{d}{d\eta}\left[(1 - \eta^2)\frac{d}{\partial\eta}S\right] + \left[\lambda - c^2\eta^2 - \frac{1}{1 - \eta^2}\right]S = 0, \qquad -1 \le \eta \le 1 \quad (1.156)$$

$$\frac{d}{d\xi}\left[(\xi^2 - 1)\frac{d}{d\xi}R\right] + \left[-\lambda + c^2\xi^2 - \frac{1}{\xi^2 - 1}\right]R = 0, \qquad 1 \le \xi < \infty,$$
$$(1.157)$$

with a separation constant λ is obtained. The angle $S(\eta) \equiv S_{1l}(c, \eta)$ and radial $R(\xi) = R_{1l}(c, \xi)$ prolate spheroidal functions with the index $m = 1$ are solutions to these equations.

The field of magnetic type (H_ξ, H_η, E_ϕ) is governed by the system of equations (1.94)–(1.96); a second order partial differential equation for the component $E_\phi(\xi, \eta)$ is readily deduced. It is also separable, with partial solutions of the form $E_\phi(\xi, \eta) = R(\xi)\, S(\eta)$ satisfying (1.156) and (1.157).

Also, the systems (1.94)–(1.96) and (1.97)–(1.99) allow separation of variables for the scalar functions $P(\xi, \eta) = h_\phi E_\phi$ and $Q(\xi, \eta) = h_\phi H_\phi$, respectively; P and Q are known as *Abraham potentials*.

1.3   Fields of Elementary Sources. Green's Functions.

In the previous section we found various solutions of the homogeneous Helmholtz equation that arises in representing solutions to the acoustic wave equation or to Maxwell's equations. As formulated in Maxwell's equations (1.25), the sources of electromagnetic fields are the currents $\vec{J}$ and charges ρ. We wish to consider compact source distributions for these equations. Useful mathematical models of real sources are the so-called *point sources*, simulating real compact sources by Dirac δ-functions with the assumption that currents or charges are concentrated at a single mathematical point of space. This representation of acoustical and electromagnetic point sources plays an important part in the construction of solutions to wave-scattering problems.

Thus, we search for solutions to the inhomogeneous Helmholtz equation with the Dirac δ-function,

$$\nabla^2 G + k^2 G = -\delta\left(\vec{r} - \vec{r'}\right); \qquad (1.158)$$

the solution $G\left(\overrightarrow{r}, \overrightarrow{r'}\right)$ is known as a Green's function in free space. (In this equation $\overrightarrow{r}$ and $\overrightarrow{r'}$ are the variable and fixed points, respectively.) It can be readily verified that in three-dimensional space the solution of (1.158) that is radially symmetric about $\overrightarrow{r} = \overrightarrow{r'}$ is given by

$$G_3\left(\overrightarrow{r}, \overrightarrow{r'}\right) = \frac{1}{4\pi} \frac{e^{ik\left|\overrightarrow{r} - \overrightarrow{r'}\right|}}{\left|\overrightarrow{r} - \overrightarrow{r'}\right|}. \tag{1.159}$$

In two-dimensional space the Green's function is

$$G_2\left(\overrightarrow{r}, \overrightarrow{r'}\right) = -\frac{i}{4} H_0^{(1)}\left(k\left|\overrightarrow{r} - \overrightarrow{r'}\right|\right). \tag{1.160}$$

As $k \to 0$ the inhomogeneous Helmholtz equation (1.158) transforms into Poisson's equation (see [1.198] of Volume I), and the solutions of inhomogeneous Helmholtz equation (1.159), (1.160) are transformed, in the three-dimensional case, to

$$G_3\left(\overrightarrow{r}, \overrightarrow{r'}\right) = \frac{1}{4\pi} \frac{1}{\left|\overrightarrow{r} - \overrightarrow{r'}\right|}$$

and, in the two-dimensional case, to

$$G_2\left(\overrightarrow{r}, \overrightarrow{r'}\right) = -\frac{i}{4\pi} \log\left(\left|\overrightarrow{r} - \overrightarrow{r'}\right|\right).$$

Let us consider the solution (1.159) under the condition that k is nonzero, but $\left|\overrightarrow{r} - \overrightarrow{r'}\right| \to 0$. Then if $\left|\overrightarrow{r} - \overrightarrow{r'}\right| \ll 1$,

$$G_3\left(\overrightarrow{r}, \overrightarrow{r'}\right) = \frac{1}{4\pi}\left\{\frac{1}{\left|\overrightarrow{r} - \overrightarrow{r'}\right|} + ik - \frac{1}{2}\left|\overrightarrow{r} - \overrightarrow{r'}\right| + O\left(\left|\overrightarrow{r} - \overrightarrow{r'}\right|^2\right)\right\}.$$
$$\tag{1.161}$$

This means that in the vicinity of the point source the electromagnetic field is of a *static* nature. In other words, the source singularity is always due to the static term that can be easily extracted from the wave function $G_3\left(\overrightarrow{r}, \overrightarrow{r'}\right)$ or $G_2\left(\overrightarrow{r}, \overrightarrow{r'}\right)$. Thus the wave process $(k > 0)$ merely modifies the static Green's function by the addition of a smooth term. This observation lies at the root of the Method of Regularisation (MoR) to be exploited intensively in this book.

Let us now reproduce in a very formal manner further useful representations of the two- and three-dimensional Green's functions in Cartesian (x, y, z), polar cylindrical (ρ, ϕ, z) and polar spherical (r, θ, ϕ) coordinates.

Since $G_3\left(\overrightarrow{r}, \overrightarrow{r'}\right)$ satisfies the homogeneous Helmholtz equation when $\overrightarrow{r} \neq \overrightarrow{r'}$, and is a symmetric function of the primed and unprimed coordinates, we may represent $G_3\left(\overrightarrow{r}, \overrightarrow{r'}\right) \equiv G_3\left(x, y, z; x', y', z'\right)$ in terms of the eigenfunctions of the Helmholtz equation (see [1.107] and [1.108]),

$$G_3\left(\overrightarrow{r}, \overrightarrow{r'}\right) =$$
$$\int_0^\infty \cos\nu\,(x - x') \int_0^\infty f(\nu, \mu)\cos\mu\,(y - y')\,e^{-\sqrt{\nu^2 + \mu^2 - k^2}\,|z - z'|}\,d\nu\,d\mu$$
$$(1.162)$$

where the condition $\mathrm{Im}\left(\sqrt{\nu^2 + \mu^2 - k^2}\right) \leq 0$ is imposed to ensure that the solution satisfies the *radiation conditions* at infinity (to be discussed in Section 1.5). Furthermore, $f(\nu, \mu)$ is an unknown function of two spectral parameters ν and μ, to be determined. Substitute this form into equation (1.158), and integrate with respect to z over a small interval $(z' - \varepsilon, z' + \varepsilon)$ about z'. Remembering the continuity of the solution at $z = z'$ and passing to the limit $\varepsilon \to 0$, we obtain

$$\frac{\partial}{\partial z} G_3\left(x, y, z; x', y', z'\right)\Big|_{z=z'-0}^{z=z'+0} = -\delta\,(x - x')\,\delta\,(y - y') \qquad (1.163)$$

or

$$2\int_0^\infty \cos\nu\,(x - x') \int_0^\infty f(\nu, \mu)\sqrt{\nu^2 + \mu^2 - k^2}\cos\mu\,(y - y')\,d\nu\,d\mu$$
$$= \delta\,(x - x')\,\delta\,(y - y'). \qquad (1.164)$$

Two applications of the Fourier cosine integral transform to equation (1.164) show that

$$f(\nu, \mu) = \frac{1}{2\pi^2}\frac{1}{\sqrt{\nu^2 + \mu^2 - k^2}}. \qquad (1.165)$$

Thus, the function $G_3\left(\overrightarrow{r}, \overrightarrow{r'}\right)$ is representable as

$$G_3\left(\overrightarrow{r}, \overrightarrow{r'}\right) =$$
$$\frac{1}{2\pi}\int_0^\infty \cos\nu\,(x - x') \int_0^\infty \frac{\cos\mu\,(y - y')}{\sqrt{\nu^2 + \mu^2 - k^2}}e^{-\sqrt{\nu^2 + \mu^2 - k^2}\,|z - z'|}\,d\nu\,d\mu \qquad (1.166)$$

where $\mathrm{Im}\left(\sqrt{\nu^2 + \mu^2 - k^2}\right) \leq 0$.

We may interpret this representation as an integrated spectrum of propagating and evanescent waves. For a lossless medium in the case when $\nu^2 + \mu^2 < k^2$, we have $\sqrt{\nu^2 + \mu^2 - k^2} = -i\sqrt{k^2 - \nu^2 - \mu^2}$. The integrand (1.166) then describes a spectrum of propagating waves, running from the plane $z = z'$ along the z-axis in both positive and negative directions. The propagation velocity of these waves v, defined by $v = \omega / \left|\sqrt{\nu^2 + \mu^2 - k^2}\right|$, is bigger than the velocity of light c. Uniform-phase wave-fronts are planar, but not coinciding with the planes $z = const$; the uniform-phase wave-fronts tilt with respect to the plane $z = const$ varies, depending upon values of ν and μ. If the uniform-phase wave-fronts of a plane wave does not coincide with the transverse plane with respect to the direction of its propagation, it is called an *inhomogeneous* plane wave. Thus, expression (1.166) can be treated when $\nu^2 + \mu^2 < k^2$ as a continuous spectrum of inhomogeneous plane waves, travelling along the z-axis with different phase speeds. If $\nu^2 + \mu^2 > k^2$ the value $\sqrt{\nu^2 + \mu^2 - k^2}$ is real valued, and the waves are evanescent, i.e., oscillations of exponentially decreasing amplitude as $|z - z'|$ increases. The uniform-phase fronts of this spectrum also do not coincide with the planes $z = const$.

It should be noted that in the derivation of the expression (1.166), the x-axis or the y-axis could equally be taken as the axis of propagation of the inhomogeneous plane waves.

If the dependence upon one of the coordinates is absent, for example the y-coordinate (so $\partial/\partial y \equiv 0$), we obtain a two-dimensional inhomogeneous Helmholtz equation to solve,

$$\frac{\partial^2 G_2}{\partial x^2} + \frac{\partial^2 G_2}{\partial z^2} + k^2 G_2 = -\delta\left(x - x'\right)\delta\left(z - z'\right). \tag{1.167}$$

Making use of the same approach the corresponding two-dimensional Green's function is representable as

$$G_2\left(x, z; x', z'\right) = \frac{1}{2\pi} \int_0^\infty \frac{\cos\nu\left(x - x'\right)}{\sqrt{\nu^2 - k^2}} e^{-\sqrt{\nu^2 - k^2}|z - z'|} d\nu \tag{1.168}$$

where $\text{Im}\left(\sqrt{\nu^2 - k^2}\right) \leq 0$.

In the same way let us find the Green's function in polar cylindrical coordinates. First we construct the Green's function $G_3\left(\rho, \phi, z; \rho', \phi', z'\right)$ that is discontinuous in coordinate z. According to equation (1.159),

$$G_3\left(\overrightarrow{r}, \overrightarrow{r'}\right) = \frac{1}{4\pi} \frac{e^{ikR}}{R} \tag{1.169}$$

where

$$R = \sqrt{(x - x')^2 + (y - y')^2 + (z - z')^2}$$
$$= \sqrt{\rho^2 - 2\rho\rho' \cos (\phi - \phi') + \rho'^2 + (z - z')^2}. \quad (1.170)$$

The inhomogeneous Helmholtz equation to be solved is

$$\frac{1}{\rho} \frac{\partial}{\partial \rho} \left(\rho \frac{\partial G_3}{\partial \rho} \right) + \frac{1}{\rho^2} \frac{\partial^2 G_3}{\partial \phi^2} + \frac{\partial^2 G_3}{\partial z^2} + k^2 G_3$$
$$= -\frac{1}{\rho} \delta (\rho - \rho') \delta (\phi - \phi') \delta (z - z'). \quad (1.171)$$

With reference to the formula (1.122) we seek a solution for $G_3 \left(\overrightarrow{r}, \overrightarrow{r'} \right)$ in the form

$$G_3 \left(\rho, \phi, z; \rho', \phi', z' \right) = \sum_{m=0}^{\infty} \left(2 - \delta_m^0 \right) \cos m \left(\phi - \phi' \right) \times$$
$$\int_0^{\infty} F_m (\nu) J_m (\nu \rho') J_m (\nu \rho) e^{-\sqrt{\nu^2 - k^2}|z - z'|} d\nu \quad (1.172)$$

where $\mathrm{Im} \left(\sqrt{\nu^2 - k^2} \right) < 0$. As before, we find

$$\frac{\partial}{\partial z} G_3 \left(\rho, \phi, z; \rho', \phi', z' \right) \Big|_{z=z'-0}^{z=z'+0} = -\frac{1}{\rho} \delta (\rho - \rho') \delta (\phi - \phi'). \quad (1.173)$$

Making use of the representations of the Dirac δ-function in cylindrical coordinates

$$\frac{1}{2\pi} \sum_{m=0}^{\infty} \left(2 - \delta_m^0 \right) \cos m \left(\phi - \phi' \right) = \delta \left(\phi - \phi' \right),$$
$$\int_0^{\infty} \rho J_m (\nu \rho) J_m (\mu \rho) d\rho = \frac{1}{\mu} \delta (\nu - \mu), \quad (1.174)$$

the desired representation is

$$G_3 \left(\rho, \phi, z; \rho', \phi', z' \right) = \frac{1}{4\pi} \sum_{m=0}^{\infty} \left(2 - \delta_m^0 \right) \cos m \left(\phi - \phi' \right) \times$$
$$\int_0^{\infty} \frac{e^{\pm \sqrt{\nu^2 - k^2}|z - z'|} J_m (\nu \rho') J_m (\nu \rho)}{\sqrt{\nu^2 - k^2}} \nu d\nu \quad (1.175)$$

where $\mathrm{Im}\left(\sqrt{\nu^2 - k^2}\right) \leq 0$.

This expression represents an infinite spectrum of plane-cylindrical waves, propagating along the z-axis in both positive and negative directions; when $\nu^2 - k^2 < 0$, the waves are propagating, but if $\nu^2 - k^2 > 0$ they are evanescent (exponentially damped). At first glance it is not obvious that the radiation condition is satisfied as $\rho \to \infty$. However taking into account the well-known relations

$$
\begin{aligned}
J_m\left(\nu\rho'\right) &= (-1)^m J_m\left(-\nu\rho'\right), \\
J_m\left(\nu\rho\right) &= \frac{1}{2}\left[H_m^{(1)}\left(\nu\rho\right) - (-1)^m H_m^{(1)}\left(-\nu\rho\right)\right], \quad (1.176)
\end{aligned}
$$

the expression (1.175) may be converted to the following form,

$$
G_3\left(\rho, \phi, z; \rho', \phi', z'\right) = \frac{1}{8\pi} \sum_{m=0}^{\infty}\left(2 - \delta_m^0\right) \cos m\left(\phi - \phi'\right) \times
$$

$$
\int_{-\infty}^{\infty} e^{-\sqrt{\nu^2-k^2}|z-z'|} \frac{\nu\, d\nu}{\sqrt{\nu^2 - k^2}}
\begin{cases}
H_m^{(1)}\left(\nu\rho'\right) J_m\left(\nu\rho\right), & \rho < \rho' \\
J_m\left(\nu\rho'\right) H_m^{(1)}\left(\nu\rho\right), & \rho > \rho',
\end{cases} \quad (1.177)
$$

where $\mathrm{Im}\left(\sqrt{\nu^2 - k^2}\right) \leq 0$. It can be seen that the expressions for spectral density of these waves are different for $\rho < \rho'$ and $\rho > \rho'$ and satisfy the requirements that the electromagnetic field be finite at $\rho = 0$ and satisfy the radiation condition at infinity ($\rho \to \infty$).

Let us now deduce another representation of the Green's function G_3 in polar cylindrical coordinates that is discontinuous in coordinate ρ. Using the same procedure, we seek the solution in the form, defined by the general solution (1.127),

$$
G_3\left(\rho, \phi, z; \rho', \phi', z'\right) = \sum_{m=0}^{\infty}\left(2 - \delta_m^0\right) \cos m\left(\phi - \phi'\right) \times
$$

$$
\int_{0}^{\infty} d\nu\, A_m\left(\nu\right) \cos \nu\left(z - z'\right)
\begin{cases}
K_m\left(\sqrt{\nu^2 - k^2}\rho'\right) I_m\left(\sqrt{\nu^2 - k^2}\rho\right), & \rho < \rho' \\
I_m\left(\sqrt{\nu^2 - k^2}\rho'\right) K_m\left(\sqrt{\nu^2 - k^2}\rho\right), & \rho > \rho'
\end{cases}
$$

$$(1.178)$$

where $A_m\left(\nu\right)$ is an unknown function to be determined from equation (1.171). By construction the form (1.178) already satisfies the continuity condition at $\rho = \rho'$ and exhibits the correct behaviour at the special points $\rho = 0$ and $\rho \to \infty$. As before, we also find that the following relation is true,

$$
\rho \frac{\partial}{\partial \rho} G_3 \left(\rho, \phi, z; \rho', \phi', z' \right) \Big|_{\rho=\rho'-0}^{\rho=\rho'+0} = -\delta \left(\rho - \rho' \right) \delta \left(\phi - \phi' \right). \tag{1.179}
$$

To derive the desired expression we use the representation of the Dirac δ-function in polar cylindrical coordinates,

$$
\frac{1}{\pi} \int_0^\infty \cos \nu \left(z - z' \right) d\nu = \delta \left(z - z' \right), \tag{1.180}
$$

and the value of the Wronskian of the pair I_m, K_m,

$$
W \left(I_m \left(z \right), K_m \left(z \right) \right) = -z^{-1}. \tag{1.181}
$$

A routine transformation produces the desired result

$$
G_3 \left(\rho, \varphi, z; \rho', \varphi', z' \right) = -\frac{1}{2\pi^2} \sum_{m=0}^\infty \left(2 - \delta_m^0 \right) \cos m \left(\varphi - \varphi' \right) \times
$$

$$
\int_0^\infty d\nu \cos \nu \left(z - z' \right) \left\{ \begin{array}{l} K_m \left(\sqrt{\nu^2 - k^2} \rho' \right) I_m \left(\sqrt{\nu^2 - k^2} \rho \right), \rho < \rho' \\ I_m \left(\sqrt{\nu^2 - k^2} \rho' \right) K_m \left(\sqrt{\nu^2 - k^2} \rho \right), \rho > \rho' \end{array} \right. \tag{1.182}
$$

where $\operatorname{Im} \left(\sqrt{\nu^2 - k^2} \right) > 0$. Using the well-known relations

$$
\begin{aligned}
I_m \left(\sqrt{\nu^2 - k^2} \rho \right) &= (-i)^m J_m \left(\sqrt{k^2 - \nu^2} \rho \right) \\
K_m \left(\sqrt{\nu^2 - k^2} \rho \right) &= \frac{\pi}{2} i^{m+1} H_m^{(1)} \left(\sqrt{k^2 - \nu^2} \rho \right)
\end{aligned}
$$

an alternative representation of equation (1.182) is

$$
G_3 \left(\rho, \varphi, z; \rho', \varphi', z' \right) = \frac{1}{4\pi i} \sum_{m=0}^\infty \left(2 - \delta_m^0 \right) \cos m \left(\varphi - \varphi' \right) \times
$$

$$
\int_0^\infty d\nu \cos \nu \left(z - z' \right) \left\{ \begin{array}{l} H_m^{(1)} \left(\sqrt{k^2 - \nu^2} \rho' \right) J_m \left(\sqrt{k^2 - \nu^2} \rho \right), \rho < \rho' \\ J_m \left(\sqrt{k^2 - \nu^2} \rho' \right) H_m^{(1)} \left(\sqrt{k^2 - \nu^2} \rho \right), \rho > \rho' \end{array} \right. \tag{1.183}
$$

where $\operatorname{Im} \left(\sqrt{k^2 - \nu^2} \right)$ must be chosen nonnegative. This representation is an infinite spectrum of cylindrical waves propagating along the radial direction with modulation in the z direction. When $\rho > \rho'$ and $\nu^2 < k^2$ the contribution is damped as is obvious from the asymptotics of the Hankel function

$$H_m^{(1)}(\mu\rho) \approx \sqrt{\frac{2}{\pi\mu\rho}} e^{i\left(\mu\rho - m\frac{\pi}{2} - \frac{\pi}{4}\right)} \tag{1.184}$$

as $\mu\rho \to \infty$, with $\mu = \sqrt{k^2 - \nu^2}$.

If there is no dependence on the z-coordinate z, we obtain the two-dimensional Green's function $G_2(\rho, \varphi; \rho', \varphi')$. We make use of the Fourier method to obtain the representation

$$G_2(\rho, \varphi; \rho', \varphi') =$$
$$\sum_{m=0}^{\infty} (2 - \delta_m^0) a_m \left\{ \begin{array}{l} H_m^{(1)}(k\rho') J_m(k\rho), \rho < \rho' \\ J_m(k\rho') H_m^{(1)}(k\rho), \rho > \rho' \end{array} \right\} \cos m(\varphi - \varphi'). \tag{1.185}$$

By construction $G_2\left(\overrightarrow{r}, \overrightarrow{r'}\right)$ is continuous at the point $\rho = \rho'$ and exhibits the correct behaviour at the special points $\rho = 0$ and $\rho \to \infty$; it remains to find the unknown coefficients a_m. Applying familiar arguments we easily find the desired representation

$$G_2(\rho, \varphi; \rho', \varphi') =$$
$$-\frac{i}{4} \sum_{m=0}^{\infty} (2 - \delta_m^0) \left\{ \begin{array}{l} H_m^{(1)}(k\rho') J_m(k\rho), \rho < \rho' \\ J_m(k\rho') H_m^{(1)}(k\rho), \rho > \rho' \end{array} \right\} \cos m(\varphi - \varphi'). \tag{1.186}$$

The result (1.186) can be obtained in an alternative way from the *addition theorem* [3] for the Hankel function of zero order

$$H_0^{(1)}\left(k\left|\overrightarrow{r} - \overrightarrow{r'}\right|\right) = H_0^{(1)}\left(k\sqrt{\rho^2 + \rho'^2 - 2\rho\rho' \cos(\varphi - \varphi')}\right)$$
$$= \sum_{m=0}^{\infty} (2 - \delta_m^0) \left\{ \begin{array}{l} H_m^{(1)}(k\rho') J_m(k\rho), \rho < \rho' \\ J_m(k\rho') H_m^{(1)}(k\rho), \rho > \rho' \end{array} \right\} \cos m(\varphi - \varphi'). \tag{1.187}$$

Finally we find the three-dimensional Green's function $G_3\left(\overrightarrow{r}, \overrightarrow{r'}\right)$ in polar spherical coordinates that is discontinuous in coordinate r. First represent equation (1.158) in spherical coordinates (r, θ, φ),

$$\frac{1}{r^2}\frac{\partial}{\partial r}\left(r^2 \frac{\partial G_3}{\partial r}\right) + \frac{1}{r^2 \sin\theta}\frac{\partial}{\partial \theta}\left(\sin\theta \frac{\partial G_3}{\partial \theta}\right) + \frac{1}{r^2 \sin^2\theta}\frac{\partial^2 G_3}{\partial \varphi^2} + k^2 G_3$$
$$= -\frac{1}{r^2 \sin\theta}\delta(r - r')\delta(\theta - \theta')\delta(\varphi - \varphi'). \tag{1.188}$$

Make use of the general solution (1.143) to construct a solution that is valid at all points of space except $\vec{r} = \vec{r}'$. This gives

$$G_3\left(\rho, \varphi, z; \rho', \varphi', z'\right) = \sum_{m=0}^{\infty} \left(2 - \delta_m^0\right) \cos m \left(\varphi - \varphi'\right) \times$$

$$\sum_{n=m}^{\infty} a_n^m P_n^m \left(\cos \theta'\right) P_n^m \left(\cos \theta\right) \begin{cases} h_n^{(1)}\left(kr'\right) j_n\left(kr\right), & r < r' \\ j_n\left(kr'\right) h_n^{(1)}\left(kr\right), & r > r'. \end{cases} \quad (1.189)$$

As usual, we may deduce from equation (1.188) that

$$r^2 \frac{\partial G_3}{\partial r}\bigg|_{r=r'-0}^{r=r'+0} = -\frac{1}{\sin \theta} \delta\left(\theta - \theta'\right) \delta\left(\varphi - \varphi'\right). \quad (1.190)$$

Using the representation of the Dirac δ–function in spherical coordinates,

$$\frac{1}{2} \sum_{n=m}^{\infty} \left(2n + 1\right) \frac{(n-m)!}{(n+m)!} P_n^m \left(\cos \theta'\right) P_n^m \left(\cos \theta\right) = \frac{1}{\sin \theta'} \delta\left(\theta - \theta'\right) \quad (1.191)$$

and taking into account the value of Wronskian of the pair $j_n, h_n^{(1)}$,

$$W\left\{j_n\left(z\right), h_n^{(1)}\left(z\right)\right\} = iz^{-2}, \quad (1.192)$$

we find

$$a_n^m = \frac{ik}{4\pi} \left(2n + 1\right) \frac{(n-m)!}{(n+m)!},$$

and obtain the desired result

$$G_3\left(\rho, \theta, \varphi; \rho', \theta', \varphi'\right) = \frac{ik}{4\pi} \sum_{m=0}^{\infty} \left(2 - \delta_m^0\right) \cos m \left(\varphi - \varphi'\right) \times$$

$$\sum_{n=m}^{\infty} \left(2n + 1\right) \frac{(n-m)!}{(n+m)!} P_n^m \left(\cos \theta'\right) P_n^m \left(\cos \theta\right) \begin{cases} h_n^{(1)}\left(kr'\right) j_n\left(kr\right), & r < r' \\ j_n\left(kr'\right) h_n^{(1)}\left(kr\right), & r > r'. \end{cases}$$

$$(1.193)$$

The location of the point source in free space can be chosen at will. Notice that if $r' = 0$, $j_n\left(kr'\right) = \delta_{0n}$, and the expression (1.193) collapses to a single term,

$$G_3 = \frac{ik}{4\pi} h_0^{(1)}\left(kr\right) = \frac{1}{4\pi} \frac{e^{ikr}}{r}; \quad (1.194)$$

this is the well-known expression for the three-dimensional Green's function in closed form.

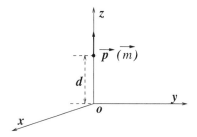

FIGURE 1.1. Vertical electric and magnetic dipoles.

1.4   Representations of Incident Electromagnetic Waves.

The results obtained in previous sections may be used to provide various representations of incident electromagnetic waves that will be used throughout this book. We represent electromagnetic waves by scalar functions or scalar potentials. Of course, alternative representations can be used, as, in essence, they must be equivalent.

Incident electromagnetic waves are simulated by point sources or plane waves. The simplest representations are obtained for vertical electric or magnetic dipoles. The term *vertical* always means that the dipole is located on the z-axis and its moment is aligned parallel to the z-axis, irrespective of the coordinate system employed (Cartesian, cylindrical or spherical) (see Figure 1.1).

Vertical electric and magnetic dipoles are described by the z-components of the electric and magnetic Hertz vectors, respectively. If the dipole moments are unity $(|\vec{p}| = |\vec{m}| = 1)$, the relevant Hertz vectors are

$$\vec{\Pi} = \vec{i}_z G_3, \quad \vec{\Pi}^{(m)} = \vec{i}_z G_3. \tag{1.195}$$

If the dipole is located at $(x', y', z') = (0, 0, d)$, it follows that

$$G_3(x, y, z; 0, 0, d) =$$
$$\frac{1}{2\pi^2} \int_0^\infty \cos \nu x \int_0^\infty \frac{\cos \mu y}{\sqrt{\nu^2 + \mu^2 - k^2}} e^{-\sqrt{\nu^2 + \mu^2 - k^2}|z - d|} d\nu d\mu \tag{1.196}$$

where $\operatorname{Im}\left(\sqrt{\nu^2 + \mu^2 - k^2}\right) \leq 0$. The electromagnetic field components are now found from formulae (1.51)–(1.54), noting that the vertical electric dipole radiates TM waves, whilst the vertical magnetic dipole radiates TE waves.

The corresponding polar cylindrical coordinate representation is derived by setting $\rho = 0$, $z = d$ in (1.175), giving the ϕ, ϕ' independent form

$$G_2\left(\rho, z; 0, d\right) \equiv G_3\left(\rho, \phi, z; 0, \phi', d\right) = \frac{1}{4\pi} \int_0^\infty \frac{J_0\left(\nu\rho\right)}{\sqrt{\nu^2 - k^2}} \nu e^{-\sqrt{\nu^2 - k^2}|z - d|} d\nu,$$

(1.197)

where $\mathrm{Im}\left(\sqrt{\nu^2 - k^2}\right) \leq 0$. It is rather instructive to deduce equation (1.197) directly from equation (1.196). Set $x = \rho\cos\phi$, $y = \rho\sin\phi$ and use the substitutions $\tau = \sqrt{\nu^2 + \mu^2}$, $\mu = \sqrt{\tau^2 - \nu^2}$ and

$$d\mu = \frac{\tau d\tau}{\sqrt{\tau^2 - \nu^2}}.$$

Then use the Dirichlet formula (see Section 1.5 of Volume I) to interchange the integration order in (1.196) and obtain

$$G_3 = \frac{1}{2\pi} \int_0^\infty \frac{\tau e^{-\sqrt{\tau^2 - k^2}|z - d|}}{\sqrt{\tau^2 - k^2}} \int_0^\tau \frac{\cos\left(\rho\sqrt{\tau^2 - \nu^2}\sin\phi\right)}{\sqrt{\tau^2 - \nu^2}} \cos\left(\rho\nu\cos\phi\right) d\nu d\tau$$

(1.198)

The inner integral is easily evaluated [30] to be

$$\int_0^\tau \frac{\cos\left(\rho\sqrt{\tau^2 - \nu^2}\sin\phi\right)}{\sqrt{\tau^2 - \nu^2}} \cos\left(\rho\nu\cos\phi\right) d\nu = \frac{\pi}{2} J_0\left(\tau\rho\right)$$

(1.199)

and we again obtain the representation (1.197).

The electromagnetic field components radiated by the vertical electric or magnetic dipole may be calculated from the formulae (1.55). An alternative way is to obtain the "generating" functions H_ϕ (for TM waves) and E_ϕ (for TE waves) and use the formulae (1.86)–(1.87). The "generating" functions are extracted from formulae (1.55),

$$H_\phi = ik\frac{\partial\Pi_z}{\partial\rho}, \quad E_\phi = -ik\frac{\partial\Pi_z^{(m)}}{\partial\rho}.$$

From the representation (1.197) we obtain the desired result

$$\left\{ \begin{array}{c} H_\phi \\ E_\phi \end{array} \right\} = \left\{ \begin{array}{c} -ik \\ ik \end{array} \right\} \frac{1}{4\pi} \int_0^\infty \frac{J_1\left(\nu\rho\right)}{\sqrt{\nu^2 - k^2}} e^{-\sqrt{\nu^2 - k^2}|z - d|} d\nu$$

(1.200)

where $\mathrm{Im}\left(\sqrt{\nu^2 - k^2}\right) \leq 0$.

Let us now find the function G_3 that describes the electromagnetic field radiated by an electric or magnetic dipole in spherical coordinates. Set

$r' = d$ and $\theta' = 0$ in (1.193), and noting that $P_n^m(1) = 0$ for all $m > 0$, the result is

$$G_3(r, \theta; d, 0) = \frac{i}{4\pi} \cdot \frac{1}{kdr} \sum_{n=0}^{\infty} (2n+1) \left\{ \begin{array}{l} \zeta_n(kd)\,\psi_n(kr)\,, r < d \\ \psi_n(kd)\,\zeta_n(kr)\,, r > d \end{array} \right\} P_n(\cos\theta)$$

(1.201)

where the functions ψ_n and ζ_n were defined in equation (1.142).

There are two descriptions of the electromagnetic field radiated by the vertical dipole. The first uses the "generating" functions H_ϕ and E_ϕ with the formulae (1.89)–(1.90). This approach is restricted to the axially symmetric case $(\partial/\partial\phi \equiv 0)$. The general case employs the two scalar functions introduced in Section 1.1, the *electric Debye potential U* and the *magnetic Debye potential V*. Both potentials satisfy the Helmholtz equation

$$\nabla^2 U + k^2 U = \nabla^2 V + k^2 V = 0,$$

(1.202)

and the related electromagnetic fields are given by the formulae (1.80). (To avoid confusion with the formulae in [41], note that we use the harmonic time dependence $\exp(-i\omega t)$ instead of $\exp(+i\omega t)$).

In terms of the "generating" functions H_ϕ or E_ϕ the electromagnetic field of the vertical electric (magnetic) dipole uses formulae (1.201) and the representation of the electric (magnetic) z-component of Hertz vector $\overrightarrow{\Pi}$ in spherical coordinates, where

$$\overrightarrow{\Pi} = \overrightarrow{i_r}\, G_3 \cos\theta + \overrightarrow{i_\theta}\,(-G_3 \sin\theta)\,;$$

(1.203)

recall that

$$G_3 = \frac{1}{4\pi} \frac{e^{ikR}}{R},$$

(1.204)

where $R = \sqrt{r^2 - 2dr\cos\theta + d^2}$. We make use of the relations

$$H_\phi = -ik \left(\mathrm{curl}\, \overrightarrow{\Pi}\right)_\phi = -\frac{ik}{r}\left[\frac{\partial}{\partial r}\left(r\Pi_\theta - \frac{\partial}{\partial\theta}\Pi_r\right)\right],$$

(1.205)

$$E_\phi = ik \left(\mathrm{curl}\, \overrightarrow{\Pi}^{(m)}\right)_\phi = \frac{ik}{r}\left[\frac{\partial}{\partial r}\left(r\Pi_\theta^{(m)}\right) - \frac{\partial}{\partial\theta}\Pi_r^{(m)}\right]$$

(1.206)

and with reference to formula (1.203) obtain

$$\left\{ \begin{array}{c} H_\phi \\ E_\phi \end{array} \right\} = \left\{ \begin{array}{c} ik \\ -ik \end{array} \right\} \left[\sin\theta \frac{\partial G_3}{\partial r} + \frac{\cos\theta}{r}\frac{\partial G_3}{\partial\theta}\right].$$

(1.207)

To transform (1.207) we need the partial derivatives

$$\frac{\partial G_3}{\partial r} = \frac{1}{4\pi}\frac{ikR-1}{R^3}e^{ikR}\left(r - d\cos\theta\right),$$

$$\frac{\partial G_3}{\partial \theta} = \frac{1}{4\pi}\frac{ikR-1}{R^3}e^{ikR}dr\sin\theta,$$

$$\frac{\partial G_3}{\partial d} = \frac{1}{4\pi}\frac{ikR-1}{R^3}e^{ikR}\left(d - r\cos\theta\right), \qquad (1.208)$$

from which it follows that

$$\sin\theta\frac{\partial G_3}{\partial r} + \frac{\cos\theta}{r}\frac{\partial G_3}{\partial \theta} = \frac{1}{d}\frac{\partial G_3}{\partial \theta}, \qquad (1.209)$$

$$\frac{\partial G_3}{\partial r} + \cos\theta\frac{\partial G_3}{\partial d} = \frac{\sin\theta}{d}\frac{\partial G_3}{\partial \theta}, \qquad (1.210)$$

$$\frac{\sin\theta}{d}\frac{\partial G_3}{\partial \theta} - \cos\theta\frac{\partial G_3}{\partial d} = \frac{\partial G_3}{\partial r}. \qquad (1.211)$$

Taking into account (1.209), equation (1.207) becomes

$$\left\{\begin{array}{c} H_\phi \\ E_\phi \end{array}\right\} = \left\{\begin{array}{c} ik \\ -ik \end{array}\right\}\frac{1}{d}\frac{\partial G_3}{\partial \theta}, \qquad (1.212)$$

and using equation (1.201) we finally deduce

$$\left\{\begin{array}{c} H_\phi \\ E_\phi \end{array}\right\} = \frac{\pm 1}{4\pi d^2 r}\sum_{n=1}^{\infty}(2n+1)\left\{\begin{array}{c} \zeta_n(kd)\,\psi_n(kr)\,,\, r < d \\ \psi_n(kd)\,\zeta_n(kr)\,,\, r > d \end{array}\right\}P_n^1(\cos\theta).$$

$$(1.213)$$

Alternatively, we may describe the electromagnetic field of a vertical (electric or magnetic) dipole in terms of Debye potentials U and V, respectively. The derivation is based on the fact that the radial (electric or magnetic) components (E_r or H_r) are defined by a single scalar function. Concretely, E_r is defined by the electric Debye potential U, and H_r is defined by the magnetic Debye potential V (see formulae [1.80]). On the other hand we can obtain independently these components, using the expansion (1.101) of the Green function and the relations

$$E_r = \left(\text{curl curl }\vec{\Pi}\right)_r, \quad H_r = \left(\text{curl curl }\vec{\Pi}^{(m)}\right)_r. \qquad (1.214)$$

The comparison leads to the desired formulae for U and V.

As both U and V satisfy the Helmholtz equation (1.202) let us expand them in spherical harmonics

$$U = \frac{1}{r} \sum_{n=0}^{\infty} a_n \left\{ \begin{array}{l} \zeta_n\left(kd\right)\psi_n\left(kr\right), r < d \\ \psi_n\left(kd\right)\zeta_n\left(kr\right), r > d \end{array} \right\} P_n\left(\cos\theta\right), \quad (1.215)$$

$$V = \frac{1}{r} \sum_{n=0}^{\infty} b_n \left\{ \begin{array}{l} \zeta_n\left(kd\right)\psi_n\left(kr\right), r < d \\ \psi_n\left(kd\right)\zeta_n\left(kr\right), r > d \end{array} \right\} P_n\left(\cos\theta\right), \quad (1.216)$$

where a_n and b_n are unknown coefficients to be determined. From (1.80) we deduce

$$E_r = \frac{1}{r^2} \sum_{n=1}^{\infty} n\left(n+1\right) a_n \left\{ \begin{array}{l} \zeta_n\left(kd\right)\psi_n\left(kr\right), r < d \\ \psi_n\left(kd\right)\zeta_n\left(kr\right), r > d \end{array} \right\} P_n\left(\cos\theta\right), \quad (1.217)$$

$$H_r = \frac{1}{r^2} \sum_{n=1}^{\infty} n\left(n+1\right) b_n \left\{ \begin{array}{l} \zeta_n\left(kd\right)\psi_n\left(kr\right), r < d \\ \psi_n\left(kd\right)\zeta_n\left(kr\right), r > d \end{array} \right\} P_n\left(\cos\theta\right). \quad (1.218)$$

Now use (1.214) and (1.203) to determine that

$$\{E_r, H_r\} = \frac{1}{dr} \frac{1}{\sin\theta} \frac{\partial}{\partial\theta} \left(\sin\theta \frac{\partial G_3}{\partial\theta} \right) =$$

$$-\frac{i}{4\pi} \frac{1}{kd^2 r^2} \sum_{n=1}^{\infty} n\left(n+1\right)\left(2n+1\right) \left\{ \begin{array}{l} \zeta_n\left(kd\right)\psi_n\left(kr\right), r < d \\ \psi_n\left(kd\right)\zeta_n\left(kr\right), r > d \end{array} \right\} P_n\left(\cos\theta\right).$$

$$(1.219)$$

The orthogonality of Legendre polynomials on $(0, \pi)$ implies that

$$\{a_n, b_n\} = -\frac{i}{4\pi} \frac{1}{kd^2}\left(2n+1\right)$$

so that

$$(U, V) = \frac{i}{4\pi} \frac{1}{ikd^2 r} \sum_{n=1}^{\infty}\left(2n+1\right) \left\{ \begin{array}{l} \zeta_n\left(kd\right)\psi_n\left(kr\right), r < d \\ \psi_n\left(kd\right)\zeta_n\left(kr\right), r > d \end{array} \right\} P_n\left(\cos\theta\right).$$

$$(1.220)$$

The electromagnetic field components radiated by a vertical electric or magnetic dipole are comparatively easy to derive.

However, the derivation of the electromagnetic field components radiated by an electric or magnetic *horizontal* dipole is a more complex problem. The term *horizontal* means that the dipole moment $\vec{m}$ is parallel to the plane

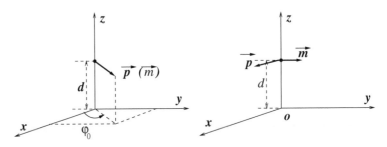

FIGURE 1.2. The horizontal electric and magnetic dipoles (left) and the Huygens source (right).

$z = const$ (see Figure 1.2 [left]). We intend to construct its representation in spherical coordinates. Having found the Debye potential for an arbitrarily oriented horizontal dipole, we then use a superposition principle to deduce the Debye potentials U_{PHS} and V_{PHS} of a point Huygens source (see Figure 1.2 [right]) that is formed by a mutually orthogonal pair of electric and magnetic dipoles, of equal dipole moments $(p = |\vec{p}| = |\vec{m}| = m)$.

We begin with the electric dipole (see Figure 1.2 [left]) that lies in the $x - z$ plane. The electric Hertz vector $\vec{\Pi}_0$ in spherical coordinates is

$$\vec{\Pi}_0 = pG_3 \left\{ \vec{i}_r \sin\theta \cos\phi + \vec{i}_\theta \cos\theta \cos\phi - \vec{i}_\phi \sin\phi \right\}. \tag{1.221}$$

Use relations (1.209)–(1.210) to obtain the radial field components

$$E_r = -\frac{1}{dr}\frac{\partial}{\partial d}\left(d\frac{\partial G_3}{\partial \theta}\right)\cos\phi, \tag{1.222}$$

$$H_r = \frac{ik}{r}\frac{\partial G_3}{\partial \theta}\sin\phi. \tag{1.223}$$

so that with reference to the expansion (1.201) we deduce the explicit spherical harmonic expansion

$$E_r = \frac{ip}{4\pi}\frac{\cos\phi}{dr^2}\sum_{n=1}^{\infty}(2n+1)\left\{\begin{array}{l}\zeta_n'(kd)\,\psi_n(kr)\,,r<d\\\psi_n'(kd)\,\zeta_n(kr)\,,r>d\end{array}\right\}P_n^1(\cos\theta)\,, \tag{1.224}$$

$$H_r = \frac{p}{4\pi}\frac{\sin\phi}{dr^2}\sum_{n=1}^{\infty}(2n+1)\left\{\begin{array}{l}\zeta_n(kd)\,\psi_n(kr)\,,r<d\\\psi_n(kd)\,\zeta_n(kr)\,,r>d\end{array}\right\}P_n^1(\cos\theta)\,. \tag{1.225}$$

Now the general solutions of the Helmholtz equation (1.202) satisfied by both Debye potentials have the forms

$$
U =
$$

$$
\frac{1}{r} \sum_{m=0}^{\infty} \left(2 - \delta_m^0\right) \cos m\phi \sum_{n=m}^{\infty} a_n^m \left\{ \begin{array}{l} \zeta_n' (kd) \psi_n (kr) , r < d \\ \psi_n' (kd) \zeta_n (kr) , r > d \end{array} \right\} P_n^m (\cos \theta)
$$

$$
+ \frac{2}{r} \sum_{m=1}^{\infty} \sin m\phi \sum_{n=m}^{\infty} b_n^m \left\{ \begin{array}{l} \zeta_n' (kd) \psi_n (kr) , r < d \\ \psi_n' (kd) \zeta_n (kr) , r > d \end{array} \right\} P_n^m (\cos \theta) , \quad (1.226)
$$

$$
V =
$$

$$
\frac{1}{r} \sum_{m=0}^{\infty} \left(2 - \delta_m^0\right) \cos m\phi \sum_{n=m}^{\infty} c_n^m \left\{ \begin{array}{l} \zeta_n (kd) \psi_n (kr) , r < d \\ \psi_n (kd) \zeta_n (kr) , r > d \end{array} \right\} P_n^m (\cos \theta)
$$

$$
+ \frac{2}{r} \sum_{m=1}^{\infty} \sin m\phi \sum_{n=m}^{\infty} f_n^m \left\{ \begin{array}{l} \zeta_n (kd) \psi_n (kr) , r < d \\ \psi_n (kd) \zeta_n (kr) , r > d \end{array} \right\} P_n^m (\cos \theta) , \quad (1.227)
$$

for suitable constants a_n^m, b_n^m, c_n^m and f_n^m. Using the differential equation for the spherical Bessel functions satisfied by $R = \psi_n$ and $R = \zeta_n$,

$$
r^2 \frac{d^2 R}{dr^2} + \left[k^2 r^2 - n (n + 1)\right] R = 0, \quad (1.228)
$$

we find the radial components, starting from (1.226)–(1.227). A comparison with (1.224)–(1.225) shows that $b_n^m = c_n^m = 0$, and $a_n^m = f_n^m = 0$ when $m \neq 1$ and

$$
a_n^1 = \frac{ip}{8\pi d} \frac{2n + 1}{n (n + 1)}, \quad f_n^1 = \frac{p}{8\pi d} \frac{2n + 1}{n (n + 1)}. \quad (1.229)
$$

Thus the Debye potentials describing the electromagnetic field of an electric horizontal dipole are

$$
U = \frac{ip}{4\pi} \frac{\cos \phi}{dr} \sum_{n=1}^{\infty} \frac{2n + 1}{n (n + 1)} \left\{ \begin{array}{l} \zeta_n' (kd) \psi_n (kr) , r < d \\ \psi_n' (kd) \zeta_n (kr) , r > d \end{array} \right\} P_n^1 (\cos \theta) \quad (1.230)
$$

and

$$
V = \frac{p}{4\pi} \frac{\sin \phi}{dr} \sum_{n=1}^{\infty} \frac{2n + 1}{n (n + 1)} \left\{ \begin{array}{l} \zeta_n (kd) \psi_n (kr) , r < d \\ \psi_n (kd) \zeta_n (kr) , r > d \end{array} \right\} P_n^1 (\cos \theta) .
$$

$$
(1.231)
$$

In a similar way the Debye potentials for the magnetic horizontal dipole are

$$U^{(m)} = \frac{m \cos \phi}{4\pi} \frac{d}{dr} \sum_{n=1}^{\infty} \frac{2n+1}{n(n+1)} \left\{ \begin{array}{l} \zeta_n(kd) \psi_n(kr), r < d \\ \psi_n(kd) \zeta_n(kr), r > d \end{array} \right\} P_n^1(\cos\theta),$$

(1.232)

$$V^{(m)} = \frac{im \sin \phi}{4\pi} \frac{d}{dr} \sum_{n=1}^{\infty} \frac{2n+1}{n(n+1)} \left\{ \begin{array}{l} \zeta_n'(kd) \psi_n(kr), r < d \\ \psi_n'(kd) \zeta_n(kr), r > d \end{array} \right\} P_n^1(\cos\theta).$$

(1.233)

When the electric dipole is located at $(0,0,-d)$ in the lower half-space $(z' < 0)$ the general term in (1.230) is multiplied by the factor $(-1)^{n+1}$, and the general term in (1.231) is multiplied by the factor $(-1)^n$. Analogously, when the magnetic horizontal dipole is located at the same point in the lower space, the general term in (1.232) is multiplied by the factor $(-1)^n$ and the general term in (1.233) is multiplied by the factor $(-1)^{n+1}$.

To obtain the Debye potentials U_{PHS} and V_{PHS}, describing the electromagnetic field of a point Huygens source, first set the dipole strengths to be equal: $p = m = \alpha$. The potentials are derived by linear superposition of the representations (1.230)–(1.233):

$$U_{PHS} = \frac{\alpha k \cos \phi}{4\pi r} \Phi(r,\theta), \quad V_{PHS} = \frac{\alpha k \sin \phi}{4\pi r} \Phi(r,\theta),$$

(1.234)

where

$$\Phi(r,\theta) = \sum_{n=1}^{\infty} \frac{2n+1}{n(n+1)} \left\{ \begin{array}{l} q_n(kd) \psi_n(kr), r < d \\ t_n(kd) \zeta_n(kr), r > d \end{array} \right\} P_n^1(\cos\theta),$$

(1.235)

and

$$q_n(kd) = \frac{1}{kd} \left[i\zeta_n'(kd) + \zeta_n(kd) \right],$$

(1.236)

$$t_n(kd) = \frac{1}{kd} \left[i\psi_n'(kd) + \psi_n(kd) \right].$$

(1.237)

If $d \to 0$, then $t_n(kd) \to 0$ if $n > 1$, and $t_1(kd) \to i\frac{2}{3}$, so that equation (1.237) simplifies to

$$\Phi(r,\theta) = i\zeta_1(kr) \sin\theta.$$

Hence the Debye potentials of a point Huygens source located at the origin are

$$U_{PHS} = i\frac{\alpha k}{4\pi} \frac{\zeta_1(kr)}{r} \sin\theta \cos\phi,$$

$$V_{PHS} = i\frac{\alpha k}{4\pi} \frac{\zeta_1(kr)}{r} \sin\theta \sin\phi.$$

(1.238)

The asymptotics

$$\zeta_n(x) = (-1)^{n+1} e^{ikx} + O(x^{-1}) \tag{1.239}$$

as $x \to \infty$ and the relations (1.80) show that in the far field

$$E_\theta = H_\phi + O\left(r^{-2}\right) = -i\frac{\alpha k^2}{4\pi}\frac{e^{ikr}}{r}\left(1 + \cos\theta\right)\cos\phi + O\left(r^{-2}\right), \tag{1.240}$$

$$E_\phi = -H_\theta + O\left(r^{-2}\right) = i\frac{\alpha k^2}{4\pi}\frac{e^{ikr}}{r}\left(1 + \cos\theta\right)\sin\phi + O\left(r^{-2}\right), \tag{1.241}$$

and $E_r = O\left(r^{-2}\right)$, $H_r = O\left(r^{-2}\right)$ as $r \to \infty$. It follows from (1.241) that the average energy flux density per oscillation period is

$$\vec{S} = \vec{i}_r \frac{1}{2} \operatorname{Re}\left(\vec{E} \times \vec{H}^*\right) = \vec{i}_r \frac{1}{2}\left(|E_\theta|^2 + |E_\phi|^2\right)$$

$$= \frac{\alpha^2 k^4}{32\pi^2}\frac{\left(1 + \cos\theta\right)^2}{r^2} + O\left(r^{-3}\right) \tag{1.242}$$

as $r \to \infty$. Thus at large distances as $r \to \infty$, the electromagnetic field of the source behaves as an outgoing spherical wave with an axisymmetric pattern that is obtained by rotating a cardioid about the z-axis.

In conclusion, let us derive the Debye potentials U_0, V_0 that describe the electromagnetic plane wave travelling along the positive z-axis. Consider an electric horizontal dipole at the point $z' = -d$, $\theta' = \pi$, and let $d \to \infty$. The asymptotic formula (1.239) shows that

$$\left\{ \begin{array}{c} U_0 \\ V_0 \end{array} \right\} = \left\{ \begin{array}{c} \cos\phi \\ \sin\phi \end{array} \right\} \frac{1}{ik^2 r} \sum_{n=1}^{\infty} i^n \frac{2n+1}{n(n+1)} \psi_n(kr) P_n^1(\cos\theta) \tag{1.243}$$

after normalisation by the factor

$$\frac{pk^2}{4\pi}\frac{e^{ikd}}{d}.$$

More generally, we may obtain representations for arbitrarily oriented Huygens sources. Suppose that a Huygens source is located on the surface of a sphere of radius d so that its location is described by the spherical coordinates (d, α, φ_0). The orthogonal dipole moments $\vec{p}$ and $\vec{m}$ are of unit strength ($|\vec{p}| = |\vec{m}| = 1$) and are tangential to the spherical surface $r = d$. The electromagnetic flux propagates along a radial line through the origin O.

It can be readily verified that the electric (respectively, magnetic) dipole field is described by the electric (respectively, magnetic) Hertz vector $\overrightarrow{\Pi}$ (respectively, $\overrightarrow{\Pi}^{(m)}$) as follows,

$$\overrightarrow{\Pi} = G_3\{[\sin\alpha\cos\theta - \cos\alpha\sin\theta\cos(\varphi - \varphi_0)]\,\overrightarrow{i}_r -$$
$$[\sin\alpha\sin\theta + \cos\alpha\cos\theta\cos(\varphi - \varphi_0)]\,\overrightarrow{i}_\theta + \cos\alpha\sin(\varphi - \varphi_0)\,\overrightarrow{i}_\varphi\}$$

$$\overrightarrow{\Pi}^{(m)} = G_3\{\sin\theta\sin(\varphi - \varphi_0)\,\overrightarrow{i}_r + \tag{1.244}$$
$$\cos\theta\sin(\varphi - \varphi_0)\,\overrightarrow{i}_\theta + \cos(\varphi - \varphi_0)\,\overrightarrow{i}_\varphi\} \tag{1.245}$$

where G_3 is defined by (1.193) with $r' = d, \theta' = \alpha, \varphi' = \varphi_0$.

Using relations of the type (1.209)-(1.211) the radial components of the electromagnetic field due to the electric dipole are given by

$$E_r^{(e)} = -\frac{1}{rd}\frac{\partial}{\partial d}\left(d\frac{\partial G_3}{\partial\alpha}\right), \quad H_r^{(e)} = -\frac{ik}{r\sin\alpha}\frac{\partial G_3}{\partial\varphi} \tag{1.246}$$

and radial components due to the magnetic dipole are given by

$$E_r^{(m)} = -\frac{ik}{r}\frac{\partial G_3}{\partial\alpha}, \quad H_r^{(m)} = -\frac{1}{rd\sin\alpha}\frac{\partial}{\partial d}\left(d\frac{\partial G_3}{\partial\varphi}\right) \tag{1.247}$$

The deduction of scalar functions U and V for the Huygens source is similar to that used above in obtaining the formulae (1.230)–(1.233). The final result is

$$U_{HS} = \frac{2k}{r}\sum_{m=1}^{\infty}\cos m(\varphi - \varphi_0) \times$$

$$\sum_{n=m}^{\infty}\frac{2n+1}{n(n+1)}\frac{(n-m)!}{(n+m)!}\left\{\begin{array}{l} t_n(kd)\zeta_n(kr), r > d \\ q_n(kd)\psi_n(kr), r < d \end{array}\right\}\tau_n^m(\cos\alpha)\,P_n^m(\cos\theta)$$

$$\tag{1.248}$$

$$V_{HS} = -\frac{2k}{r}\sum_{m=1}^{\infty}m\sin m(\varphi - \varphi_0) \times$$

$$\sum_{n=m}^{\infty}\frac{2n+1}{n(n+1)}\frac{(n-m)!}{(n+m)!}\left\{\begin{array}{l} t_n(kd)\zeta_n(kr), r > d \\ q_n(kd)\psi_n(kr), r < d \end{array}\right\}\pi_n^m(\cos\alpha)\,P_n^m(\cos\theta),$$

$$\tag{1.249}$$

where now the values $q_n(kd)$ and $t_n(kd)$ are defined by the formulae

$$t_n(kd) = \frac{1}{kd}\{-i\psi_n'(kd) + \psi_n(kd)\}, \quad q_n(kd) = \frac{1}{kd}\{-i\zeta_n'(kd) + \zeta_n(kd)\} \tag{1.250}$$

and also

$$\tau_n^m(\cos x) = \frac{d}{dx} P_n^m(\cos x), \quad \pi_n^m(\cos x) = \frac{1}{\sin x} P_n^m(\cos x) \qquad (1.251)$$

It should be noted that in the formulae (1.248)–(1.249) we have suppressed a factor $1/4\pi$.

We may readily deduce from (1.248)–(1.249) the scalar functions U_0 and V_0 describing the electromagnetic plane wave at oblique incidence. As $d \to \infty$,

$$q_n = \frac{2}{kd}(-i)^{n+1} e^{ikd}\left\{1 + O\left((kd)^{-1}\right)\right\}$$

and

$$U_0 = \frac{2}{ik^2 r} \sum_{m=1}^{\infty} \cos m\,(\varphi - \varphi_0) \times$$

$$\sum_{n=m}^{\infty} (-i)^n \frac{2n+1}{n(n+1)} \frac{(n-m)!}{(n+m)!} \psi_n\,(kr)\, \tau_n^m(\cos \alpha)\, P_n^m(\cos \theta) \qquad (1.252)$$

$$V_0 = -\frac{2}{ik^2 r} \sum_{m=1}^{\infty} m \sin m\,(\varphi - \varphi_0) \times$$

$$\sum_{n=m}^{\infty} (-i)^n \frac{2n+1}{n(n+1)} \frac{(n-m)!}{(n+m)!} \psi_n\,(kr)\, \pi_n^m(\cos \alpha)\, P_n^m(\cos \theta) \qquad (1.253)$$

where we suppress the factor $2k^3 e^{ikd}/kd$.

Setting $\alpha = \pi$ and using obvious results

$$\tau_n^m(-1) = \pi_n^m(-1) = 0, \quad m > 1$$

$$\tau_n^1(-1) = -\pi_n^1(-1) = (-1)^n \frac{n(n+1)}{2},$$

we see that formulae (1.252)–(1.253) coincide with the formulae (1.243) describing an electromagnetic plane wave travelling along the positive z-axis.

1.5   Formulation of Wave Scattering Theory for Structures with Edges.

In the preceding sections we have provided basic descriptions of the acoustic and electromagnetic fields in free space. There is an extensive literature (see [16], [38]) to which the interested reader is referred for deeper developments of this classical subject. Our main interest is the interaction of travelling acoustic or electromagnetic waves with bodies of varying acoustic or electromagnetic properties and of varying shape.

The interaction between waves and obstacles (or *scatterers*) causes disturbances to incident or primary wave fields, generally referred to as diffraction phenomena. A very general definition identifies any deviation of the wave field, apart from that resulting from the elementary application of geometrical optics, as a *diffraction phenomenon*. It is well known that the higher frequency the better the postulates of geometrical optics apply to wave-scattering, with one single but very important stipulation that can be quantified by the ratio of the characteristic dimension l of a scatterer and the wavelength λ of the incident wave.

When $\lambda/l \ll 1$ low-frequency scattering that is a perturbation of the incident wave occurs and is referred to as *Rayleigh scattering*. When the wavelength λ is comparable to the characteristic dimension of the obstacle ($\lambda \sim l$) one or several diffraction phenomena dominate; this region is also called the *resonance region*. The high-frequency region or quasi-optical region is characterised by $\lambda \ll l$. To a greater or lesser extent, Rayleigh-scattering can be studied by various perturbation methods, and high-frequency scattering can be studied by well-developed high-frequency approximate techniques. The most difficult diffraction problems fall in the resonance region where there are no suitable approximate models, and accurate or rigorous approaches are therefore preferable.

Independently of the specific scattering mechanisms or the assumed frequency regime, the perfectly general and rigorous formulation of the diffraction problem for scatterers of any characteristic dimension is uniform. Although the incident field satisfies the Helmholtz equation or Maxwell's equations as appropriate, as does the total field resulting from interaction with the scatterer, several conditions must be imposed to ensure that the total field (however calculated) exists and is unique. These include boundary conditions, (Sommerfeld's) radiation conditions, and in the case of scatterers with sharp edges, a boundedness condition on the scattered energy. Let us discuss these in turn.

First of all, the total field U^t is decomposed as a sum of the incident field U^0 and scattered (or diffracted) field U^{sc} so that

$$U^t = U^0 + U^{sc};$$

the total and scattered fields must be solutions of the appropriate wave equation in the exterior of the scatterer. (The incident field is a solution of the appropriate wave equation in the whole space.) Furthermore, U^t and U^{sc} and their partial derivatives are continuous everywhere in the space exterior to, and onto the surface of the scatterer, except possibly at

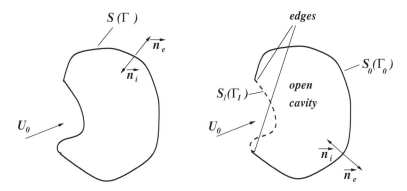

FIGURE 1.3. A closed scatterer (left) and open cavity structure (right).

singular points on the scatterer surface (where the unit normal vector to the surface does not exist or does not vary continuously).

Next so-called boundary conditions apply. We formulate them briefly, making a distinction between smooth obstacles and those with cavities and edges (Figure 1.3). We restrict ourselves to the acoustic case. As a point of notation, the symbol S always refers to a closed smooth surface (corresponding to a three-dimensional obstacle) and Γ always refers to a closed smooth contour (corresponding to a two-dimensional obstacle). If an aperture is opened in the closed surface S, the remaining *open* (infinitely thin) surface will be denoted S_0 and the removed portion (or *aperture* surface) will be denoted S_1; $S = S_0 \cup S_1$. Likewise if a portion Γ_1 of the closed curve Γ is removed, the remaining *open* contour will be denoted Γ_0; $\Gamma = \Gamma_0 \cup \Gamma_1$.

For closed scatterers (Figure 1.3 [left]) the boundary conditions satisfied by the total velocity potential U^t are as follows. If the surface is acoustically rigid or *hard,* the fluid velocity vanishes at the surface. This is characterised by the *hard boundary condition*

$$\left. \frac{\partial U^t}{\partial n} \right|_{S(\Gamma)} = 0 \tag{1.254}$$

where $\vec{n}$ is a unit normal to the surface S or contour Γ as appropriate. For external boundary value problems, the unit normal vector is chosen in the outward direction $\vec{n} = \vec{n}_e$; for the internal boundary value problem, the unit normal vector is chosen in the inward direction $\vec{n} = \vec{n}_i$. If the surface is acoustically *soft,* the pressure vanishes on the surface. This is characterised by the *soft boundary condition*

$$\left. U^t \right|_{S(\Gamma)} = 0 \tag{1.255}$$

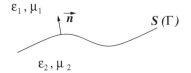

FIGURE 1.4. The boundary separating two media with differing electromagnetic properties.

that applies for both external and internal boundary value problems.

For open thin-walled obstacles the distinction between external and internal boundary value problems disappears; the cavity region is connected to the exterior region through apertures (S_1) or slots (Γ_1) (Figure 1.3 [right]). Both problems now are mixed and we refer to them as *mixed boundary value problems for the Helmholtz equation*.

Let us formulate the mixed boundary conditions. When S_0 (or Γ_0) is acoustically hard the total fields U_e^t and U_i^t in the exterior and interior regions, respectively, satisfy

$$\frac{\partial U_e^t}{\partial n} = \frac{\partial U_i^t}{\partial n} = 0 \text{ on } S_0\,(\Gamma_0)\,, \tag{1.256}$$

$$U_e^t = U_i^t \text{ on } S_1\,(\Gamma_1)\,, \tag{1.257}$$

where $\vec{n} \equiv \vec{n}_e = -\vec{n}_i$. Thus the field is continuous across the aperture surface and obeys the hard boundary condition at all interior and exterior points of S_0 (or Γ_0). However there is a discontinuity in the field value in moving from the interior to the exterior through a point on S_0 (or Γ_0). When S_0 (or Γ_0) is acoustically soft, the total solution satisfies the boundary conditions

$$U_i^t = U_e^t = 0 \text{ on } S_0\,(\Gamma_0)\,, \tag{1.258}$$

$$\frac{\partial U_i^t}{\partial n} = \frac{\partial U_e^t}{\partial n} \text{ on } S_1\,(\Gamma_1)\,. \tag{1.259}$$

Thus the normal derivative is continuous across the aperture, but there is a discontinuity in its value in moving from the interior to the exterior through a point on S_0 (or Γ_0).

The formulation of the electromagnetic boundary conditions is as follows. In the general case (see Figure 1.4) the boundary surface S (or contour Γ) separates two media with different electromagnetic parameters $\varepsilon_1,\ \mu_1,\ \sigma_1$, and $\varepsilon_2,\ \mu_2,\ \sigma_1$; denote the corresponding electromagnetic fields by $\vec{E}_1, \vec{H}_1$ and $\vec{E}_2, \vec{H}_2$, respectively. Assume that there are no extraneous charges and currents on S (or Γ).

Then boundary conditions are derived from the observation that the normal components of electric and magnetic flux are continuous whereas the tangential components of the electric and magnetic field intensity are continuous across S (or Γ). However the normal components of the electric and magnetic field intensity are in general discontinuous across S (or Γ). The conditions are

$$\varepsilon_1 \vec{E}_1 \cdot \vec{n} = \varepsilon_2 \vec{E}_2 \cdot \vec{n} \tag{1.260}$$
$$\vec{E}_1 \times \vec{n} = \vec{E}_2 \times \vec{n} \tag{1.261}$$
$$\mu_1 \vec{H}_1 \cdot \vec{n} = \mu_2 \vec{H}_2 \cdot \vec{n} \tag{1.262}$$
$$\vec{H}_1 \times \vec{n} = \vec{H}_2 \times \vec{n}. \tag{1.263}$$

These boundary conditions (1.260)–(1.263) hold when the conductivities of both media are finite so that the existence of surface currents is impossible. If the surface S (or contour Γ) enclosing the interior region labelled by index 2 is ideally conducting, the boundary conditions (1.260)–(1.263) are replaced by

$$\vec{n} \times \vec{E}_1 = 0 \tag{1.264}$$
$$\vec{E}_1 \cdot \vec{n} = q \tag{1.265}$$
$$\vec{n} \times \vec{H}_1 = \vec{J} \tag{1.266}$$
$$\vec{H}_1 \cdot \vec{n} = 0 \tag{1.267}$$

where q and $\vec{J}$ are the surface (or line) charges and currents, respectively, induced on $S(\Gamma)$.

With reference to Figure 1.3 let us now formulate the boundary conditions of mixed type for the open ideally conducting surface S_0 (or contour Γ_0), with aperture surface S_1 (or slot Γ_1). Let $\left(\vec{E}_i, \vec{H}_i\right)$ and $\left(\vec{E}_e, \vec{H}_e\right)$ denote the electromagnetic fields in the interior and exterior regions, respectively. Then

$$\vec{n} \times \vec{E}_i = \vec{n} \times \vec{E}_e = 0 \quad \text{on } S_0 \tag{1.268}$$
$$\vec{n} \times \vec{H}_i = \vec{n} \times \vec{H}_e \quad \text{on } S_1 \tag{1.269}$$
$$\vec{n} \times \left(\vec{H}_i - \vec{H}_e\right) = \vec{J} \quad \text{on } S_0 \tag{1.270}$$
$$\vec{n} \cdot \left(\vec{E}_i - \vec{E}_e\right) = q \quad \text{on } S_1 \tag{1.271}$$

where $\vec{J}$ and q are the net currents and charges at each point of S_0 or S_1. When an open cavity is excited by an incident electromagnetic wave, the last two conditions are superfluous, because the values $\vec{J}$ and q are due to the induced electromagnetic field that is to be found.

The boundary conditions for scalar functions U and V follow from (1.268)–(1.269). It should be noted that the formulation of boundary conditions in this form is always correct for both closed and open surfaces. However the analogous form of boundary conditions is not valid for the scalar functions U and V. Let us consider, for example, the open spherical surface (a metallic spherical cap) described in spherical coordinates (r, θ, φ) by $r = a$, $\theta \in (0, \theta_0)$, $\varphi \in (0, 2\pi)$, and irradiated by a plane wave. In the general case of oblique incidence (see formulae [1.252]–[1.253]), use of one of the boundary conditions (1.268)–(1.269) leads to a pair of first order differential equations

$$\frac{d}{d\theta} F_m(\theta) - i \frac{m^2}{\sin \theta} G_m(\theta) = 0, \quad \theta \in (0, \theta_0) \tag{1.272}$$

$$\frac{1}{\sin \theta} F_m(\theta) - i \frac{d}{d\theta} G_m(\theta) = 0, \quad \theta \in (0, \theta_0), \tag{1.273}$$

where the functions F_m and G_m arise from the expansion of the Debye potentials in the form (with $\varphi_0 = 0$)

$$U = U^0 + U^{sc} = \sum_{m=1}^{\infty} \cos m\varphi F_m(\theta)$$

$$V = V^0 + V^{sc} = \sum_{m=1}^{\infty} m \sin m\varphi G_m(\theta). \tag{1.274}$$

Equations (1.272)–(1.273) are equivalent to a second order differential equation for either F_m or G_m; the function G_m satisfies

$$\frac{d}{d\theta} \left[\sin \theta \frac{d}{d\theta} G_m(\theta) \right] - \frac{m^2}{\sin \theta} G_m(\theta) = 0, \quad \theta \in (0, \theta_0) \tag{1.275}$$

and the function F_m is obtained from (1.273). The general solution to (1.275) employing two arbitrary constants $C_m^{(1)}$ and $C_m^{(2)}$ is

$$G_m(\theta) = C_m^{(1)} \tan^m \frac{\theta}{2} + C_m^{(2)} \cot^m \frac{\theta}{2}, \quad \theta \in (0, \theta_0). \tag{1.276}$$

The requirement of solution boundedness forces $C_m^{(2)} \equiv 0$, so that

$$G_m(\theta) = C_m^{(1)} \tan^m \frac{\theta}{2}, \quad F_m(\theta) = i m C_m^{(1)} \tan^m \frac{\theta}{2}, \quad \theta \in (0, \theta_0) \tag{1.277}$$

The format of the solution (1.277) reflects the coupling of two types of waves when the spherical surface is open ($\theta_0 \neq \pi$). When $\theta_0 = \pi$, the requirement of solution boundedness further forces $C_m^{(1)} \equiv 0$. In this case the boundary conditions simplify to

$$G_m(\theta) = F_m(\theta) = 0, \quad \theta \in (0, \pi), \tag{1.278}$$

so that boundary conditions for an ideally conducting sphere take the well-known form

$$\frac{\partial}{\partial r}(rU)\bigg|_{r=a} = 0, \ V|_{r=a} = 0, \ \ \theta \in (0, \pi) \tag{1.279}$$

Hence, the combination of boundary conditions (1.279) for an open spherical shell is prohibited, and in this case it is necessary to use the conditions of *mixed* type (1.277).

The same analysis is valid if the spherical cap is defined by the interval $\theta \in (\theta_0, \pi)$, provided we replace the functions $\tan^m \frac{1}{2}\theta$ by $\cot^m \frac{1}{2}\theta$, the latter being regular at the point $\theta = \pi$.

The constants $C_m^{(1)}$ and $C_m^{(2)}$ are known as *polarisation constants*. The argument above is quite general and is employed in the formulation and analysis of vectorial scattering problems for other structures such as the circular disc, the hollow finite cylinder, etc.

In order to find physically reasonable solutions the so-called *Sommerfeld radiation conditions* must be imposed. In three-dimensional problems, for any scattered scalar field ψ that satisfies the Helmholtz equation these state that $r\psi$ is bounded for all r, i.e.,

$$|r\psi| < K \tag{1.280}$$

for some constant K, and

$$r\left(\frac{\partial \psi}{\partial r} - ik\psi\right) \to 0 \tag{1.281}$$

uniformly with respect to direction as $r \to \infty$. In two-dimensional problems conditions (1.280) and (1.281) are replaced by

$$\left|\sqrt{r}\psi\right| < K \tag{1.282}$$

and

$$\sqrt{r}\left(\frac{\partial \psi}{\partial r} - ik\psi\right) \to 0 \tag{1.283}$$

uniformly with respect to direction as $r \to \infty$, respectively. These conditions mean that in two- (respectively, three-) dimensional space the scattered field must behave as an outgoing cylindrical (respectively, spherical) wave at very large distances from the scatterer. The minus sign in both formulae is replaced by a plus sign if the time harmonic dependence is changed from $\exp(-i\omega t)$ to $\exp(+i\omega t)$.

The corresponding conditions for the three-dimensional electromagnetic case are

$$\left| r \overrightarrow{E} \right| < K, \quad \left| r \overrightarrow{H} \right| < K \tag{1.284}$$

and

$$r \left(\overrightarrow{E} + Z_0 \overrightarrow{r} \times \overrightarrow{H} \right) \to 0, \quad r \left(\overrightarrow{H} - Z_0^{-1} \overrightarrow{r} \times \overrightarrow{E} \right) \to 0 \tag{1.285}$$

uniformly with respect to direction as $r \to \infty$; in the two-dimensional case the factor r occurring in (1.284) and (1.285) is replaced by $\sqrt{r}$.

For smooth closed scatterers, the enforcement of these boundary and radiation conditions is adequate to ensure that a unique solution to the scattering problem at hand exists (see [41], [93]). However if the scattering surface has singular points, or is open, further conditions to guarantee uniqueness must be imposed.

The distinctive feature of open thin-walled cavities is the presence of sharp edges. In their vicinity of edges some components of the electromagnetic field exhibit singular behaviour of the form $\rho^{-\frac{1}{2}}$, where ρ is the distance measured from the point of field observation (at an off-body point) to the nearest point on the edge. This applies also to an on-surface point not located on the edge of the scatterer. If ρ is small enough (in the sense $k\rho \ll 1$), then in the neighbourhood of an edge the electromagnetic field is of a static character (see, for example [1.161]). In fact, the singularity order of the field near an edge is determined by the static part of the Helmholtz operator $L = \nabla^2 + k^2$. Hence the discussion of edge conditions developed for *potential* problems in Part I (Section 1.3) [1] is pertinent for wave-scattering problems. Thus, a suitable *edge condition* to be imposed is the *boundedness* of scattered acoustic or electromagnetic energy within every arbitrarily chosen finite volume V of space that may include the edges. Thus, in the case of acoustic scattering by an obstacle with edges we demand that

$$\frac{1}{2} \iiint_V \left\{ |\nabla U|^2 + k^2 |U|^2 \right\} dV < \infty, \tag{1.286}$$

and in the electromagnetic case we require that

$$\frac{1}{2} \iiint_V \left\{ \left| \vec{E} \right|^2 + \left| \vec{H} \right|^2 \right\} dV < \infty, \tag{1.287}$$

where the symmetrised form (1.58)–(1.59) of Maxwell's equations has been employed. These conditions provide the correct choice of the solution class for the acoustic or electromagnetic field, that in turn provides the correct order of the singularity in the field components near the edges. A proof of uniqueness is given by Jones [41] (chapter 9).

If the finite energy condition is not imposed, a multiplicity of nonzero solutions to the Helmholtz equation that vanish on the scatterer and satisfy the radiation conditions may be found. Consider the soft half-plane given by $y = 0$, $x \le 0$ (and z arbitrary). In cylindrical polars (ρ, ϕ, z), the two-dimensional field

$$U = H_{\frac{1}{2}}^{(1)} (k\rho) \sin \frac{1}{2} (\phi - \pi)$$

is such a solution that is regular everywhere off the half-plane. Obviously any constant multiple is also such a solution. Similar solutions can be constructed for the rigid half-plane, or in the electromagnetic case, for the perfectly conducting half-plane. As Jones [41] notes, this non-uniqueness cannot be attributed to the infinite extent of the scatterer: two solutions have been found for the circular disc.

In summary a wave-scattering problem posed for a scatterer incorporating cavities and edges is guaranteed a solution that is unique provided it satisfies the Helmholtz equation or Maxwell's equations, appropriate boundary conditions, Sommerfeld's radiation condition and the boundedness condition on the scattered energy. It is unnecessary to impose the last condition if the scatterer is a smooth closed obstacle; it is easy to prove that the scattered energy is always finite in any finite volume containing the scatterer. However it is essential to enforce the bounded energy condition for scatterers with sharp edges in order to ensure uniqueness of the solution.

1.6 Single- or Double-Layer Surface Potentials and Dual Series Equations.

In many ways the justification of various representations of the solution to a wave-scattering problem coincides with the analogous representations employed in potential theory. These were discussed at some length in Section 1.7 of Volume I [1]. Thus we may represent the solution to the Helmholtz equation by a single- or double-layer surface potential, in particular when the scatterer possesses edges. To avoid some duplication of argument we simply state the final results.

In acoustics we represent the total field by a total velocity potential U^t that is decomposed into a sum of an incident velocity potential U^0 and a scattered velocity potential U^{sc},

$$U^t = U^0 + U^{sc}. \tag{1.288}$$

Under the soft acoustic condition on S_0 a first-kind boundary value problem for the Helmholtz equation is obtained. Let $U^{(i)}$ and $U^{(e)}$ denote the desired potentials in the interior and exterior regions, respectively. In the three-dimensional case

$$U^{sc}\left(\vec{r}\right) = \iint_{S_0} \left[\frac{\partial U^{(i)}}{\partial n} - \frac{\partial U^{(e)}}{\partial n}\right] G_3\left(\vec{r}, \vec{r'}\right) ds \qquad (1.289)$$

where $\vec{n} \equiv \vec{n}_e = -\vec{n}_i$ is the outward unit normal to S_0 and

$$G_3\left(\vec{r}, \vec{r'}\right) = \frac{1}{4\pi} \frac{e^{ik\left|\vec{r} - \vec{r'}\right|}}{\left|\vec{r} - \vec{r'}\right|}$$

is the free space Green's function; in the two-dimensional case, the Green's function is replaced by

$$G_2\left(\vec{r}, \vec{r'}\right) = -\frac{i}{4\pi} H_0^{(1)}\left(k\left|\vec{r} - \vec{r'}\right|\right)$$

and

$$U^{sc}\left(\vec{r}\right) = -\int_{\Gamma_0} \left[\frac{\partial U^{(i)}}{\partial n} - \frac{\partial U^{(e)}}{\partial n}\right] G_2\left(\vec{r}, \vec{r'}\right) d\Gamma \qquad (1.290)$$

where the integral is taken over the open contour Γ_0.

Under the hard boundary condition on S_0 (or Γ_0) we use the representation

$$U\left(\vec{r}\right) = -\iint_{S_0} \left[U^{(e)}\left(\vec{r'}\right) - U^{(i)}\left(\vec{r'}\right)\right] \frac{\partial}{\partial n} G_3\left(\vec{r}, \vec{r'}\right) ds \qquad (1.291)$$

in three-dimensional space, and the representation

$$U\left(\vec{r}\right) = -\int_{\Gamma_0} \left[U^{(e)}\left(\vec{r'}\right) - U^{(i)}\left(\vec{r'}\right)\right] \frac{\partial}{\partial n} G_2\left(\vec{r}, \vec{r'}\right) d\Gamma \qquad (1.292)$$

in the two-dimensional case.

Following the argument developed in Section 1.7 of Part I [1], we introduce the *jump functions*

$$\sigma_D\left(\vec{r'}\right) = \frac{\partial U^{(i)}}{\partial n} - \frac{\partial U^{(e)}}{\partial n} \qquad (1.293)$$

$$\sigma_N\left(\vec{r'}\right) = U^{(e)}\left(\vec{r'}\right) - U^{(i)}\left(\vec{r'}\right) \qquad (1.294)$$

so that the integral formulae for the first-kind boundary value problem become

$$U^{sc}\left(\overrightarrow{r}\right) = -\iint_{S_0} \sigma_D\left(\overrightarrow{r'}\right) G_3\left(\overrightarrow{r}, \overrightarrow{r'}\right) ds \qquad (1.295)$$

$$U^{sc}\left(\overrightarrow{r}\right) = -\int_{\Gamma_0} \sigma_D\left(\overrightarrow{r'}\right) G_2\left(\overrightarrow{r}, \overrightarrow{r'}\right) d\Gamma \qquad (1.296)$$

whilst those for the second-kind boundary value problem become

$$U^{sc}\left(\overrightarrow{r}\right) = -\iint_{S_0} \sigma_N\left(\overrightarrow{r'}\right) \frac{\partial}{\partial n} G_3\left(\overrightarrow{r}, \overrightarrow{r'}\right) ds \qquad (1.297)$$

$$U^{sc}\left(\overrightarrow{r}\right) = -\int_{\Gamma_0} \sigma_N\left(\overrightarrow{r'}\right) \frac{\partial}{\partial n} G_2\left(\overrightarrow{r}, \overrightarrow{r'}\right) d\Gamma. \qquad (1.298)$$

The representation for $U^{sc}\left(\overrightarrow{r}\right)$ defined by (1.295) or (1.296) is the velocity potential associated with a simple or *single-layer distribution* on S or Γ, respectively; the representation defined by (1.297) or (1.298) is the velocity potential of a *double-layer* distribution on S or Γ, respectively.

The first-kind boundary value problem (prescribing the value of U on S_0 [or Γ_0]) specifies that $U^t|_{S_0} = 0$ and gives rise to the following Fredholm integral equations of the first kind for the unknown single-layer distribution σ_D,

$$\iint_{S_0} \sigma_D\left(\overrightarrow{r'}\right) G_3\left(\overrightarrow{r_s}, \overrightarrow{r'}\right) ds = U^0\left(\overrightarrow{r_s}\right), \; \overrightarrow{r_s} \in S_0 \qquad (1.299)$$

$$\int_{\Gamma_0} \sigma_D\left(\overrightarrow{r'}\right) G_2\left(\overrightarrow{r_s}, \overrightarrow{r'}\right) d\Gamma = U^0\left(\overrightarrow{r_s}\right), \; \overrightarrow{r_s} \in \Gamma_0. \qquad (1.300)$$

The second-kind boundary value problem specifies that $\left.\frac{\partial U^t}{\partial n}\right|_{S_0(\Gamma_0)} = 0$ and gives rise to the following Fredholm integral equations of the first kind for the unknown double-layer distribution σ_N,

$$\iint_{S_0} \sigma_N\left(\overrightarrow{r'}\right) \frac{\partial^2}{\partial n_s \partial n'} G_3\left(\overrightarrow{r_s}, \overrightarrow{r'}\right) ds = \frac{\partial U^0}{\partial n_s}, \; \overrightarrow{r_s} \in S_0 \qquad (1.301)$$

$$\int_{\Gamma_0} \sigma_N\left(\overrightarrow{r'}\right) \frac{\partial^2}{\partial n_s \partial n'} G_2\left(\overrightarrow{r_s}, \overrightarrow{r'}\right) d\Gamma = \frac{\partial U^0}{\partial n_s}, \; \overrightarrow{r_s} \in \Gamma_0 \qquad (1.302)$$

where $\overrightarrow{n_s}$ denotes the outward-pointing unit normal at $\overrightarrow{r_s}$.

These integral equations provide a means of calculating the surface layer distribution that in turn determine the scattered field at any off-body point, including points in the far field (the Green's function in equations [1.295]–[1.298] can be replaced by a simpler asymptotic form that transforms these equations that have the explicit format of outwardly radiating waves).

In special geometries the solution to the Helmholtz equation can be formulated in terms of dual (or triple) series equations or dual (or triple) integral equations; this approach is analogous to that of the potential theory situation described in Chapter 1 (page 50) of Part I. Let us now establish the equivalence of the dual series (and integral) equations approach and the method of single- and double-layer potentials in solving mixed boundary value problems for the Helmholtz equation.

Consider by way of illustration two simple examples of acoustic scattering by a soft spherical cap and by a soft circular disc. For the sake of simplicity the incident potential U^0 is due to a plane wave travelling along the z-axis.

Consider first the spherical cap S_0 that is supposed to occupy the region $r = a$, $\theta \in (0, \theta_0)$, $\phi \in (0, 2\pi)$. Let S_1 denote the complementary part of the spherical surface (the aperture) defined by $r = a$, $\theta \in (\theta_0, \pi)$, $\phi \in (0, 2\pi)$. Seek a solution by the method of separation of variables in the form

$$U^{sc}(r, \theta) = \sum_{n=0}^{\infty} a_n \left\{ \begin{array}{l} h_n^{(1)}(ka) \, j_n(kr), r < a \\ j_n(ka) \, h_n^{(1)}(kr), r > a \end{array} \right\} P_n(\cos \theta) \qquad (1.303)$$

where the unknown coefficients a_n are to be determined and, as always, the total potential has been decomposed as the sum

$$U^t = U^0 + U^{sc}. \qquad (1.304)$$

Imposition of the boundary conditions

$$U^t(a+0, \theta) = U^t(a-0, \theta), \quad \theta \in (0, \theta_0) \qquad (1.305)$$

$$\left. \frac{\partial U^t}{\partial r} \right|_{r=a+0} = \left. \frac{\partial U^t}{\partial r} \right|_{r=a-0}, \quad \theta \in (\theta_0, \pi) \qquad (1.306)$$

leads to the dual series equations for the unknown coefficients

$$\sum_{n=0}^{\infty} a_n j_n(ka) \, h_n^{(1)}(ka) \, P_n(\cos \theta) = -U^0(a, \theta), \theta \in (0, \theta_0) \,(1.307)$$

$$\sum_{n=0}^{\infty} a_n P_n(\cos \theta) = 0, \qquad \theta \in (\theta_0, \pi) \,(1.308)$$

where the incident potential has the expansion in spherical harmonics

$$U^0(a, \theta) = -e^{ika \cos \theta} = -\sum_{n=0}^{\infty} i^n j_n(ka) \, P_n(\cos \theta). \qquad (1.309)$$

Our aim is to show that the Fredholm equation of the first kind (1.299) is equivalent to the dual series equations (1.307)–(1.308). Because the jump function

$$\psi\left(\theta',\phi'\right) = \left.\frac{\partial U}{\partial r}\right|_{r=a-0} - \left.\frac{\partial U}{\partial r}\right|_{r=a+0} = \begin{cases} \sigma_D\left(\theta',\phi'\right) & \text{on } S_0 \\ 0 & \text{on } S_1 \end{cases} \quad (1.310)$$

vanishes on S_1, the surface of integration in equation (1.299) may be extended to the whole of the spherical surface $S = S_0 \cup S_1$ (given by $r = a$), so that

$$a^2 \int_0^{2\pi} d\phi' \int_0^\pi d\theta' \sin\theta' \psi\left(\theta',\phi'\right) G_3\left(a,\theta,\phi;a,\theta',\phi'\right) = e^{ika\cos\theta},$$

$$\theta \in (0,\theta_0) \quad (1.311)$$

where the kernel $G_3\left(a,\theta,\phi;a,\theta',\phi'\right)$ is given by formula (1.193) with $r = r' = a$. Expand the function $\psi\left(\theta',\phi'\right)$ in surface spherical harmonics

$$\psi\left(\theta',\phi'\right) = \sum_{s=0}^\infty \left(2-\delta_s^0\right)\cos s\phi' \sum_{p=s}^\infty a_p^s P_p^s\left(\cos\theta'\right)$$

$$+ 2\sum_{s=1}^\infty \sin s\phi' \sum_{p=s}^\infty b_p^s P_p^s\left(\cos\theta'\right) \quad (1.312)$$

where the coefficients a_p^s and b_p^s are to be determined. Insert the representations (1.312) and (1.193), evaluated at $r = r' = a$, into integral equation (1.311) and integrate to obtain

$$ika^2\left\{\sum_{m=0}^\infty \left(2-\delta_s^0\right)\cos m\phi \sum_{n=m}^\infty a_n^m j_n\left(ka\right) h_n^{(1)}\left(ka\right) P_n^m\left(\cos\theta\right)\right.$$

$$\left.+2\sum_{m=1}^\infty \sin m\phi \sum_{n=m}^\infty b_n^m j_n\left(ka\right) h_n^{(1)}\left(ka\right) P_n^m\left(\cos\theta\right)\right\}$$

$$= \sum_{n=0}^\infty i^n\left(2n+1\right) j_n\left(ka\right) h_n^{(1)}\left(ka\right); \quad (1.313)$$

this is valid for $\theta \in (0,\theta_0)$ and $\phi \in (0,2\pi)$. By virtue of the orthogonality of the complete set of functions $\{\cos m\phi\}_{m=0}^\infty$ and $\{\sin m\phi\}_{m=1}^\infty$ on $(0,2\pi)$ it follows that $a_n^m = b_n^m = 0$ when $m \geq 1$ and the equation (1.313) simplifies to the first member (1.308) of the dual series pair,

$$\sum_{n=0}^\infty a_n j_n\left(ka\right) h_n^{(1)}\left(ka\right) P_n\left(\cos\theta\right) = -U^0\left(a,\theta\right), \quad \theta \in (0,\theta_0) \quad (1.314)$$

where we identify $ika^2 a_n^0 = -a_n$. The second member of the dual series pair is directly obtained from definition (1.310), which implies that $\psi\left(\theta', \phi'\right) = 0$ when $\theta \in (\theta_0, \pi)$ so that

$$\sum_{n=0}^{\infty} a_n P_n (\cos \theta) = 0, \quad \theta \in (\theta_0, \pi) . \tag{1.315}$$

It is now obvious that equations (1.314)–(1.315) obtained from the single-layer formulation are completely equivalent to the dual series equations (1.307)–(1.308) obtained by the method of separation of variables.

Now consider the circular disc S_0 that is supposed to occupy the region $z = 0$, $\rho < a$, $\phi \in (0, 2\pi)$. Let S_1 denote the complementary part of the disc in the plane $z = 0$ defined by $\rho > a$, $\phi \in (0, 2\pi)$. Seek a solution by the method of separation of variables in the form

$$U^{sc} (\rho, z) = \int_0^\infty g(\nu) J_0(\nu\rho) e^{-\sqrt{\nu^2 - k^2}|z|} d\nu \tag{1.316}$$

where $\text{Im}\left(\sqrt{\nu^2 - k^2}\right) \leq 0$ and the unknown function $g(\nu)$ is to be determined. As usual the total velocity potential U^t is decomposed as the sum

$$U^t = U^0 + U^{sc} \tag{1.317}$$

where the incident potential is described by

$$U^0 = e^{ikz} . \tag{1.318}$$

Enforcement of the boundary conditions

$$U^t (\rho, +0) = U^t (\rho, -0), \quad \rho < a \tag{1.319}$$

$$\left.\frac{\partial U^t}{\partial z}\right|_{z=+0} = \left.\frac{\partial U^t}{\partial z}\right|_{z=-0}, \quad \rho > a \tag{1.320}$$

leads to the dual integral equations

$$\int_0^\infty g(\nu) J_0(\nu\rho)\, d\nu = -1, \quad \rho < a \tag{1.321}$$

$$\int_0^\infty \sqrt{\nu^2 - k^2}\, g(\nu) J_0(\nu\rho)\, d\nu = 0, \quad \rho > a \tag{1.322}$$

for the function g.

The jump function

$$\psi\left(\rho', \phi'\right) = \left.\frac{\partial U}{\partial z}\right|_{z=+0} - \left.\frac{\partial U}{\partial z}\right|_{z=-0} = \begin{cases} \sigma_D\left(\rho', \phi'\right), & \text{on } S_0 \\ 0, & \text{on } S_1 \end{cases} \tag{1.323}$$

vanishes on S_1; so we may extend the surface of integration in equation (1.299) to the whole plane $(z = 0)$ so that

$$\int_0^{2\pi} d\phi' \int_0^{\pi} d\rho' \rho' \psi \left(\rho', \phi'\right) G_3 \left(\rho, \phi, 0; \rho', \phi', 0\right) = 1, \quad \rho < a \qquad (1.324)$$

where the kernel $G_3 \left(\rho, \phi, 0; \rho', \phi', 0\right)$ is given by formula (1.175) with $z = z' = 0$,

$$G_3 \left(\rho, \phi, 0; \rho', \phi', 0\right) =$$

$$\frac{1}{4\pi} \sum_{m=0}^{\infty} \left(2 - \delta_m^0\right) \cos m \left(\phi - \phi'\right) \int_0^{\infty} \frac{J_m \left(\nu \rho'\right) J_m \left(\nu \rho\right)}{\sqrt{\nu^2 - k^2}} \nu d\nu \qquad (1.325)$$

where $\mathrm{Im} \left(\sqrt{\nu^2 - k^2}\right) \leq 0$. Now represent the function $\psi \left(\rho', \phi'\right)$ by the Fourier series

$$\psi \left(\rho', \phi'\right) = \sum_{s=0}^{\infty} \left(2 - \delta_s^0\right) \cos s\phi' \int_0^{\infty} G_s \left(\mu\right) J_s \left(\mu \rho'\right) d\mu$$

$$+ 2 \sum_{s=1}^{\infty} \sin s\phi' \int_0^{\infty} F_s \left(\mu\right) J_s \left(\mu \rho'\right) d\mu. \qquad (1.326)$$

Insert the representations (1.325)–(1.326) into integral equation (1.324) to obtain

$$\frac{1}{2} \sum_{m=0}^{\infty} \left(2 - \delta_m^0\right) \cos m\phi \int_0^{\infty} G_m \left(\nu\right) \frac{J_m \left(\nu \rho\right)}{\sqrt{\nu^2 - k^2}} d\nu$$

$$+ \frac{1}{2} \sum_{m=1}^{\infty} \sin m\phi \int_0^{\infty} F_m \left(\nu\right) \frac{J_m \left(\nu \rho\right)}{\sqrt{\nu^2 - k^2}} d\nu = 1. \qquad (1.327)$$

It follows that $F_m \left(\nu\right) = G_m \left(\nu\right) \equiv 0$ when $m \geq 1$ and equation (1.327) simplifies to

$$\frac{1}{2} \int_0^{\infty} G_0 \left(\nu\right) \frac{J_0 \left(\nu \rho\right)}{\sqrt{\nu^2 - k^2}} d\nu = 1, \quad \rho < a. \qquad (1.328)$$

The second member of the dual integral equation pair follows from the definition (1.323) requiring that $\psi \left(\rho, \phi\right) = 0$ when $\rho > a$ so that

$$\int_0^{\infty} G_0 \left(\nu\right) J_0 \left(\nu \rho\right) d\nu = 0, \quad \rho > a. \qquad (1.329)$$

The identification

$$\frac{1}{2} G_0 \left(\nu\right) = -\sqrt{\nu^2 - k^2} g \left(\nu\right)$$

shows that the equations (1.328) and (1.329) coincide with the dual integral equations (1.321)–(1.322) previously obtained by the method of separation of variables.

Throughout this book we will mostly use the method of separation of variables to obtain dual series equations and dual integral equations relevant to certain canonical scattering problems. However as these examples make clear this approach is completely equivalent to integral equation methods based on single- or double-layer distributions.

1.7   A Survey of Methods for Scattering.

In this section we provide a brief survey of the techniques available to solve scattering problems from arbitrarily shaped structures. The study of canonical scatterers has played an important role in the development of general methods for calculating the scattering from arbitrarily shaped obstacles, in at least two ways. First these studies allow us to identify various scattering mechanisms and quantify the likely effect in other scattering objects. Secondly canonical scatterers provide solutions of undisputed accuracy against which the results of computation with more general purpose algorithms may be compared. These *benchmark* solutions allow us to assess the relative accuracy of the various choices, such as surface discretisation, that must be made in the implementation of any numerical scattering algorithm. The classic text [13] provides an excellent survey of known results for a variety of mainly *closed* canonical surfaces (such as the sphere). One purpose of this book is to provide *benchmark* solutions for the class of scatterers characterised by edges and cavities that are either empty or enclose another scattering object. The canonical scatterers to be studied are typified by the spherical cavity, i.e., the surface obtained by removal of a circular aperture in the spherical surface.

As mentioned in earlier sections, the ratio of the wavelength to a characteristic dimension of the scatterer is very important in determining what scattering mechanisms are dominant and it is usual to divide the entire frequency band into three parts, the low frequency or long wavelength (Rayleigh) regime, the intermediate or resonance regime, where the scatterer around one or a few wavelengths in size, and the high frequency or optical (short wavelength) regime. At low frequency, the scattered field may be regarded as a perturbation of a corresponding static problem and an expansion may be sought as a series of powers of ka, where k is the wavenumber and a a typical diameter of the scatterer; the first term is usually satisfactory for approximate purposes. In the resonance regime, integral equation approaches and differential equation approaches provide a basis for numerical methods, whilst at high frequency, ray tracing techniques are often deployed.

If an integral equation is employed, the unknown surface pressure or velocity (as appropriate) is determined by an integral equation over the scattering surface with a suitable kernel derived from a Green's function; the equation is usually solved by methods such as Galerkin's method, the least squares method, the method of weighted residuals or collocation methods, all of which give rise to a finite system of linear algebraic equations for the coefficients of the selected basis functions representing the desired surface pressure (or velocity). In the electrical engineering literature the application of the method of weighted residuals is known as the method of moments. It was popularised by Harrington [31], and is described fully in [71]; many examples of its applications are given in the compilation [64]. Whilst these focus on the time harmonic context, time dependent integral equations have also been employed for wide-band calculations [81].

If a time harmonic differential equation approach is employed, the total field is determined in the vicinity of the scatterer (and beyond); the equation may be solved by a spatial finite difference replacement of the total field employing field values at points of a mesh surrounding the obstacle; the mesh is terminated to one of finite extent by enforcing some artificial condition (several variants have been explored in the literature) related to the radiation condition on the terminating surface (see [22], [10] and more recently, [8]); the unknown field values are obtained from the solution of a system of linear algebraic equations; if the fully time dependent wave equation is employed, with temporal as well as spatial finite difference replacements for the total field quantities on the same type of truncated mesh, then at each time step the total field quantities may be updated in terms of the quantities calculated at previous time steps; the calculation is pursued until a steady state response or a quiescent state is achieved, from which the time harmonic response may be determined by an appropriate discrete Fourier transform. These methods are usually termed finite difference frequency domain (FDFD) methods and finite difference time domain (FDTD) methods, respectively. A fourth order method is examined in [72]. A more sophisticated representation of the total field in the mesh of finite extent surrounding the scatterer leads to finite element methods (FEM), producing again systems of linear algebraic equations for the coefficients of the basis functions representing the total field throughout the mesh region [121].

High frequency methods assume that the (total or scattered) field behaves in a ray-like fashion; in free space, the rays travel in straight lines; rules for their reflection and diffraction from scattering surfaces distinguish the various approximate methods. Physical optics (PO) is the simplest such method in which the total field (or its normal derivative, as appropriate)

on the illuminated side of the scattering surface is determined precisely
as though the surface were planar (with the same incident illumination);
the scattered field is readily expressed as a closed integral over the surface (which may be calculated numerically or approximately by stationary
phase arguments). The limitations of PO become apparent at angles well
away from normal incidence, or when edge or corner contributions are significant. More sophisticated treatments rely upon exact calculations of the
high frequency scattering from the edge of a semi-infinite half plane or from
wedges or from cones in similar semi-infinite structures; these are incorpo-
rated as diffraction coefficients modifying the amplitude and phase of the
ray as it strikes the edge or corner. This is the basis for the well-known
geometrical theory of diffraction (GTD) developed by Keller [46] and others [35] and later developments such as the uniform theory of diffraction
(UTD) and the physical theory of diffraction (PTD): see also [42].

Cavity scatterers present one of the principal difficulties encountered by
GTD and its variants: rays may be trapped inside the cavity, i.e., some
of the paths traced out according to the rules of whichever method is employed become totally confined to the cavity region and lead to unreliable
estimates for the total as well as scattered fields. Methods employed at
lower frequency also have their difficulties. If an integral equation based
algorithm is employed, the rate of convergence of the computed solution
as a function of the discretisation number (number of basis elements or
unknowns used to represent the desired field or surface quantity) is usually
much poorer than for convex, or at least closed, scatterers, and in some
cases it may be impossible to have any surety of the accuracy of the computed solution. This may be attributable to the inherent instability of the
integral equation used: if it is of first kind, then, as finer discretisations
are employed, the condition number of the associated system of equations
grows rapidly, and although the theoretical accuracy of solution representation is apparently increased, actual computational accuracy degrades often
to a point where the computed solution is unreliable. This remark holds,
notwithstanding the beneficial stabilising influence of a singularity that is
usually present in the kernel of such integral equations. Differential equation based algorithms encounter similar difficulties for cavity structures.

The validation of putative methods for calculating the scattering from
objects incorporating edges and cavities, which may themselves enclose a
variety of scatterers, depends entirely on comparison with the results of
other proven approaches, whether analytical, computational or experimental. It is our purpose to provide a set of benchmark solutions to a selection
of canonical scattering problems to meet this need.

2

Acoustic Diffraction from a Circular Hole in a Thin Spherical Shell.

The closed sphere is the simplest three-dimensional scatterer of finite extent. Both acoustically hard and soft spherical surfaces have been intensively studied in all wavelength bands: the low frequency or long wavelength (Rayleigh) regime, the intermediate or resonance regime, where the scatterer around one or a few wavelengths in size, and the high frequency or optical (short wavelength) regime. At low frequency, the scattered field may be expanded in a series of powers of ka, where k is the wavenumber and a the spherical radius; the first term is usually satisfactory for approximate purposes. In the resonance regime, the scattered field may be expressed as a convergent series of spherical harmonics, known as the *Mie series* [41], [13]; this series is satisfactory for calculations on spheres up to several wavelengths. The poor convergence at higher frequency is ameliorated by the use of transformations such as the *Watson transformation* to a rapidly convergent series [41].

Thus the sphere provides a test bed for approximate methods that may be developed to treat various classes of scatterers in these regimes, as discussed in Section 1.7. Opening apertures in the closed sphere creates a cavity and associated scattering mechanisms that are completely absent in the closed structure; it therefore provides a significantly more useful and testing benchmark for such approximate methods.

In this chapter we examine perhaps the simplest three-dimensional cavity structure: the thin spherical shell with a circular aperture. The shell may be acoustically soft or hard, and is illuminated by an acoustic plane wave. The symmetry inherent in the configuration allows us to represent the scattered and total fields (inside and outside the cavity) in spherical harmonics; enforcement of boundary conditions on the scattering surface and continuity conditions across the circular aperture leads to a pair of dual series equations for the unknown spherical harmonic coefficients. Dual series equations of this type were extensively analysed in Chapter 2 of Volume I [1]. The standard transformation to a second kind Fredholm matrix equation is exploited in Section 2.1, so that solutions of guaranteed accuracy are obtained.

The closed spherical structure supports a standing spherical wave oscillation in the interior; when a small circular aperture is opened in the shell, it is possible to excite similar oscillations. It is a particular example of a *Helmholtz acoustic resonator,* in which the lowest frequency oscillation or *Helmholtz mode* is of interest. The dependence of the resonant frequency (at which the peak amplitude oscillations occur) and Q-factor upon aperture size is discussed in Section 2.2.

Large amplitude oscillations may be excited at frequencies near to those frequencies at which interior resonance of the closed sphere occurs. They are characterised by the so called quasi-eigenvalues of the cavity – complex numbers at which the regularised matrix equation is singular. These are discussed in Section 2.3.

Two measures of the energy in the scattered field are discussed in Section 2.4 – the total cross-section (measuring the total scattered energy) and the sonar cross-section measuring the energy scattered backwards, opposite to the direction of travel of the incident acoustic wave. Cross-sections for both acoustically soft and hard shells are examined in three frequency ranges: the Rayleigh regime, the resonance regime $(1 < ka \leq 20)$ and the high frequency regime. The low frequency solution is shown to be in accord with well-known solutions published in the literature. If the aperture is small, the resonant regime is characterised by *resonance doublets* (approximating the derivative of a delta function) superimposed on a more smoothly varying response typical of the closed sphere; the locations of the doublets are easily related to the quasi-eigenvalues of the cavity. If the aperture is opened so fully that the shell becomes a shallow curved disc, the accurately computed solution may be compared to that derived from a physical optics approach. Finally the cross-sections at short wavelength (the quasi-optical regime) results are presented for shells of size $ka \leq 320$. By way of illustration, an aperture of more than 20 wavelengths in a spherical shell of around 100 wavelengths is examined.

In Section 2.5, the force exerted on the open hard shell by the incident wave is examined as a function of frequency; this force is generated by the pressure difference between the interior and exterior of the cavity.

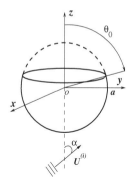

FIGURE 2.1. A spherical shell illuminated by a plane wave incident at angle α.

In Section 2.6, the scalar spherical reflector antenna is examined and the distribution of acoustic energy in the focal region determined; in section 2.6, the same structure in transmitting mode is examined. The results of this chapter are in some respects developments of the earlier investigations on this subject (see [19], [92], [62], [63], [34], [96] and [102]).

2.1 Plane W ave Diffraction from a Soft or Hard Spherical Cap.

Consider an acoustic plane wave of unit amplitude propagating at an angle α to the positive direction of z-axis (see Figure 2.1). The wave impacts a spherical cap (or spherical cavity) described in spherical coordinates by $r = a$, $\theta \in (0, \theta_0)$, $\varphi \in (0, 2\pi)$; the cap is part of a spherical surface centred at the origin and subtends an angle θ_0 at the origin. The cap is supposed to be acoustically soft or acoustically hard (rigid). The problem is to find the acoustic field scattered by the structure.

The velocity potential of the incident wave is

$$U^i \equiv U^i(r, \theta, \varphi) = \exp(ikr \cos \psi), \qquad (2.1)$$

where for a suitable φ_0,

$$\cos \psi = \cos \alpha \cos \theta + \sin \alpha \sin \theta \cos(\varphi - \varphi_0);$$

because of the problem symmetry, we may assume that $\varphi_0 = 0$ without loss of generality. Let U^s denote the scattered velocity potential. In accordance with the principle of linear superposition, the total velocity potential U^t is the sum of incident and scattered potentials, i.e.,

$$U^t = U^i + U^s. \qquad (2.2)$$

A rigorous statement of the problem to be solved that is based entirely on a theorem guaranteeing uniqueness [41] is as follows. We state it first for the acoustically hard cap.

Problem 1 *Find the potential U^t that satisfies the following conditions. (1) The function U^t satisfies the Helmholtz equation*

$$(\triangle + k^2)U^t = 0 \tag{2.3}$$

at all points off the shell. (2) U^t and its partial derivatives are continuous and finite everywhere, including at all points on the spherical surface $r = a$, but the value of U^t may change discontinuously across the shell. (3) The so-called mixed boundary conditions to be enforced on two regions of the spherical surface $r = a$, the cap and the complementary cap formed by the remainder of the spherical surface are

$$\frac{\partial}{\partial r}U^t|_{r=a-0} = \frac{\partial}{\partial r}U^t|_{r=a+0} = 0, \theta \in (0, \theta_0),$$
$$U^t|_{r=a-0} = U^t|_{r=a+0}, \theta \in (\theta_0, \pi) \tag{2.4}$$

for all $\varphi \in (0, 2\pi)$. (4) The scattered field satisfies the radiation conditions (1.280)–(1.281) that implies it must be an outgoing spherical wave at infinity; as $r \to \infty$, it has the asymptotic form

$$U^s = F(\theta, \varphi)\frac{e^{ikr}}{r} + O\left(r^{-2}\right) \tag{2.5}$$

where the scattered field pattern $F(\theta, \varphi)$ is a function dependent only on angular direction. (5) The energy of the scattered field is bounded, meaning that the integral over any finite volume V of space (possibly including the edges of spherical cap)

$$W = \frac{1}{4}\iiint_V \left\{ |\text{grad } U^s|^2 + k^2 |U^s|^2 \right\} dV < \infty \tag{2.6}$$

is finite.

The conditions (2) and (3) imply the following continuity condition for normal derivatives of the total velocity U^t

$$\frac{\partial}{\partial r}U^t|_{r=a-0} = \frac{\partial}{\partial r}U^t|_{r=a+0}, \quad \theta \in (0, \pi), \varphi \in (0, 2\pi). \tag{2.7}$$

In connection with the final condition (2.6), it is convenient to use the interior of sphere $(r \leq a)$ as the finite volume of integration. Continuity conditions do not apply at singular points of the surface, namely the edge;

in the vicinity of such points, the field singularity is determined by (2.6). This determines the solution class and provides the correct singularity order for the scattered field. The acoustically soft problem may be formulated as follows.

Problem 2 *Find the potential U^t that satisfies the following conditions. (1) The function U^t satisfies the Helmholtz equation (2.3) at all points off the shell. (2) U^t is continuous and finite everywhere, including at all points on the spherical surface $r = a$,*

$$U^t|_{r=a-0} = U^t|_{r=a+0}, \theta \in (0, \pi); \varphi \in (0, 2\pi). \tag{2.8}$$

The partial derivatives are continuous off the shell, but change discontinuously across the shell surface. (3) The mixed boundary conditions to be enforced on two regions of the spherical surface $r = a$, the cap and the complementary cap formed by the remainder of the spherical surface are

$$
\begin{aligned}
U^t|_{r=a-0} &= U^t|_{r=a+0} = 0, \theta \in (0, \theta_0), \\
\frac{\partial U^t}{\partial r}\Big|_{r=a-0} &= \frac{\partial U^t}{\partial r}\Big|_{r=a+0}, \theta \in (\theta_0, \pi)
\end{aligned}
\tag{2.9}
$$

for all $\varphi \in (0, 2\pi)$. (4) The scattered field satisfies the radiation conditions (1.280)–(1.281). (5) The energy integral defined by (2.6) is bounded, for any finite volume V of space.

As for the rigid shell, continuity conditions do not apply at singular points of the surface, namely the edge; in the vicinity of such points, the field singularity is determined by (2.6).

Expand the incident field (2.1) in spherical wave functions

$$U^i = \sum_{m=0}^{\infty} (2 - \delta_m^0) \cos m\varphi \times$$

$$\sum_{n=m}^{\infty} i^n (2n+1) \frac{(n-m)!}{(n+m)!} j_n(kr) P_n^m(\cos \alpha) P_n^m(\cos \theta). \tag{2.10}$$

We also employ spherical harmonics to seek solutions of the Helmholtz equation (2.3) that satisfy the radiation condition (2.5) and the hard or soft boundary conditions, (2.7) or (2.8), as appropriate. In the hard case,

$$U^t = U^i + \sum_{m=0}^{\infty} (2 - \delta_m^0) \cos m\varphi \sum_{n=m}^{\infty} i^n (2n+1) \frac{(n-m)!}{(n+m)!} a_n^m \times$$

$$P_n^m(\cos \alpha) P_n^m(\cos \theta) \begin{cases} j_n(kr), & r < a \\ j_n'(ka) h_n^{(1)}(kr)/h_n^{(1)'}(ka), & r > a \end{cases} \tag{2.11}$$

where a_n^m are the unknown Fourier coefficients to be determined. Enforcement of the mixed boundary conditions (2.4) leads to the dual series equations

$$\sum_{n=m}^{\infty} i^n (2n+1) \frac{(n-m)!}{(n+m)!} j_n'(ka)[a_n^m + 1] P_n^m(\cos \alpha) P_n^m(\cos \theta) = 0,$$

$$\theta \in (0, \theta_0) \quad (2.12)$$

$$\sum_{n=m}^{\infty} i^n (2n+1) \frac{(n-m)!}{(n+m)!} \frac{a_n^m}{h_n^{(1)\prime}(ka)} P_n^m(\cos \alpha) P_n^m(\cos \theta) = 0,$$

$$\theta \in (\theta_0, \pi), \quad (2.13)$$

valid for each $m = 0, 1, 2, \ldots$.

In the soft case, we seek a solution in the form

$$U^t = U^i + \sum_{m=0}^{\infty} (2 - \delta_m^0) \cos m\varphi \sum_{n=m}^{\infty} i^n (2n+1) \frac{(n-m)!}{(n+m)!} b_n^m \times$$

$$P_n^m(\cos \alpha) P_n^m(\cos \theta) \begin{cases} j_n(kr), & r < a \\ j_n(ka) h_n^{(1)}(kr)/h_n^{(1)}(ka), & r > a \end{cases} \quad (2.14)$$

where b_n^m are the unknown Fourier coefficients to be determined; enforcement of the mixed boundary conditions (2.9) leads to a similar set of dual series equations

$$\sum_{n=m}^{\infty} i^n (2n+1) \frac{(n-m)!}{(n+m)!} j_n(ka)[b_n^m + 1] P_n^m(\cos \alpha) P_n^m(\cos \theta) = 0,$$

$$\theta \in (0, \theta_0) \quad (2.15)$$

$$\sum_{n=m}^{\infty} i^n (2n+1) \frac{(n-m)!}{(n+m)!} \frac{b_n^m}{h_n^{(1)}(ka)} P_n^m(\cos \alpha) P_n^m(\cos \theta) = 0,$$

$$\theta \in (\theta_0, \pi). \quad (2.16)$$

valid for each $m = 0, 1, 2, \ldots$.

The solution of these two sets of dual series equations relies upon the results expounded in Chapter 2 of Volume I; they are solved in an identical manner. We outline the details for the hard case, and merely state the result for the soft case. Define new coefficients x_n^m by

$$x_n^m = i^n \frac{\hat{P}_n^m(\cos \alpha)}{h_n^{(1)\prime}(ka)} a_n^m. \quad (2.17)$$

Set $z = \cos\theta$, $z_0 = \cos\theta_0$ and reformulate equations (2.12) and (2.13) in terms of the orthonormal associated Legendre functions $\hat{P}_n^m(\cos\theta)$; the result is

$$\sum_{n=m}^{\infty} x_n^m \hat{P}_n^m(z) = 0, \qquad z \in (-1, z_0) \qquad (2.18)$$

$$\sum_{n=m}^{\infty} (2n+1)x_n^m \left\{ \frac{j_n'(ka)h_n^{(1)'}(ka)}{2n+1} \right\} \hat{P}_n^m(z) =$$

$$= -\sum_{n=m}^{\infty} i^n j_n'(ka)\hat{P}_n^m(\cos\alpha)\hat{P}_n^m(z), \quad z \in (z_0, 1). \quad (2.19)$$

The asymptotics of the spherical Bessel functions [3] show that

$$\frac{j_n'(ka)h_n^{(1)'}(ka)}{2n+1} = \frac{i}{4(ka)^3}\left\{1 + O(n^{-2})\right\} \qquad (2.20)$$

as $n \to \infty$, so that the asymptotically small parameter ε_n defined by

$$\varepsilon_n = 1 + 4i(ka)^3(2n+1)^{-1}j_n'(ka)h_n^{(1)'}(ka) \qquad (2.21)$$

has the property $\varepsilon_n = O(n^{-2})$, as $n \to \infty$. We may now extract the singular part of the dual series equations by writing them in the form as

$$\sum_{n=m}^{\infty} x_n^m \hat{P}_n^m(z) = 0, \qquad z \in (-1, z_0) \qquad (2.22)$$

$$\sum_{n=m}^{\infty} (2n+1)x_n^m \hat{P}_n^m(z) = \sum_{n=m}^{\infty} (2n+1)(x_n^m \varepsilon_n + \alpha_n^m)\hat{P}_n^m(z),$$

$$z \in (z_0, 1), \quad (2.23)$$

where $\alpha_n^m = 4i(ka)^3(2n+1)^{-1}i^n j_n'(ka)\hat{P}_n^m(\cos\alpha)$.

The relationship between the orthonormal associated Legendre functions $\hat{P}_n^m$ and the Jacobi polynomials $\hat{P}_{n-m}^{(m,m)}$ (see Appendix B.3 of Volume I) allows us to rewrite (2.22) and (2.23) in the equivalent form

$$\sum_{s=0}^{\infty} x_{s+m}^m \hat{P}_s^{(m,m)}(z) = 0, \qquad z \in (-1, z_0) \qquad (2.24)$$

$$\sum_{s=0}^{\infty} (s + m + \frac{1}{2}) x_{s+m}^m \hat{P}_s^{(m,m)}(z) =$$

$$\sum_{s=0}^{\infty} (s + m + \frac{1}{2})(x_{s+m}^m \varepsilon_{s+m} + \alpha_{s+m}^m) \hat{P}_s^{(m,m)}(z), \quad z \in (z_0, 1). \quad (2.25)$$

Enforcement of the finite energy condition (2.6) provides a definition of the solution space for the unknown coefficients x_{s+m}^m:

$$W = \pi(ka)^2 \sum_{m=0}^{\infty} (2 - \delta_m^0) \times$$

$$\sum_{s=0}^{\infty} |x_{s+m}^m|^2 \left\{ j_{s+m}(ka) j_{s+m}'(ka) + j_{s+m}^2(ka) \right\} |h_{s+m}^{(1)'}(ka)|^2 < \infty. \quad (2.26)$$

Thus we have reduced the original dual series equations (2.12) and (2.13) to the pair (2.24) and (2.25), subject to (2.26). Such systems were extensively considered in Sections 2.2 and 2.4 of Volume I, where it was shown that the infinite system of the linear algebraic equations (i.s.l.a.e.) may be transformed to a second kind Fredholm matrix equation, with a matrix operator that is a completely continuous perturbation of the identity operator, in the functional space l_2. Defining

$$\{X_s^m, A_s^m\} = (s + m + \frac{1}{2})^{\frac{1}{2}} \{x_{s+m}^m, \alpha_{s+m}^m\}, \quad (2.27)$$

the transformed system is (for each fixed $m = 0, 1, 2, \ldots$),

$$X_l^m - \sum_{s=0}^{\infty} X_s^m \varepsilon_{s+m} \hat{Q}_{sl}^{(m+\frac{1}{2}, m-\frac{1}{2})}(z_0) = \sum_{s=0}^{\infty} A_s^m \hat{Q}_{sl}^{(m+\frac{1}{2}, m-\frac{1}{2})}(z_0) \quad (2.28)$$

where $l = 0, 1, 2, \ldots$, and $\hat{Q}_{sl}^{(\alpha,\beta)}$ denotes the normalised incomplete scalar product defined in Appendix B.6 of Volume I.

This regularised system of equations has several notable advantages (theoretical and numerical) over the original system of dual series equations, and these are discussed in Appendix C.3 of Volume I. In particular we wish to show that it has a unique solution (the existence of which is already assured by the Fredholm alternative). It is convenient to let $\overrightarrow{X^m}$ denote the vector $(X_0^m, X_1^m, \ldots)$, E denote the diagonal matrix $\operatorname{diag}(\varepsilon_m, \varepsilon_{m+1}, \ldots)$ and Q denote the matrix with elements $\hat{Q}_{sl}^{(m+\frac{1}{2}, m-\frac{1}{2})}(z_0)$, where s, l take values $0, 1, 2, \ldots$. Then the system (2.28) has a unique solution if and only if the homogeneous form of this system (with the right hand side set to zero) has only the zero solution, i.e.,

$$\overrightarrow{X^m} - \overrightarrow{X^m} EQ = \overrightarrow{0}. \quad (2.29)$$

Thus, denoting the complex conjugate by a star and the transpose of $\overrightarrow{X^m}$ by $\left(\overrightarrow{X^m}\right)^t$,

$$\overrightarrow{X^m} E^* \left(\overrightarrow{X^{m*}}\right)^t = \overrightarrow{X^m} E Q E^* \left(\overrightarrow{X^{m*}}\right)^t. \tag{2.30}$$

The right hand side of (2.30) is real, because the matrix Q is real and symmetric. Thus

$$\sum_{s=0}^{\infty} |X_s^m| \operatorname{Im} \varepsilon_{s+m} = 0. \tag{2.31}$$

From the definition (2.21), $\operatorname{Im} \varepsilon_{s+m} \geq 0$ for all $m = 0, 1, 2, \ldots$, and for any fixed ka, at most one of the numbers $\operatorname{Im} \varepsilon_{s+m}$ is zero. If none of the numbers $\operatorname{Im} \varepsilon_{s+m}$ vanish, Equation (2.29) has only the zero solution. If one of the numbers $\operatorname{Im} \varepsilon_{s+m}$ vanish, when say, $m = l$, all the components of $\overrightarrow{X^m}$ except X_l^m vanish; but then the homogeneous form of Equations (2.24)–(2.25) (with $\alpha_{s+m}^m = 0$) are untenable unless $z_0 = -1$ (because a polynomial cannot vanish on the subinterval $(-1, z_0)$). Thus (2.29) has only the zero solution unless the aperture in the cavity closes, in which case nontrivial solutions correspond to eigenoscillations of the closed sphere. In any case, the system describing the *open* cavity structure has a unique solution.

This argument may be adapted to all the regularised i.s.l.a.e. that are derived in later sections and chapters, so that unique solutions are assured. Some remarks about the practical computation of the coefficients in (2.28) will be made once the corresponding i.s.l.a.e. for the soft case has been derived.

A similar procedure may be applied to dual series equations (2.15) and (2.16) arising in the soft case. Define

$$b_n^m = (-i)^n (2n+1) \frac{h_n^{(1)}(ka)}{\hat{P}_n^m(\cos \alpha)} y_n^m, \tag{2.32}$$

$$\beta_n^m = -i^{n+1} ka j_n(ka) \hat{P}_n^m(\cos \alpha).$$

The results of Chapter 2, Volume I may be applied to obtain the i.s.l.a.e. (for each fixed $m = 0, 1, 2, \ldots$),

$$Y_l^m - \sum_{s=0}^{\infty} Y_s^m \mu_{s+m} \hat{Q}_{sl}^{(m-\frac{1}{2}, m+\frac{1}{2})}(z_0) = \sum_{s=0}^{\infty} B_s^m \hat{Q}_{sl}^{(m-\frac{1}{2}, m+\frac{1}{2})}(z_0) \tag{2.33}$$

where $l = 0, 1, 2, \ldots$; the rescaled quantities are

$$\{Y_s^m, B_s^m\} = (s + m + \frac{1}{2})^{\frac{1}{2}} \{y_{s+m}^m, \beta_{s+m}^m\}, \tag{2.34}$$

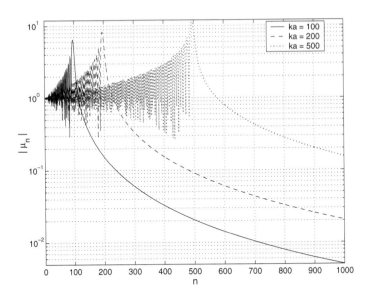

FIGURE 2.2. Asymptotic behaviour of μ_n for various wavenumbers.

and the asymptotically small parameter μ_n defined by

$$\mu_n = 1 - ika(2n+1)j_n(ka)h_n^{(1)}(ka) \qquad (2.35)$$

has the property $\mu_n = O\left(n^{-2}\right)$ as $n \to \infty$.

The numerical computation of the coefficients in (2.28) or (2.33) relies on the truncation of the i.s.l.a.e. to a finite number N_{tr} of equations and as noted in Appendix C.3 of Volume I, reliable bounds on the error can be obtained (see, for example, [44]). In wave-scattering problems, the choice of truncation number can be directly related to the size of the scatterer in wavelengths, or equivalently, by the parameter kL, where L is a typical dimension of the scatterer. For the sphere it is natural to use $L = a$. A practical choice is guided by the asymptotic behaviour of the parameters ε_n or μ_n. The behaviour of μ_n for three different values of ka is shown in Figure 2.2. It is clearly seen that the parameter rapidly decays when n exceeds ka, so that ka may be regarded as the "cut-off" value, so that N_{tr} should be chosen somewhat in excess of ka.

In the practical calculations on spherical cavity systems reported in this and later chapters, a value $N_{tr} = ka + 12$ was used; this produced three to four significant digits in the computed Fourier coefficients (in (2.28) and the other i.s.l.a.e.).

We may now constructively prove that the solution obtained to either of the regularised systems (2.28) or (2.33) satisfies all the conditions of the wave-scattering problem and consequently is unique.

By definition, the solution to the sound-hard case (see [2.11]) satisfies the Helmholtz equation (2.3), the continuity condition (2.7) and the radiation condition (2.5); likewise the solution to the sound-soft case (see [2.14]) satisfies the Helmholtz equation, the continuity condition (2.8) and the radiation condition. It remains to prove that the solution satisfies the mixed boundary conditions, (2.4) or (2.9) as appropriate, and the edge condition (2.6) as well.

First regroup in a seemingly very formal way the terms in (2.28), (2.27):

$$X_l^m = \sum_{s=0}^{\infty} (A_s^m + X_s^m \varepsilon_{s+m}) \hat{Q}_{sl}^{(m+\frac{1}{2}, m-\frac{1}{2})}(z_0), \tag{2.36}$$

$$Y_l^m = \sum_{s=0}^{\infty} (B_s^m + Y_s^m \mu_{s+m}) \hat{Q}_{sl}^{(m-\frac{1}{2}, m+\frac{1}{2})}(z_0). \tag{2.37}$$

In the sound-hard case, the jump in the total potential across spherical surface $r = a$ is

$$U^t|_{r=a-0} - U^t|_{a+0} =$$
$$\frac{2i}{(ka)^2} \sum_{m=0}^{\infty} (2 - \delta_m^0)(1 - z^2)^{\frac{m}{2}} \cos m\varphi \ K_m(z, z_0) \tag{2.38}$$

where

$$K_m(z, z_0) = \sum_{l=0}^{\infty} (l + m + \frac{1}{2})^{-\frac{1}{2}} X_l^m \hat{P}_l^{(m,m)}(z);$$

the normal derivative of velocity potential on the spherical surface has value

$$\frac{\partial U^t}{\partial r}|_{r=a} = \frac{ik}{(ka)^3} \sum_{m=0}^{\infty} (2 - \delta_m^0)(1 - z^2)^{\frac{m}{2}} \cos(m\varphi) \ F_m(z, z_0) \tag{2.39}$$

where

$$F_m(z, z_0) = \sum_{l=0}^{\infty} (l + m + \frac{1}{2})^{\frac{1}{2}} \{(1 - \varepsilon_{l+m}) X_l^m - A_l^m\} \hat{P}_l^{(m,m)}(z).$$

In the sound-soft case, the jump in the normal derivative of the total potential across spherical surface $r = a$ is

$$\frac{\partial}{\partial r}U^t|_{r=a-0} - \frac{\partial}{\partial r}U^t|_{r=a+0} =$$

$$- i\frac{4k}{(ka)^2}\sum_{m=0}^{\infty}(2 - \delta_m^0)(1 - z^2)^{\frac{m}{2}}\cos m\varphi\, S_m(z, z_0) \quad (2.40)$$

where

$$S_m(z, z_0) = \sum_{l=0}^{\infty}(l + m + \frac{1}{2})^{\frac{1}{2}}Y_l^m\,\hat{P}_l^{(m,m)}(z);$$

the value of the velocity potential on the spherical surface is

$$U^t|_{r=a} = \frac{2i}{ka}\sum_{m=0}^{\infty}(2 - \delta_m^0)(1 - z^2)^{\frac{m}{2}}\cos m\varphi\, G_m(z, z_0) \quad (2.41)$$

where

$$G_m(z, z_0) = \sum_{l=0}^{\infty}(l + m + \frac{1}{2})^{-\frac{1}{2}}\left\{B_l^m - (1 - \mu_{l+m})Y_l^m\right\}\hat{P}_l^{(m,m)}(z).$$

Now use the following property (see Appendix B.6 of Volume I) concerning the incomplete scalar product,

$$\hat{Q}_{sl}^{(\alpha,\beta)}(-z_1) = \delta_{sl} - (-1)^{s-l}\hat{Q}_{sl}^{(\beta,\alpha)}(z_1),$$

to transform equations (2.28) and (2.33). Setting $\theta_1 = \pi - \theta_0$ and $z_1 = \cos\theta_1 = -z_0$ we obtain

$$(1 - \varepsilon_{l+m})\tilde{X}_l^m + \sum_{s=0}^{\infty}\tilde{X}_s^m\varepsilon_{s+m}\hat{Q}_{sl}^{(m-\frac{1}{2},m+\frac{1}{2})}(z_1)$$

$$= \tilde{A}_l^m - \sum_{s=0}^{\infty}\tilde{A}_s^m\hat{Q}_{sl}^{(m-\frac{1}{2},m+\frac{1}{2})}(z_1) \quad (2.42)$$

$$(1 - \mu_{l+m})\tilde{Y}_l^m + \sum_{s=0}^{\infty}\tilde{Y}_s^m\mu_{s+m}\hat{Q}_{sl}^{(m+\frac{1}{2},m-\frac{1}{2})}(z_1)$$

$$= \tilde{B}_l^m - \sum_{s=0}^{\infty}\tilde{B}_s^m\hat{Q}_{sl}^{(m+\frac{1}{2},m-\frac{1}{2})}(z_1) \quad (2.43)$$

where $\left\{ \tilde{X}_l^m, \tilde{Y}_l^m, \tilde{A}_l^m, \tilde{B}_l^m \right\} = (-1)^l \left\{ X_l^m, Y_l^m, A_l^m, B_l^m \right\}$; equivalently

$$(1 - \varepsilon_{l+m})\tilde{X}_l^m - \tilde{A}_l^m = -\sum_{s=0}^{\infty}(\tilde{A}_s^m + \tilde{X}_s^m \varepsilon_{s+m})\hat{Q}_{sl}^{(m-\frac{1}{2},m+\frac{1}{2})}(z_1) \quad (2.44)$$

$$(1 - \mu_{l+m})\tilde{Y}_l^m - \tilde{B}_l^m = -\sum_{s=0}^{\infty}(\tilde{B}_s^m + \tilde{Y}_s^m \mu_{s+m})\hat{Q}_{sl}^{(m+\frac{1}{2},m-\frac{1}{2})}(z_1). \quad (2.45)$$

The substitution $z = -u$ in (2.39) and (2.41) leads to

$$F_m(u, z_1) = \sum_{l=0}^{\infty}(l + m + \frac{1}{2})^{\frac{1}{2}}\left\{ (1 - \varepsilon_{l+m})\tilde{X}_l^m - \tilde{A}_l^m \right\} \hat{P}_l^{(m,m)}(u) \quad (2.46)$$

$$G_m(u, z_1) = \sum_{l=0}^{\infty}(l + m + \frac{1}{2})^{-\frac{1}{2}}\left\{ \tilde{B}_l^m - (1 - \mu_{l+m})\tilde{Y}_l^m \right\} \hat{P}_l^{(m,m)}(u).$$
$$(2.47)$$

Substitute (2.36), (2.46), (2.47) into the corresponding expressions and change the summation order to obtain

$$K_m(z, z_0) = \sum_{s=0}^{\infty}(A_s^m + X_s^m \varepsilon_{s+m})\Phi_s^m(z, z_0) \quad (2.48)$$

$$S_m(z, z_0) = \sum_{s=0}^{\infty}(B_s^m + Y_s^m \mu_{s+m})R_s^m(z, z_0)$$

$$F_m(u, z_1) = -\sum_{s=0}^{\infty}(\tilde{A}_s^m + \tilde{X}_s^m \varepsilon_{s+m})R_s^m(u, z_1)$$

$$G_m(u, z_1) = \sum_{s=0}^{\infty}(\tilde{B}_s^m + \tilde{Y}_s^m \mu_{s+m})\Phi_s^m(u, z_1)$$

where

$$\Phi_s^m(x, x_0) = \sum_{l=0}^{\infty}(l + m + \frac{1}{2})^{-\frac{1}{2}}\hat{Q}_{sl}^{(m+\frac{1}{2},m-\frac{1}{2})}(x_0)\hat{P}_l^{(m,m)}(x) \quad (2.49)$$

$$R_s^m(x, x_0) = \sum_{l=0}^{\infty}(l + m + \frac{1}{2})^{\frac{1}{2}}\hat{Q}_{sl}^{(m-\frac{1}{2},m+\frac{1}{2})}(x_0)\hat{P}_l^{(m,m)}(x) \quad (2.50)$$

The functions $\Phi_s^m(x, x_0)$ and $R_s^m(x, x_0)$ are closely related. If we substitute the formula (see Appendix B.6 of Volume I)

$$\hat{Q}_{sl}^{(m-\frac{1}{2},m+\frac{1}{2})}(x_0) =$$

$$(1-x_0^2)^{\frac{m}{2}}\hat{P}_s^{(m-\frac{1}{2},m+\frac{1}{2})}(x_0)\frac{\hat{P}_l^{(m+\frac{1}{2},m-\frac{1}{2})}(x_0)}{l+m+\frac{1}{2}} + \frac{s+m+\frac{1}{2}}{l+m+\frac{1}{2}}\hat{Q}_{sl}^{(m+\frac{1}{2},m-\frac{1}{2})}(x_0)$$

$$(2.51)$$

into (2.50) we obtain

$$R_s^m(x, x_0) = (s+m+\frac{1}{2})\Phi_s^m(x, x_0) +$$

$$(1-x_0^2)^{\frac{m}{2}}\hat{P}_s^{(m-\frac{1}{2},m+\frac{1}{2})}(x_0)\sum_{l=0}^{\infty}(l+m+\frac{1}{2})^{-\frac{1}{2}}\hat{P}_l^{(m+\frac{1}{2},m-\frac{1}{2})}(x_0)\hat{P}_l^{(m,m)}(x).$$

$$(2.52)$$

The series appearing in (2.52) has the discontinuous value

$$\sum_{l=0}^{\infty}(l+m+\frac{1}{2})^{-\frac{1}{2}}\hat{P}_l^{(m+\frac{1}{2},m-\frac{1}{2})}(x_0)\hat{P}_l^{(m,m)}(x) =$$

$$= \pi^{-\frac{1}{2}}(1+x)^{-m}(1-x_0)^{-m-\frac{1}{2}}(x-x_0)^{-\frac{1}{2}}H(x-x_0) \quad (2.53)$$

so that the function $R_s^m(x, x_0)$ is represented as

$$R_s^m(x, x_0) = (s+m+\frac{1}{2})\Phi_s^m(x, x_0) +$$

$$\pi^{-\frac{1}{2}}\frac{(1+x_0)^{\frac{m}{2}}}{(1-x_0)^{\frac{m+1}{2}}}(1+x)^{-m}\frac{\hat{P}_s^{(m-\frac{1}{2},m+\frac{1}{2})}(x_0)}{(x-x_0)^{\frac{1}{2}}}H(x-x_0). \quad (2.54)$$

Thus the analytical properties of $R_s^m(x, x_0)$ are readily deduced from those of $\Phi_s^m(x, x_0)$.

It can be shown that

$$\Phi_s^m(x, x_0) = \pi^{\frac{1}{2}}(1+x)^{-m}G_s^m(x, x_0)H(x-x_0) \quad (2.55)$$

where

$$G_s^m(x, x_0) = \int_{x_0}^{x}(1+u)^{m-\frac{1}{2}}\frac{\hat{P}_s^{(m+\frac{1}{2},m-\frac{1}{2})}(u)}{(x-u)^{\frac{1}{2}}}du. \quad (2.56)$$

Thus the function $R_s^m(x, x_0)$ has the final representation

$$R_s^m(x, x_0) = \pi^{\frac{1}{2}}(1+x)^{-m}(s+m+\frac{1}{2})G_s^m(x, x_0)H(x-x_0) +$$

$$\pi^{-\frac{1}{2}}\frac{(1+x_0)^{\frac{m}{2}}}{(1-x_0)^{\frac{m+1}{2}}}(1+x)^{-m}\frac{\hat{P}_s^{(m-\frac{1}{2}, m+\frac{1}{2})}(x_0)}{(x-x_0)^{\frac{1}{2}}}H(x-x_0). \quad (2.57)$$

Let us compare the definition (2.56) of the functions $G_s^m(x, x_0)$ with the integral representation for the normalised Jacobi polynomials

$$\hat{P}_s^{(m,m)}(x) = \frac{(s+m+\frac{1}{2})^{\frac{1}{2}}}{\pi^{\frac{1}{2}}(1+x)^m}\int_{-1}^{x}(1+u)^{m-\frac{1}{2}}\frac{\hat{P}_s^{(m+\frac{1}{2}, m-\frac{1}{2})}(u)}{(x-u)^{\frac{1}{2}}}du. \quad (2.58)$$

Observing that

$$G_s^m(x, -1) = \pi^{\frac{1}{2}}\frac{(1+x)^m}{(s+m+\frac{1}{2})^{\frac{1}{2}}}\hat{P}_s^{(m,m)}(x)$$

$$= \pi^{\frac{1}{2}}(s+m+\frac{1}{2})^{-\frac{1}{2}}\left(\frac{1+x}{1-x}\right)^{\frac{m}{2}}\hat{P}_s^m(x), \quad (2.59)$$

we may regard the functions $G_s^m(x, x_0)$ as generalisations of the functions $\hat{P}_s^m$. The functions $\hat{P}_s^m(x)$ are eigenfunctions of the whole sphere; with some care, the functions $G_s^m(x, x_0)$ may be called *spherical cap eigenfunctions*.

Finally it is clear from the representations (2.55) and (2.57) that the solution obtained to the sound-hard and the sound-soft problems *automatically* satisfy the mixed boundary conditions and the edge condition.

2.2   Rigorous Theory of the Spherical Helmholtz Resonator.

The Helmholtz resonator is the archetypical system exhibiting resonance with distributed parameters. Although this system has been well studied in the literature for many years (see for example, [75] and [65]), our purpose is to provide some formulae that are more accurate than previously reported. We exploit the mathematically rigorous description of the previous section, concerning diffraction by a circular hole in an acoustically hard spherical shell, to construct a rigorous theory of the spherical Helmholtz resonator.

In a simple form, this classical acoustic resonator may be realised as a vessel of a certain volume V coupled to free space by a tube or column of length l and cross-sectional area S (see Figure 2.3).

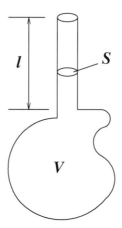

FIGURE 2.3. The Helmholtz resonator.

From a physical point of view it is reasonable to postulate that all the kinetic energy of the system is concentrated in the column of air that moves in the tube as a plunger. On the other hand, the potential energy of the system is stored through the process of elastic deformation of the air contained in the vessel. With these assumptions, a simple model can be developed for the acoustical resonator in which the resonant frequency f_0 does not depend on the vessel shape nor on the cross-section of the tube, being given by the simple formula

$$f_0 = \frac{v_s}{2\pi} \left(\frac{S}{l \cdot V} \right)^{\frac{1}{2}} \qquad (2.60)$$

where v_s is the velocity of sound in air.

If the acoustic resonator is coupled to free space by a small aperture one may suppose that all the kinetic energy, or at least its major part, is concentrated in the air moving near the aperture. The resonant frequency f_0 now depends on the aperture shape, but, as previously, may be expected to be independent of vessel shape.

When the aperture is circular, of radius r, the formula (2.60) becomes

$$f_0 = \frac{v_s}{2\pi} \left(\frac{2r}{V} \right)^{\frac{1}{2}}. \qquad (2.61)$$

In particular, if the vessel is spherical, of radius a and volume $V = \frac{4}{3}\pi a^3$, and the circular aperture subtends a small angle $\theta_1 (\ll 1)$ at the centre of the sphere, then $r = a \sin \theta_1 \simeq a\theta_1$ and relative wave-number at resonance is

$$(ka)_0 = \frac{2\pi f_0 a}{v_s} = \left(\frac{3\theta_1}{2\pi}\right)^{\frac{1}{2}}. \tag{2.62}$$

These and similar physically plausible models of the distribution of the sound energy near the aperture have been developed by many authors to determine more exactly the resonant frequency of the acoustic resonator and the Q-factor of this Helmholtz mode. The most accurate formula appears to be due to the author of [62]; the dominant terms in the improved estimate give

$$(ka)_0 = \left(\frac{3\theta_1}{2\pi}\right)^{\frac{1}{2}} \left\{1 + \frac{9}{20\pi}\theta_1\right\}. \tag{2.63}$$

The analytical treatment of the previous section allows us to deduce a more accurate formula for the Helmholtz mode. We set $m = 0$ in equation (2.42), so that

$$(1 - \varepsilon_l)\tilde{X}_l^0 + \sum_{s=0}^{\infty} \tilde{X}_s^0 \varepsilon_s Q_{sl} = \tilde{A}_l^0 - \sum_{s=0}^{\infty} \tilde{A}_s^0 Q_{sl}, \tag{2.64}$$

where $l = 0, 1, \ldots$, and

$$Q_{sl} \equiv \hat{Q}_{sl}^{(-\frac{1}{2},\frac{1}{2})}(\cos\theta_1) = \frac{1}{\pi}\left[\frac{\sin(s-l)\theta_1}{s-l} + \frac{\sin(s+l+1)\theta_1}{s+l+1}\right], \tag{2.65}$$

and

$$\tilde{A}_l^0 = 2i^{l+1}(l + \frac{1}{2})^{-\frac{1}{2}}(ka)^3 j_l'(ka)\hat{P}_l(\cos\alpha). \tag{2.66}$$

On the supposition that the parameters ka and θ_1 are small ($\ll 1$), we may solve (2.64) by the method of successive approximations.

Because resonance occurs when $(ka)^2 \sim \theta_1$ (and both parameters $(ka)^2$ and θ_1 are small), we modify this method and separate the first row (or *resonance row*) corresponding to $l = 0$, so that equation (2.64) takes the form

$$[1 - \varepsilon_0(1 - Q_{00})]\tilde{X}_0^0 = \tilde{A}_0^0 - \sum_{s=0}^{\infty} \tilde{A}_s^0 Q_{s0} - \sum_{s=1}^{\infty} \tilde{X}_s^0 \varepsilon_s Q_{s0} \tag{2.67}$$

and when $l > 0$,

$$(1 - \varepsilon_l)\tilde{X}_l^0 = \tilde{A}_l^0 - \sum_{s=0}^{\infty} \tilde{A}_s^0 Q_{sl} - \tilde{X}_0^0 \varepsilon_0 Q_{0l} - \sum_{s=1}^{\infty} \tilde{X}_s^0 \varepsilon_s Q_{sl}. \tag{2.68}$$

A triple application of the method of successive approximations to the equation (2.68) leads to the following approximate expression for $\tilde{X}_l^0$:

$$\tilde{X}_s^0 = \frac{\tilde{A}_s^0}{1 - \varepsilon_s} -$$

$$\frac{\varepsilon_0 \tilde{X}_0}{1 - \varepsilon_s} \left\{ Q_{0s} - \sum_{p=1}^{\infty} \frac{\varepsilon_p}{1 - \varepsilon_p} Q_{ps} Q_{0p} + \sum_{p=1}^{\infty} \frac{\varepsilon_p}{1 - \varepsilon_p} Q_{ps} \sum_{n=1}^{\infty} \frac{\varepsilon_n}{1 - \varepsilon_n} Q_{0n} Q_{np} \right\}.$$

$$(2.69)$$

Substitution of (2.69) into equation (2.67) and some manipulation leads to

$$\tilde{X}_0^0 \left\{ 1 - \varepsilon_0 [1 - Q_{00} + \sum_{s=1}^{\infty} \frac{\varepsilon_s}{1 - \varepsilon_s} Q_{s0} Q_{0s} - \sum_{s=1}^{\infty} \frac{\varepsilon_s}{1 - \varepsilon_s} Q_{s0} \sum_{p=1}^{\infty} \frac{\varepsilon_p}{1 - \varepsilon_p} Q_{ps} Q_{0p} \right.$$

$$\left. + \sum_{s=1}^{\infty} \frac{\varepsilon_s}{1 - \varepsilon_s} Q_{s0} \sum_{p=1}^{\infty} \frac{\varepsilon_p}{1 - \varepsilon_p} Q_{ps} \sum_{n=1}^{\infty} \frac{\varepsilon_n}{1 - \varepsilon_n} Q_{0n} Q_{np}] \right\} =$$

$$= \tilde{A}_0^0 (1 - Q_{00}) - \sum_{s=1}^{\infty} \frac{\tilde{A}_s^0}{1 - \varepsilon_s} Q_{s0}. \quad (2.70)$$

The characteristic equation that determines the Helmholtz mode approximately is obtained by equating the expression in curly brackets to zero.

It can be readily seen that the characteristic equation which determines the Helmholtz mode exactly is representable, at least in a formal sense, as

$$1 - \varepsilon_0 \{ 1 - Q_{00} + \sum_{s=1}^{\infty} \frac{\varepsilon_s}{1 - \varepsilon_s} Q_{s0} \{ Q_{0s} - \sum_{p=1}^{\infty} \frac{\varepsilon_p}{1 - \varepsilon_p} Q_{ps} \{ Q_{0p} -$$

$$\sum_{n=1}^{\infty} \frac{\varepsilon_n}{1 - \varepsilon_n} Q_{np} \{ Q_{0n} - ... \} \} \} \} = 0. \quad (2.71)$$

The more delicate issue of convergence may be addressed by estimating the size of the terms. The expressions Q_{nm} quantify the perturbation due to the hole. It is clear that the characteristic equation (2.71) cannot be solved exactly. Approximate equations are derived from (2.71) by retaining a finite number of perturbation terms. The precise number of terms is closely related to the desired number of corrections to the initial approximate value $(ka)_0 \simeq (3\theta_1/2\pi)^{\frac{1}{2}}$. From a practical point of view it is sufficient to keep one or two corrections. Here we use the approximate characteristic equation with three perturbation terms

$$1-\varepsilon_0\left\{1-Q_{00}+\sum_{s=1}^{\infty}\frac{\varepsilon_s}{1-\varepsilon_s}Q_{s0}Q_{0s}-\sum_{s=1}^{\infty}\frac{\varepsilon_s}{1-\varepsilon_s}Q_{s0}\sum_{p=1}^{\infty}\frac{\varepsilon_s}{1-\varepsilon_s}Q_{ps}Q_{0s}\right\}$$

$$= 0. \quad (2.72)$$

Recalling that ka is small, we may expand the asymptotically small parameter ε_n in powers of ka. When $n > 0$,

$$\mathrm{Re}(\varepsilon_n) = \frac{1}{(2n+1)^2} + \frac{4\left(2n^2+2n-1\right)}{(2n-1)(2n+1)^2(2n+3)}(ka)^2$$

$$+ \frac{16n^2+19n+48}{2(2n-3)(2n-1)(2n+1)^2(2n+3)(2n+5)}(ka)^4$$

$$+ \frac{73n^2+48n+1080}{3(2n-5)(2n-3)(2n-1)(2n+1)^2(2n+3)(2n+5)(2n+7)}(ka)^6$$

$$+ O\left((ka/n)^8\right), \quad (2.73)$$

and

$$\mathrm{Im}(\varepsilon_n) = \frac{(ka)^{2n+1}}{2n+1}\left\{2^{n+1}\frac{n\Gamma(n+1)}{\Gamma(2n+2)}\right\}^2 \times$$

$$\left\{1-\frac{n+2}{n(2n+3)}(ka)^2+\frac{n^3+6n^2+10n+5}{n^2(2n+3)^2(2n+5)}(ka)^4+O\left(\frac{(ka)^6}{n^3}\right)\right\};$$

$$(2.74)$$

also

$$\mathrm{Re}(\varepsilon_0) = 1 + \frac{4}{3}(ka)^2 + \frac{8}{15}(ka)^4 - \frac{8}{35}(ka)^6 + \frac{16}{567}(ka)^8 + O\left((ka)^{10}\right)$$

$$(2.75)$$

and

$$\mathrm{Im}(\varepsilon_0) = \frac{4}{9}(ka)^5 - \frac{4}{45}(ka)^7 + \frac{4}{525}(ka)^9 + O\left((ka)^{11}\right). \quad (2.76)$$

After substitution of (2.73)–(2.76) into (2.72), the remainder has order $O((ka)^8)$ and $O(\theta_1^4)$ in the parameters ka and θ_1, respectively. After some algebra one may deduce an approximate characteristic equation for the square of the resonant wavenumber $x = (ka)^2$ to be

$$2\theta - \theta^2 + \left(\frac{1}{2} + \frac{\pi^2}{6}\right)\theta^3 - \frac{4}{3}x\left\{1 - 2\theta + \left(1 + \frac{\pi^2}{12} + \frac{35}{36}\right)\theta^2\right\}$$

$$- \frac{8}{15}x^2(1 - 2\theta) + \frac{8}{35}x^3$$

$$- i\left\{\frac{3}{4}x^{\frac{3}{2}}\theta^2 - \frac{4}{45}x^{\frac{7}{2}} + \frac{4}{9}x^{\frac{5}{2}}(1 - 2\theta)\right\} + O(x^4, \theta_1^4) = 0, \quad (2.77)$$

where $\theta = \theta_1/\pi$. Newton's method may be used to solve (2.77) for x. Let $f(x)$ denote the terms on the left hand side of that equation, ignoring the $O(x^4, \theta_1^4)$ terms. We calculate the successive iterates

$$x^{(n+1)} = x^{(n)} - \frac{f(x^{(n)})}{f'(x^{(n)})} \quad (2.78)$$

commencing with $x^{(0)} = 0$. Thus

$$x^{(1)} = \frac{3}{2}\theta[1 + O(\theta)], \quad (2.79)$$

$$x^{(2)} = \frac{3\theta}{2}\left\{1 + \frac{9}{10}\theta - i\frac{1}{3}\left(\frac{3\theta}{2}\right)^{\frac{3}{2}}\right\}, \quad (2.80)$$

and

$$x^{(3)} = \frac{3\theta}{2}\left\{1 + \frac{9}{10}\theta + \frac{919}{1575}\theta^2 - i\frac{1}{3}\left(\frac{3\theta}{2}\right)^{\frac{3}{2}}\left[1 + \frac{5}{4}\theta\right]\right\}. \quad (2.81)$$

Extraction of the square root from (2.81) produces the real and imaginary parts of resonant wavenumber for the Helmholtz mode:

$$(ka)_H = \sqrt{\frac{3\theta}{2}}\left\{1 + \frac{9}{20}\theta + \frac{9601}{50400}\theta^2 - i\frac{1}{6}\left(\frac{3\theta}{2}\right)^{\frac{3}{2}}\left[1 + \frac{5}{4}\theta\right]\right\}. \quad (2.82)$$

Obviously in the context of an approximation the fraction $\frac{9601}{50400}$ may be replaced by $\frac{4}{21}$.

2.3 Quasi-Eigenoscillations: Spectrum of the Open Spherical Shell.

It is well known that the spectrum of the eigenoscillations of the closed spherical cavity is multiply degenerate: to each eigenvalue (except the lowest), there are several distinct (linearly independent) modes of oscillation corresponding to that eigenvalue.

Actually, the set of eigenfunctions is representable as follows. Consider first the acoustically hard case. For each $n = 0, 1, 2, \ldots$, the eigenvalues of the closed cavity are the roots $\varkappa_{nl}$ of the characteristic (dispersion) equation

$$j_n'(\varkappa_{nl}) = 0 \tag{2.83}$$

ordered in ascending magnitude $\varkappa_{n1} < \varkappa_{n2} < \varkappa_{n3} \ldots$; associated with each eigenvalue $\varkappa_{nl}$ are the $(2n + 1)$ eigenoscillations

$$\psi_{l,n,m}^{(r)}(r, \theta, \varphi) = j_n\left(\varkappa_{nl}\frac{r}{a}\right) P_n^m(\cos\theta) \begin{Bmatrix} \cos m\varphi, m = 0, 1, \ldots n \\ \sin m\varphi, m = 1, 2, \ldots n \end{Bmatrix}. \tag{2.84}$$

In the acoustically soft case, the eigenvalues of the closed cavity are the roots ν_{nl} of the equation

$$j_n(\nu_{nl}) = 0 \tag{2.85}$$

ordered in ascending magnitude $\nu_{n1} < \nu_{n2} < \nu_{n3} \ldots$; associated with each eigenvalue ν_{nl} are the $(2n + 1)$ eigenoscillations

$$\psi_{l,n,m}^{(s)}(r, \theta, \varphi) \sim j_n\left(\nu_{nl}\frac{r}{a}\right) P_n^m(\cos\theta) \begin{Bmatrix} \cos m\varphi, m = 0, 1, \ldots n \\ \sin m\varphi, m = 1, 2, \ldots n \end{Bmatrix}. \tag{2.86}$$

It will be remarked that l is the number of zeros as r ranges over the interval $(0, a)$, $n - m$ is the number of zeros as θ ranges over the interval $(0, \pi)$ and $2m$ is the number of zeros as φ ranges over the interval $(0, 2\pi)$.

It may be anticipated that the loss of symmetry caused by opening an aperture in the walls of the closed resonator breaks the spectrum degeneracy, splitting each multiple eigenvalue into several nondegenerate eigenvalues. Thus if the thin spherical shell is punctured by a circular hole the spectrum of closed spherical cavity is perturbed in this fashion. To analyse this phenomenon quantitatively, one must find nontrivial solutions to a certain homogeneous problem for the punctured spherical shell and so determine the perturbed spectrum.

Practically we have already solved this problem in the previous section. In fact, the relevant homogeneous problems are obtained by setting $\tilde{A}_l^m \equiv 0$ and $\tilde{B}_l^m \equiv 0$ in equations (2.42) and (2.43), respectively, yielding

$$(1 - \varepsilon_{l+m})\tilde{X}_l^m + \sum_{s=0}^{\infty} \tilde{X}_s^m \varepsilon_{s+m} \hat{Q}_{sl}^{(m-\frac{1}{2}, m+\frac{1}{2})}(z_1) = 0 \tag{2.87}$$

and

$$(1 - \mu_{l+m})\tilde{Y}_l^m + \sum_{s=0}^{\infty} \tilde{Y}_s^m \mu_{s+m} \hat{Q}_{sl}^{(m+\frac{1}{2}, m-\frac{1}{2})}(z_1) = 0 \tag{2.88}$$

where $m = 1, 2, \ldots$.

Nontrivial solutions of either equation arise at certain *complex* values of ka. The coupling of the inner cavity region $(r \le a)$ to the outer region $(a < r < \infty)$ allows energy to dissipate through the hole, i.e., radiation losses occur, and real eigenvalues associated with the *closed* sphere become complex. These complex roots are termed *quasi-eigenvalues* and are associated with *quasi-eigenoscillations*.

In fact, there are two possible spectral parameters to consider in treating (2.87) or (2.88). One is aperture size, described by the value $z_1 = \cos \theta_1$. More intrinsic to the geometry, the other spectral parameter is the relative wave number ka; the set of values where nontrivial solutions for equations (2.87), (2.88) occur form a discrete set (complex-valued). The characteristic equations to be solved are

$$\det \{A\} = 0 \tag{2.89}$$

where, in the acoustically hard case, the matrix A is defined to be

$$A = A^{(r)} = \left\{ (1 - \varepsilon_{l+m}) \delta_{ls} + \varepsilon_{s+m} \hat{Q}_{sl}^{(m-\frac{1}{2}, m+\frac{1}{2})}(z_1) \right\}_{l,s=0}^{\infty}, \tag{2.90}$$

and in the soft case, it is defined to be

$$A = A^{(s)} = \left\{ (1 - \mu_{l+m}) \delta_{ls} + \mu_{s+m} \hat{Q}_{sl}^{(m+\frac{1}{2}, m-\frac{1}{2})}(z_1) \right\}_{l,s=0}^{\infty}. \tag{2.91}$$

In general, when the aperture is not necessarily small, the equations (2.89) can only be solved numerically. However if the aperture is small $(1 - z_1 = \varepsilon \ll 1)$ the equations (2.89) may be replaced by an approximate form and solved analytically.

From its definition, $\hat{Q}_{sl}^{(\alpha, \beta)}(z_1) \to 0$ as $z_1 \to 1$; it vanishes at $z_1 = 1$. We suppose that $z_1 \simeq 1$ and that all the terms of form $\hat{Q}_{sl}^{(\alpha, \beta)}(z_1)$ are small. Retain in (2.90), (2.91) terms of the form $\hat{Q}_{sl}^{(\alpha, \beta)}(z_1)$ and their products. Suppressing the details one deduces the approximate equations

$$\det \left\{ A^{(r)} \right\} \simeq \prod_{l=0}^{\infty} \{ 1 - \varepsilon_{l+m} F_l^m \} = 0 \tag{2.92}$$

where

$$F_l^m = 1 - \hat{Q}_{l,l}^{(m-\frac{1}{2}, m+\frac{1}{2})}(z_1) +$$

$$\sum_{\substack{s=0 \\ s \ne l}}^{\infty} \frac{\varepsilon_{s+m}}{1 - \varepsilon_{s+m}} \hat{Q}_{s,l}^{(m-\frac{1}{2}, m+\frac{1}{2})}(z_1) \hat{Q}_{l,s}^{(m-\frac{1}{2}, m+\frac{1}{2})}(z_1),$$

and

$$\det\left\{A^{(s)}\right\} \simeq \prod_{l=0}^{\infty}\left\{1 - \mu_{l+m}G_l^m\right\} = 0 \tag{2.93}$$

where

$$G_l^m = 1 - \hat{Q}_{l,l}^{(m+\frac{1}{2},m-\frac{1}{2})}(z_1) +$$

$$\sum_{\substack{s=0 \\ s\neq l}}^{\infty} \frac{\mu_{s+m}}{1 - \mu_{s+m}} \hat{Q}_{s,l}^{(m+\frac{1}{2},m-\frac{1}{2})}(z_1)\hat{Q}_{l,s}^{(m+\frac{1}{2},m-\frac{1}{2})}(z_1).$$

Confining ourselves to the very first correction term to the unperturbed eigenvalues $\varkappa_{nl}$ and ν_{nl} we simplify the approximate dispersion equations and extract only the imaginary parts of the quadratic products involving $\hat{Q}_{s,l}^{(\alpha,\beta)}(z_1)$. It is obvious from the definitions of ε_n (2.21) and μ_n (2.35) that

$$\operatorname{Im}\left\{\frac{\varepsilon_{s+m}}{1 - \varepsilon_{s+m}}\right\} = \frac{2s + 2m + 1}{4(ka)^3|h_{s+m}^{(1)\prime}(ka)|^2} \tag{2.94}$$

$$\operatorname{Im}\left\{\frac{\mu_{s+m}}{1 - \mu_{s+m}}\right\} = -\frac{1}{ka(2s + 2m + 1)|h_{s+m}^{(1)}(ka)|^2}. \tag{2.95}$$

Thus the approximate characteristic equations are

$$1 - \varepsilon_{l+m}[1 - \hat{Q}_{l,l}^{(m-\frac{1}{2},m+\frac{1}{2})}(z_1) +$$

$$\frac{i}{4(ka)^3}\sum_{\substack{s=0 \\ s\neq l}}^{\infty}\frac{2s + 2m + 1}{|h_{s+m}^{(1)\prime}(ka)|^2}\hat{Q}_{s,l}^{(m-\frac{1}{2},m+\frac{1}{2})}(z_1)\hat{Q}_{l,s}^{(m-\frac{1}{2},m+\frac{1}{2})}(z_1)] = 0 \tag{2.96}$$

and

$$1 - \mu_{l+m}[1 - \hat{Q}_{l,l}^{(m+\frac{1}{2},m-\frac{1}{2})}(z_1) -$$

$$\frac{i}{ka}\sum_{\substack{s=0 \\ s\neq l}}^{\infty}\frac{\hat{Q}_{s,l}^{(m+\frac{1}{2},m-\frac{1}{2})}(z_1)\cdot\hat{Q}_{l,s}^{(m+\frac{1}{2},m-\frac{1}{2})}(z_1)}{(2s + 2m + 1)|h_{s+m}^{(1)}(ka)|^2}] = 0. \tag{2.97}$$

We defer investigation of the case that corresponds to the Helmholtz resonator to the next section; this case corresponds to the values $l = 0$ and

$m = 0$ in (2.96). Now transform (2.96), (2.97) to a more convenient form by re-labelling the indices $n = l + m$ and $p = s + m$ to obtain

$$1 - \varepsilon_n [1 - \hat{Q}_{n-m,n-m}^{(m-\frac{1}{2},m+\frac{1}{2})}(z_1) +$$

$$\frac{i}{4(ka)^3} \sum_{\substack{p=m \\ p \neq n}}^{\infty} \frac{2p+1}{|h_p^{(1)\prime}(ka)|^2} \hat{Q}_{p-m,n-m}^{(m-\frac{1}{2},m+\frac{1}{2})}(z_1) \hat{Q}_{n-m,p-m}^{(m-\frac{1}{2},m+\frac{1}{2})}(z_1)] = 0 \quad (2.98)$$

and

$$1 - \mu_n [1 - \hat{Q}_{n-m,n-m}^{(m+\frac{1}{2},m-\frac{1}{2})}(z_1) -$$

$$\frac{i}{ka} \sum_{\substack{p=m \\ p \neq n}}^{\infty} \frac{\hat{Q}_{p-m,n-m}^{(m+\frac{1}{2},m-\frac{1}{2})}(z_1) \cdot \hat{Q}_{n-m,p-m}^{(m+\frac{1}{2},m-\frac{1}{2})}(z_1)}{(2p+1)|h_p^{(1)}(ka)|^2}] = 0. \quad (2.99)$$

These equations may be solved by Newton's method. Suppressing the simple calculations, final results for the perturbations of the eigenvalues are

$$\varkappa_{nl}^m / \varkappa_{nl} = 1 + \frac{1}{2} \frac{(n+\frac{1}{2})}{[\varkappa_{nl}^2 - n(n+1)]} \hat{Q}_{n-m,n-m}^{(m-\frac{1}{2},m+\frac{1}{2})}(z_1) -$$

$$i\frac{1}{8} \frac{(n+\frac{1}{2})}{\varkappa_{nl}^3 [\varkappa_{nl}^2 - n(n+1)]} \sum_{\substack{p=m \\ p \neq n}}^{\infty} \frac{2p+1}{|h_p^{(1)\prime}(\varkappa_{nl})|^2} \hat{Q}_{p-m,n-m}^{(m-\frac{1}{2},m+\frac{1}{2})}(z_1) \hat{Q}_{n-m,p-m}^{(m-\frac{1}{2},m+\frac{1}{2})}(z_1)$$

$$(2.100)$$

and

$$\nu_{nl}^m / \nu_{nl} = 1 - \frac{1}{2} \left(n + \frac{1}{2}\right)^{-1} \hat{Q}_{n-m,n-m}^{(m+\frac{1}{2},m-\frac{1}{2})}(z_1) -$$

$$i\frac{1}{2} \frac{1}{(n+\frac{1}{2}) \nu_{nl}} \sum_{\substack{p=m \\ p \neq n}}^{\infty} \frac{\hat{Q}_{p-m,n-m}^{(m+\frac{1}{2},m-\frac{1}{2})}(z_1) \cdot \hat{Q}_{n-m,p-m}^{(m+\frac{1}{2},m-\frac{1}{2})}(z_1)}{(2p+1)|h_p^{(1)}(\nu_{nl})|^2}. \quad (2.101)$$

These formulae provide the perturbation corresponding to each index m as it ranges over the values $0, 1, 2, \ldots n$; the indices n and l are fixed in these formulae. Notice that the perturbed spectrum arising from the small circular hole has a double degeneracy at each value of $m = 1, 2, \ldots n$, whereas the spectrum of the closed cavity has a $(2m + 1)$-fold degeneracy. The impact of the hole is a reduction of spherical symmetry to cylindrical symmetry.

For the acoustically hard case the real parts of the quasi-eigenvalues have a positive shift compared to the unperturbed eigenvalues; for the soft case this shift is negative. In both cases the imaginary part of the perturbed quasi-eigenvalues $(\varkappa_{nl}^m, \nu_{nl}^m)$ is negative. The negativity testifies to the *dissipation* of energy by the system and provides finite nonzero Q-factors for each perturbed mode.

To obtain more practical results, express the values $\varkappa_{nl}^m$ and ν_{nl}^m in terms of the aperture size defined by the polar angle θ_1. It can be shown that

$$\hat{Q}_{sl}^{(\alpha,\beta)}(z_1) = \frac{(1-z_1)^{\alpha+1}(1+z_1)^{\beta}}{\{(l+\alpha+1)(\beta+l)\}^{\frac{1}{2}}} \hat{P}_s^{(\alpha,\beta)}(1)\hat{P}_l^{(\alpha+1,\beta-1)}(1)$$

$$+ O\left((1-z_1)^{\alpha+2}\right). \quad (2.102)$$

Now substitute the values of the normalised Jacobi polynomials into the formula (2.102) to obtain the major term of the expansion in powers of θ_1. The result is

$$\hat{Q}_{sl}^{(\alpha,\beta)}(z_1) \simeq \frac{(\theta_1/2)^{2\alpha+2}}{\Gamma(\alpha+1)\Gamma(\alpha+2)} \times$$

$$\left\{ \frac{\Gamma(\alpha+s+1)\Gamma(\alpha+l+1)\Gamma(\alpha+\beta+s+1)}{\Gamma(s+1)\Gamma(l+1)} \right\}^{\frac{1}{2}} \times \quad (2.103)$$

$$\left\{ \frac{\Gamma(\alpha+\beta+l+1)(\alpha+\beta+2s+1)(\alpha+\beta+2l+1)}{\Gamma(\beta+s+1)\Gamma(\beta+l+1)} \right\}^{\frac{1}{2}}. \quad (2.104)$$

As a corollary, the following approximations hold:

$$\hat{Q}_{n-m,n-m}^{(m-\frac{1}{2},m+\frac{1}{2})}(\cos\theta_1) \simeq 2\frac{(\theta_1/2)^{2m+1}}{\Gamma(m+\frac{1}{2})\Gamma(m+\frac{3}{2})} \cdot \frac{\Gamma(n+m+1)}{\Gamma(n-m+1)}$$

and

$$\hat{Q}_{n-m,n-m}^{(m+\frac{1}{2},m-\frac{1}{2})}(\cos\theta_1) \simeq \frac{1}{2}\frac{(\theta_1/2)^{2m+3}}{\Gamma(m+\frac{3}{2})\Gamma(m+\frac{5}{2})}(2n+1)^2\frac{\Gamma(n+m+1)}{\Gamma(n-m+1)}.$$

Furthermore the following approximations for products of the incomplete scalar product hold:

$$\hat{Q}_{p-m,n-m}^{(m-\frac{1}{2},m+\frac{1}{2})}(\cos\theta_1)\hat{Q}_{n-m,p-m}^{(m-\frac{1}{2},m+\frac{1}{2})}(\cos\theta_1) = \left\{\hat{Q}_{p-m,n-m}^{(m-\frac{1}{2},m+\frac{1}{2})}(\cos\theta_1)\right\}^2$$

$$\simeq 4\frac{(\theta_1/2)^{4m+2}}{\left[\Gamma(m+\frac{1}{2})\Gamma(m+\frac{3}{2})\right]^2} \cdot \frac{\Gamma(n+m+1)}{\Gamma(n-m+1)} \cdot \frac{\Gamma(p+m+1)}{\Gamma(p-m+1)}$$

and

$$\hat{Q}_{p-m,n-m}^{(m+\frac{1}{2},m-\frac{1}{2})}(\cos\theta_1)\hat{Q}_{n-m,p-m}^{(m+\frac{1}{2},m-\frac{1}{2})}(\cos\theta_1) = \left\{\hat{Q}_{p-m,n-m}^{(m+\frac{1}{2},m-\frac{1}{2})}(\cos\theta_1)\right\}^2 \simeq$$

$$\frac{1}{4}\frac{(\theta_1/2)^{4m+6}}{\left[\Gamma(m+\frac{3}{2})\Gamma(m+\frac{5}{2})\right]^2}(2n+1)^2(2p+1)^2\frac{\Gamma(n+m+1)}{\Gamma(n-m+1)}\cdot\frac{\Gamma(p+m+1)}{\Gamma(p-m+1)}.$$

With these approximations, we may deduce the Q-factor of the quasi-eigenoscillation of the spherical cavity with a small aperture ($\theta_1 \ll 1$). In the acoustically hard case, the Q-factor of the perturbed mode characterised by indices n, l and m is

$$Q_{nl}^m = -\frac{2\,\mathrm{Re}(\varkappa_{nl}^m)}{\mathrm{Im}(\varkappa_{nl}^m)} \simeq 32\left\{\Gamma(m+\frac{1}{2})\Gamma(m+\frac{3}{2})\right\}^2(2/\theta_1)^{4m+2}\times$$

$$\varkappa_{nl}^3\left[\varkappa_{nl}^2 - n(n+1)\right]\left\{(2n+1)^2 H_n^m\sum_{\substack{p=m\\p\neq n}}^{\infty}\frac{(2p+1)^2}{|h_p^{(1)\prime}(\varkappa_{nl})|^2}H_p^m\right\}^{-1}; \quad (2.105)$$

in the acoustically soft case, the Q-factor of the perturbed mode (characterised by indices n, l and m) is

$$Q_{nl}^m = -\frac{2\,\mathrm{Re}(\nu_{nl}^m)}{\mathrm{Im}(\nu_{nl}^m)} = 32\left\{\Gamma(m+\frac{3}{2})\Gamma(m+\frac{5}{2})\right\}^2\left(\frac{2}{\theta_1}\right)^{4m+6}\times$$

$$\nu_{nl}\left\{(2n+1)^2 H_n^m\sum_{\substack{p=m\\p\neq n}}^{\infty}\frac{(2p+1)^2}{|h_p^{(1)}(\nu_{nl})|^2}H_p^m\right\}^{-1}. \quad (2.106)$$

In these formulae, $H_n^m = \dfrac{2}{2n+1}\dfrac{(n+m)!}{(n-m)!}$.

2.4   Total and Sonar Cross-Sections.

What is the impact of an open spherical shell on a propagating acoustic plane wave? The results obtained in previous sections provide a basis for analytical and numerical answers to this question. Previous sections studied resonance phenomena. In the scattering context, resonance (or near-resonance phenomena) occurs when the frequency of the incident wave is such that the relative wave number ka coincides with the real part of a quasi-eigenvalue associated with a quasi-eigenoscillation of the open spherical cavity. In this section we compute various measures of the acoustic energy scattered to the far field zone, i.e., at large distances from the scatterer.

Recalling that at large distances from scattering objects (when $kr \gg ka$) the scattered velocity potential takes the form

$$U^s = \frac{e^{ikr}}{r} f(\theta, \varphi) + O\left(r^{-2}\right) \tag{2.107}$$

as $r \to \infty$, we may employ the well-known asymptotics [3]

$$h_n^{(1)\prime}(kr) = (-i)^{n+1} \frac{e^{ikr}}{kr} + O((kr)^{-2}), \text{ as } kr \to \infty$$

in equations (2.11) and (2.14) to determine the *scattering pattern* $f(\theta, \varphi)$. In the hard case,

$$f(\theta, \varphi) = f^h(\theta, \varphi) = \frac{2}{ik} \sum_{m=0}^{\infty} (2 - \delta_m^0) \cos m\varphi \times$$

$$\sum_{s=0}^{\infty} (-i)^{s+m} (s + m + \frac{1}{2})^{-\frac{1}{2}} X_s^m \hat{j}_{s+m}'(ka) \hat{P}_{s+m}^m (\cos \theta) \tag{2.108}$$

whereas in the soft case,

$$f(\theta, \varphi) = f^s(\theta, \varphi) = \frac{4}{ik} \sum_{m=0}^{\infty} (2 - \delta_m^0) \cos m\varphi \times$$

$$\sum_{s=0}^{\infty} (-i)^{s+m} (s + m + \frac{1}{2})^{\frac{1}{2}} Y_s^m j_{s+m}(ka) \hat{P}_{s+m}^m (\cos \theta). \tag{2.109}$$

Recall that the rescaled variables X_s^m and Y_s^m were defined by equations (2.27), (2.17), (2.34) and (2.32).

A measure of the total energy scattered by the scattering object is the *total cross-section*, defined by

$$\sigma_T = \int_\Omega |f|^2 d\Omega, \tag{2.110}$$

where $d\Omega = \sin\theta d\theta d\varphi$ is the solid angle element and the integration is taken over the complete sphere Ω; equivalently

$$\sigma_T = \int_0^{2\pi} \int_0^\pi |f(\theta, \varphi)|^2 \sin\theta d\theta d\varphi. \tag{2.111}$$

Thus, the total cross-sections of the acoustically hard shell and of the acoustically soft shell (with circular aperture) are

$$\sigma_T^h = \frac{8\pi}{k^2} \sum_{m=0}^{\infty} \sum_{s=0}^{\infty} (s + m + \frac{1}{2})^{-1} |X_s^m|^2 [j'_{s+m}(ka)]^2 \qquad (2.112)$$

and

$$\sigma_T^s = \frac{32\pi}{k^2} \sum_{m=0}^{\infty} \sum_{s=0}^{\infty} (s + m + \frac{1}{2}) |Y_s^m|^2 [j_{s+m}(ka)]^2, \qquad (2.113)$$

respectively.

A measure of the energy scattered by the shell in the direction determined by the angles θ and φ is the *differential scattering cross-section* or *bistatic sonar cross-section* $\sigma = \sigma(\theta, \varphi)$ defined by (see, for example, [13] or [77])

$$\sigma(\theta, \varphi) = \lim_{r \to \infty} 4\pi r^2 \frac{|U^s(r, \theta, \varphi)|^2}{|U^i(r, \theta, \varphi)|^2} = 4\pi |f(\theta, \varphi)|^2. \qquad (2.114)$$

(Note that the normalising magnitude of the incident field in the present discussion is unity.) The bistatic cross-section will be denoted $\sigma^h(\theta, \varphi)$ or $\sigma^s(\theta, \varphi)$ according to whether the shell is acoustically hard or soft. It depends upon the direction of the incident field.

The monostatic sonar cross-section (or simply the sonar cross-section) measures the energy scattered by the shell in the direction from which the incident wave came, i.e., it measures the energy in the *backscattered* direction, defined by $\theta = \pi - \alpha, \varphi = 0$ in (2.114). It is denoted σ; thus $\sigma = \sigma(\pi - \alpha, 0)$. The monostatic sonar cross-section of the acoustically hard shell is

$$\sigma^h = \frac{16\pi}{k^2} \times$$

$$\left| \sum_{m=0}^{\infty} (-i)^m (2 - \delta_m^0) \sum_{s=0}^{\infty} i^s (s + m + \frac{1}{2})^{-\frac{1}{2}} X_s^m j'_{s+m}(ka) \hat{P}_{s+m}^m(\cos \alpha) \right|^2 \qquad (2.115)$$

and that of the acoustically soft shell is

$$\sigma^s = \frac{64\pi}{k^2} \times$$

$$\left| \sum_{m=0}^{\infty} (-i)^m (2 - \delta_m^0) \sum_{s=0}^{\infty} i^s (s + m + \frac{1}{2})^{\frac{1}{2}} Y_s^m j_{s+m}(ka) \hat{P}_{s+m}^m(\cos \alpha) \right|^2. \qquad (2.116)$$

In general these cross-sections $\sigma_T^{(h,s)}, \sigma^{(h,s)}$ must be computed numerically. However at low frequency, some analytical approximations are possible. In this regime, referred to as the *Rayleigh scattering* regime, the shell is rather smaller than a wavelength ($ka \ll 1$) and, in common with other scattering structures, this is the simplest scenario in which analytical scattering formulae may be extracted.

2.4.1 Rayleigh Scattering.

The soft shell provides a simple example of how the equations derived in (section 2.1) may be exploited to deduce analytical expressions for the cross-sections σ_T^s and σ^s. In the limiting case $ka \ll 1$, the parameter μ_n, defined by (2.35), has the behaviour

$$\operatorname{Re}\mu_n = -\frac{2}{(2n-1)(2n+3)}(ka)^2 + O((ka)^4), \qquad (2.117)$$

$$\operatorname{Im}\mu_n = -\frac{2^n n!}{(2n)!(2n+1)!}(ka)^{2n+1}\left\{1 - \frac{(ka)^2}{2n+3} + O((ka)^4)\right\}. \qquad (2.118)$$

When $ka = 0$, the system (2.33) provides the exact solution of the corresponding potential problem (see Chapter 3 of Volume I [1]):

$$Y_l^m = \sum_{s=0}^{\infty} B_s^m \hat{Q}_{sl}^{(m-\frac{1}{2},m+\frac{1}{2})}(z_0). \qquad (2.119)$$

When $ka \ll 1$, the solution to equation (2.33) is a perturbation of the static solution (2.119); it may be constructed by the method of successive approximations as an expansion in powers of ka.

Let us restrict our calculation to the first three powers of ka so that the remainder is $O((ka)^4)$. To this order of approximation, the calculations of σ_T^s and σ^s require only the lowest term ($m = 0$). Application of the method of successive approximations to (2.37) shows that (recall that $z_0 = \cos\theta_0$)

$$\begin{aligned}
Y_l^0 = &-\frac{i}{2}ka\left[\hat{Q}_{0l}^{(-\frac{1}{2},\frac{1}{2})}(z_0)\right] \\
&+\frac{1}{2}(ka)^2\left[P_1(\cos\alpha)\hat{Q}_{1l}^{(-\frac{1}{2},\frac{1}{2})}(z_0) - \hat{Q}_{00}^{(-\frac{1}{2},\frac{1}{2})}(z_0)\hat{Q}_{0l}^{(-\frac{1}{2},\frac{1}{2})}(z_0)\right] \\
&-\frac{i}{2}(ka)^3\left[\hat{Q}_{10}^{(-\frac{1}{2},\frac{1}{2})}(z_0)P_1(\cos\alpha) - [\hat{Q}_{00}^{(-\frac{1}{2},\frac{1}{2})}(z_0)]^2 - \frac{1}{6}\right]\hat{Q}_{0l}^{(-\frac{1}{2},\frac{1}{2})}(z_0) \\
&-\frac{i}{2}(ka)^3\left[-\frac{1}{3}P_2(\cos\alpha)\hat{Q}_{2l}^{(-\frac{1}{2},\frac{1}{2})}(z_0) + \Phi_l(z_0)\right] + O((ka)^4) \qquad (2.120)
\end{aligned}$$

where

$$\Phi_l(z_0) = -2 \sum_{s=0}^{\infty} \frac{\hat{Q}_{sl}^{(-\frac{1}{2},\frac{1}{2})}(z_0)\hat{Q}_{s0}^{(-\frac{1}{2},\frac{1}{2})}(z_0)}{(2s-1)(2s+3)} =$$

$$= -2\frac{\hat{Q}_{0l}^{(-\frac{1}{2},\frac{1}{2})}(z_0)}{(2l-1)(2l+3)} - \frac{1}{4\pi}\left[\frac{2}{3}\cos\frac{3}{2}\theta_0 + 2\cos\frac{\theta_0}{2}\right] \times$$

$$\left[\frac{\sin(l+\frac{3}{2})\theta_0}{l+\frac{3}{2}} + \frac{\sin(l-\frac{1}{2})\theta_0}{l-\frac{1}{2}}\right]. \quad (2.121)$$

Let Q_{sl} denote the incomplete scalar product

$$Q_{sl} \equiv \hat{Q}_{sl}^{(-\frac{1}{2},\frac{1}{2})}(\cos\theta_0) = \frac{1}{\pi}\left[\frac{\sin(s-l)\theta_0}{s-l} + \frac{\sin(s+l+1)\theta_0}{s+l+1}\right],$$

and recall that the Legendre polynomials satisfy $P_1(\cos\alpha) = \cos\alpha$ and $P_2(\cos\alpha) = \frac{3}{2}\cos^2\alpha - \frac{1}{2}$. Substitution of (2.120) into the expressions (2.113) leads to the low-frequency approximate formulae for the total and monostatic cross-sections,

$$\frac{\sigma_T^s}{\pi a^2} = 4Q_{00}^2 + 4(ka)^2\left\{\frac{1}{3}[Q_{01}^2 + Q_{20}Q_{00} - 2Q_{00}^2] - Q_{00}^4\right\} +$$

$$4(ka)^2\left\{2Q_{00}\Phi_0 + [Q_{10}^2 - Q_{20}Q_{00}]\cos^2\alpha\right\} + O((ka)^4) \quad (2.122)$$

and

$$\frac{\sigma^s}{\pi a^2} = 4Q_{00}^2 + 4(ka)^2\left\{-\frac{2}{3}Q_{00}(Q_{00} + Q_{20}) - Q_{00}^4\right\} +$$

$$4(ka)^2\left\{2Q_{00}\Phi_0 + 2(2Q_{10}^2 - Q_{00}Q_{11} - Q_{00}Q_{20})\cos^2\alpha\right\} + O((ka)^4). \quad (2.123)$$

When $\theta_0 = \pi$, formulae (2.122) and (2.123) provide well-known expressions (see, for example, [13]) for the cross-sections of the closed soft sphere. The angle α is set to 0 or π, and the elementary property $\hat{Q}_{sl}^{(-\frac{1}{2},\frac{1}{2})}(-1) = \delta_{sl}$ may be used to show that

$$\frac{\sigma_T^s}{\pi a^2} = 4\left\{1 - \frac{1}{3}(ka)^2 + O((ka)^4)\right\}; \quad (2.124)$$

$$\frac{\sigma^s}{\pi a^2} = 4\left\{1 - \frac{7}{3}(ka)^2 + O((ka)^4)\right\}. \quad (2.125)$$

Another limiting case occurs when $\theta_0 \ll 1$, and the spherical cap is very nearly a circular disc of radius $a_d = a\theta_0$. In this case,

$$Q_{sl} \simeq \frac{1}{\pi} \left\{ 2\theta_0 - \frac{1}{3} \left[s(s+1) + l(l+1) + \frac{1}{2} \right] \theta_0^3 \right\} \qquad (2.126)$$

and one can use (2.122), (2.123) to obtain the major term of the cross-sections for the circular disc,

$$\frac{\sigma_T^s}{\pi a_d^2} = \frac{16}{\pi^2} [1 + O(\theta_0^2)], \qquad (2.127)$$

$$\frac{\sigma^s}{\pi a_d^2} = \frac{16}{\pi^2} [1 + O(\theta_0^2)]. \qquad (2.128)$$

More accurate values of the normalised cross-sections $\dfrac{\sigma_T^s}{\pi a^2}$ and $\dfrac{\sigma^s}{\pi a^2}$ can be obtained by using more terms of the expansion of Y_s^m in powers of ka (see [49] for a discussion of normal incidence on the disc).

In contrast to the soft shell case described by (2.33), the corresponding equation (2.28) for the acoustically hard shell cannot be solved to provide a uniformly valid solution with respect to parameter θ_0 at low frequencies ($ka \ll 1$). The reason is the absence of a solution in static limit ($ka = 0$). This is traceable to the behaviour of the asymptotically parameter: when $ka \ll 1$, and $n \geq 1$,

$$\mathrm{Re}\,\varepsilon_n = (2n+1)^{-2} + 4\frac{2n^2 + 2n - 1}{(2n-1)(2n+1)^2(2n+3)}(ka)^2 + O((ka)^4), \qquad (2.129)$$

$$\mathrm{Im}\,\varepsilon_n = \frac{4}{2n+1} \left\{ \frac{2^n n \cdot n!}{(2n+1)!} \right\}^2 (ka)^{2n+1} \left\{ 1 + O((ka)^2) \right\}, $$

so that

$$\lim_{ka \to 0} \varepsilon_n = (2n+1)^{-2}. \qquad (2.130)$$

An analytical treatment of (2.28) is possible only when some restriction is placed on the angular parameter, so that $ka \ll 1$, and simultaneously $\theta_0 \ll 1$ or $\theta_1 = \pi - \theta_0 \ll 1$. To provide an analytical treatment when θ_0 has an arbitrary value (and $ka \ll 1$), we introduce the new parameter

$$\varepsilon_n^* = \varepsilon_n - (2n+1)^{-2} \qquad (2.131)$$

that satisfies

$$\lim_{ka \to 0} \varepsilon_n^* = 0$$

and reformulate equation (2.28).

In the transformation of the dual series equations (2.24) and (2.25) to the regularised system (2.28), the following system is obtained

$$\sum_{s=0}^{\infty} X_s^m \hat{P}_s^{(m+\frac{1}{2}, m-\frac{1}{2})}(z) =$$

$$\begin{cases} 0, & z \in (-1, z_0) \\ \sum_{s=0}^{\infty} [X_s^m \varepsilon_{s+m} + A_s^m] \hat{P}_s^{(m+\frac{1}{2}, m-\frac{1}{2})}(z), & z \in (z_0, 1) \end{cases} \quad (2.132)$$

(applying appropriate inner products to this equation directly leads to (2.28)). Introduce the parameter ε_n^* and replace the equation defined on $z \in (z_0, 1)$ by

$$\sum_{s=0}^{\infty} X_s^m \hat{P}_s^{(m+\frac{1}{2}, m-\frac{1}{2})}(z) - \frac{1}{4} \sum_{s=0}^{\infty} X_s^m (s + m + \frac{1}{2})^{-2} \hat{P}_s^{(m+\frac{1}{2}, m-\frac{1}{2})}(z)$$

$$= F_m(z), \qquad z \in (z_0, 1) \quad (2.133)$$

where

$$F_m(z) = \sum_{s=0}^{\infty} [X_s^m \varepsilon_{s+m}^* + A_s^m] \hat{P}_s^{(m+\frac{1}{2}, m-\frac{1}{2})}(z). \quad (2.134)$$

The key to further progress is to observe that the function

$$G_m(z) \overset{def}{=} \sum_{s=0}^{\infty} X_s^m (s + m + \frac{1}{2})^{-2} \hat{P}_s^{(m+\frac{1}{2}, m-\frac{1}{2})}(z) \quad (2.135)$$

obeys a certain second order differential equation. This idea was discussed at the end of Section 2.1 of Volume I. It may be checked that by employing the differentiation formulae for Jacobi polynomials (see Volume I) the equation (2.133) may be replaced by the differential equation

$$(1 - z^2)G_m''(z) - [1 + (2n + 2)z] \cdot G_m'(z) - m(m + 1)G_m(z)$$

$$= -F_m(z), \qquad z \in (z_0, 1). \quad (2.136)$$

One may readily observe that the homogeneous variant of this differential equation coincides with differential equations for Jacobi polynomials $P_n^{(m+\frac{1}{2}, m-\frac{1}{2})}(z)$ of degree $n = -m$ and $n = -m - 1$.

Here we deduce the formulae for the cross-sections when $\theta_1 \ll 1$, $ka \sim \theta_1^2 \ll 1$; this relates to excitation of the Helmholtz mode. By keeping very few terms in the expansion of the Fourier coefficients X_{s+m}^m, the results of Section 2 of this chapter allow us to state approximate analytical formulae. Using a Taylor series expansion in ka near $(ka)_0$, where

$$(ka)_0 = \left(\frac{3\theta_1}{2\pi}\right)^{\frac{1}{2}} \left[1 + \frac{9}{20}\frac{\theta_1}{\pi} + O(\theta_1^2)\right],$$

one deduces the following approximate values for the cross-sections σ_T^h and σ^h of the hard spherical shell with a small aperture $(\theta_1 \ll 1)$:

$$\frac{\sigma_T^h}{\pi a^2} = \left[\frac{1}{3} + \frac{\frac{4}{9}(ka)^4}{[(ka)^2 - (ka)_0^2]^2 + \frac{1}{9}(ka)^{10}}\right](ka)^4 \left[1 + O((ka)^2, (ka)_0^2)\right];$$

$$(2.137)$$

$$\frac{\sigma^h}{\pi a^2} = \left[\frac{1}{3} + \frac{\frac{4}{9}(ka)^4}{[(ka)^2 - (ka)_0^2]^2 + \frac{1}{9}(ka)^{10}}\right](ka)^4 \left[1 + O((ka)^2, (ka)_0^2)\right].$$

$$(2.138)$$

If $\theta_1 = 0$, then $(ka)_0 = 0$, and the result (2.137) coincides with the well-known value of the total cross-section for a hard sphere (see [13])

$$\frac{\sigma_T^h}{\pi a^2} = \frac{7}{9}(ka)^4[1 + O((ka)^2)].$$

$$(2.139)$$

When the acoustic resonator is excited at a frequency corresponding to $ka = (ka)_0$, the total cross-section given by (2.137) is

$$\sigma_T^h = \frac{\lambda_0^2}{\pi},$$

$$(2.140)$$

where $\lambda_0 = 2\pi/k_0$ denotes wavelength. Notice that in the low frequency limit, $\sigma_T^h \gg 1$.

Accurate numerical calculations for the total cross-section σ_T^h of the hard shell are shown in Figure 2.4; at the three sizes of circular aperture displayed $(\theta_1 = 5°, 10°, 30°)$, the results computed from (2.112) and (2.115) using the solution of the system (2.28) very nearly coincide with the approximate analytical formulae stated above. The plot of the corresponding sonar cross-sections is very similar.

Now we turn to the less well-studied frequency band in which the dimension of the scatterer is comparable to a wavelength: $\lambda \sim a$. This so-called *resonance region* or *diffraction region* does not have a precise upper limit though the usual bound is typically $ka \lesssim 20$.

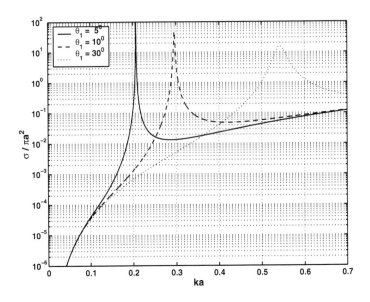

FIGURE 2.4. Total cross-section of the hard open spherical shell, $\alpha = 0°$.

2.4.2   Resonance Region.

When ka is close to the real part of the quasi-eigenvalues $\varkappa_{nl}^{m}$ or ν_{nl}^{m} (defined by [2.100] or [2.101]), so that $|ka - \varkappa_{nl}^{m}| \ll 1$ or $|ka - \nu_{nl}^{m}| \ll 1$, the values of the corresponding cross-sections $(\sigma_{T}^{h}, \sigma^{h}$ in the hard case or $\sigma_{T}^{s}, \sigma^{s}$ in the soft case) vary rapidly in such an interval.

From the physical point of view, at resonance the cavity accumulates energy from the incoming wave. In the vicinity of the resonant frequency, deep nulls occur in the sonar cross-sections $(\sigma^{h}$ or $\sigma^{s})$ if the Q-factor is high enough. In fact the phenomenon is more complicated. Deep nulls exist side by side with resonance peaks also lying at frequencies close to the real part of quasi-eigenvalues ($\mathrm{Re}(\varkappa_{nl}^{m})$ or $\mathrm{Re}(\nu_{nl}^{m})$ as appropriate).

Let us illustrate these features with the specific example of a soft open shell with a circular aperture of angle $\theta_{1} = 15°$. The normalised sonar cross-section $\sigma^{s}/\pi a^{2}$ is shown as a function of ka in Figure 2.5, with an expanded view in the vicinity of the first resonance in Figure 2.6(left). The backscattering strongly depends on the quasi-eigenoscillations excited in the cavity. This seemingly chaotic picture can be understood in terms of the spectrum of the quasi-eigenoscillations, the real part of which is readily observed at the values of the wavenumber at which the spikes representing very rapid variation of the cross-section occur. The expanded view of Figure

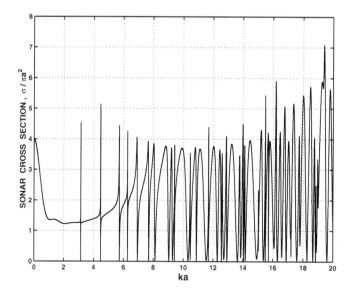

FIGURE 2.5. Soft open spherical shell, $\theta_0 = 15°$, $\alpha = 0°$.

2.6(left) shows the typical variation that we shall term a *resonance doublet* or *double extremum*.

The impact of the lowest quasi-eigenmodes on the magnitude of the cross-section $\sigma^s/\pi a^2$ is easily interpreted as there is no competition from modes excited at adjacent frequencies. However the real parts of quasi-eigenvalues associated with different modes become increasingly closely packed and less well separated at higher frequencies, and the scatterer's cross-section is affected by this mode competition. Complex phenomena such as so-called *hybrid modes* or *mode suppression* may appear at higher frequencies.

Leaving aside such special features, let us concentrate on the main feature, the resonance doublet. Consider a plane wave normally incident ($\alpha = 0°$) on the circular aperture of a soft open spherical shell (a similar treatment holds for other angles α of incidence). The angle subtended by the aperture at the centre is small: $\theta_1/\pi \ll 1$. Normalise the far field backscattered spherical wave amplitude given by (2.109) so that

$$A = \frac{ik}{4} f^s(\pi, 0) = \sum_{s=0}^{\infty} \left(s + \frac{1}{2} \right) Y_s^0 j_s(ka)(-i)^s. \qquad (2.141)$$

At wavenumbers ka well removed from the quasi-eigenvalues $\nu_{pk}^{(0)}$ the major contribution to the value of A is obtained by neglecting the presence of the small aperture and using the corresponding closed sphere backscattered

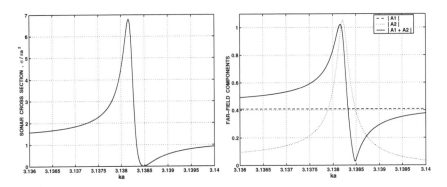

FIGURE 2.6. Soft open spherical shell, $\theta_0 = 15°$, $\alpha = 0°$.

amplitude A_1. Since

$$Y_s^0 = -\frac{(-i)^s}{2h_s^{(1)}(ka)} + O(\theta_1^3), \tag{2.142}$$

$$A \simeq A_1 = -\frac{1}{4}\sum_{s=0}^{\infty}(2s+1)(-1)^s\frac{j_s(ka)}{h_s^{(1)}(ka)}. \tag{2.143}$$

If the parameter ka approaches one of the values $\mathrm{Re}\,\nu_{pk}^{(0)}$ the situation dramatically changes. The difference $A_2 = A - A_1$ measures the effect of the aperture. It may be readily shown analytically that if $|ka - \nu_{pk}^{(0)}| \sim \theta_1^3$ the magnitude of aperture radiation A_2 previously neglected is comparable in magnitude to that of A_1.

The resonance doublet arises from the interference of two spherical waves arriving in the far-field zone with comparable amplitudes. Whilst the phase variation of A_1 within a narrow frequency band is negligibly small the same is not true for that of A_2. Its rapid variation creates the possibility for constructive and destructive interference at two very closely spaced frequencies. To illustrate this phenomenon consider the lowest quasi-eigenvalue, located near $\nu_{01} = \pi$. Accurate numerical calculations of the magnitude of A_1 and A_2 are shown in Figure 2.6(right) in the frequency band $|ka - \nu_{01}^{(0)}| \sim \theta_1^3$. Between 3.138 and 3.139, the phase of A_2 changes abruptly by π, whilst that of A_1 is almost unchanged. When ka is about 3.1385 the absolute values of A_1 and A_2 are very nearly equal, but their phase difference is very nearly equal to π. Cancellation in the sum $A = A_1 + A_2$ occurs (destructive interference) and a deep null in the sonar cross-section is observed near this point. On the other hand, the peak value of the sonar cross-section at $ka = 3.1382$ results from the dominant contribution A_2 to the

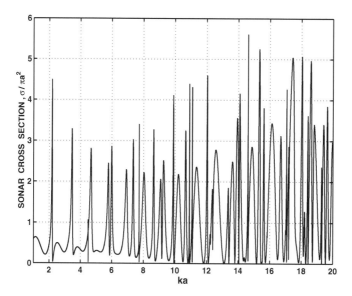

FIGURE 2.7. Sonar cross-section of a hard open spherical shell with $\theta_1 = 15°$, at incident angle $\alpha = 0°$.

sum $A = A_1 + A_2$. The same behaviour occurs at wavenumbers ka in the vicinity of the real part of other quasi-eigenvalues.

A similar phenomenon occurs for the acoustically hard spherical cavity: calculations of the sonar cross-section of the scatterer with the same circular aperture ($\theta_1 = 15°$) are shown in Figure 2.7. The impact of this phenomenon is also visible in total cross-section calculations. Results for the acoustically hard and soft cavities are displayed in Figures 2.8 and 2.9, respectively.

The strength of the coupling of energy into the cavity from the incident wave is very dependent upon the incident angle α. Although the location (in frequency) of rapid variation in sonar cross-sections can be correlated with the real part of suitable quasi-eigenvalues, the precise magnitude is not easy to predict analytically as α varies. However the overall reduction in peak values of the sonar cross-section at incidence angle $\alpha = 180°$ (shown in Figure 2.10) is obviously due to the nonoptimal excitation of the spherical cavity: the aperture is strongly shadowed.

2.4.3 High Frequency Regime.

Let us now consider the frequency dependence of $\sigma^s/\pi a^2$ at larger aperture sizes. In the resonance regime the ratio of the aperture diameter $D_a =$

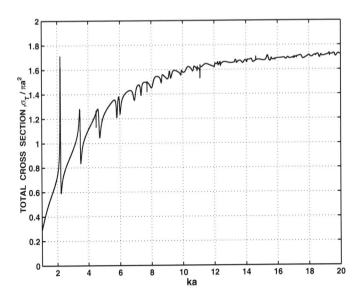

FIGURE 2.8. Total cross-section of a hard open spherical shell with $\theta_1 = 15°$, at incident angle $\alpha = 0°$.

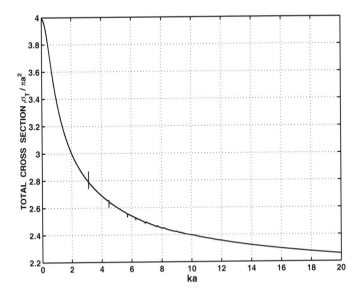

FIGURE 2.9. Total cross-section of a soft open spherical shell with $\theta_1 = 15°$, at incident angle $\alpha = 0°$.

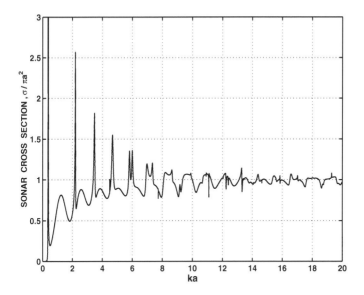

FIGURE 2.10. Sonar cross-section of the hard open spherical shell with $\theta_1 = 15°$, at incident angle $\alpha = 180°$.

$2a \sin \theta_1$ to wavelength λ varies from 0.258 $(ka \sim 3.14)$ to 1.647 $(ka = 20)$, i.e., $D_a \sim \lambda$.

We now consider the case when the aperture diameter is many wavelengths $D_a \gg \lambda$. Extend the numerical calculation of the cross-sections of the hard open shell (with aperture parameter $\theta_1 = 15°$) from $ka = 20$ (see Figure 2.7) to $ka = 320$ $(D_a/\lambda = 26.35)$; the results are displayed in Figures 2.11 and 2.12. At the highest wavenumber the diameter of the sphere is $2a \simeq 101\lambda$; the upper frequency range is *quasi-optical*.

Let us compare the cross-sections for direct illumination $(\alpha = 0)$ of the hole, with those calculated when the hole is deeply shadowed $(\alpha = 180°)$.

As might be expected, there is a significant contrast between the two cases of incidence. When the aperture is directly illuminated $(\alpha = 0°)$, the acoustic plane wave passes through the aperture with little disturbance, reflects from the inner surface of the cavity and passes out through the aperture again with little disturbance. As a result backscattering increases with frequency, and the total sonar cross-section steadily grows (see Figure 2.11). This simple explanation accounts for the envelope observed in these results. The numerous oscillations inside this envelope are due to the multiple reflections of rays from the inner surface of the cavity. At certain frequencies the acoustic field is "locked" inside the cavity, resulting in a null in the value of $\sigma^h(ka)$. At other individual values of ka the acoustic

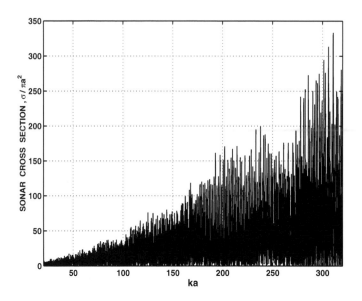

FIGURE 2.11. Sonar cross-section of a hard open spherical shell with $\theta_1 = 15°$, at incident angle $\alpha = 0°$.

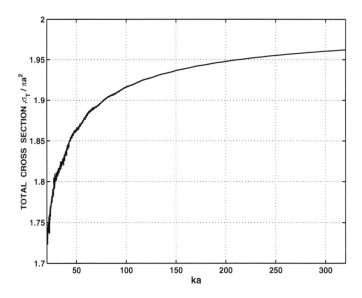

FIGURE 2.12. Total cross-section of a hard open spherical shell with $\theta_1 = 15°$, at incident angle $\alpha = 0°$.

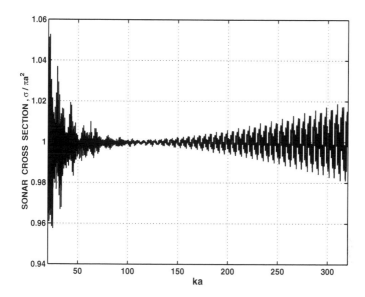

FIGURE 2.13. Sonar cross-section of a hard open spherical shell with $\theta_1 = 15°$, at incident angle $\alpha = 180°$.

field is reflected from the cavity completely, resulting in a peak value for $\sigma^h(ka)$. This complicated picture is difficult to predict analytically.

When the hole is in shadow ($\alpha = 180°$), the results are more or less predictable. The average value of the sonar cross-section (see Figure 2.13) is very close to that of the closed hard sphere; the numerous small amplitude oscillations about the average value $\sigma^h/\pi a^2 = 1$ is the only indication about of the existence of the aperture lying in the deep shadow. The total cross-section is nearly identical with that under direct illumination (see Figure 2.11).

Our solution of wave scattering from open spherical shells is uniformly valid with respect to the polar angle θ_0. Thus we may carry out an exhaustive study of the backscatter from slightly curved circular discs or shallow spherical caps. In particular, we wish to find the deviation of the standard physical optics (PO) approximation from our accurately computed solution.

On the illuminated surface of the scatterer, the physical optics approach (see [88]) approximates, in the hard case, the velocity potential U by $2U^i$ and, in the soft case, approximates its normal derivative $\dfrac{\partial U}{\partial n'}$ (in the soft case) by $2\dfrac{\partial U^0}{\partial n'}$ where U^i is the velocity potential (2.1) of incident plane wave; on the shadowed side of the surface these quantities are approximated by zero.

The velocity potentials for the scattered field at an arbitrarily taken point of space $\overrightarrow{r}$ are then calculated from

$$U^s(\overrightarrow{r}) = -\frac{1}{2\pi} \iint_S \frac{e^{ikR}}{R} \frac{\partial}{\partial n'} U^i(\overrightarrow{r'}) ds \quad \text{(soft body)} \quad (2.144)$$

$$U^s(\overrightarrow{r}) = \frac{1}{2\pi} \iint_S U^i(\overrightarrow{r'}) \frac{\partial}{\partial n'} \left(\frac{e^{ikR}}{R}\right) ds \quad \text{(hard body)} \quad (2.145)$$

where $\overrightarrow{r'}$ refers to the body surface and

$$R = \left\{r^2 + a^2 + 2ar\left[\cos\theta\cos\theta' + \sin\theta\sin\theta'\cos(\phi - \phi')\right]\right\}^{\frac{1}{2}}.$$

Substituting (2.144), (2.145) into (2.114) and carrying out the elementary integration, one obtains for normal incidence ($\alpha = 0°, 180°$) the expressions for monostatic sonar cross-sections:

$$\frac{\sigma^s}{\pi a_d^2} = \csc^2\theta_0 \left\{ \begin{array}{ll} |S_s(ka, \theta_0)|^2, & \alpha = 0° \\ |S_s^*(ka, \theta_0)|^2, & \alpha = 180° \end{array} \right\} \quad (2.146)$$

where

$$S_s(ka, \theta_0) = 1 - \cos\theta_0 e^{-i2ka(1-\cos\theta_0)} + \frac{i}{2ka}\left[1 - e^{-i2ka(1-\cos\theta_0)}\right];$$
$$(2.147)$$

and

$$\frac{\sigma^h}{\pi a_d^2} = \csc^2\theta_0 \left\{ \begin{array}{ll} |P_h(ka, \theta_0)|^2, & \alpha = 0° \\ |P_h^*(ka, \theta_0)|^2, & \alpha = 180° \end{array} \right\} \quad (2.148)$$

where

$$P_h(ka, \theta_0) = 1 - e^{-i2ka(1-\cos\theta_0)}. \quad (2.149)$$

In these formulae $a_d = a\sin\theta_0$ is aperture size of the spherical cap and the asterisk indicates complex conjugate values. It is notable that the cross-sections are the same for $\alpha = 0°$ and $\alpha = 180°$.

It can be readily shown if $ka \to \infty$, and $\theta_0 \to 0$ so that the product $ka\theta_0$ is a large but finite value, then both formulae give the well-known high-frequency limit for circular discs,

$$\frac{\sigma^s}{\pi a_d^2} = (ka_d)^2; \frac{\sigma^h}{\pi a_d^2} = (ka_d)^2.$$

The maximum depth of the spherical cap is $h = a(1 - \cos\theta_0)$. Since $2ka(1 - \cos\theta_0) = 4\pi h/\lambda$, then the maximum depth plays a decisive role in the backscattering magnitude. Thus

$$\frac{\sigma^h}{\pi a_d^2} = \csc^2\theta_0 \left|1 - \exp\left(-i4\pi\frac{h}{\lambda}\right)\right|^2 = 4\csc^2\theta_0 \sin^2\left(2\pi\frac{h}{\lambda}\right), \qquad (2.150)$$

so that cross-sectional nulls occur when $\dfrac{h}{\lambda} = \dfrac{n}{2} (n = 0, 1, ...)$ whilst cross-sectional maxima occur when $\dfrac{h}{\lambda} = \dfrac{n}{2} + \dfrac{1}{4}$. The same behaviour happens for the soft spherical cap though the values of h/λ at which nulls and peaks occur are slightly shifted by an amount $O((ka)^{-1})$.

Accurate calculations for the cross-section $\sigma^s/\pi a_d^2$ of a soft spherical cap with $\theta_0 = 15°$ over the range $0 < ka_d \leq 50$ are compared with the PO approximation in Figure 2.14. The true cross-section $\sigma^s_{180}/\pi a_d^2$ of the illuminated convex surface ($\alpha = 180°$) is different from the true cross-section $\sigma^s_0/\pi a_d^2$ of the illuminated concave surface ($\alpha = 0°$), though the PO estimates are identical. The peaks and nulls occur very close to the approximate values of h/λ determined by the PO approach. The PO estimate (2.146) falls between the two accurately computed cross-sections and is very close to their geometric mean $(\sigma^s_0\sigma^s_{180})^{\frac{1}{2}}/\pi a_d^2$. The same assertions are true for the normalised cross-section $\sigma^h/\pi a_d^2$ of a hard spherical cap with $\theta_0 = 15°$; the cross-section over the same range $(0 < ka_d \leq 50)$ is shown in Figure 2.15.

As the angular parameter decreases, the accuracy of the PO estimate improves for both soft and hard caps, as may be verified by further calculations over the same frequency range $0 < ka_d \leq 50$; moreover the PO estimate is very close to the geometric mean of the two accurately computed cross-sections. The PO estimate for scatterers of comparable size (in wavelength dimensions) is better for more nearly planar structures. This is to be expected because the guiding principle of physical optics is to treat each locality on the scatterer as a planar reflector.

2.5 The Mechanical Force Factor.

The acoustic pressure variation p_1 is related to the velocity potential $U = U(\overrightarrow{r}, t)$ by

$$p_1 = \rho_0 \frac{\partial}{\partial t} U(\overrightarrow{r}, t) \qquad (2.151)$$

where ρ_0 is the gas density. Thus the net pressure Δp exerted on an open hard spherical shell is proportional to the jump in velocity potential from

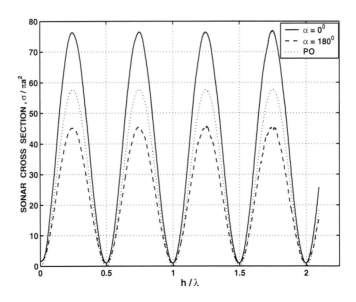

FIGURE 2.14. Soft spherical cap, $\theta_0 = 15°$. The PO estimate is approximately the geometric mean of the accurate cross-sections for $\alpha = 0°$ and $180°$.

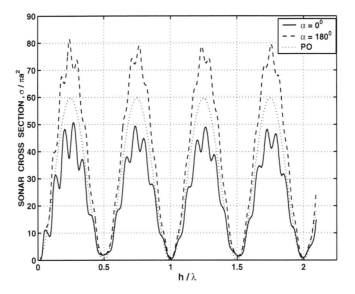

FIGURE 2.15. Hard spherical cap, $\theta_0 = 15°$. The PO estimate is approximately the geometric mean of the accurate cross-sections for $\alpha = 0°$ and $180°$.

the interior to the exterior:

$$\Delta p = -i\omega \rho_1 \left\{ U(a-0,\theta,\phi) - U(a+0,\theta,\phi) \right\}. \tag{2.152}$$

(Across the aperture, there is no jump.) The total force F_α on the shell, acting along the direction of plane wave propagation, is obtained by integration over the surface of the shell of the force component $\Delta F_\alpha = \Delta p_\alpha dS$ exerted on an element of area dS, where $\Delta p_\alpha = \Delta p \cos \psi$; it equals

$$F_\alpha = \iint_S \Delta p_\alpha dS. \tag{2.153}$$

Thus for suitable constants a_1^0 and a_1^1, the mechanical force is

$$F_\alpha = i\frac{4\pi\omega\rho_0}{k^2 h_1^{(1)\prime}(ka)} \left\{ a_1^0 \cos^2\alpha + a_1^1 \sin^2\alpha \right\}. \tag{2.154}$$

For the closed spherical shell $(\theta_0 = \pi)$, $a_1^0 = a_1^1 = -1$ and

$$F_\alpha \equiv F_\alpha^0 = -i\frac{4\pi\omega\rho_0}{k^2 h_1^{(1)\prime}(ka)}. \tag{2.155}$$

Define the *mechanical force factor* η by

$$\eta = \frac{|F_\alpha|}{|F_\alpha^0|};$$

thus, in terms of the coefficients X_1^0, X_1^1 defined by equation (2.27),

$$\eta = \frac{2}{3} \left| X_1^0 \cos\alpha + \sqrt{2} X_1^1 \sin\alpha \right| \cdot \left| h_1^{(1)\prime}(ka) \right|. \tag{2.156}$$

This basic formula allows us to distinguish the mechanical impact of the incident wave on the hard closed sphere from that on the thin-shelled hard cavity.

One conclusion is obvious: when the plane wave is propagating along the aperture plane $(\alpha = \pi/2)$ the open shell does not operate as a Helmholtz resonator at the Helmholtz mode wavenumber $ka = (ka)_0$, because the resonant contribution of X_1^0 is suppressed $(\cos\alpha = 0)$. It can be readily shown (see Section 2.2) that in the Helmholtz mode $(ka = \mathrm{Re}(ka)_H)$, the mechanical force factor is

$$\eta \cong \left(\frac{2\pi}{\theta_1} - 1 \right) |\cos\alpha|. \tag{2.157}$$

Some numerical calculations of η are presented in Figure 2.16; they are in good agreement with (2.157). Notice that values of η under diametrically

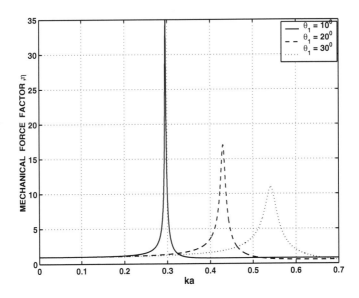

FIGURE 2.16. Mechanical force factor for various hard open shells at incident angle $\alpha = 0°$.

opposite directions of incidence are the same. In this low-frequency region the resonant behaviour of η is due to excitation of the Helmholtz mode.

The behaviour of η in the resonance region ($ka < 20$) is dominated by the resonant spikes in values of X_1^0, X_1^1 that occur precisely at wavenumbers $ka = \varkappa_{l10}$ and $ka = \varkappa_{l11}$, respectively. There is notable variation in the value of η at various angles of incidence. The efficiency of excitation depends directly on incidence angle α: the best performance of η occurs when $45° \leq \alpha \leq 90°$; the worst performance occurs when the aperture is shadowed ($\alpha = 180°$). See Figures 2.17 and 2.18 for the results of illumination of the cavity of aperture angle $\theta_1 = 15°$ from the illuminated ($\alpha = 0°$) and shadowed ($\alpha = 180°$) sides, respectively. The response at various illumination angles for the more open structure with aperture angle $\theta_1 = 60°$ is shown in Figure 2.19. The Q-factor of cavity excited quasi-eigenoscillations decreases as aperture size increases. The same is true for values of η at resonant frequencies.

When the open spherical shell becomes a spherical cap ($\theta_0 \leq 90°$) the behaviour of η changes, especially over a wider frequency range (up to quasi-optical). This may be explained in terms of interference phenomena. Although it is not so evident for the hemisphere ($\theta_0 = 90°$), because of the multiple reflections of rays (see Figure 2.20), the regular behaviour of η becomes more obvious for shallower caps. There is a simple explanation in

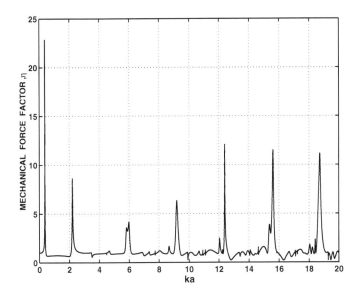

FIGURE 2.17. Mechanical force factor for the hard open shell ($\theta_0 = 15°$) at incident angle $\alpha = 0°$.

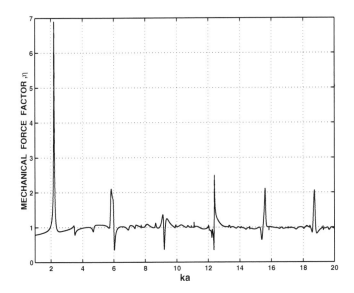

FIGURE 2.18. Mechanical force factor for the hard open shell ($\theta_0 = 15°$) at incident angle $\alpha = 180°$.

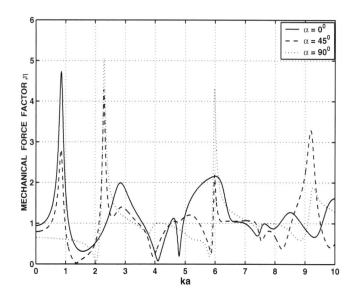

FIGURE 2.19. Mechanical force factor for the hard open spherical shell ($\theta_1 = 60°$) at various angles of incidence.

geometrical optics terms: for caps of angle dimension $\theta_0 < 45°$ the only reflection of normally incident rays is from the specular point lying at the centre of the cap.

The ratio of maximal depth $h = a(1 - \cos\theta_0)$ of the cap to wavelength λ is the crucial parameter. In the results plotted in Figure 2.21 for caps of angle $\theta_0 = 30°$ and $15°$ an interference phenomenon is clearly visible: maxima (minima, respectively) of η occur at depths h equal to an even (odd, respectively) number of half-wavelengths.

2.6   The Focal Region of a Spherical Reflector Antenna.

The design of reflector antennas is guided by a fundamental geometric feature of the parabola, that any ray which emanates from the focus is reflected from the interior of the parabolic surface in a direction parallel to the axis of symmetry; conversely any incoming ray parallel to the axis is reflected by the surface so that it passes through the focal point. Moreover any collection of such incoming rays parallel to the axis arrives, after reflection, at the focal point in phase, having travelled the same distance irrespective of the point of reflection. The same observations are true for

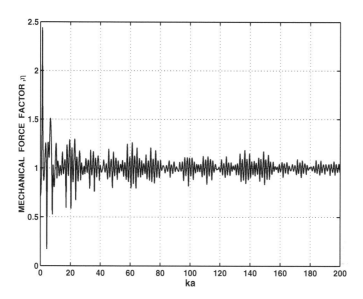

FIGURE 2.20. Mechanical force factor for the hard semi-spherical shell $(\theta_0 = 90°)$, at incident angle $\alpha = 0°$.

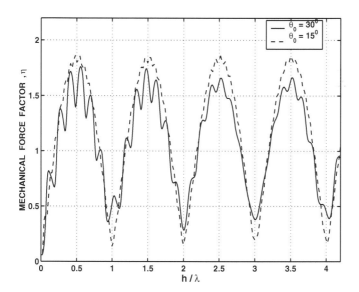

FIGURE 2.21. Mechanical force factor for various hard open spherical shells, at incident angle $\alpha = 0°$.

the paraboloid obtained by revolution of the parabola about its axis of symmetry. Constructive interference of the incident field at the focal point concentrates energy in the focal region where a receiver or sensor may be placed for signal reception. This principle applies equally to acoustic and electromagnetic waves.

This description of the antenna's function in terms of rays relies on the physical nature of wave motion at high frequencies, and the traditional geometric optics (GO) approach has enjoyed some success in analysing the field distribution around reflector antennas that are of large dimension in wavelengths. Various reflector surfaces have been considered in the literature, principally the paraboloid and the spherical reflector and their variants.

In this section we investigate the acoustic field distribution in the near-field zone of an axisymmetric receiving reflector antenna that is spherically shaped. The aperture D is acoustically large ($D \gg \lambda$). The GO approach encounters some difficulties in accounting for the contribution from the rim of the finite extent reflector. Furthermore, GO analysis breaks down in the focal region because of the collapse of a ray tube to a point (this difficulty is generally encountered with caustics). Although various corrections may be available via techniques, such as physical optics (PO) or the geometrical theory of diffraction (GTD), it is difficult to estimate their accuracy *a priori*. The accurate full-wave solution of the canonical scattering problem described in Section 2.1 – the spherical cap – provides one benchmark to test out the accuracy of these high frequency asymptotic techniques.

The usefulness of this proposed benchmark solution at high frequencies is, at first sight, surprising. Since it obtained by a technique that inverts the singular (or static) part of the Helmholtz operator, the solution of the regularised system might be expected to be most effective and efficient at low frequencies (Rayleigh scattering) and moderately effective in the resonance regime. However one purpose of this section is to demonstrate the effectiveness of this approach even in the quasi-optical regime.

Consider the case when the acoustic plane wave is normally incident on the spherical cap (i.e., the reflector). Setting $\alpha = 0$ in equations (2.28) and (2.33), the system simplifies so that only those equations with parameter $m = 0$ are retained:

$$Y_l^0 - \sum_{s=0}^{\infty} Y_s^0 \mu_s Q_{sl}(\theta_0) = \sum_{s=0}^{\infty} B_s^0 Q_{sl}(\theta_0) \qquad (2.158)$$

$$X_l^0 - \sum_{s=0}^{\infty} X_s^0 \varepsilon_s R_{sl}(\theta_0) = \sum_{s=0}^{\infty} A_s^0 R_{sl}(\theta_0) \qquad (2.159)$$

where

$$Q_{sl}(\theta_0) \equiv \hat{Q}_{sl}^{(-\frac{1}{2},\frac{1}{2})}(\cos\theta_0) = \frac{1}{\pi}\left[\frac{\sin(s-l)\theta_0}{s-l} + \frac{\sin(s+l+1)\theta_0}{s+l+1}\right] \quad (2.160)$$

and

$$R_{sl}(\theta_0) \equiv \hat{Q}_{sl}^{(\frac{1}{2},-\frac{1}{2})}(\cos\theta_0) = \frac{1}{\pi}\left[\frac{\sin(s-l)\theta_0}{s-l} - \frac{\sin(s+l+1)\theta_0}{s+l+1}\right]. \quad (2.161)$$

First of all let us consider accurately the influence of edges on the acoustic equivalent of the surface current density distribution on the spherical reflector. For the hard reflector, the appropriate surface density, denoted J^h, is proportional to the total velocity potential U on the spherical surface $(r = a)$; for the soft reflector this value (denoted J^s) is proportional to the normal derivative of U on the surface $r = a$.

On the illuminated side of the surface, the physical optics approximations for J^h and J^s are, respectively,

$$J_{po}^h = 2U^i \quad (2.162)$$

and

$$J_{po}^s = 2\frac{\partial U^i}{\partial n}; \quad (2.163)$$

on the shadowed side J_{po}^h and J_{po}^s are set to zero. Denote the illuminated and shadowed sides of the spherical cap by S^+ and S^-, respectively. Throughout U^t will denote the velocity potential of the total field; $U^t = U^i + U^s$ is its decomposition as a sum of the incident plane wave potential U^i and scattered velocity potential U^s.

In fact, the acoustic field penetrates the shadowed region so that J^h and J^s take nonzero values. It is therefore useful to consider *two-sided surface current densities* defined in the hard case by

$$J^h = \begin{cases} U^t(a-0,\theta,\phi) & \text{on } S^+, \\ U^t(a+0,\theta,\phi) & \text{on } S^-, \end{cases} \quad (2.164)$$

and in the soft case by

$$J^s = \begin{cases} \dfrac{\partial}{\partial r}U^t\bigg|_{r=a-0} & \text{on } S^+, \\[4mm] \dfrac{\partial}{\partial r}U^t\bigg|_{r=a+0} & \text{on } S^-. \end{cases} \quad (2.165)$$

In addition, we define associated *jump functions* by

$$j^h = U^t(a - 0, \theta, \phi) - U^t(a + 0, \theta, \phi) \tag{2.166}$$

in the hard case, and by

$$j^s = \left. \frac{\partial}{\partial r} U^t \right|_{r=a-0} - \left. \frac{\partial}{\partial r} U^t \right|_{r=a+0} \tag{2.167}$$

in the soft case.

The densities $(2.164) - (2.167)$ may be expressed in terms of the Fourier coefficients X_s^0 and Y_s^0. Thus the two-sided surface current densities are

$$J^h = e^{ika\cos\theta} + \begin{cases} 2 \sum\limits_{n=0}^{\infty} j_n(ka) h_n^{(1)\prime}(ka) X_n^0 P_n(\cos\theta), \text{ on } S^+, \\ 2 \sum\limits_{n=0}^{\infty} j_n'(ka) h_n^{(1)}(ka) X_n^0 P_n(\cos\theta), \text{ on } S^-, \end{cases} \tag{2.168}$$

and

$$J^s = ik\cos\theta\, e^{ika\cos\theta} +$$

$$\begin{cases} 2k \sum\limits_{n=0}^{\infty} j_n'(ka) h_n^{(1)}(ka)(2n+1) Y_n^0 P_n(\cos\theta), \text{ on } S^+, \\ 2k \sum\limits_{n=0}^{\infty} j_n(ka) h_n^{(1)\prime}(ka)(2n+1) Y_n^0 P_n(\cos\theta), \text{ on } S^-, \end{cases} \tag{2.169}$$

and the jump functions are

$$j^h = \frac{2i}{(ka)^2} \sum_{n=0}^{\infty} X_n^0 P_n(\cos\theta), \tag{2.170}$$

$$j^s = \frac{-2ki}{(ka)^2} \sum_{n=0}^{\infty} (2n+1) Y_n^0 P_n(\cos\theta). \tag{2.171}$$

Using the systems (2.158) and (2.159), we deduce

$$j^h = \frac{2i}{(ka)^2} \sum_{s=0}^{\infty} \left(X_s^0 \varepsilon_s + A_s^0 \right) \sum_{n=0}^{\infty} R_{sn}(\theta_0) P_n(\cos\theta), \tag{2.172}$$

$$j^s = \frac{-2ki}{(ka)^2} \sum_{s=0}^{\infty} \left(Y_s^0 \mu_s + B_s^0 \right) \sum_{n=0}^{\infty} (2n+1) Q_{sn}(\theta_0) P_n(\cos\theta). \tag{2.173}$$

It is obvious from the Definitions (2.49) and (2.50) that

$$\sum_{n=0}^{\infty} R_{sn}(\theta_0) P_n(\cos\theta) = \bar{\Phi}_s^0(\cos\theta, \cos\theta_0),$$

$$\sum_{n=0}^{\infty} (2n+1) Q_{sn}(\theta_0) P_n(\cos\theta) = 2R_s^0(\cos\theta, \cos\theta_0). \qquad (2.174)$$

In accordance with (2.55) and (2.57) these expressions vanish when $\theta > \theta_0$; when $\theta < \theta_0$, we use equations (2.56) and (2.57) to deduce that

$$j^h = \frac{2\pi^{\frac{1}{2}} i}{(ka)^2} \sum_{s=0}^{\infty} \left(X_s^0 \varepsilon_s + A_s^0 \right) G_s^0(\cos\theta, \cos\theta_0), \qquad (2.175)$$

$$j^s = \frac{-2ki}{(ka)^2} \left\{ \pi^{\frac{1}{2}} \sum_{s=0}^{\infty} \left(Y_s^0 \mu_s + B_s^0 \right) \left(s + \frac{1}{2} \right) G_s^0(\cos\theta, \cos\theta_0) + \right.$$
$$\left. \left(\frac{2}{\pi}\right)^{\frac{1}{2}} \csc\left(\frac{\theta_0}{2}\right) (\cos\theta - \cos\theta_0)^{-\frac{1}{2}} \sum_{s=0}^{\infty} \left(Y_s^0 \mu_s + B_s^0 \right) \hat{P}_s^{\left(-\frac{1}{2},\frac{1}{2}\right)}(\cos\theta_0) \right\}. $$
$$ (2.176) $$

The representations (2.175) and (2.176) have beneficial features. The relatively slowly converging series in (2.170) and (2.171), with convergence rate of $O(s^{-\frac{3}{2}})$ and $O(s^{-\frac{1}{2}})$, respectively, are replaced by the more rapidly convergent series (2.175) and (2.176), possessing convergence rates of $O(s^{-\frac{7}{2}})$ and $O(s^{-\frac{5}{2}})$, respectively.

The utility of these representations depends upon the ease in calculating $G_s^0(\cos\theta, \cos\theta_0)$. To this end, observe that

$$G_s^0(\cos\theta, \cos\theta_0) = \left(\frac{2}{\pi}\right)^{\frac{1}{2}} \int_\theta^{\theta_0} \frac{\sin\left(s + \frac{1}{2}\right)\phi}{\sqrt{\cos\phi - \cos\theta_0}} d\phi \qquad (2.177)$$

and consider the rescaled function

$$G_s(\theta, \theta_0) = \pi^{-\frac{1}{2}} G_s^0(\cos\theta, \cos\theta_0) = \frac{\sqrt{2}}{\pi} \int_\theta^{\theta_0} \frac{\sin\left(s + \frac{1}{2}\right)\phi}{\sqrt{\cos\phi - \cos\theta_0}} d\phi. \qquad (2.178)$$

Notice that (see Equation [2.59]) $G_s(\theta, \pi) = P_s(\cos\theta)$. Upon substituting $z = e^{i\phi}$ an alternative representation is the contour integral

$$G_s(\theta, \theta_0) = \frac{2}{\pi} \operatorname{Im} \left\{ \int_{e^{i\theta}}^{e^{i\theta_0}} \frac{z^s dz}{\sqrt{z^2 - 2z\cos\theta + 1}} \right\} \qquad (2.179)$$

in the complex plane along a circular arc connecting the points $e^{i\theta}$ and $e^{i\theta_0}$. The calculation of the integral (2.179) is easily effected by use of the recurrence formula

$$
G_{s+1}(\theta, \theta_0) = \frac{2\sqrt{2}}{\pi(s+1)} \cos\left(s + \frac{1}{2}\right)\theta_0 \cdot \sqrt{\cos\theta - \cos\theta_0}
$$
$$
+ \frac{2s+1}{s+1} \cos\theta \cdot G_s(\theta, \theta_0) - \frac{s}{s+1} G_{s-1}(\theta, \theta_0), \quad (2.180)
$$

where $s = 0, 1, 2, \ldots$, and is initialised by the value

$$
G_0(\theta, \theta_0) = \frac{2}{\pi} \arctan\left[\frac{\sqrt{2}\sin\frac{1}{2}\theta_0\sqrt{\cos\theta - \cos\theta_0} + \cos\theta - \cos\theta_0}{\sqrt{2}\cos\frac{1}{2}\theta_0\sqrt{\cos\theta - \cos\theta_0} + \sin\theta_0}\right].
$$
$$
(2.181)
$$

The recurrence formula (2.180) is a generalisation of the recurrence relation for Legendre polynomials: if setting $\theta_0 = \pi$ in both formulae (2.180) and (2.181) we obtain

$$
(s+1)P_{s+1}(x) - (2s+1)xP_s(x) + sP_{s-1}(x) = 0
$$

and

$$
G_0(\theta, \pi) = 1 = P_0^0(\cos\theta).
$$

In terms of the notation (2.178), the final form of the surface current density jump functions is

$$
j^h = \frac{2i}{(ka)^2} \sum_{s=0}^{\infty} \left[X_s^0 \varepsilon_s + A_s^0\right] G_s(\theta, \theta_0), \quad (2.182)
$$

$$
j^s = \frac{-2ki}{(ka)^2} \left\{ \sum_{s=0}^{\infty} \left(Y_s^0 \mu_s + B_s^0\right) G_s(\theta, \theta_0) + \right.
$$
$$
\left. \frac{4}{\pi} \csc(\theta_0)(\cos\theta - \cos\theta_0)^{-\frac{1}{2}} \sum_{s=0}^{\infty} \left(Y_s^0 \mu_s + B_s^0\right) \cos\left(s + \frac{1}{2}\right)\theta_0 \right\}. \quad (2.183)
$$

Let us present some illustrative calculations to compare the PO estimate of surface current density on the hard spherical dish with the exact result given by the method of regularisation (MoR). The relevant formulae are (2.162), (2.168) and (2.175).

Two values of the parameter θ_0, 15° and 35.13°, are chosen. The latter value corresponds to that of the spherical reflector antenna (SRA) installed

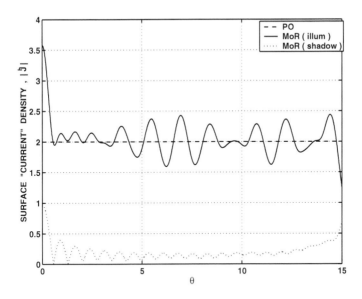

FIGURE 2.22. Acoustic surface current density on the hard spherical reflector, $D/\lambda = 20$, $\theta_0 = 15°$.

in the Arecibo observatory; this SRA is the largest reflector of its type in the world. A useful design parameter is D/λ, where $D = 2a \sin \theta_0$ denotes aperture dimension and λ wavelength. Thus the relative wave number is $ka = \pi(D/\lambda)/\sin \theta_0$. The maximal depth of the spherical dish is $h = a(1 - \cos \theta_0)$; the ratio of maximal depth to aperture dimension is $h^* = h/D = \frac{1}{2} \tan \frac{1}{2}\theta_0$; for the two angles of interest, $15°$ and $35.08°$, this ratio takes the values 0.0658 and 0.1580, respectively.

Consider first the shallower dish ($\theta_0 = 15°$). The absolute value of surface current density $|J|$ along the dish is plotted in Figures 2.22, 2.23 and 2.24 for aperture sizes $D/\lambda = 20, 50$ and 200, respectively. Oscillations of $|J|$ about the PO value ($|J| = 2$) decrease, for the most part of the dish, as D/λ increases; this is true for both illuminated and shadowed sides.

The main difference between the approximate PO results and the accurate MoR results is visible in the regions adjacent to the dish centre and rim; it is most important near the centre. As D/λ increases the shape of $|J|$ becomes similar to a *pulse signal* or *train of spatially separated pulses*. It is not so difficult to suppose that this is due to the edges; powerful secondary radiation from the edges produces this *quasi-singular* response.

The phase distribution for $D/\lambda = 20$ is plotted in Figure 2.25. At first glance the PO estimates seem satisfactory, at least over most of the dish. However, as for the amplitude distribution, there are notable deviations

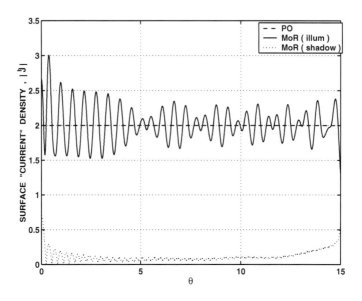

FIGURE 2.23. Acoustic surface current density on the hard spherical reflector, $D/\lambda = 50$, $\theta_0 = 15°$.

in the central region. For a spherical dish of size $D/\lambda = 200$ the maximal phase deviation, about $0.07^c \simeq 4°$, occurs in the middle of the dish; see Figure 2.26.

Our second set of numerical results concerns a deeper dish modelling the Arecibo antenna. Its physical dimensions are radius of curvature $a = 870$ feet and aperture dimension $D = 1000$ feet; thus the angle parameter is $\theta_0 = \arcsin (D/2a) = 35.13°$. Its operating frequency is about 430 MHz, corresponding to a wavelength λ equal to 0.697 metre or 2.28 feet; thus $D/\lambda \simeq 437$.

For $D/\lambda = 437$, the magnitude of surface current density is plotted in Figure 2.27; the behaviour of the distribution is very similar to that observed for the shallower dish, as is indeed the case for smaller values of D/λ. The corresponding phase distribution is also very similar to that observed for the shallower dish, exhibiting a very smooth quadratic variation.

Let us consider more closely the surface current on the shadowed part of the spherical dish. According to physical optics, the surface current density vanishes in the shadow region. However in reality, the incident wave penetrates the shadow region. We may subdivide the shadowed surface into three parts. In the *transient zone* $(20° < \theta \leq 35.13°)$ the current progressively decreases. In the *deep shadow zone* the current is stable and tiny. Finally, there is a narrow zone around the central point $(\theta = 0°)$ where

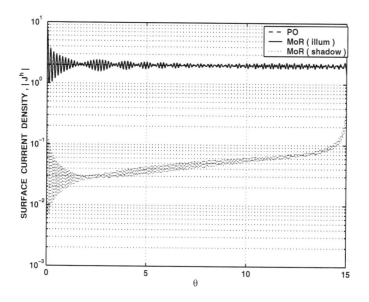

FIGURE 2.24. Acoustic surface current density on the hard spherical reflector, $D/\lambda = 200$, $\theta_0 = 15°$.

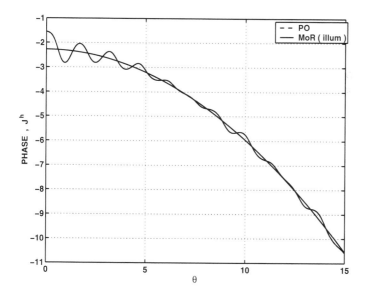

FIGURE 2.25. Hard spherical reflector with $D/\lambda = 20$, $\theta_0 = 15°$. Phase distribution of the acoustic surface current.

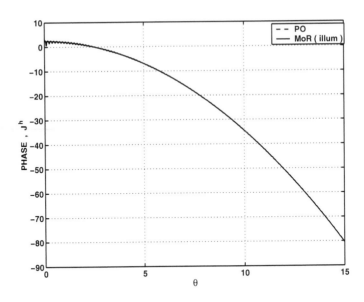

FIGURE 2.26. Hard spherical reflector with $D/\lambda = 200$, $\theta_0 = 15°$. Phase distribution of the acoustic surface current.

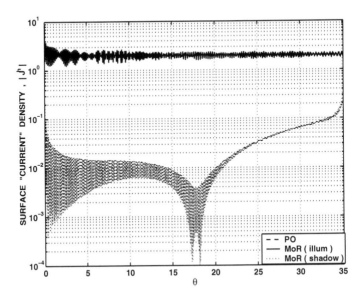

FIGURE 2.27. Acoustic surface current distribution on the hard spherical reflector with $D/\lambda = 437$, $\theta_0 = 35°$.

there is a *bright spot* indicating appreciable penetration of the acoustic field.

In conclusion we remark that calculations of the jump function given by Formulae (2.166) and (2.175) employing the special function $G_s^{(0)}(\theta, \theta_0)$ associated with the spherical segment give extremely accurate results for a wide range of the parameter (D/λ); these results agree well with those computed by other methods (such as integral equation methods).

Our comparison of PO results with the accurate results obtained by MoR illustrates certain inadequacies of physical optics in estimating the contribution of edge scattering. Although our results concern a specific scatterer, they highlight a generic difficulty encountered in physical optics. Also it will be recalled that the reflector was uniformly illuminated by a plane wave. From a physical point of view, it may be anticipated that edge scattering will play a less significant role if the reflector is *nonuniformly* illuminated, so that there is reduced illumination at the edges. This aspect is considered in the next section.

2.7   The Transmitting Spherical Reflector Antenna.

The final section of this chapter is devoted to an accurate analysis of the acoustic transmitting spherical reflector antenna, based on the method developed in Section 2.1. Some results have appeared in [115].

As mentioned in the previous section, the performance of a reflector antenna is often assessed by asymptotic high-frequency techniques such as Physical Optics (PO) combined with Geometrical Theory of Diffraction (GTD). The Method of Moments (MoM) is also used on integral equation based analyses of reflectors that are small to moderately sized in terms of wavelength. The merits and limitations of both approaches are well known and were briefly discussed in Section 1.7.

In spite of their flexibility, neither PO nor GTD are uniformly accurate with respect to spatial direction, and both fail to characterize smaller reflectors. MoM algorithms for the full-wave integral equation based approaches become computationally expensive for larger reflectors, due to the large matrices generated. Besides, not every MoM approximation scheme is convergent to the exact result as the number of equations is increased, in the sense that the computation error cannot be progressively minimised.

The feed field of such reflectors is normally simulated via a Gaussian beam or a spherical-wave expansion multiplied with an angular window function. Usually such simulations ignore the fact that such a feed field model does not satisfy the Helmholtz equation exactly, although the radiated or scattered field is found as a solution of the full-wave integral equation. The so-called *complex point source* beam, or a combination of such beams, has therefore been proposed as a model of the feed field. These satisfy the Helmholtz equation exactly at every point in the physical observation space. In [7], [89] this concept was combined with PO and GTD for a characterization of spherical-wave scalar beam scattering from a circular aperture. This concept was further developed in [68] and [69], which contain practical and useful results.

In this section we combine this feed model with the numerically accurate MoR, to analyse soft and hard spherical reflector antennas. We begin with a brief description of the complex point source concept. Our treatment closely follows that of Jones [42].

2.7.1 *Complex Point Source.*

The function $\exp(ikR)/R$, where $R^2 = (x - x_0)^2 + (y - y_0)^2 + (z - z_0)^2$, is a solution of Helmholtz's equation in (x, y, z) for any fixed (x_0, y_0, z_0) even if x_0, y_0, z_0 are complex. Now suppose that $x_0 = y_0 = 0$ and $z = ib$ with b positive real, so that

$$R^2 = \rho^2 + (z - ib)^2$$

where $\rho = (x^2 + y^2)^{\frac{1}{2}}$. Then the real values of (x, y, z) at which R vanishes are given by $z = 0, \rho = b$. Thus R is a multiple-valued function in (x, y, z) space. To render it single valued introduce a cut in $z = 0, \rho \leq b$ with R specified to be $z - ib$ on $\rho = 0$ when $z > 0$ and to be $-(z - ib)$ when $z < 0$. Values elsewhere are determined by continuity. Such a definition is consistent with the customary one for R when $b = 0$.

If $\rho^2 \ll z^2 + b^2$ and $z > 0$,

$$\frac{e^{ikR}}{R} \approx \frac{1}{z - ib} \exp\left\{ik\left(z - ib + \frac{1}{2}\frac{\rho^2}{z - ib}\right)\right\}$$

$$\approx \frac{1}{(z^2 + b^2)^{\frac{1}{2}}} \exp\left\{ikz\left(1 + \frac{1}{2}\frac{\rho^2}{z^2 + b^2}\right) - \frac{1}{2}\frac{kb\rho^2}{z^2 + b^2} + kb + i\tan^{-1}\frac{b}{z}\right\}$$

$$(2.184)$$

The range of observation points that satisfy $\rho^2 \ll z^2 + b^2$ define the *paraxial* region. In that part of the paraxial region where $z^2 \ll b^2$ the wave propagates parallel to the z axis with little distortion of the phase front, but decays exponentially in the direction perpendicular to the z axis, falling to $1/e$ of its value at $\rho = 0$ when $\rho = (2b/k)^{\frac{1}{2}}$. Thus, for $z \ll b^2$, a well-collimated beam is formed in the paraxial region (see Figure 2.28 which is closely based on [38]).

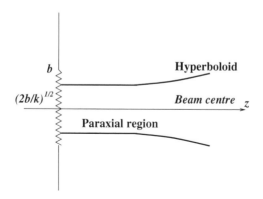

FIGURE 2.28. Gaussian Beam.

For larger values of z, the $1/e$ points lie on a hyperboloid that is asymptotic to the hyperbolic cone $\frac{1}{2}kb\rho^2 = z^2$. Fields of this type are known as *Gaussian beams*. The plane $z = 0$ is known as the *beam waist* and $(2b/k)^{\frac{1}{2}}$ as the *spot size* at the beam waist. Thus, a Gaussian beam is an approximate representation in the paraxial region of an exact solution of Helmholtz's equation generated by a source at a complex point.

Outside the paraxial region the Gaussian approximation is no longer valid. Nevertheless, any known solutions for real point sources that can be continued analytically to sources at complex points will provide solutions for excitation by Gaussian beams. In a later Section we will use the concept of a complex point source to construct the so-called *Complex Point Huygens Source*.

2.7.2 Regularised Solution and Far-Field Characteristics.

The transmission problem may be formulated as in section 2.1, except that the incident field is no longer a plane wave but a complex source point located axisymmetrically. Thus consider a zero-thickness, soft or hard, spherical reflector of radius a and angular width $2\theta_0$, symmetrically excited by the field of a complex point source beam located at a point with spherical coordinates $(r_s, 0, 0)$ where the radial coordinate is a complex value: $r_s = r_0 + ib$. Both the incident and scattered acoustic fields are axisymmetric and thus independent of ϕ: $U^0 = U^0(r, \theta), U^s = U^s(r, \theta)$. The geometry of the problem is shown in Figure 2.29. The incident scalar (acoustic) wave field is

$$U^0(r, \theta) = \exp(ikR)/R \qquad (2.185)$$

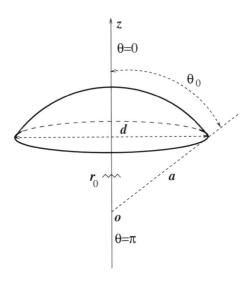

FIGURE 2.29. Geometry of the transmitting SRA.

where $R = (r^2 - 2rr_s \cos\theta + r_s^2)^{\frac{1}{2}}$, and k is the (real-valued) free-space wave number. The scattered field $U^s(r, \theta)$ is the solution of the Helmholtz equation, with a boundary condition of either hard or soft type (see [2.4] or [2.9]) enforced on the reflector surface M (given by $r = a$, $\theta \in (0, \theta_0)$, $\phi \in (0, 2\pi)$). In addition, to ensure uniqueness, the scattered field must satisfy (i) a finite energy condition, equivalent to the edge condition $U^s \sim O\left(\rho^{\frac{1}{2}}\right)$, $\partial U^s/\partial r \sim O\left(\rho^{-\frac{1}{2}}\right)$, as the distance from the dish rim $\rho \to 0$, and (ii) the outgoing radiation condition at infinity.

For the hard spherical reflector, the scattered field has the form

$$U_h^s = ik \sum_{n=0}^{\infty} x_n^h \left\{ \begin{array}{l} h_n^{(1)\prime}(ka)j_n(kr), r < a \\ j_n'(ka)h_n^{(1)}(kr), r > a \end{array} \right\} P_n(\cos\theta), \qquad (2.186)$$

whilst for the soft spherical reflector, it takes the form

$$U_s^s = ik \sum_{n=0}^{\infty} (2n+1)x_n^s \left\{ \begin{array}{l} h_n^{(1)}(ka)j_n(kr), r < a \\ j_n(ka)h_n^{(1)}(kr), r > a \end{array} \right\} P_n(\cos\theta). \qquad (2.187)$$

From the results of Section 2.1 the unknown coefficients x_m^h and x_m^s satisfy the following second kind Fredholm matrix equations

$$x_m^h - \sum_{n=0}^{\infty} x_n^h \varepsilon_n \hat{Q}_{nm}^{(\frac{1}{2},-\frac{1}{2})}(z_0) = \sum_{n=0}^{\infty} B_n^h \hat{Q}_{nm}^{(\frac{1}{2},-\frac{1}{2})}(z_0), \qquad (2.188)$$

$$x_m^s - \sum_{n=0}^{\infty} x_n^s \mu_s \hat{Q}_{nm}^{(-\frac{1}{2},\frac{1}{2})}(z_0) = \sum_{n=0}^{\infty} B_n^s \hat{Q}_{nm}^{(-\frac{1}{2},\frac{1}{2})}(z_0), \qquad (2.189)$$

where $m = 0, 1, 2, \ldots$, $z_0 = \cos\theta_0$, and ε_n and μ_n are defined by formulae (2.21) and (2.35), respectively. In addition,

$$B_n^h = 4i(ka)^3 j_n(kr_s) h_n^{(1)\prime}(ka), \qquad (2.190)$$
$$B_n^s = -ika(2n+1) j_n(kr_s) h_n^{(1)}(ka). \qquad (2.191)$$

The normalised incomplete scalar products $\hat{Q}_{nm}^{(\pm\frac{1}{2},\mp\frac{1}{2})}(z_0)$ have the simple form given by (2.160) and (2.161).

Due to the well-known asymptotics

$$h_n^{(1)}(x) = (-i)^{n+1} e^{ix}/x + O\left(x^{-2}\right), \qquad (2.192)$$

as $x \to \infty$, the total acoustic field in the far-field zone has the asymptotic form

$$U_{tot}^{s,h}(r,\theta) = \frac{e^{ikr}}{r} \psi^{s,h}(\theta) + O\left(r^{-2}\right), \qquad (2.193)$$

where the far field patterns ψ^s, ψ^h are given by

$$\psi^s(\theta) = e^{-ikr_s\cos\theta} + \sum_{n=0}^{\infty} (-i)^n (2n+1) x_n^s j_n(ka) P_n(\cos\theta) \qquad (2.194)$$

and

$$\psi^h(\theta) = e^{-ikr_s\cos\theta} + \sum_{n=0}^{\infty} (-i)^n x_n^h j_n'(ka) P_n(\cos\theta). \qquad (2.195)$$

Due to the completeness and orthogonality of the Legendre functions, the integral over the unit sphere defining the time-average far-zone power flux $P_{tot}^{(s,h)}$ may be performed analytically. The free-space radiated power and free-space directivity (or gain) of the complex point source described above are

$$P_0 = \frac{2\pi}{k^2} \frac{\sinh 2kb}{kb}, \quad D_0 = \frac{1}{4\pi} \frac{2kbe^{2kb}}{\sinh 2kb}. \qquad (2.196)$$

respectively. The overall directivity should be compared with D_0. In the soft case, the normalised far field power flux is

$$P_{tot}^s / P_0 = 1 +$$

$$\frac{2kb}{\sinh 2kb} \sum_{n=0}^{\infty} (2n+1) j_n (ka) \left\{ j_n (ka) \left| x_n^s \right|^2 + 2 \operatorname{Re} \left[x_n^s j_n (kr_s) \right] \right\}, \quad (2.197)$$

and in the hard case, it is

$$P_{tot}^h / P_0 = 1 +$$

$$\frac{2kb}{\sinh 2kb} \sum_{n=0}^{\infty} j_n' (ka) \left\{ (2n+1)^{-1} j_n' (ka) \left| x_n^h \right|^2 + 2 \operatorname{Re} \left[x_n^h j_n (kr_s) \right] \right\}.$$

$$(2.198)$$

For spherical reflectors with no losses, the directivity coincides with the gain of the antenna, being

$$G_{tot}^s = \frac{4\pi}{k^2 P_{tot}^s} \left| e^{ikr_s} + \sum_{n=0}^{\infty} i^n (2n+1) x_n^s j_n (ka) \right|^2 \quad (2.199)$$

$$G_{tot}^h = \frac{4\pi}{k^2 P_{tot}^h} \left| e^{ikr_s} + \sum_{n=0}^{\infty} i^n x_n^h j_n' (ka) \right|^2 \quad (2.200)$$

in the soft and hard case, respectively.

2.7.3   Numerical Results.

We have already seen that the complex point source generates a Gaussian beam. The importance of edge contributions to scattering from the reflector was highlighted in the previous section. One device to minimise their effect is to *taper* the beam source, i.e., the location and waist of the beam source are chosen so that the edges are weakly illuminated whilst the central portion of the reflector is fully illuminated.

Consider the spherical reflector antenna with aperture $D = 2a \sin \theta_0$. Both the central point and edges of the reflector are in the far-field zone of the source and so the acoustic energy density is very nearly proportional to $w (a, \theta) = k^2 \left| U^0 (a, \theta) \right|$ at the dish surface. The level of incident energy is determined by the ratio $w (a, \theta_0) / w (a, 0)$; the edge taper N is the value of this ratio in decibels (dB). Thus if N is specified, the value of kb is determined by

$$10 \log \left(w (a, \theta_0) / w (a, 0) \right) + N = 0. \quad (2.201)$$

Uniform illumination corresponds to $kb = 0$. In fact, because the distance $d_t = a - r_0$ from an acoustic monopole to the central point of the dish differs from the distance $d_e = \left(a^2 - 2ar_0 \cos\theta_0 + r_0^2\right)^{\frac{1}{2}}$ to the edges, the intrinsic edge taper is not exactly zero; for example, if the monopole is placed at the geometrical focus $(r_0 = 0.5a)$, the edge taper N takes values -0.55dB and -2.37dB corresponding values of θ_0 equal to $15°$ and $35.13°$, respectively. These angular values are used in subsequent calculations in this section. However we shall ignore this relatively small difference between the true value of N and zero, and ask the reader to bear in mind that on subsequent plots an edge taper of 0 dB is to be interpreted as uniform illumination of the reflector.

A rule of thumb employed by antenna designers states that the radiation features of a spherical (or other) reflector antenna are very similar to those of a parabolic reflector provided two reflector surfaces differ by less than $\lambda/16$. For the shallower SRA (with $\theta_0 = 15°$) this rule constrains its size to $D/\lambda \leq 53.5$. However the deeper reflector described in the previous section, the Arecibo reflector antenna (with $\theta_0 = 35.13°$ and $D/\lambda = 437$), is well outside the bounds of this rule. It is of great interest to provide an accurate prediction of the performance of a nonparabolic reflector antenna of these dimensions. We present some calculations for both shallow and deep reflectors with edge taper N equal to 0 dB (i.e., uniform illumination), -5 dB and -10.8 dB.

The dependence of gain G on feed location along the z axis for the shallower SRA of angular measure $\theta_0 = 15°$ is shown in Figure 2.30 $(D/\lambda = 20)$ and Figure 2.31 $(D/\lambda = 50)$. The performance is very sensitive to the edge taper. At the lowest value of edge taper $(N = -10.8$ dB$)$ the gain is comparable to the ideal theoretical limit $D_{\max} = (\pi D/\lambda)^2$ for these aperture sizes (see [76]). The ideally attainable value $D_{\max}$ arises when the acoustic field distribution is uniform in both amplitude and phase, with an edge taper N around -10.8 (-11dB is optimal).

The dependence of gain G on normalised source location (r_0/a) is oscillatory, which is inconvenient in practical application.

There are two observations to be made. First there is no big difference in the gain G amongst the various edge tapers that attain a -10.8 dB level at the edge. The second is the increasing divergence of G from $D_{\max}$ for the deeper reflector as D/λ increases. The maximum value of gain $G_{\max}$ is about $0.907 D_{\max}$ when $D/\lambda = 20$, but this decreases to $0.656 D_{\max}$ and $0.225 D_{\max}$ as D/λ increases to 50 and 200 (see Figure 2.32), respectively. This explained by increasing spherical aberration introducing nonuniformities into the acoustic field distribution in the aperture plane. The obvious conclusion, well known to antenna designers, is that no completely satisfactory performance of a deep reflector antenna is possible with a single

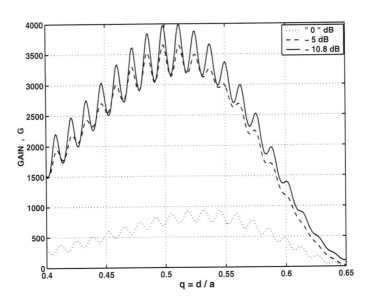

FIGURE 2.30. Gain of a hard spherical reflector, $\theta_0 = 15°$, $D/\lambda = 20$.

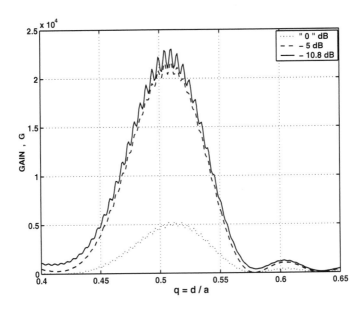

FIGURE 2.31. Gain of a hard spherical reflector, $\theta_0 = 15°$, $D/\lambda = 50$.

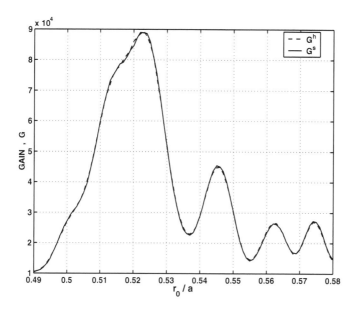

FIGURE 2.32. Gains G^h and G^s of a hard SRA and a soft SRA ($\theta_0 = 35.13°$), $D/\lambda = 200$.

feed point, and better performance is obtained only by using a distributed feed.

Whilst gain is important, the level of sidelobes (peaks in the off-axis radiation pattern) must be low enough for satisfactory engineering performance, with successive sidelobes progressively decreasing as one moves off-axis. Normalised radiation patterns were calculated for the shallow soft SRA ($\theta_0 = 15°$) with an optimal location of the feed r_0 along the optical axis; these locations are $r_0/a = 0.5055$ and $r_0/a = 0.5069$ when $D/\lambda = 20$ and 50, respectively. In both cases the edge taper $N = -10.8$ dB.

The results are shown in Figures 2.33 and 2.34. The value of G is close to the theoretical limit, and the first sidelobe is about -25 dB below the main beam. This is excellent for application.

Let us now examine the deep SRA ($\theta_0 = 35.13°$), locating, as before, the feed at those points of optical axis where the maximum value of gain occur, using an edge taper N equal to -10.8 dB. The optimal source locations are $r_0/a = 0.534$, 0.538 and 0.5222 when D/λ takes values 20, 50 and 200, respectively. The simple idea, of locating the feed at that point of axis corresponding to the maximum of G, fails to produce a satisfactory sidelobe distribution for the deep SRA. Instead of a Gaussian beam-like radiation the pattern shows undesirable widening that is more obvious for

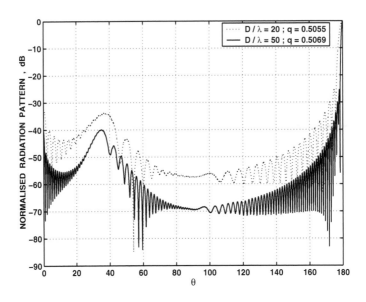

FIGURE 2.33. Normalised radiation pattern for soft spherical reflectors with $\theta_0 = 15°$ and $D/\lambda = 20$ or 50.

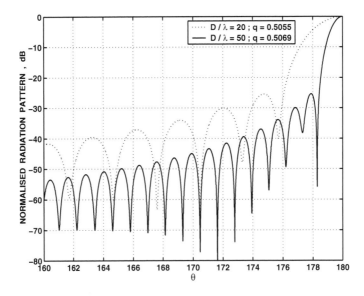

FIGURE 2.34. Normalised radiation pattern for soft spherical reflectors with $\theta_0 = 15°$ and $D/\lambda = 20$ or 50.

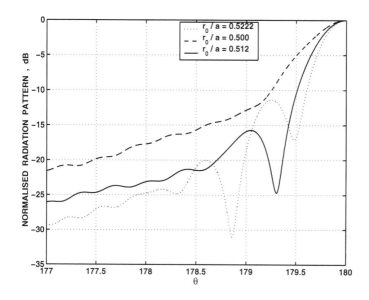

FIGURE 2.35. Normalised radiation patterns for the soft spherical reflector with $\theta_0 = 35.13°$ with $D/\lambda = 200$; various source locations.

the largest value of D/λ (200). The distortion of the main beam is the result of a nonuniform acoustic field distribution over the aperture plane. A better and nearly ideal shape of the main beam is obtained by shifting the feed nearer to the geometrical focus at $r_0/a = 0.5$; directivity (or gain) falls (see Figure 2.35). The value of G drops significantly. In spite of plausibly shaped radiation patterns the directivity (or gain) when the feed is located at the point $r_0/a = 0.512$ is very much reduced (to about 17%).

A characteristic feature of the deep large aperture SRA is the oscillatory dependence of gain G on feed location r_0/a (see Figure 2.32). This is directly related to the acoustic energy density distribution on the optical axis when the SRA is in the receiving regime. This is an instance of the reciprocity theorem.

As is evident from Figure 2.32, location of a single feed near to, but not at the point where G is maximal, improves the radiation pattern at the expense of directivity. The situation can only be improved by employing distributed feeds or line feeds.

The results plotted in Figure 2.32 also show that there is negligible difference in the performance of hard or soft spherical reflectors when they are large. For small and moderate size reflectors, although the dependence of the gain (denoted G^h or G^s according as the reflector is hard or soft) on source location r_0/a is of a highly oscillatory character, the geometric mean $G = \left(G^h G^s\right)^{\frac{1}{2}}$ is smoothly varying. Our calculations reveal that this value G is closely related to the value of gain (or directivity) for the corresponding electromagnetic case when the SRA is perfectly conducting (see Chapter 4).

3

Acoustic Diffraction from Various Spherical Cavities.

In this chapter the approach of Chapter 2 is extended to topologically more complex scatterers. The first is a thin spherical shell with two circular holes (a *spherical barrel*): topologically this object is doubly connected. The second scatterer is the complementary surface, a spherical shell with an equatorial slot (see Figure 3.1). In both cases, the interest lies in the coupling of wave energy from the exterior free space region to the interior cavity region, via the circular apertures (in the first case) or via a slot (in the second case).

From a general point of view, one might expect that scattering theory for more complex structures would be significantly more difficult than that for the basic scatterers studied in the previous chapter (the spherical shell with a circular hole).

It is perhaps surprising that substantive scattering analysis is possible for spherical shells either with two equal and symmetrically located circular holes or with an equatorial slot. That these problems are solvable is possible by exploiting the method developed for solving the corresponding electrostatic problems in Volume I. Any departure from symmetry in the location of the aperture(s) increases the mathematical complexity of the scattering analysis; although it is possible to exploit the potential theory developed for an asymmetrically located slot and develop a corresponding theory for wave scattering, the final form of the solution is rather too bulky a solution to be included in the present volume.

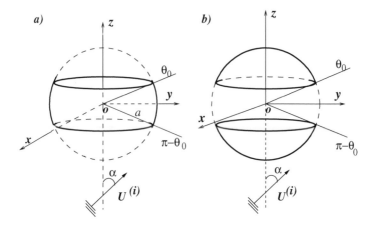

FIGURE 3.1. Plane wave illumination of a) a spherical barrel and b) a sphere with an equatorial slot.

Our examination of these cavity structures parallels that of the previous chapter for the single aperture cavity. In Sections 3.1 and 3.2, plane wave scattering from these more complex cavities is considered, for both soft and hard surfaces. Their behaviour as Helmholtz resonators is examined in Section 3.3, and the quasi-eigenoscillations of these cavities is described in Section 3.4. The total and sonar cross-sections and the mechanical force factor are investigated in Section 3.5. The results highlight the differences between these structures and the topologically simpler structure of the previous chapter.

3.1   The Hard Spherical Barrel and Soft Slotted Spherical Shell.

The investigation carried out in Chapter 2 is readily adapted to the present structures of interest. We may formulate the equations to be solved, by modifying (2.24) and (2.25). Thus plane wave illumination (defined by [2.10]) of the acoustically hard spherical barrel leads to the symmetrical triple series equations for $m = 0, 1, 2, \ldots,$

$$\sum_{s=0}^{\infty} x_{s+m}^{m} \hat{P}_{s}^{(m,m)}(z) = 0, \qquad z \in (-1, -z_0) \tag{3.1}$$

$$\sum_{s=0}^{\infty}(s+m+\frac{1}{2})\left\{x_{s+m}^m(1-\varepsilon_{s+m})-\alpha_{s+m}^m\right\}\hat{P}_s^{(m,m)}(z)=0,$$

$$z \in (-z_0, z_0) \quad (3.2)$$

$$\sum_{s=0}^{\infty}x_{s+m}^m\hat{P}_s^{(m,m)}(z)=0, \qquad z \in (z_0, 1) \qquad (3.3)$$

where the notation retains the same meaning as in Section 2.1, except that the constant θ_0 describes the angular size of each circular aperture in the symmetrical structure (the barrel is defined by $r = a$, $\theta_0 < \theta < \pi - \theta_0$, $\varphi \in (0, 2\pi)$), and $z_0 = \cos\theta_0$.

The scattering of the same plane wave by an acoustically soft spherical shell with an equatorial slot leads in a similar way to the following triple series equations for $m = 0, 1, 2, \ldots$,

$$\sum_{s=0}^{\infty}\left\{y_{s+m}^m(1-\mu_{s+m})-\beta_{s+m}^m\right\}\hat{P}_s^{(m,m)}(z)=0, \quad z \in (-1, -z_0) \qquad (3.4)$$

$$\sum_{s=0}^{\infty}(s+m+\frac{1}{2})y_{s+m}^m\hat{P}_s^{(m,m)}(z)=0, \qquad z \in (-z_0, z_0) \qquad (3.5)$$

$$\sum_{s=0}^{\infty}\left\{y_{s+m}^m(1-\mu_{s+m})-\beta_{s+m}^m\right\}\hat{P}_s^{(m,m)}(z)=0, \quad z \in (z_0, 1) \qquad (3.6)$$

where the notation retains the same meaning as in Section 2.1, including the constant θ_0 that describes the angular size of the upper cap (which is identical to the lower cap) in the symmetrical structure (the slot is defined by $r = a$, $\theta_0 < \theta < \pi - \theta_0$, $\varphi \in (0, 2\pi)$), and $z_0 = \cos\theta_0$.

Mathematically, both systems of equations are the same: the solution of the second is easily deduced from the first. Systems of this type were generally discussed in Section 2.4 of Volume I, where they were designated as *Type A triple series equations*. Let us briefly describe the regularisation process applied especially to the system (3.1)–(3.3) of present interest.

Using the symmetry property of Jacobi polynomials

$$\hat{P}_s^{(m,m)}(-z) = (-1)^s \hat{P}_s^{(m,m)}(z),$$

the system may be split into the following two decoupled systems for the even and odd index coefficients,

$$\sum_{s=0}^{\infty}x_{2s+m}^m\hat{P}_{2s}^{(m,m)}(z)=0, \qquad z \in (-1, -z_0) \qquad (3.7)$$

$$\sum_{s=0}^{\infty}(s+\frac{m}{2}+\frac{1}{4})\left\{x_{2s+m}^{m}(1-\varepsilon_{2s+m})-\alpha_{2s+m}\right\}\hat{P}_{2s}^{(m,m)}(z)=0,$$

$$z\in(-z_0,0)\quad(3.8)$$

and

$$\sum_{s=0}^{\infty}x_{2s+1+m}^{m}\hat{P}_{2s+1}^{(m,m)}(z)=0,\qquad z\in(-1,-z_0)\qquad(3.9)$$

$$\sum_{s=0}^{\infty}(s+\frac{m}{2}+\frac{3}{4})\left\{x_{2s+1+m}^{m}(1-\varepsilon_{2s+1+m})-\alpha_{2s+1+m}^{m}\right\}\hat{P}_{2s+1}^{(m,m)}(z)=0,$$

$$z\in(-z_0,0).\quad(3.10)$$

The next step uses the Gegenbauer polynomial relationship (see [2.131] in Chapter 2 of Volume I),

$$\hat{P}_{2s+l}^{(m,m)}(z)=\left[\frac{h_s^{(m,l-\frac{1}{2})}}{h_{2s+l}^{(m,m)}}\right]^{\frac{1}{2}}\frac{\Gamma(s+1)\Gamma(2s+m+1+l)}{\Gamma(2s+1+l)\Gamma(s+m+1)}z^l\,\hat{P}_s^{(m,l-\frac{1}{2})}(2z^2-1),$$

$$(3.11)$$

where $h_n^{(\alpha,\beta)}\equiv\left\|P_n^{(\alpha,\beta)}\right\|^2$ denotes the squared norm of Jacobi polynomials (see Appendix, Volume I).

Upon setting $u=2z^2-1=\cos 2\theta$ and $u_0=2z_0^2-1=\cos 2\theta_0$, we obtain the following two sets of dual series equations on the interval $(-1,1)$,

$$\sum_{s=0}^{\infty}(s+\frac{m}{2}+\frac{1}{4})\left\{\tilde{x}_{2s+m}^{m}(1-\varepsilon_{2s+m})-\tilde{\alpha}_{2s+m}^{m}\right\}\hat{P}_s^{(m,-\frac{1}{2})}(u)=0,$$

$$u\in(-1,u_0)\quad(3.12)$$

$$\sum_{s=0}^{\infty}\tilde{x}_{2s+m}^{m}\,\hat{P}_s^{(m,-\frac{1}{2})}(u)=0,\,u\in(u_0,1)\qquad(3.13)$$

and

$$\sum_{s=0}^{\infty}(s+\frac{m}{2}+\frac{3}{4})\left\{\tilde{x}_{2s+1+m}^{m}(1-\varepsilon_{2s+1+m})-\tilde{\alpha}_{2s+1+m}^{m}\right\}\hat{P}_s^{(m,\frac{1}{2})}(u)=0,$$

$$u\in(-1,u_0)\quad(3.14)$$

$$\sum_{s=0}^{\infty} \tilde{x}_{2s+1+m}^m \hat{P}_s^{(m,\frac{1}{2})}(u) = 0, \, u \in (u_0, 1) \tag{3.15}$$

where

$$\begin{Bmatrix} \tilde{x}_{2s+m+l} \\ \tilde{\alpha}_{2s+m+l} \end{Bmatrix} = \begin{Bmatrix} h_s^{(m,l-\frac{1}{2})} \\ h_{2s+l}^{(m,m)} \end{Bmatrix}^{\frac{1}{2}} \frac{\Gamma(s+1)\Gamma(2s+m+1+l)}{\Gamma(2s+1+l)\Gamma(s+m+1)} \begin{Bmatrix} x_{2s+m+l} \\ \alpha_{2s+m+l} \end{Bmatrix}. \tag{3.16}$$

In order to apply the Abel integral transform technique to the systems (3.12), (3.13) and (3.14), (3.15) introduce new variables and coefficients so that

$$X_{2s+l}^m = \left\{ \frac{\Gamma(s+l+1)\Gamma(s+m+1)}{\Gamma(s+l+\frac{1}{2})\Gamma(s+m+\frac{1}{2})} \right\}^{\frac{1}{2}} \tilde{x}_{2s+l+m}, \tag{3.17}$$

$$A_{2s+l}^m = (s + \frac{m+l}{2} + \frac{1}{4}) \left\{ \frac{\Gamma(s+m+\frac{1}{2})\Gamma(s+l+\frac{1}{2})}{\Gamma(s+m+1)\Gamma(s+l+1)} \right\}^{\frac{1}{2}} \tilde{\alpha}_{2s+l+m}. \tag{3.18}$$

These rescalings produce the systems

$$\sum_{s=0}^{\infty} \left\{ (s + \frac{m}{2} + \frac{1}{4}) \frac{\Gamma(s+m+\frac{1}{2})\Gamma(s+\frac{1}{2})}{\Gamma(s+m+1)\Gamma(s+1)} X_{2s}^m (1 - \varepsilon_{2s+m}) - A_{2s}^m \right\} \times$$

$$\hat{P}_s^{(m-\frac{1}{2},u)}(u) = 0, \quad u \in (-1, u_0) \tag{3.19}$$

$$\sum_{s=0}^{\infty} X_{2s}^m \hat{P}_s^{(m-\frac{1}{2},u)}(u) = 0, \quad u \in (u_0, 1) \tag{3.20}$$

and

$$\sum_{s=0}^{\infty} \left\{ (s + \frac{m}{2} + \frac{3}{4}) \frac{\Gamma(s+m+\frac{1}{2})\Gamma(s+\frac{3}{2})}{\Gamma(s+m+1)\Gamma(s+2)} X_{2s+1}^m (1 - \varepsilon_{2s+1+m}) - A_{2s+1}^m \right\}$$

$$\times \hat{P}_s^{(m-\frac{1}{2},1)}(u) = 0, \, u \in (-1, u_0), \tag{3.21}$$

$$\sum_{s=0}^{\infty} X_{2s+1}^m \hat{P}_s^{(m-\frac{1}{2},1)}(u) = 0, \quad u \in (u_0, 1). \tag{3.22}$$

We now identify the parameters

$$q_{sm}^{(l)} = 1 - (s + \frac{m+l}{2} + \frac{1}{4}) \frac{\Gamma(s+m+\frac{1}{2})}{\Gamma(s+m+1)} \cdot \frac{\Gamma(s+m+\frac{1}{2})}{\Gamma(s+m+1)}; \tag{3.23}$$

as $s \to \infty$, they are asymptotically small: $q_{sm}^{(l)} = O(s^{-2})$ as $s \to \infty$. Thus we may rewrite the dual series equations in a standard form

$$\sum_{s=0}^{\infty} X_{2s+l}^m \hat{P}_s^{(m-\frac{1}{2},l)}(u) = \begin{cases} F_1(u), & u \in (-1, u_0) \\ 0, & u \in (u_0, 1) \end{cases} \tag{3.24}$$

where

$$F_1(u) = \sum_{s=0}^{\infty} \left\{ \left[(1 - q_{sm}^{(l)}) \varepsilon_{2s+l+m} + q_{sm}^{(l)} \right] X_{2s+l}^m + A_{2s+l}^m \right\} \hat{P}_s^{(m-\frac{1}{2},l)}(u)$$

and $l = 0$ or 1, corresponding to the even or odd index systems, respectively. The process is completed by exploiting the orthonormality of the Jacobi polynomials to transform these equations to the following second kind Fredholm matrix equations in which the matrix operator is a completely continuous perturbation of the identity matrix in l_2:

$$(1 - q_{nm}^{(l)})(1 - \varepsilon_{2n+l+m}) X_{2n+l}^m +$$

$$+ \sum_{s=0}^{\infty} \left\{ (1 - q_{sm}^{(l)}) \varepsilon_{2s+l+m} + q_{sm}^{(l)} \right\} X_{2s+l}^m \hat{Q}_{sn}^{(m-\frac{1}{2},l)}(u_0) =$$

$$= A_{2n+l}^m - \sum_{s=0}^{\infty} A_{2s+l}^m \hat{Q}_{sn}^{(m-\frac{1}{2},l)}(u_0), \tag{3.25}$$

where $l = 0$ or 1. The solution of this system thus obtained is uniformly valid with respect to angle size θ_0 and acoustical size ($\sim a/\lambda$).

In some cases it is more convenient to re-formulate the systems (3.25) in terms of the complementary angle $\theta_1 = \frac{\pi}{2} - \theta_0$: geometrically (see Figure 3.1) θ_1 is the half-angle subtended by the barrel at the origin. It is evident that $u_1 = \cos(2\theta_1) = -u_0$. The alternative solution form is (where $l = 0$ or 1),

$$\hat{X}_{2n+l}^m - \sum_{s=0}^{\infty} \left\{ (1 - q_{sm}^{(l)}) \varepsilon_{2s+l+m} + q_{sm}^{(l)} \right\} \hat{X}_{2s+l}^m \hat{Q}_{sn}^{(l,m-\frac{1}{2})}(u_1)$$

$$= \sum_{s=0}^{\infty} \hat{A}_{2s+l}^m \hat{Q}_{sn}^{(l,m-\frac{1}{2})}(u_1), \tag{3.26}$$

where $\left\{ \hat{X}_{2s+l}^m, \hat{A}_{2s+l}^m \right\} = (-1)^s \left\{ X_{2s+l}^m, A_{2s+l}^m \right\}$.

The solution of the system (3.4)–(3.6) is constructed in a similar way. Corresponding to (3.24) one obtains (where $l = 0$ or 1 corresponding to even or odd index systems as appropriate),

$$\sum_{s=0}^{\infty} Y_{2s+l}^m \hat{P}_s^{(m-\frac{1}{2},l)}(u) =$$

$$\begin{cases} \displaystyle\sum_{s=0}^{\infty} Y_{2s+l}^m q_{sm}^{(l)} \hat{P}_s^{(m-\frac{1}{2},l)}(u), & u \in (-1, u_0) \\ \displaystyle\sum_{s=0}^{\infty} \left[Y_{2s+l}^m \mu_{2s+l+m} + B_{2s+l}^m \right] \hat{P}_s^{(m-\frac{1}{2},l)}(u), & u \in (u_0, 1) \end{cases}, \quad (3.27)$$

where

$$Y_{2s+l}^m = 2^{-\frac{m}{2}+\frac{l}{2}+\frac{1}{4}} \left\{ \frac{\Gamma(s+l+1)\Gamma(2s+2m+l+1)\Gamma(s+1)}{\Gamma(s+m+\frac{1}{2})\Gamma(s+m+l+\frac{1}{2})\Gamma(2s+1+l)} \right\}^{\frac{1}{2}} y_{2s+l} \quad (3.28)$$

and

$$B_{2s+l}^m = 2^{-\frac{m}{2}+\frac{l}{2}+\frac{1}{4}} \left(s + \frac{m+l}{2} + \frac{1}{4}\right) \frac{\Gamma(s+l+\frac{1}{2})}{\Gamma(s+m+1)} \times$$

$$\left\{ \frac{\Gamma(s+m+\frac{1}{2})\Gamma(2s+l+2m+1)\Gamma(s+1)}{\Gamma(s+l+1)\Gamma(s+m+l+\frac{1}{2})\Gamma(2s+1+l)} \right\}^{\frac{1}{2}} \beta_{2s+m+l}. \quad (3.29)$$

From the system (3.27) one can obtain the following regularised system in the form of an i.s.l.a.e. (with a matrix operator that is a completely continuous perturbation of the identity in l_2),

$$Y_{2n+l}^m\left(1 - q_{nm}^{(l)}\right) + \sum_{s=0}^{\infty} Y_{2s+l}^m\left(q_{sm}^{(l)} - \mu_{2s+l+m}\right)\hat{Q}_{sn}^{(m-\frac{1}{2},l)}(u_0)$$

$$= \sum_{s=0}^{\infty} B_{2s+l}^m \hat{Q}_{sn}^{(m-\frac{1}{2},l)}(u_0), \quad (3.30)$$

where $l = 0$ or 1.

In terms of the parameter $u_1 = \cos(2\theta_1)$, where θ_1 is the slot semi-width, the system (3.30) takes the form (with $l = 0$ or 1)

$$\left(1 - \mu_{2n+l+m}\right)\hat{Y}_{2n+l}^m - \sum_{s=0}^{\infty} \hat{Y}_{2s+l}^m\left(q_{sm}^{(l)} - \mu_{2s+l+m}\right)\hat{Q}_{sn}^{(l,m-\frac{1}{2})}(u_1)$$

$$= \hat{B}_{2n+l}^m - \sum_{s=0}^{\infty} \hat{B}_{2s+l}^m \hat{Q}_{sn}^{(l,m-\frac{1}{2})}(u_1). \quad (3.31)$$

The regularised systems obtained here will be used in the following sections for analytic and numerical studies of the scattering of acoustic plane waves by the class of scatterers described above. Uniqueness of solutions to these systems, and of those encountered in susbsequent sections in this chapter, is established by adapting the uniqueness argument given in Section 2.1; the truncation number for numerical studies is chosen according to the criterion set out in that section.

3.2   The Soft Spherical Barrel and Hard Slotted Spherical Shell.

When a plane acoustic wave strikes an acoustically soft spherical barrel or an acoustically hard spherical shell with an equatorial slot, the scattered field may be determined from the solution of a set of symmetrical triple series equations related to, but possessing an important difference from the triple series equations discussed in the preceding section. The relevant system is a set of *Type B triple series equations*, of the sort discussed in Section 2.4.2 of Volume I [1]. As indicated in that section, some additional operations are needed to transform the series equations into a standard form for regularisation.

In fact we may omit writing down the appropriate triple series equations and start from a system of dual series derived from (3.12)–(3.15) by simply interchanging the subintervals on which each series equation is defined. Thus, employing the same notation as in the preceding section, scattering of a plane wave from an acoustically hard slotted spherical shell gives rise to the following system, for the even $(l = 0)$ or odd $(l = 1)$ index coefficients:

$$\sum_{s=0}^{\infty} \tilde{x}^m_{2s+l+m} \, \hat{P}^{(m,l-\frac{1}{2})}_s(u) = 0, \qquad u \in (-1, u_0), \qquad (3.32)$$

$$\sum_{s=0}^{\infty} \left(s + \frac{m+l}{2} + \frac{1}{4} \right) \left\{ \tilde{x}^m_{2s+l+m}(1 - \varepsilon_{2s+l+m}) - \tilde{\alpha}^m_{2s+l+m} \right\} \hat{P}^{(m,l-\frac{1}{2})}_s(u)$$
$$= 0, \qquad u \in (u_0, 1). \quad (3.33)$$

The parameter θ_0 is the angle subtended at the origin by each spherical cap, and the index $m = 0, 1, 2, \ldots$.

Similarly, the plane wave diffraction problem for an acoustically soft spherical barrel also leads to the following system, for the even $(l = 0)$ or odd ($l = 1$) index coefficients:

$$\sum_{s=0}^{\infty} \left\{ \tilde{y}_{2s+l+m}^m (1 - \mu_{2s+l+m}) - \tilde{\beta}_{2s+l+m}^m \right\} \hat{P}_s^{(m,l-\frac{1}{2})}(u) = 0,$$

$$u \in (-1, u_0) \quad (3.34)$$

$$\sum_{s=0}^{\infty} \left(s + \frac{m+l}{2} + \frac{1}{4} \right) \tilde{y}_{2s+l+m}^m \hat{P}_s^{(m,l-\frac{1}{2})}(u) = 0, \qquad u \in (u_0, 1). \quad (3.35)$$

(Again the notation is as introduced in the previous section.)

The regularisation of hard and soft systems is similar; so we concentrate on the systems defined by equations (3.32) and (3.33). The solution for odd index ($l = 1$) coefficients $\left(\tilde{x}_{2s+1+m}^m \right)$ is relatively straightforward and is constructed in accordance with the general scheme described in Section 2.4 of Volume I. Thus one obtains

$$\sum_{s=0}^{\infty} X_{2s+1}^m \hat{P}_s^{(m+\frac{1}{2},0)}(u) = \begin{cases} 0, & u \in (-1, u_0) \\ F_2(u), & u \in (u_0, 1) \end{cases} \quad (3.36)$$

where

$$F_2(u) = \sum_{s=0}^{\infty} \left\{ \left[(1 - p_{sm}^{(1)}) \varepsilon_{2s+1+m} + p_{sm}^{(1)} \right] X_{2s+1}^m + A_{2s+1}^m \right\} \hat{P}_s^{(m+\frac{1}{2},0)}(u),$$

and

$$X_{2s+1}^m = \left\{ \frac{\Gamma(s + \frac{3}{2})\Gamma(s + m + \frac{3}{2})}{\Gamma(s+1)\Gamma(s+m+1)} \right\}^{\frac{1}{2}} \tilde{x}_{2s+1+m}^m \quad (3.37)$$

and

$$A_{2s+1}^m = \left(s + \frac{m}{2} + \frac{3}{4} \right) \left\{ \frac{\Gamma(s+1)\Gamma(s+m+1)}{\Gamma(s+\frac{3}{2})\Gamma(s+m+\frac{3}{2})} \right\}^{\frac{1}{2}} \tilde{\alpha}_{2s+1+m}^m \quad (3.38)$$

and

$$p_{sm}^{(1)} = 1 - \left(s + \frac{m}{2} + \frac{3}{4} \right) \cdot \frac{\Gamma(s+1)\Gamma(s+m+1)}{\Gamma(s+\frac{3}{2})\Gamma(s+m+\frac{3}{2})}. \quad (3.39)$$

The parameter $p_{sm}^{(1)}$ is asymptotically small: $p_{sm}^{(1)} = O(s^{-2})$, as $s \to \infty$. The system (3.36) is readily transformed to the following set of regularised i.s.l.a.e. (with matrix operator that is a compact perturbation of the identity in l_2), one for each index $l = 0, 1, 2, \ldots$,

$$X_{2l+1}^m - \sum_{s=0}^{\infty} \left\{ (1 - p_{sm}^{(1)}) \varepsilon_{2s+1+m} + p_{sm}^{(1)} \right\} X_{2s+1}^m \hat{Q}_{sl}^{(m+\frac{1}{2},0)}(u_0)$$

$$= \sum_{s=0}^{\infty} A_{2s+1}^m \hat{Q}_{sl}^{(m+\frac{1}{2},0)}(u_0), \quad (3.40)$$

where $m = 0, 1, 2, \ldots$.

On the other hand, the even index system defined by setting $l = 0$ in (3.34), (3.35) gives rise to a system in which the parameters fall outside the range of applicability of the method described in Section 2.1 of Volume I [1]. Specifically, the indices (α, β) of the relevant Jacobi polynomials do not satisfy $\alpha, \beta \geq -\frac{1}{2}$. To avoid mathematically erroneous transforms in the standard scheme of regularisation, it is necessary to increase the relevant indices by some means, such as using Rodrigues' formula as described in Chapter 2 of Volume I [1]. Thus set $l = 0$ and rearrange equations (3.34), (3.35) in the form

$$\sum_{s=1}^{\infty} \tilde{x}_{2s+m}^m \hat{P}_s^{(m,-\frac{1}{2})}(u) = -\tilde{x}_m^m \left[h_0^{(m,-\frac{1}{2})} \right]^{-\frac{1}{2}}, \quad u \in (-1, u_0), \quad (3.41)$$

$$\sum_{s=1}^{\infty} \left(s + \frac{m}{2} + \frac{1}{4} \right) \left\{ \tilde{x}_{2s+m}^m (1 - \varepsilon_{2s+m}) - \tilde{\alpha}_{2s+m}^m \right\} \hat{P}_s^{(m,-\frac{1}{2})}(u)$$

$$= -\left(\frac{m}{2} + \frac{1}{4} \right) \left\{ \tilde{x}_m^m (1 - \varepsilon_m) - \tilde{\alpha}_m^m \right\} \left\{ h_0^{(m,-\frac{1}{2})} \right\}^{-\frac{1}{2}}, \quad u \in (u_0, 1). \quad (3.42)$$

Rodrigues' formula for the orthonormal Jacobi polynomials occurring in these equations states

$$(1 - u)^m (1 + u)^{-\frac{1}{2}} \hat{P}_s^{(m,-\frac{1}{2})}(u) =$$

$$-\left\{ s\left(s + m + \frac{1}{2} \right) \right\}^{-\frac{1}{2}} \frac{d}{du} \left\{ (1 - u)^{m+1} (1 + u)^{\frac{1}{2}} \hat{P}_{s-1}^{(m+1,\frac{1}{2})}(u) \right\}. \quad (3.43)$$

To make the deduction more transparent let us first consider the regularisation of equations (3.41) and (3.42) with the value of m equal to 0. This value corresponds to normal incidence of the plane wave ($\alpha = 0$). Applying Rodrigues' formula (3.43) to the initial equations (3.41), (3.42) produces

$$\sum_{s=1}^{\infty} [s(s + \frac{1}{2})]^{-\frac{1}{2}} \tilde{x}_{2s}^0 \hat{P}_{s-1}^{(1,\frac{1}{2})}(u) = 2\tilde{x}_0^0 \left\{ h_0^{(0,-\frac{1}{2})} \right\}^{-\frac{1}{2}} (1 - u)^{-1}, \quad u \in (-1, u_0)$$

$$(3.44)$$

$$(1-u)^2 \sum_{s=1}^{\infty} \frac{\left(s+\frac{1}{4}\right)}{\left\{\left(s-\frac{1}{2}\right)s\left(s+\frac{1}{2}\right)(s+1)\right\}^{\frac{1}{2}}} \left\{\tilde{x}_{2s}^0(1-\varepsilon_{2s}) - \tilde{\alpha}_{2s}^0\right\} \hat{P}_{s-1}^{\left(1,\frac{1}{2}\right)}(u)$$

$$= -\frac{1}{2} \frac{\left\{\tilde{x}_0^0(1-\varepsilon_0) - \tilde{\alpha}_0^0\right\}}{\left\{h_0^{\left(0,-\frac{1}{2}\right)}\right\}^{\frac{1}{2}}} \left\{2\sqrt{2}\left[\sqrt{2} - (1+u)^{\frac{1}{2}}\right] - (1-u)\right\},$$

$$u \in (u_0, 1). \quad (3.45)$$

The standard scheme of regularisation is now applicable and produces the equation

$$(1-u)^{\frac{3}{2}} \sum_{s=1}^{\infty} X_{2s} \hat{P}_{s-1}^{\left(\frac{3}{2},0\right)}(u) = \begin{cases} F_1(u), & u \in (-1, u_0) \\ F_2(u), & u \in (u_0, 1), \end{cases} \quad (3.46)$$

where

$$F_1(u) = (2\pi)^{-\frac{1}{2}} \left\{h_0^{\left(0,-\frac{1}{2}\right)}\right\}^{-\frac{1}{2}} \tilde{x}_0^0, \quad (3.47)$$

$$F_2(u) = (1-u)^{\frac{3}{2}} \sum_{s=1}^{\infty} \left\{[p_s(1-\varepsilon_{2s}) + \varepsilon_{2s}] X_{2s} + A_{2s}\right\} \hat{P}_{s-1}^{\left(\frac{3}{2},0\right)}(u) +$$

$$\frac{\tilde{x}_0^0(1-\varepsilon_0) - \tilde{\alpha}_0^0}{\pi^{\frac{1}{2}} \left\{h_0^{\left(0,-\frac{1}{2}\right)}\right\}^{\frac{1}{2}}} \left\{(1-u)^{\frac{1}{2}} - \sqrt{2}\ln\left[\frac{\sqrt{2} + (1-u)^{\frac{1}{2}}}{(1+u)^{\frac{1}{2}}}\right]\right\}, \quad (3.48)$$

the rescaled variables are

$$X_{2s} = \frac{\Gamma\left(s+\frac{1}{2}\right)}{\Gamma(s+1)} \tilde{x}_{2s}^0, \qquad A_{2s} = \frac{s+\frac{1}{4}}{s\left(s+\frac{1}{2}\right)} \cdot \frac{\Gamma(s+1)}{\Gamma\left(s+\frac{1}{2}\right)} \tilde{\alpha}_{2s}^0, \quad (3.49)$$

and the asymptotically small parameter

$$p_s = 1 - \frac{s+\frac{1}{4}}{s\left(s+\frac{1}{2}\right)} \left[\frac{\Gamma(s+1)}{\Gamma\left(s+\frac{1}{2}\right)}\right]^2, \quad (3.50)$$

satisfies $p_s = O(s^{-2})$, as $s \to \infty$.

Let $F(u)$ denote the left hand side of equation (3.46). This equation shows that the function F has piecewise continuous representation on the complete interval $[-1, 1]$, of the form

$$F(u) = \begin{cases} F_1(u), & u \in (-1, u_0) \\ F_2(u), & u \in (u_0, 1). \end{cases} \quad (3.51)$$

Moreover the function $F(u)$ has a series expansion in terms of the complete orthonormal set $\left\{\hat{P}_{s-1}^{(\frac{3}{2},0)}\right\}_{s=1}^{\infty}$.

It is obvious that for arbitrary values of $\tilde{x}_0^0$ the coefficients $\{X_{2s}\}_{s=1}^{\infty}$ belong to l_2 (i.e., are square summable). In fact the coefficients have a faster rate of decay than mere square summability implies: $X_{2s} = O(s^{-2})$, as $s \to \infty$. This follows from the condition (2.6). Thus the function F is continuous, and the apparent arbitrariness in the choice of $\tilde{x}_0^0$ is eliminated by enforcing continuity at the point $u = u_0$, by demanding $F_1(u_0) = F_2(u_0)$. Thus

$$\tilde{x}_0^0 = \frac{\tilde{\alpha}_0}{\sqrt{2\pi}} \frac{1}{1 - \varepsilon_0} \cdot \frac{\varkappa(ka, u_0) - 1}{\varkappa(ka, u_0)} +$$

$$\frac{\left\{h_0^{(0,-\frac{1}{2})}\right\}^{\frac{1}{2}}}{(2\pi)^{\frac{1}{2}}} \frac{(1 - u_0)^{\frac{3}{2}}}{\varkappa(ka, u_0)} \sum_{s=1}^{\infty} \{[p_s(1 - \varepsilon_{2s}) + \varepsilon_{2s}] X_{2s} + A_{2s}\} \hat{P}_{s-1}^{(\frac{3}{2},0)}(u_0)$$

$$\tag{3.52}$$

where

$$\varkappa(ka, u_0) = 1 - \frac{1}{\pi}(1 - \varepsilon_0) \left\{ \left(\frac{1 - u_0}{2}\right)^{\frac{1}{2}} - \ln\left[\frac{\sqrt{2} + (1 - u_0)^{\frac{1}{2}}}{(1 + u_0)^{\frac{1}{2}}}\right] \right\}.$$

Suppressing some bulky transforms, the final format of the regularised system is

$$y_{2l} - \sum_{s=1}^{\infty} y_{2s} \left[p_s(1 - \varepsilon_{2s}) + \varepsilon_{2s}\right] R_{s-1,l-1}(ka, u_0)$$

$$= 2^{-\frac{1}{4}} \tilde{\alpha}_0 \frac{\hat{Q}_{l-1}^{(\frac{3}{2},0)}(u_0)}{\varkappa(ka, u_0)} + \sum_{s=1}^{\infty} b_{2s} R_{s-1,l-1}(ka, u_0) \tag{3.53}$$

where $\{y_{2n}, b_{2n}\} = \left\{n(n + \frac{1}{2})\right\}^{\frac{1}{2}} \{X_{2n}, A_{2n}\}$ and

$$R_{s-1,l-1}(ka, u_0) = \hat{Q}_{s-1,l-1}^{(\frac{1}{2},1)}(u_0) - 2\sqrt{2}\frac{1 - \varepsilon_0}{\varkappa(ka, u_0)} \hat{Q}_{s-1}^{(\frac{3}{2},0)}(u_0)\hat{Q}_{l-1}^{(\frac{3}{2},0)}(u_0),$$

$$\hat{Q}_{n-1}^{(\frac{3}{2},0)}(u_0) = \frac{1}{\sqrt{\pi}} \left(\frac{1 - u_0}{2}\right)^{\frac{3}{2}} \cdot \frac{\hat{P}_{n-1}^{(\frac{3}{2},0)}(u_0)}{[n(n + \frac{1}{2})]^{\frac{1}{2}}}. \tag{3.54}$$

With this notation, the value of $\tilde{x}_0^0$ is

$$\tilde{x}_0^0 = \frac{\tilde{\alpha}_0^0}{1 - \varepsilon_0} \cdot \frac{\varkappa(ka, u_0) - 1}{\varkappa(ka, u_0)} +$$

$$\frac{2^{\frac{7}{4}}}{\varkappa(ka, u_0)} \sum_{s=1}^{\infty} \{[p_s(1 - \varepsilon_{2s}) + \varepsilon_{2s}] y_{2s} + b_{2s}\} \hat{Q}_{s-1}^{(\frac{3}{2}, 0)}(u_0). \quad (3.55)$$

It follows from (3.53) and (3.54) that the coefficient sequence is square summable, i.e., $\{y_{2n}\}_{n=1}^{\infty} \in l_2$.

For further investigation of the acoustic resonator it is convenient to formulate the final system (3.53) in terms of $u_1 = -u_0 = \cos 2\theta_1$, where θ_1 is the slot half-angle. Setting $\{y_{2l}^*, b_{2l}^*\} = (-1)^l \{y_{2l}, b_{2l}\}$, the new coefficients satisfy

$$\{1 - p_l(1 - \varepsilon_{2l}) - \varepsilon_{2l}\} y_{2l}^* + \sum_{s=1}^{\infty} \{p_s(1 - \varepsilon_{2s}) + \varepsilon_{2s}\} y_{2s}^* \bar{\Phi}_{s-1,l-1}(ka, u_1)$$

$$= -2^{-\frac{1}{4}} \tilde{\alpha}_0^0 \frac{\hat{Q}_{l-1}^{(0,\frac{3}{2})}(u_1)}{\xi(ka, u_1)} + b_{2l}^* - \sum_{s=1}^{\infty} b_{2s}^* \bar{\Phi}_{s-1,l-1}(ka, u_1), \quad (3.56)$$

where $l = 1, 2, \ldots$, and

$$\xi(ka, u_1) = \varkappa(ka, -u_1) =$$

$$1 - \frac{1}{\pi\sqrt{2}}(1 - \varepsilon_0)\left\{(1 + u_1)^{\frac{1}{2}} - \ln\left[\frac{\sqrt{2} + (1 + u_1)^{\frac{1}{2}}}{(1 - u_1)^{\frac{1}{2}}}\right]\right\}, \quad (3.57)$$

$$\bar{\Phi}_{s-1,l-1}(ka, u_1) = \hat{Q}_{s-1,l-1}^{(1,\frac{1}{2})}(u_1) + 2\sqrt{2}\frac{1 - \varepsilon_0}{\xi(ka, u_1)}\hat{Q}_{s-1}^{(0,\frac{3}{2})}(u_1)\hat{Q}_{l-1}^{(0,\frac{3}{2})}(u_1). \quad (3.58)$$

With this notation, the value of $\tilde{x}_0^0$ is

$$\tilde{x}_0^0 = \frac{\tilde{\alpha}_0^0}{1 - \varepsilon_0} \cdot \frac{\xi(ka, u_1) - 1}{\xi(ka, u_1)} -$$

$$\frac{2^{\frac{7}{4}}}{\xi(ka, u_1)} \sum_{s=1}^{\infty} \{[p_s(1 - \varepsilon_{2s}) + \varepsilon_{2s}] y_{2s}^* + b_{2s}^*\} \hat{Q}_{s-1}^{(0,\frac{3}{2})}(u_1). \quad (3.59)$$

The construction of the regularised system from (3.41) and (3.42) for arbitrary values of m is obtained with the same procedure outlined above. For the sake of brevity and lucidity it is omitted.

3.3 Helmholtz Resonators: Barrelled or Slotted Spherical Shells.

In Section 2.2 we determined a comparatively accurate approximation for the relative wavenumber $(ka)_0$ of the Helmholtz mode for the classical acoustic resonator formed by the spherical cavity with a small circular aperture. As noted there have been several previous studies of this subject by many authors.

Our objective in this section is to determine the relative wavenumber $(ka)_0$ of the Helmholtz mode for more complex acoustic resonators. These results seem to be new; at least they are based on the mathematically rigorous solutions obtained in first two sections of this chapter. The first resonator to be considered is the acoustically hard spherical shell with two small and equal circular holes, co-axially located (see Figure 3.1a). From a physical point of view it has many similarities with the spherical resonator with a single circular aperture. The second resonator to be considered is the hard spherical shell with a narrow (equatorial) slot, symmetrically located.

The difference between this resonator and the classical one manifests itself in the nonclassical concentration of acoustic kinetic energy in a toroidal region near the slot. The classical formula (2.61) does not predict even the dominant term of the Helmholtz mode. Nevertheless both resonators exhibit a clear spatial distribution of potential and kinetic energy that is typical of resonant systems. We now demonstrate these assertions by producing approximate analytic expressions for both resonators in the Helmholtz mode.

First consider the hard spherical barrel. The characteristic equation is deduced by setting $m = l = 0$ in (3.25), and following the argument in Section 2.2. To avoid awkward calculations, a smaller number of perturbation terms than employed in (2.72) will be used, and we consider the approximate characteristic equation

$$
(1 - q_{00}^{(0)})(1 - \varepsilon_0) + \left[(1 - q_{00}^{(0)})\varepsilon_0 + q_{00}^{(0)}\right] \times
$$
$$
\left\{ \hat{Q}_{00}^{(-\frac{1}{2},0)}(u_0) - \sum_{s=1}^{\infty} \frac{(1 - q_{s0}^{(0)})\varepsilon_{2s} + q_{s0}^{(0)}}{(1 - q_{s0}^{(0)})(1 - \varepsilon_{2s})} \hat{Q}_{s0}^{(-\frac{1}{2},0)}(u_0)\hat{Q}_{0s}^{(-\frac{1}{2},0)}(u_0) \right\} = 0.
$$

$$(3.60)$$

As before, we assume $(ka)^2 \sim \theta_0$ and retain all terms that satisfy this assumption. The neglected terms in the approximate equation (3.60) are $O(ka^6, \theta_0^3)$. Insert the exact values $q_{00}^{(0)} = 1 - \pi/4$, $\hat{Q}_{00}^{(-\frac{1}{2},0)}(u_0) = \sin\theta_0$ and use the expansion (see [2.75])

$$\varepsilon_0 = 1 + \frac{4}{3}(ka)^2 + \frac{8}{15}(ka)^4 + i\frac{4}{9}(ka)^5 + O((ka)^6)$$

to obtain

$$\sum_{s=1}^{\infty} \frac{(1 - q_{s0}^{(0)})\varepsilon_{2s} + q_{s0}^{(0)}}{(1 - q_{s0}^{(0)})(1 - \varepsilon_{2s})} \hat{Q}_{s0}^{(-\frac{1}{2},0)}(u_0)\hat{Q}_{0s}^{(-\frac{1}{2},0)}(u_0) =$$

$$\frac{4\theta_0^2}{\pi}\left\{\frac{1}{8}\sum_{s=1}^{\infty}\frac{1}{s(2s+1)} + \sum_{s=1}^{\infty}\left[1 - (s + \frac{1}{4})\frac{\Gamma^2(s + \frac{1}{2})}{\Gamma^2(s+1)}\right]\right\} + O(\theta_0^3, (ka)^2\theta_0^2).$$

$$(3.61)$$

The series on the right hand side of (3.61) may be evaluated. The first is

$$\sum_{s=1}^{\infty}\frac{1}{s(2s+1)} = 2 - 2\ln 2.$$

Observing that the complete elliptic integral of the first kind has a power series expansion in terms of the modulus k [23],

$$K(k) = \frac{1}{2}\sum_{m=0}^{\infty}\left[\frac{\Gamma(m+\frac{1}{2})}{\Gamma(m+1)}\right]^2 k^{2m},$$

the second series may be calculated to be

$$\sum_{s=1}^{\infty}\left[1 - (s+\frac{1}{4})\frac{\Gamma^2(s+\frac{1}{2})}{\Gamma^2(s+1)}\right] = \frac{1}{4}(\pi - 3). \qquad (3.62)$$

Thus the approximate characteristic equation for the Helmholtz mode is

$$(ka)^2 - \frac{3\theta_0}{\pi} + \frac{3}{\pi^2}(\pi - 2 - \ln 2)\theta_0^2 + \frac{2}{5}(ka)^4 - (ka)^2\theta_0 + i\frac{1}{3}(ka)^5$$

$$+ O((ka)^6, \theta_0^3) = 0. \quad (3.63)$$

An approximate solution of (3.63) is readily generated by Newton's method:

$$(ka)_0 = \left(\frac{3\theta_0}{\pi}\right)^{\frac{1}{2}}\left\{1 + \frac{1 + 5\ln 2}{10}\frac{\theta_0}{\pi} - i\frac{1}{6}\left(\frac{3\theta_0}{\pi}\right)^{\frac{3}{2}} + O(\theta_0^2)\right\}. \qquad (3.64)$$

If one doubles the parameter θ_0 in expression (3.64) the value of the dominant term equals that of the classical acoustic resonator. This confirms that the physical processes taking place in both resonators are similar in origin.

Turning to the slotted acoustic resonator (see Figure 3.1b) we may notice that in deriving (3.56) we have essentially deduced the approximate characteristic equation. In fact, the equation

$$\xi(ka, u_1) = 0 \qquad (3.65)$$

gives the dominant term in the full characteristic equation. If the slot is logarithmically small $([\ln(2/\theta_1)]^{-1} \ll 1)$ one may deduce from (3.65) a rough estimate for $(ka)_0$ to be

$$\mathrm{Re}(ka)_0 = \frac{1}{2}(3\pi)^{\frac{1}{2}}\left[\ln\left(2/\theta_1\right)\right]^{-\frac{1}{2}}; \qquad \mathrm{Im}(ka)_0 = 0. \qquad (3.66)$$

Actually (3.65) can be solved approximately when θ_1 is small and the error is easily estimated.

Now consider the function $\bar{\Phi}_{s-1,l-1}(ka, u_1)$ given by (3.58). Due to equation (2.102), the first term of this function is negligibly small $\left(O(\theta_1^4)\right)$ in comparison with the second $\left(O\left((\ln 2/\theta_1)^{-1}\right)\right)$ and may be neglected when $\theta_1 \ll 1$, so that the characteristic equation degenerates to one that can be solved exactly. First, write the approximate equations (see [3.56], [3.59]) in the form

$$y_{2l}^* - \{[p_l(1 - \varepsilon_{2l}) + \varepsilon_{2l}]\,y_{2l}^* + b_{2l}^*\} +$$

$$2\sqrt{2}\,\frac{1 - \varepsilon_0}{\xi(ka, u_1)}\hat{Q}_{l-1}^{(0,\frac{3}{2})}(u_1) \sum_{s=1}^{\infty} \{[p_s(1 - \varepsilon_{2s}) + \varepsilon_{2s}]\,y_{2s}^* + b_{2s}^*\}\,\hat{Q}_{s-1}^{(0,\frac{3}{2})}(u_1)$$

$$= -2^{-\frac{1}{4}}\tilde{\alpha}_0^0\,\frac{\hat{Q}_{l-1}^{(0,\frac{3}{2})}(u_1)}{\xi(ka, u_1)}, \qquad (3.67)$$

$$\tilde{x}_0^0 + \frac{2^{\frac{7}{4}}}{\xi(ka, u_1)}\sum_{s=1}^{\infty} \{[p_s(1 - \varepsilon_{2s}) + \varepsilon_{2s}]\,y_{2s}^* + b_{2s}^*\}\,\hat{Q}_{s-1}^{(0,\frac{3}{2})}(u_1)$$

$$= \frac{\tilde{\alpha}_0^0}{1 - \varepsilon_0}\,\frac{\xi(ka, u_1) - 1}{\xi(ka, u_1)}. \qquad (3.68)$$

Now multiply both parts of (3.68) by the factor $\hat{Q}_{l-1}^{(0,\frac{3}{2})}(u_1)$ and sum over the index l to obtain

$$\sum_{l=1}^{\infty} y_{2l}^*\hat{Q}_{l-1}^{(0,\frac{3}{2})}(u_1) + \left[2\sqrt{2}\,\frac{1 - \varepsilon_0}{\xi(ka, u_1)}G(u_1) - 1\right]F(ka, u_1)$$

$$= -2^{-\frac{1}{4}}\tilde{\alpha}_0^0\,\frac{G(u_1)}{\xi(ka, u_1)}, \qquad (3.69)$$

$$\tilde{x}_0^0 + \frac{2^{\frac{7}{4}}}{\xi(ka, u_1)} F(ka, u_1) = \frac{\tilde{\alpha}_0^0}{1 - \varepsilon_0} \frac{\xi(ka, u_1) - 1}{\xi(ka, u_1)}, \tag{3.70}$$

where

$$F(ka, u_1) = \sum_{n=1}^{\infty} \{[p_n(1 - \varepsilon_{2n}) + \varepsilon_{2n}] y_{2n}^* + b_{2n}^*\} \hat{Q}_{n-1}^{(0, \frac{3}{2})}(u_1), \tag{3.71}$$

$$G(u_1) = \sum_{n=1}^{\infty} \left\{ \hat{Q}_{n-1}^{(0, \frac{3}{2})}(u_1) \right\}^2. \tag{3.72}$$

Thus

$$\sum_{l=1}^{\infty} y_{2l}^* \hat{Q}_{l-1}^{(0, \frac{3}{2})}(u_1) = -\frac{1}{\sqrt{\pi}} \left(\frac{1 - u_0}{2} \right)^{\frac{3}{2}} \sum_{l=1}^{\infty} X_{2l} \hat{P}_l^{(\frac{3}{2}, 0)}(u_0)$$

$$= 2^{-\frac{7}{4}} [\xi(ka, u_1) - 1] \tilde{x}_0^0 - 2^{-\frac{7}{4}} \frac{\xi(ka, u_1) - 1}{1 - \varepsilon_0} \tilde{\alpha}_0^0. \tag{3.73}$$

We thus obtain a pair of simultaneous algebraic equations for the unknown $\tilde{x}_0^0$ and $F(ka, u_1)$,

$$2^{-\frac{7}{4}} [\xi(ka, u_1) - 1] \tilde{x}_0^0 + \left[2\sqrt{2} \frac{1 - \varepsilon_0}{\xi(ka, u_1)} G(u_1) - 1 \right] F(ka, u_1) \tag{3.74}$$

$$= 2^{-\frac{7}{4}} \frac{[\xi(ka, u_1) - 1] \xi(ka, u_1) - 2\sqrt{2}(1 - \varepsilon_0) G(u_1)}{(1 - \varepsilon_0) \xi(ka, u_1)} \tilde{\alpha}_0^0, \tag{3.75}$$

$$\tilde{x}_0^0 + \frac{2^{\frac{7}{4}}}{\xi(ka, u_1)} F(ka, u_1) = \frac{\tilde{\alpha}_0^0}{1 - \varepsilon_0} \frac{\xi(ka, u_1) - 1}{\xi(ka, u_1)}, \tag{3.76}$$

with solution

$$F(ka, u_1) = \frac{\Delta_F}{\Delta}, \qquad \tilde{x}_0^0 = \frac{\Delta_{\tilde{x}_0^0}}{\Delta}, \tag{3.77}$$

where

$$\Delta = \frac{2\xi(ka, u_1) - [1 + 2\sqrt{2}(1 - \varepsilon_0) G(u_1)]}{\xi(ka, u_1)},$$

$$\Delta_F = -2^{-\frac{7}{4}} \tilde{\alpha}_0^0 \frac{\xi(ka, u_1) - [1 + 2\sqrt{2}(1 - \varepsilon_0) G(u_1)]}{(1 - \varepsilon_0) \xi(ka, u_1)}, \tag{3.78}$$

$$\Delta_{\tilde{x}_0^0} = \frac{2[\xi(ka, u_1) - 1] - 2\sqrt{2}(1 - \varepsilon_0) G(u_1)}{(1 - \varepsilon_0) \xi(ka, u_1)} \tilde{\alpha}_0^0.$$

Explicitly the solution is

$$F(ka, u_1) = \frac{-2^{-\frac{7}{4}} \left(\xi(ka, u_1) - \left[1 + 2\sqrt{2}\, (1 - \varepsilon_0)\, G(u_1) \right] \right)}{(1 - \varepsilon_0) \left\{ 2\xi(ka, u_1) - \left[1 + 2\sqrt{2}\, (1 - \varepsilon_0)\, G(u_1) \right] \right\}} \tilde{\alpha}_0^0, \quad (3.79)$$

$$\tilde{x}_0^0 = \frac{2 \left[\xi(ka, u_1) - 1 \right] - 2\sqrt{2}\, (1 - \varepsilon_0)\, G(u_1)}{(1 - \varepsilon_0) \left\{ 2\xi(ka, u_1) - \left[1 + 2\sqrt{2}\, (1 - \varepsilon_0)\, G(u_1) \right] \right\}} \tilde{\alpha}_0^0. \quad (3.80)$$

where, from (3.72) and (3.54),

$$G(u_1) = \frac{2^{-\frac{5}{2}}}{\pi} \left(\frac{1 + u_1}{2} \right)^3 \sum_{n=1}^{\infty} \frac{2n + \frac{1}{2}}{n(n + \frac{1}{2})} \left\{ P_{n-1}^{(0, \frac{3}{2})}(u_1) \right\}^2. \quad (3.81)$$

The utility of this solution depends upon a suitable approximation for G that is now derived. Use both the integral representations of Abel type for Jacobi polynomials contained in Section 1.5 of Volume I [1]. One can transform (3.81) to

$$G(u_1) = \frac{2^{-\frac{7}{2}}}{\pi} \left[\frac{3 + u_1}{4} - \frac{1 - u_1}{2} \left(\frac{1 + u_1}{2} \right)^{\frac{1}{2}} \right] \times$$

$$\left\{ (3 + u_1) \ln \left[\frac{\sqrt{2} + (1 + u_1)^{\frac{1}{2}}}{(1 - u_1)^{\frac{1}{2}}} \right] - 2 \left(\frac{1 + u_1}{2} \right)^{\frac{1}{2}} \right\} -$$

$$\frac{2^{-\frac{7}{2}}}{\pi} \left(\frac{1 + u_1}{2} \right)^3 \sum_{n=0}^{\infty} \frac{2n + \frac{5}{2}}{(n + \frac{1}{2})(n + 1)(n + \frac{3}{2})(n + 2)} \left\{ P_n^{(0, \frac{3}{2})}(u_1) \right\}^2. \quad (3.82)$$

The major term in (3.82) is that which contains the logarithmic function; $G(\cos(2\theta_1)) = O(\ln(2/\theta_1))$. The transform of G from (3.81) to (3.82) extracts the singular logarithmic part exactly. The remainder is the smooth function given by the absolutely convergent series in (3.82). If we were to use the Jacobi polynomial expansion near $u_1 = 1$,

$$P_n^{(0, \frac{3}{2})}(u_1) \simeq 1 - \frac{1}{2}n(n + \frac{5}{2})(1 - u),$$

then

$$\begin{aligned} S(u_1) &= \sum_{n=0}^{\infty} \frac{2n + \frac{5}{2}}{(n + \frac{1}{2})(n + 1)(n + \frac{3}{2})(n + 2)} \left\{ P_n^{(0, \frac{3}{2})}(u_1) \right\}^2 \\ &= 2 + O \left(\theta_1^2 \ln (\theta_1^{-1}) \right). \end{aligned} \quad (3.83)$$

Unfortunately, the last result is inadequate for deducing a more accurate value of $(ka)_0$ (see [3.66]) and a better approximation is required. Exact summation of (3.83) seems impossible so we proceed as follows. If $1-u_1 \ll 1$ the function $S(u_1)$ may be approximated by

$$S(u_1) = -2 + 2 \sum_{k=0}^{\infty} \frac{2k + \frac{5}{2}}{(k + \frac{1}{2})(k + 1)(k + \frac{3}{2})(k + 2)} P_k^{(0,\frac{3}{2})}(u_1) + O((1 - u_1)).$$

$$(3.84)$$

Using the partial fraction decomposition

$$\frac{2k + \frac{5}{2}}{(k + \frac{1}{2})(k + 1)(k + \frac{3}{2})(k + 2)} = 2 \left\{ \frac{1}{k + \frac{1}{2}} - \frac{1}{k + 1} - \frac{1}{k + \frac{3}{2}} - \frac{1}{k + 2} \right\}$$

and the well-known formula (see Appendix B.3, Volume I) for the generating function

$$\sum_{k=0}^{\infty} t^k P_k^{(0,\frac{3}{2})}(u_1) = \frac{2^{\frac{3}{2}}}{R(1 + t + R)^{\frac{3}{2}}}$$

$$(3.85)$$

where

$$R = \sqrt{1 - 2u_1 t + t^2},$$

one obtains the integral

$$S(u_1) =$$

$$- 2 + 16\sqrt{2} \int_0^1 \frac{(1 - x)(1 - x^2)dx}{\sqrt{1 - 2u_1x^2 + x^4} \left\{ 1 + x^2 + \sqrt{1 - 2u_1x^2 + x^4} \right\}^{\frac{3}{2}}}.$$

$$(3.86)$$

Noting that

$$\left\{ 1 + x^2 + \sqrt{1 - 2u_1x^2 + x^4} \right\}^{-\frac{3}{2}} =$$

$$2^{-\frac{3}{2}} - 3 \cdot 2^{-\frac{7}{2}} \frac{x^2}{1 - x^2}(1 - u_1) + O((1 - u_1)^2) \quad (3.87)$$

the function S has the representation

$$S(u_1) = -2 + 8(I_0 - I_1 - I_2 + I_3) - 6(I_2 - I_3)(1 - u_1) + O((1 - u_1))$$

$$(3.88)$$

where $I_i(u_1) \equiv I_i$ $(i = 0, 1, 2, 3)$ are the following integrals tabulated in [30] and [14]:

$$I_0 = \int_0^1 \frac{dx}{\sqrt{1 - 2tx^2 + x^4}} = \frac{1}{2}K(\cos\theta_1)$$

$$I_1 = \int_0^1 \frac{x\,dx}{\sqrt{1 - 2tx^2 + x^4}} = \frac{1}{2}\ln\left[\frac{1 + \sin\theta_1}{\sin\theta_1}\right]$$

$$I_2 = \int_0^1 \frac{x^2\,dx}{\sqrt{1 - 2tx^2 + x^4}} = \frac{1}{2}K(\cos\theta_1) - E(\cos\theta_1) + \sin\theta_1$$

$$I_3 = \int_0^1 \frac{x^3\,dx}{\sqrt{1 - 2tx^2 + x^4}}$$

$$= -\frac{1}{2} + \frac{1}{2}\cos(2\theta_1)\ln\left[\frac{1 + \sin\theta_1}{\sin\theta_1}\right] + \sin\theta_1. \tag{3.89}$$

Finally we have, when $1 - u_1 \ll 1$,

$$S(u_1) = 2 - 4\sin^2\theta_1 \ln(\operatorname{cosec}\theta_1) + O(\sin^2\theta_1), \tag{3.90}$$

with the correspondingly accurate result

$$G(u_1) = \frac{2^{-\frac{3}{2}}}{\pi}\left\{\ln 2 - 1 + \ln(\operatorname{cosec}\theta_1) - \sin^2\theta_1 \ln(\operatorname{cosec}\theta_1)\right\} + O(\sin^2\theta_1). \tag{3.91}$$

Observing formulae (3.79) and (3.80) the characteristic equation of crudest accuracy (3.65) may be replaced by the more accurate

$$2\xi(ka, u_1) - 1 - 2\sqrt{2}(1 - \varepsilon_0)G(u_1) = 0. \tag{3.92}$$

Inserting (3.91) it is transformed to

$$1 - \frac{1}{\pi}(1 - \varepsilon_0)\left\{1 - \ln 2 - \ln(\operatorname{cosec}\theta_1) - \sin^2\theta_1 \ln(\operatorname{cosec}\theta_1)\right\} = 0. \tag{3.93}$$

In spite of the fact that the only distinction between (3.93) and (3.65) is the addition of one term, the corrected equation enables us to determine the relative wavenumber $(ka)_0$ of the Helmholtz mode for the slotted acoustic resonator with an essentially higher accuracy.

In contrast to the characteristic equation for the barrelled acoustic resonator (see the beginning of this section), equation (3.93) incorporates mixed terms in parameter θ_1, as products of $\sin^2 \theta_1$ and $\ln(\mathrm{cosec}\,\theta_1)$. This creates a problem. Powers of ka, logarithmic terms in θ_1, and mixed terms in θ_1 and products of powers of ka and terms containing θ_1 must be properly balanced. A clear prompt is given by the formula (3.66). The value of the parameter ka near $(ka)_0$ is comparable to $\varepsilon = [\ln(\mathrm{cosec}\,\theta_1)]^{-\frac{1}{2}}$, i.e., $ka \sim \varepsilon$.

In accordance with (2.75), $1 - \varepsilon_0$ has the expansion, when $ka \ll 1$,

$$1 - \varepsilon_0 = \sum_{n=2}^{N} (ka)^n a_n + O((ka)^{N+1}) \tag{3.94}$$

where the power N is to be specified later; the coefficients a_n are known ($a_3 = 0$). Now rewrite (3.93) in terms of the parameter ε

$$1 - \frac{1}{\pi} \left(\sum_{n=2}^{N} a_n \varepsilon^n \right) \left\{ C - \varepsilon^{-2} - \sin^2 \theta_1 \cdot \varepsilon^{-2} \right\} = 0, \tag{3.95}$$

where $C = 1 - \ln 2$. Fix the parameter $\theta_1 \ll 1$. Choose γ so that

$$\sin^2 \theta_1 = \varepsilon^\gamma; \tag{3.96}$$

thus

$$\gamma = -\frac{2}{\varepsilon^2 \ln \varepsilon}. \tag{3.97}$$

Use the approximation

$$\sin^2 \theta_1 \sim \varepsilon^M \tag{3.98}$$

where $M = entier(\gamma)$ and substitute in the equation (3.95):

$$1 - \frac{1}{\pi} \left(\sum_{n=2}^{N} a_n \varepsilon^n \right) \left\{ C - \varepsilon^{-2} - \varepsilon^{M-2} + O(\varepsilon^M) \right\} = 0. \tag{3.99}$$

The latter can be reduced to an equation determined correctly to order $O(\varepsilon^M)$ by choosing $N = M + 1$, so that

$$1 - \frac{1}{\pi} \left[a_2 + a_3 + \sum_{n=2}^{M-1} (C a_n - a_{n+2}) \varepsilon^n + O(\varepsilon^M) \right] = 0. \tag{3.100}$$

It is important to note that in the range of $0 < \theta_1 < 5°$, the value of M is no more than 10, so that all terms corresponding to those in (2.75) can be taken into account in constructing a realistic dispersion equation. Retaining a reasonable number of terms the approximate dispersion equation is (with $ka = x$),

$$1 - \frac{4}{3\pi}x^2\varepsilon^{-2} + \frac{4}{3\pi}Cx^2 - \frac{8}{15\pi}x^4\varepsilon^{-2} - i\frac{4}{9\pi}x^5\varepsilon^{-2} + \frac{8}{15\pi}Cx^4$$
$$+ \frac{8}{35\pi}x^6\varepsilon^{-2} + i\frac{4}{9\pi}Cx^5 + i\frac{4}{45\pi}x^7\varepsilon^{-2} + O(\varepsilon^6, x^6) = 0. \quad (3.101)$$

As before an approximate root obtained by Newton's method is

$$(ka)_0 = \frac{(3\pi)^{\frac{1}{2}}}{2}\varepsilon\left\{1 + \frac{1}{2}\left(C - \frac{3\pi}{10}\right)\varepsilon^2 + \frac{1}{2}\left(\frac{3}{4}C^2 - \frac{9\pi}{20}C + \frac{711}{2800}\pi^2\right)\varepsilon^4\right.$$
$$\left. - i\frac{(3\pi)^{\frac{1}{2}}}{16}\varepsilon^3\left[1 - \left(\frac{3\pi}{2} - 3C + \frac{3\pi^2}{20}\right)\varepsilon^2\right]\right\} + O(\varepsilon^6). \quad (3.102)$$

The Q-factor of the slotted spherical acoustic resonator is

$$Q = -\frac{2\,\mathrm{Re}(ka)_0}{\mathrm{Im}(ka)_0} = \frac{16}{(3\pi)^{\frac{1}{2}}\varepsilon^3}[1 + O(\varepsilon^2)]. \quad (3.103)$$

It is clear that one can find the value of $(ka)_0$ more accurately by calculating more terms in the expansion of $1 - \varepsilon_0$ as powers of ka.

To end this section we state the following approximate formulae that will be used to analyse the total cross-section of the Helmholtz resonator:

$$F(ka, u_1) = 0, \quad (3.104)$$

$$x_0^{(0)} = \frac{\tilde{\alpha}_0^0}{1 - \varepsilon_0}\left\{1 - \frac{1}{2\xi(ka, u_1) - 1 - 2\sqrt{2}(1 - \varepsilon_0)G(u_1)}\right\}$$
$$= \frac{\tilde{\alpha}_0^0}{1 - \varepsilon_0}\left[1 - \frac{1}{\xi(ka, u_1)}\right]. \quad (3.105)$$

3.4   Quasi-Eigenoscillations of the Spherical Cavity.

In this section we determine the quasi-eigenoscillations of the spherical cavity with two equal and symmetrically located circular holes or with a symmetrically located equatorial slot. The procedure is very similar to that developed in Section 2.3, so that an abbreviated treatment is given. There is a notable difference from the corresponding results obtained for the spherical shell with a single hole. When the scatterer has two apertures or an

equatorial slot the scattering process can be represented as two independent processes. From equations (3.25) and (3.31) it may be seen that the
even and odd spherical wave harmonics comprising the entire acoustic field
are scattered independently, without mutual coupling; the corresponding
approximate characteristic equations are also independent.

Employing the same procedure as used in Section 2.3 leads to approximate characteristic equations for the quasi-eigenoscillations of the spherical
barrel or the slotted shell. However for the sake of brevity, we restrict attention to the acoustically hard barrel and the soft slotted shell.

The quasi-eigenoscillations are indexed by an index l taking values 0 for
even modes and values 1 for odd modes index and by an index m taking
values $0, 1, 2, \ldots$. The approximate characteristic equations for the quasi-
eigenoscillations of the acoustically hard barrelled shell are (for $l = 0, 1$ and
$m = 0, 1, 2, \ldots$),

$$
1 + \tilde{q}_{nm}^{(l)} - (\varepsilon_{2n+l+m} + \tilde{q}_{nm}^{(l)}) \times
$$

$$
\left\{
1 - \hat{Q}_{nn}^{(m-\frac{1}{2},l)}(u_0) + \sum_{\substack{s=0 \\ s \neq n}}^{\infty} \frac{\varepsilon_{2s+l+m} + \tilde{q}_{sm}^{(l)}}{1 - \varepsilon_{2s+l+m}} \hat{Q}_{sn}^{(m-\frac{1}{2},l)}(u_0) \hat{Q}_{ns}^{(m-\frac{1}{2},l)}(u_0)
\right\}
$$

$$
= 0, \quad (3.106)
$$

where $\tilde{q}_{nm}^{(l)} = q_{nm}^{(l)} / \left(1 - q_{nm}^{(l)}\right)$; the corresponding equations for the acoustically soft slotted shell are

$$
1 - q_{nm}^{(l)} - \left(\mu_{2n+l+m} - 1 + q_{nm}^{(l)}\right) \times
$$

$$
\left\{
1 - \hat{Q}_{nn}^{(l,m-\frac{1}{2})}(u_1) - \sum_{\substack{s=0 \\ s \neq n}}^{\infty} \frac{\mu_{2s+l+m} + q_{sm}^{(l)}}{1 - \mu_{2s+l+m}} \hat{Q}_{sn}^{(l,m-\frac{1}{2})}(u_1) \hat{Q}_{ns}^{(l,m-\frac{1}{2})}(u_1)
\right\}
$$

$$
= 0. \quad (3.107)
$$

The notation is defined as in Section 3.1.

For the acoustically hard slotted shell, we restrict attention to the axially-
symmetric case discussed in Section 3.2, corresponding to setting the index
m equal to 0. The approximate characteristic equation for higher quasi-

eigenvalues of odd-type is

$$1 - \left[(1 - p_{l0}^{(1)})\varepsilon_{2l+1} + p_{l0}^{(1)}\right] \times$$

$$\left\{1 - \hat{Q}_{ll}^{(0,\frac{1}{2})}(u_1) + \sum_{\substack{s=0 \\ s \neq n}}^{\infty} \frac{(1 - p_{s0}^{(1)})\varepsilon_{2s+1} + p_{s0}^{(1)}}{1 - \left[(1 - p_{s0}^{(1)})\varepsilon_{2s+1} + p_{s0}^{(1)}\right]} \hat{Q}_{sl}^{(0,\frac{1}{2})}(u_1)\hat{Q}_{ls}^{(0,\frac{1}{2})}(u_1)\right\}$$

$$= 0, \quad (3.108)$$

whilst that for even-type is

$$1 - [p_l(1 - \varepsilon_{2l}) + \varepsilon_{2l}] \times$$

$$\left\{1 - T_l(ka, u_1) + T_l(ka, u_1) \sum_{\substack{s=1 \\ s \neq n}}^{\infty} \frac{p_s(1 - \varepsilon_{2s}) + \varepsilon_{2s}}{1 - p_s(1 - \varepsilon_{2s}) - \varepsilon_{2s}} T_s(ka, u_1)\right\}$$

$$= 0, \quad (3.109)$$

where

$$T_p(ka, u_1) = 2\sqrt{2}\frac{1 - \varepsilon_0}{\xi(ka, u_1)} \left\{\hat{Q}_{p-1}^{(0,\frac{3}{2})}(u_1)\right\}^2. \quad (3.110)$$

Let us analyse the equations (3.106)–(3.109) commencing with the first. In fact, four separate cases of equation (3.106) should be considered, depending on the two possible values of l $(0, 1)$ and whether the azimuthal index m is even or odd. For the sake of clarity these four cases are stated below.

When $l = 0$ and m is even $(m = 2m')$, the equation reads

$$1 + \tilde{q}_{p-m',2m'}^{(0)} - (\varepsilon_{2p} + \tilde{q}_{p-m',2m'}^{(0)}) \cdot \{1 - \hat{Q}_{p-m',p-m'}^{(2m'-\frac{1}{2},0)}(u_0) +$$

$$\frac{i}{4(ka)^3} \sum_{\substack{k=m' \\ k \neq p}}^{\infty} \frac{(4k+1)(1 + \tilde{q}_{p-m',2m'}^{(0)})}{\left|h_{2k}^{(1)'}(ka)\right|^2} \hat{Q}_{k-m',p-m'}^{(2m'-\frac{1}{2},0)}(u_0)\hat{Q}_{p-m',k-m'}^{(2m'-\frac{1}{2},0)}(u_0)\}$$

$$= 0, \quad (3.111)$$

where we set $2n + 2m' = 2p$ and $2s + 2m' = 2k$.

When $l = 0$ and m is odd ($m = 2m' + 1$), the equation becomes

$$1 + \tilde{q}^{(0)}_{p-m',2m'+1} - \left(\varepsilon_{2p+1} + \tilde{q}^{(0)}_{p-m',2m'+1}\right) \cdot \left\{1 - \hat{Q}^{(2m'+\frac{1}{2},0)}_{p-m',p-m'}(u_0) + \right.$$

$$\left. \frac{i}{4(ka)^3} \sum_{\substack{k=m' \\ k\neq p}}^{\infty} \frac{(4k+3)\left(1 + \tilde{q}^{(0)}_{p-m',2m'+1}\right)}{\left|h^{(1)'}_{2k+1}(ka)\right|^2} \hat{Q}^{(2m'+\frac{1}{2},0)}_{k-m',p-m'}(u_0)\hat{Q}^{(2m'+\frac{1}{2},0)}_{p-m',k-m'}(u_0)\right\}$$

$$= 0 \quad (3.112)$$

where we set $2n + 2m' = 2p + 1$ and $2s + 2m' + 1 = 2k + 1$.

When $l = 1$ and m is even ($m = 2m'$), the equation reads

$$1 + \tilde{q}^{(1)}_{p-m',2m'} - \left(\varepsilon_{2p+1} + \tilde{q}^{(1)}_{p-m',2m'}\right) \cdot \left\{1 - \hat{Q}^{(2m'-\frac{1}{2},1)}_{p-m',p-m'}(u_0) + \right.$$

$$\left. \frac{i}{4(ka)^3} \sum_{\substack{k=m' \\ k\neq p}}^{\infty} \frac{(4k+3)\left(1 + \tilde{q}^{(1)}_{p-m',2m'}\right)}{\left|h^{(1)'}_{2k+1}(ka)\right|^2} \hat{Q}^{(2m'-\frac{1}{2},1)}_{k-m',p-m'}(u_0)\hat{Q}^{(2m'-\frac{1}{2},1)}_{p-m',k-m'}(u_0)\right\}$$

$$= 0 \quad (3.113)$$

where $2n + 1 + 2m' = 2p + 1$ and $2s + 1 + 2m' = 2k + 1$.

Finally, when $l = 1$ and m is odd ($m = 2m' + 1$), the equation becomes

$$1 + \tilde{q}^{(1)}_{p-m',2m'+1} - \left(\varepsilon_{2p+2} + \tilde{q}^{(1)}_{p-m',2m'+1}\right) \cdot \left\{1 - \hat{Q}^{(2m'+\frac{1}{2},1)}_{p-m',p-m'}(u_0) + \right.$$

$$\left. \frac{i}{4(ka)^3} \sum_{\substack{k=m' \\ k\neq p}}^{\infty} \frac{(4k+5)\left(1 + \tilde{q}^{(1)}_{p-m',2m'+1}\right)}{\left|h^{(1)'}_{2k+2}(ka)\right|^2} \hat{Q}^{(2m'+\frac{1}{2},1)}_{k-m',p-m'}(u_0)\hat{Q}^{(2m'+\frac{1}{2},1)}_{p-m',k-m'}(u_0)\right\}$$

$$= 0. \quad (3.114)$$

The roots of these equations are perturbations of the roots $\varkappa_{nl}$ of the dispersion equation (2.83) for the interior of the closed spherical cavity with acoustically hard walls. In the order of their succession, the solutions of these equations are, respectively,

$$\varkappa^{(2m')}_{2p,l} / \varkappa_{2p,l} = 1 +$$

$$\frac{1}{2}\left(2p + \frac{1}{2}\right)\left(1 + \tilde{q}^{(0)}_{p-m',2m'}\right)\left[\varkappa^2_{2p,l} - 2p(2p+1)\right]^{-1} \hat{Q}^{(2m'-\frac{1}{2},0)}_{p-m',p-m'}(u_0)$$

$$- i\frac{1}{8}\left(2p + \frac{1}{2}\right)\left(1 + \tilde{q}^{(0)}_{p-m',2m'}\right)\varkappa^{-3}_{2p,l}\left[\varkappa^2_{2p,l} - 2p(2p+1)\right]^{-1} \times$$

$$\sum_{\substack{k=m' \\ k\neq p}}^{\infty} \frac{(4k+1)(1+\tilde{q})}{\left|h^{(1)'}_{2k}(\varkappa_{2p,l})\right|^2} \hat{Q}^{(2m'-\frac{1}{2},0)}_{k-m',p-m'}(u_0)\hat{Q}^{(2m'-\frac{1}{2},0)}_{p-m',k-m'}(u_0), \quad (3.115)$$

$$\varkappa^{(2m'+1)}_{2p+1,l}/\varkappa_{2p+1,l} = 1 +$$

$$\frac{1}{2}(2p+\frac{3}{2})(1+\tilde{q}^{(0)}_{p-m',2m'+1})\left[\varkappa^2_{2p+1,l} - (2p+1)(2p+2)\right]^{-1}\hat{Q}^{(2m'+\frac{1}{2},0)}_{p-m',p-m'}(u_0)$$

$$- i\frac{1}{8}(2p+\frac{3}{2})(1+\tilde{q}^{(0)}_{p-m',2m'+1})\varkappa^{-3}_{2p+1,l}\left[\varkappa^2_{2p+1,l} - (2p+1)(2p+2)\right]^{-1} \times$$

$$\sum_{\substack{k=m'\\k\neq p}}^{\infty} \frac{(4k+3)(1+\tilde{q})}{\left|h^{(1)\prime}_{2k+1}(\varkappa_{2p+1,l})\right|^2}\hat{Q}^{(2m'+\frac{1}{2},0)}_{k-m',p-m'}(u_0)\hat{Q}^{(2m'+\frac{1}{2},0)}_{p-m',k-m'}(u_0), \quad (3.116)$$

$$\varkappa^{(2m')}_{2p+1,l}/\varkappa_{2p+1,l} = 1 +$$

$$\frac{1}{2}(2p+\frac{3}{2})(1+\tilde{q}^{(1)}_{p-m',2m'})\left[\varkappa^2_{2p+1,l} - (2p+1)(2p+2)\right]^{-1}\hat{Q}^{(2m'-\frac{1}{2},1)}_{p-m',p-m'}(u_0)$$

$$- \frac{i}{8}(2p+\frac{3}{2})(1+\tilde{q}^{(1)}_{p-m',2m'})\varkappa^{-3}_{2p+1,l}\left[\varkappa^2_{2p+1,l} - (2p+1)(2p+2)\right]^{-1} \times$$

$$\sum_{\substack{k=m'\\k\neq p}}^{\infty} \frac{(4k+3)(1+\tilde{q})}{\left|h^{(1)\prime}_{2k+1}(\varkappa_{2p+1,l})\right|^2}\hat{Q}^{(2m'-\frac{1}{2},1)}_{k-m',p-m'}(u_0)\hat{Q}^{(2m'-\frac{1}{2},1)}_{p-m',k-m'}(u_0), \quad (3.117)$$

$$\varkappa^{(2m'+1)}_{2p+2,l}/\varkappa_{2p+2,l} = 1 +$$

$$\frac{1}{2}(2p+\frac{5}{2})(1+\tilde{q}^{(1)}_{p-m',2m'+1})\left[\varkappa^2_{2p+2,l} - (2p+2)(2p+3)\right]^{-1}\hat{Q}^{(2m'+\frac{1}{2},1)}_{p-m',p-m'}(u_0)$$

$$- \frac{i}{8}(2p+\frac{5}{2})(1+\tilde{q}^{(1)}_{p-m',2m'+1})\varkappa^{-3}_{2p+2,l}\left[\varkappa^2_{2p+2,l} - (2p+1)(2p+3)\right]^{-1} \times$$

$$\sum_{\substack{k=m'\\k\neq p}}^{\infty} \frac{(4k+5)(1+\tilde{q})}{\left|h^{(1)\prime}_{2k+2}(\varkappa_{2p+2,l})\right|^2}\hat{Q}^{(2m'+\frac{1}{2},1)}_{k-m',p-m'}(u_0)\hat{Q}^{(2m'+\frac{1}{2},1)}_{p-m',k-m'}(u_0). \quad (3.118)$$

Without exception the real parts of these roots are positively shifted compared to the unperturbed roots; and all roots have negative imaginary parts. It is logical to ask what change in the quasi-eigenvalues would be produced if a second circular aperture, of the same size, were opened in a spherical cavity with a single circular hole Let us answer this question by considering one of the formulae above, for instance, formula (3.115).

To compare (3.115) with (2.100), we set $n = 2p, m = 2m'$. The only terms to be compared, under the condition that $\theta_0 \ll 1$, are

$$\hat{Q}^{(2m'-\frac{1}{2},2m'+\frac{1}{2})}_{2p-2m',2p-2m'}(\cos\theta_0) \text{ and } \hat{Q}^{(2m'-\frac{1}{2},0)}_{p-m',p-m'}(\cos 2\theta_0).$$

A surprisingly clear and accurate asymptotic equality holds when $\theta_0 \ll 1$: from equation (2.102) and the duplication formula for the Gamma function we deduce

$$\left(1 + \tilde{q}^{(1)}_{p-m',2m'}\right) \hat{Q}^{(2m'-\frac{1}{2},0)}_{p-m',p-m'}(\cos 2\theta_0) \approx 2\hat{Q}^{(2m'-\frac{1}{2},2m'+\frac{1}{2})}_{2p-2m',2p-2m'}(\cos\theta_0),$$
(3.119)

and

$$\hat{Q}^{(2m'-\frac{1}{2},2m'+\frac{1}{2})}_{2p-2m',2p-2m'}(\cos\theta_0) \approx 2\frac{(\theta_0/2)^{4m'+1}}{\Gamma(2m'+\frac{1}{2})\Gamma(2m'+\frac{3}{2})}\frac{\Gamma(2p+2m'+1)}{\Gamma(2p-2m'+1)}.$$
(3.120)

Analogous statements are true for the other equations (3.116)–(3.118).

Thus opening an additional hole of the same size in the cavity wall doubles the positive shift (from the unperturbed eigenvalue of closed spherical cavity) in the real part of the quasi-eigenvalue, compared to that observed for the single aperture cavity. In other words the first perturbation term arises from the separate contribution of each hole. It can be readily shown that taking into account one more term in (3.106) and solving the corresponding approximate characteristic equation produces a second correction term reflecting the "splitting" impact of both holes.

Moreover, the shift in the imaginary part of $\varkappa^{(2m')}_{2p,l}$ is also double that of the corresponding value for the cavity with a single hole. We suppress the justification of this assertion, merely noting that it depends upon the asymptotically correct approximation (for $\theta_0 \ll 1$)

$$\hat{Q}^{(\alpha,\beta)}_{sn}(\cos\theta_0)\hat{Q}^{(\alpha,\beta)}_{sn}(\cos\theta_0) \approx \hat{Q}^{(\alpha,\beta)}_{ss}(\cos\theta_0)\hat{Q}^{(\alpha,\beta)}_{nn}(\cos\theta_0).$$

These features of the quasi-eigenvalues of the quasi-eigenoscillations for the hard barrelled acoustic resonator are in clear agreement with physical expectations.

In the same way we may derive formulae for the spectrum of the soft spherical cavity with a narrow equatorial slot. The roots of approximate characteristic equations are perturbations of the roots ν_{nl} of the dispersion equation (2.85) for the interior of the closed spherical cavity with acoustically soft walls. The perturbation results corresponding to formulae (3.115)–(3.118) are as follows. With $l = 0$ and m even ($m = 2m'$),

$$\nu^{(2m')}_{2p,l}/\nu_{2p,l} = 1 - \frac{1}{2}\left(2p + \frac{1}{2}\right)^{-1}\hat{Q}^{(0,2m'-\frac{1}{2})}_{p-m',p-m'}(u_1)(1 - \tilde{q}^{(0)}_{p-m',2m'})$$

$$- i\frac{1}{2}\left(2p + \frac{1}{2}\right)^{-1}\frac{1 - \tilde{q}^{(0)}_{p-m',2m'}}{\nu_{2p,l}} \times$$

$$\sum_{\substack{k=m'\\k\neq p}}^{\infty} \frac{1 - \tilde{q}_{k-m',2m'}}{(4k+1)\left|h^{(1)}_{2k}(\nu_{2p,l})\right|^2}\hat{Q}^{(0,2m'-\frac{1}{2})}_{k-m',p-m'}(u_1)\hat{Q}^{(0,2m'-\frac{1}{2})}_{p-m',k-m'}(u_1);$$
(3.121)

with $l = 0$ and m odd ($m = 2m' + 1$),

$$\nu_{2p+1,l}^{(2m'+1)}/\nu_{2p+1,l} = 1 - \frac{1}{2}(2p + \frac{3}{2})^{-1}(1 - \tilde{q}_{p-m',2m'+1}^{(0)})\hat{Q}_{p-m',p-m'}^{(0,2m'+\frac{1}{2})}(u_1)$$

$$-i\frac{1}{2}(2p + \frac{3}{2})^{-1}\frac{1 - \tilde{q}_{p-m',2m'+1}^{(0)}}{\nu_{2p+1,l}} \times$$

$$\sum_{\substack{k=m' \\ k \neq p}}^{\infty} \frac{1 - \tilde{q}_{k-m,2m'+1}^{(0)}}{(4k+3)\left|h_{2k+1}^{(1)}(\nu_{2p+1,l})\right|^2}\hat{Q}_{k-m',p-m'}^{(0,2m'+\frac{1}{2})}(u_1)\hat{Q}_{p-m',k-m'}^{(0,2m'+\frac{1}{2})}(u_1); \quad (3.122)$$

with $l = 1$ and m even ($m = 2m'$),

$$\nu_{2p+1,l}^{(2m')}/\nu_{2p+1,l} = 1 - \frac{1}{2}(2p + \frac{3}{2})^{-1}(1 - \tilde{q}_{p-m',2m'}^{(1)})\hat{Q}_{p-m',p-m'}^{(0,2m'-\frac{1}{2})}(u_1)$$

$$-i\frac{1}{2}(2p + \frac{3}{2})^{-1}\frac{1 - \tilde{q}_{p-m',2m'}^{(1)}}{\nu_{2p+1,l}} \times$$

$$\sum_{\substack{k=m' \\ k \neq p}}^{\infty} \frac{1 - \tilde{q}_{k-m,2m'}^{(1)}}{(4k+3)\left|h_{2k+1}^{(1)}(\nu_{2p+1,l})\right|^2}\hat{Q}_{k-m',p-m'}^{(0,2m'-\frac{1}{2})}(u_1)\hat{Q}_{p-m',k-m'}^{(0,2m'-\frac{1}{2})}(u_1); \quad (3.123)$$

and finally, with $l = 1$ and m odd ($m = 2m' + 1$),

$$\nu_{2p+2,l}^{(2m'+1)}/\nu_{2p+2,l} = 1 - \frac{1}{2}(2p + \frac{5}{2})^{-1}(1 - \tilde{q}_{p-m',2m'+1}^{(1)})\hat{Q}_{p-m',p-m'}^{(0,2m'+\frac{1}{2})}(u_1)$$

$$-i\frac{1}{2}(2p + \frac{5}{2})^{-1}\frac{1 - \tilde{q}_{p-m',2m'+1}^{(1)}}{\nu_{2p+2,l}} \times$$

$$\sum_{\substack{k=m' \\ k \neq p}}^{\infty} \frac{1 - \tilde{q}_{k-m,2m'+1}^{(1)}}{(4k+5)\left|h_{2k+2}^{(1)}(\nu_{2p+2,l})\right|^2}\hat{Q}_{k-m',p-m'}^{(0,2m'+\frac{1}{2})}(u_1)\hat{Q}_{p-m',k-m'}^{(0,2m'+\frac{1}{2})}(u_1). \quad (3.124)$$

In contradistinction to the hard barrelled resonator, the real parts of the quasi-eigenvalues of the slotted soft resonator are shifted negatively compared to the eigenvalues of closed soft sphere. Another distinctive feature is the absence of any strong dependence in the perturbation upon the azimuthal index m. The shift in value is proportional to θ_1^2, slightly varying from one value of m to another.

This is clearly indicated by the approximate formula (2.102) together with the behaviour of the parameter $\tilde{q}_{p-m',2m'+s}^{(s)}$ ($s = 0, 1$): for all m it is very much smaller than unity. Thus the Q-factor of these oscillations is practically the same for all values of m' (or m).

3.5 Total and Sonar Cross-Sections; Mechanical Force Factor.

Just as for the spherical shell with a single hole (Chapter 2), numerical analysis of the basic characteristics of more complicated scatterers is best considered in three regimes (low frequency, resonance and quasi-optical ranges). The total cross-section σ_T^h and the sonar cross-section σ^h of the acoustically hard spherical shell may be calculated in the manner of Section 2.4.

The cross-sections of the spherical shell with an equatorial slot are

$$
\frac{\sigma_T^h}{\pi a^2} = \frac{4\sqrt{2}}{(ka)^2} \left\{ |\tilde{x}_0^0|^2 \, j_1^2 \, (ka) + \sum_{s=1}^{\infty} \frac{|y_{2s}|}{s\left(s+\frac{1}{2}\right)} \left[\frac{\Gamma(s+1)}{\Gamma\left(s+\frac{1}{2}\right)} \right]^2 j_{2s}'^2 \, (ka) + \right.
$$

$$
\left. \frac{1}{2} \sum_{s=0}^{\infty} \left(s+\frac{1}{2}\right)^{-2} \left[\frac{\Gamma(s+1)}{\Gamma\left(s+\frac{1}{2}\right)} \right]^2 |X_{2s+1}^0|^2 \left[j_{2s+1}' \, (ka) \right]^2 \right\} \qquad (3.125)
$$

and

$$
\frac{\sigma^h}{\pi a^2} = \frac{4\sqrt{2}}{(ka)^2} \left| \tilde{x}_0^0 j_0' \, (ka) + 2 \sum_{s=1}^{\infty} (-1)^s \left(s+\frac{1}{4}\right)^{\frac{1}{2}} \frac{\Gamma(s+1)}{\Gamma\left(s+\frac{1}{2}\right)} \frac{y_{2s} j_{2s}' \, (ka)}{\left[s\left(s+\frac{1}{2}\right)\right]^{\frac{1}{2}}} \right.
$$

$$
\left. + i\sqrt{2} \sum_{s=0}^{\infty} (-1)^s \frac{\left(s+\frac{3}{4}\right)^{\frac{1}{2}}}{s+\frac{1}{2}} \frac{\Gamma(s+1)}{\Gamma\left(s+\frac{1}{2}\right)} X_{2s+1}^0 j_{2s+1}' \, (ka) \right|^2 . \qquad (3.126)
$$

The cross-sections of the spherical shell with two holes are

$$
\frac{\sigma_T^h}{\pi a^2} = \frac{4\sqrt{2}}{(ka)^2} \sum_{s=0}^{\infty} \left[\frac{\Gamma\left(s+\frac{1}{2}\right)}{\Gamma(s+1)} \right]^2 \times
$$

$$
\left\{ |X_{2s}^0|^2 \left[j_{2s}' \, (ka) \right]^2 + \frac{1}{2} \frac{s+\frac{1}{2}}{s+1} |X_{2s+1}^0|^2 \left[j_{2s+1}' \, (ka) \right]^2 \right\} \qquad (3.127)
$$

$$
\frac{\sigma^h}{\pi a^2} = \frac{8\sqrt{2}}{(ka)^2} \left| \sum_{s=0}^{\infty} (-1)^s \frac{\Gamma\left(s+\frac{1}{2}\right)}{\Gamma(s+1)} \times \right.
$$

$$
\left. \left\{ \left(2s+\frac{1}{2}\right)^{\frac{1}{2}} X_{2s}^0 j_{2s}' \, (ka) + \frac{i}{\sqrt{2}} \sqrt{\left(2s+\frac{3}{2}\right) \frac{s+\frac{1}{2}}{s+1}} X_{2s+1}^0 j_{2s+1}' \, (ka) \right\} \right|^2 .
$$

$$
(3.128)
$$

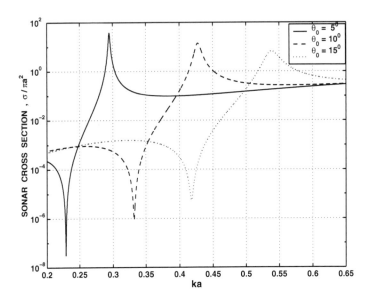

FIGURE 3.2. Sonar cross-sections of various hard spherical barrels at normal incidence.

The mechanical force factor η is considered only for the spherical barrel and is given by

$$\eta = 2^{-\frac{3}{4}} \frac{\pi^{\frac{1}{2}}}{3} \left| X_1^0 \right| \cdot \left| h_1^{(1)\prime}(ka) \right|. \tag{3.129}$$

First consider the hard spherical shell with two holes. In the low-frequency band, the sonar cross-sections for three aperture sizes are displayed in Figure 3.2; the total cross-sections are very similar. Both cross-sections are very similar to those of the spherical shell with a single hole (see Section 2.4). The only differences, for structures with the same aperture size, are a positive shift in the relative wavenumber of the Helmholtz mode for the doubly connected shell and a decrease in peak value. This is in accordance with the analytical formula (3.64).

It is worth noting that resonance peaks of the mechanical force factor η do not occur at the frequency of the Helmholtz mode. This is distinctively different from the behaviour of the single aperture cavity. A formal explanation is as follows. According to (3.129), the value of η is proportional to $\left| X_1^0 \right|$. The solution developed in Section 3.1 produced decoupled sets of equations for the odd and even index coefficients. Resonance features attributed to the even coefficients do not affect the value of $\left| X_1^0 \right|$.

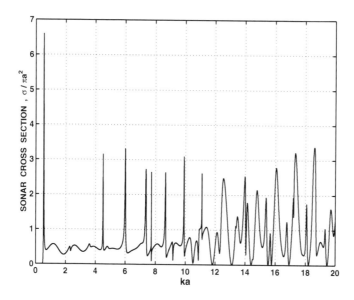

FIGURE 3.3. Normalised sonar cross-section of the hard spherical barrel ($\theta_0 = 15°$) at normal incidence.

A physical explanation is perhaps more enlightening. At the point $ka = \mathrm{Re}\,(ka)_0$ of resonance, the distribution of acoustic pressure on the shell is completely symmetrical because of shell symmetry and symmetry of the Helmholtz oscillation mode. Thus integration over the shell, of the component of acoustic pressure at each point on the shell in the direction of plane wave propagation, necessarily gives a zero value for the mechanical force factor η. Thus we omit its calculation in the low-frequency range.

The frequency dependence of the normalised sonar cross-section $(\sigma^h / \pi a^2)$ in the resonant regime $(0 < ka \leq 20)$ is examined for three apertures sizes $(\theta_0 = 15°, 30°, 45°)$; results are displayed in Figures 3.3, 3.4 and 3.5. A comparison of corresponding plots for the hard shell with one or two holes (see for instance Figure 2.7 and Figure 3.3) reveals that rather fewer resonance peaks occur for the doubly connected shell. In fact the resonance peaks for this structure occur at values of ka corresponding to quasi-eigenoscillations with *even number of acoustic field variations* as a function of the angle θ, i.e., $ka = \varkappa_{2s,l}^{(0)}$ $(s = 0, 1, ...)$. In other words the resonant contribution of quasi-eigenoscillations with an odd number of field variations (in θ) is largely suppressed. An explanation of this phenomena relies on the alignment of the shell geometry with the spatial distribution of the quasi-eigenoscillations with even number of field variations in θ.

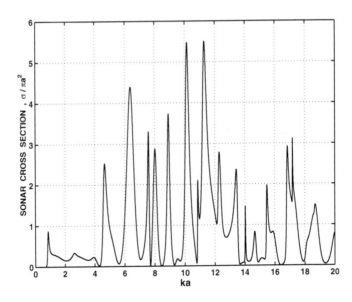

FIGURE 3.4. Normalised sonar cross-section of the hard spherical barrel $(\theta_0 = 30°)$ at normal incidence.

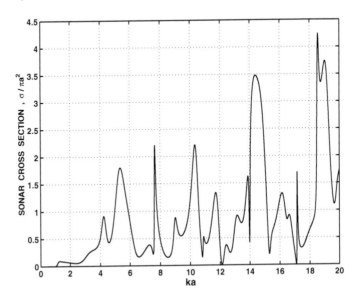

FIGURE 3.5. Normalised sonar cross-section of the hard spherical barrel $(\theta_0 = 45°)$ at normal incidence.

As the aperture widens, the Q-factor of quasi-eigenoscillations decreases. The half-width of the resonant peaks increases and strong resonant backscattering features become smeared out.

The frequency dependence of mechanical force factor η is simpler. Resonant peaks of this value (see Figures 3.6 and 3.7) accurately correspond to the quasi-eigenvalues occurring at $ka = \varkappa_{1l}^{(0)}$. Some additional peaks result from coupling of the dominant quasi-eigenoscillation with other quasi-eigenoscillations at neighbouring frequencies.

As the aperture dimension θ_0 increases beyond $45°$ the spherical cavity with two holes becomes what might be termed a short spherical hollow cylinder. The value $\theta_0 = 45°$ provides a point of demarcation between two types of cavity structures that are distinctively different at somewhat larger or smaller values of θ_0. The justification for the introduction of this terminology is the appearance of new effects in the deep quasi-optical band.

From a geometrical point of view, the spherical hollow cylinder is described by three parameters: shell radius a, height $h = 2a \sin \theta_1$ (where $\theta_1 = \pi/2 - \theta_0$) and maximal deviation of the spherical cylinder from the right circular cylinder, given by $d = a\left(1 - \cos \theta_1\right)$. The ratios of these parameters to wavelength are of great importance in scattering theory, and may be expressed as

$$h/\lambda = \frac{ka \sin \theta_1}{\pi}, \qquad d/\lambda = \frac{ka}{2\pi}\left(1 - \cos \theta_1\right).$$

A "rule-of-thumb" generally accepted in scattering theory asserts that if $d/\lambda < 1/16$ there is negligible difference in the scattering features, i.e., in this case there is negligible difference whether the generatrix of cylinder is straight or curved. With this in mind, we consider extremely short and hollow cylinders (defined by angle $\theta = 2°$) over a wide frequency range $0.1 \leq ka \leq 500$; over this range d/λ never exceeds $1/16$.

Cross-sectional calculations are plotted in Figures 3.8 and 3.9. The magnitude of scattering by these cylindrical scatterers is exceedingly small compared to that from a hard sphere of the same wavelength dimension. Nevertheless the frequency dependence is clearly regular, the distance Δka between neighbouring maxima being about π, corresponding to a half wavelength.

The period of the envelope of short-period oscillations is related to the value h/λ. The example displayed in Figure 3.10 shows that the envelope has period $h/2\lambda$, with maxima and minima occurring when $h = n\lambda/2$ and $h = (2n + 1)\lambda/4$ $(n = 1, 2, \ldots)$, respectively.

The frequency dependence of the mechanical force factor is much clearer. Ignoring the first extrema, maxima and minima of η occur when $h = \left(n + \frac{1}{2}\right)\lambda$ and $h = n\lambda$ $(n = 1, 2, \ldots)$, respectively.

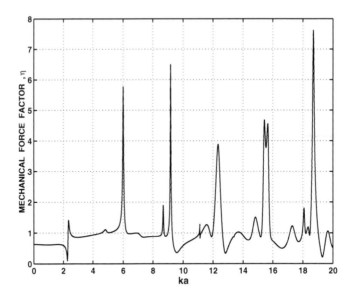

FIGURE 3.6. Mechanical force factor for the hard spherical barrel ($\theta_0 = 15°$) at normal incidence.

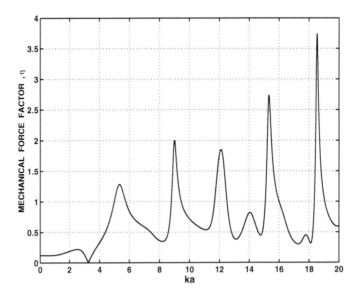

FIGURE 3.7. Mechanical force factor for the hard spherical barrel ($\theta_0 = 45°$) at normal incidence.

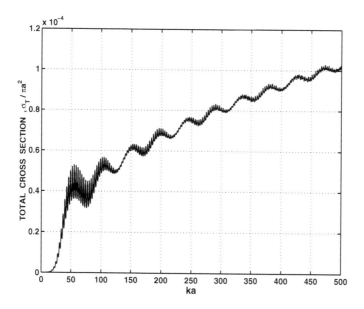

FIGURE 3.8. Normalised total cross-section of the "spherical" hollow cylinder ($\theta_0 = 2°$).

As the parameter θ_1 increases new phenomena appear. The behaviour of the short spherical hollow cylinder at higher frequencies acquires some features of an open spherical barrel-like resonator. In Figures 3.11 and 3.12 are displayed some results for the structures with $\theta_1 = 10°$.

Since the quantity η is essentially a near-field characteristic, the occurrence of high Q-factor oscillations in the cavity of open barrel-like resonator will be clearly visible in the frequency dependence of η.

Consider Figure 3.12. As ka, or h/λ, increases, increasingly sharp peaks, with repetition period about $\Delta ka = \pi$, are observed, especially when $h/\lambda \gg 1$. This feature is very characteristic for open spherical barrel-like resonators.

For the structure with parameter $\theta_1 = 15°$, other phenomena appear when the frequency range is extended to $ka = 2000$; large-scale oscillations with period Δka about 800 occur. As θ_1 increases, the period of these large-scaled oscillations reduces, to about $\Delta ka \simeq 150$ for $\theta_1 = 30°$ (see Figure 3.13) and $\Delta ka \simeq 50$ for $\theta_1 = 45°$. Simple explanations of this phenomena in terms of structural parameters are difficult, and we restrict ourselves to accurate numerical calculations.

The behaviour of value η (see Figure 3.14) demonstrates that in the quasi-optical range the spherical barrel generates or selects resonant frequencies in the manner of the classical open barrel-like resonator.

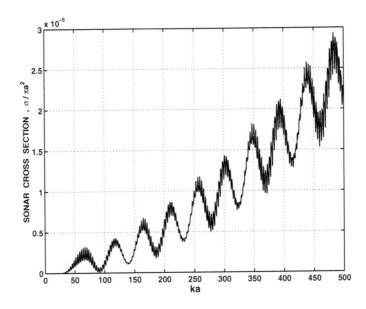

FIGURE 3.9. Normalised sonar cross-section of the "spherical" hollow cylinder ($\theta_0 = 2°$).

The final part of this section examines the slotted hard spherical shell. We will term this structure slotted if $\theta_1 < 45°$, but if $\theta_0 \leq 45°$ we consider it as two spherical caps or a two-mirrored open spherical resonator (at high frequency). First consider the low-frequency range. The normalised sonar cross-section $\sigma/\pi a_d^2$ is displayed given in Figure 3.15 with values of θ_1 equal to $10^{-3}, 10^{-2}$ and 10^{-1} (degrees). Although such angular sizes are not practicable, it is notable that, even at these extremely small angular values, the half-width of resonant peaks are relatively wide because of the logarithmic dependence of Q-factor on the polar angle θ_1 (see Section 3.3). More realistic values of the parameter ($\theta_1 = 1°, 5°, 10°$) were used to calculate the sonar cross-sections plotted in Figure 3.16. A comparison with the corresponding plots for the hard spherical shell with one or two holes reveals that the resonant frequency (of the Helmholtz mode) is significantly shifted towards the transition zone between the Rayleigh scattering zone and the resonance region. For the slot with angular semi-width of $\theta_1 = 10°$, resonance occurs near $ka \simeq 1.08$. The behaviour of the corresponding total cross-section is similar in each case.

Let us compare scattering features of the slotted spherical shell with the spherical cavity having two holes in the resonance region. We use the same values of angular semi-width θ_1 for the slotted sphere as were used

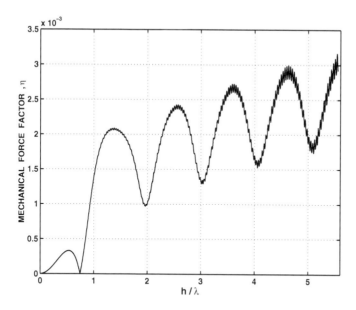

FIGURE 3.10. Mechanical force factor for the "spherical" hollow cylinder $(\theta_0 = 2°)$.

for angular semi-widths θ_0 of the circular holes in the doubly connected spherical shell.

Various cross-section results are presented in Figures 3.17 and 3.18 for $\theta_1 = 15°$ and in Figure 3.19 for $\theta_1 = 45°$. The first distinctive feature in the scattering by the slotted shell is the large resonant total cross-section value due to excitation in the spherical cavity of the lowest order quasi-eigenoscillation at $ka = \varkappa_{11}^0$. In contrast to the spherical shell with two holes where the excitation of quasi-eigenoscillations with odd number of field variations in θ were suppressed, all oscillations of an odd or even character are excited.

The complex responses of hybrid quasi-eigenoscillations that are due to the interaction of two (or more) closely spaced quasi-eigenvalues is a special topic in the *spectral theory of open structures*. The mathematical methods developed in this book are basically sufficient to study these problems in further detail, but since this topic is not central to this book, it will not be pursued further.

We now describe below some interesting scattering features of this scatterer in quasi-optical range. The upper wave size is limited so that the maximal distance between the central points of the spherical caps is $2a/\lambda = 100$; the parameter θ_0 varies.

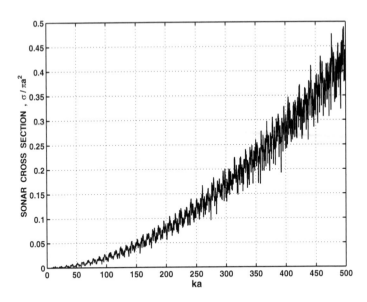

FIGURE 3.11. Sonar cross-section of the "spherical" hollow cylinder ($\theta_1 = 10°$).

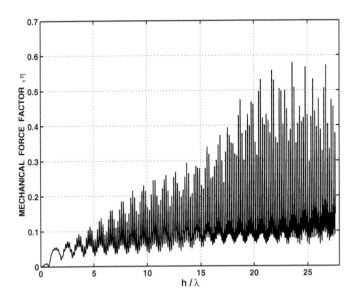

FIGURE 3.12. Mechanical force factor for the "spherical" hollow cylinder ($\theta_1 = 10°$).

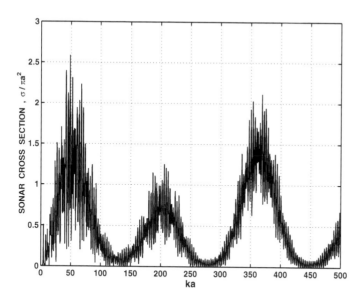

FIGURE 3.13. Sonar cross-section of the "spherical" hollow cylinder ($\theta_1 = 30°$).

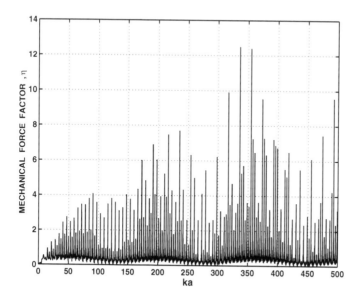

FIGURE 3.14. Mechanical force factor for the "spherical" hollow cylinder ($\theta_1 = 30°$).

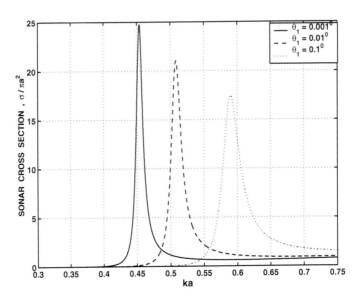

FIGURE 3.15. Sonar cross-section of the slotted spherical shell, normal incidence.

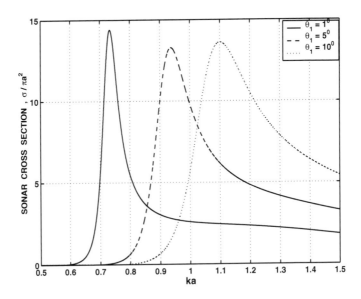

FIGURE 3.16. Sonar cross-section of the hard slotted spherical shell, normal incidence.

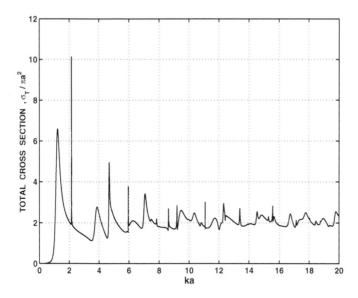

FIGURE 3.17. Total cross-section of the hard slotted spherical shell ($\theta_1 = 15°$), at normal incidence.

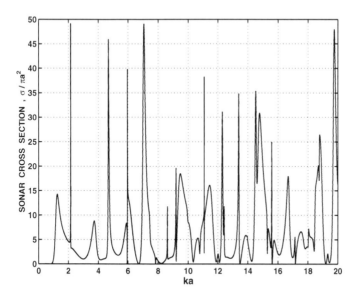

FIGURE 3.18. Sonar cross-section of the hard slotted spherical shell ($\theta_1 = 15°$), at normal incidence.

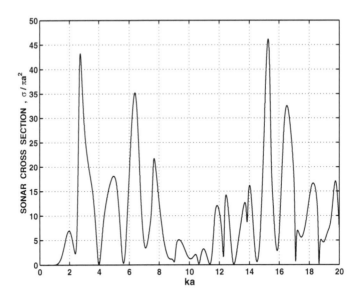

FIGURE 3.19. Sonar cross-section of the hard slotted spherical shell ($\theta_1 = 45°$), at normal incidence.

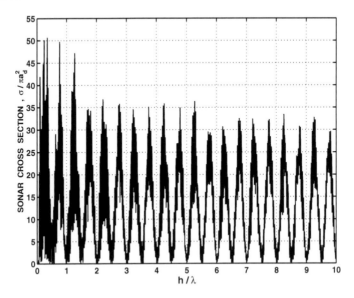

FIGURE 3.20. Sonar cross-section of two coaxial spherical caps ($\theta_0 = 30°$), at normal incidence.

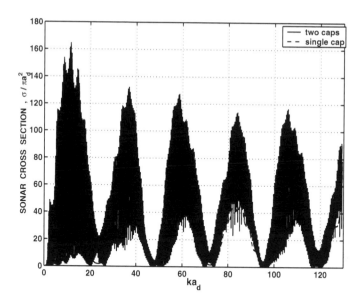

FIGURE 3.21. Comparison of the sonar cross-sections of one and two caps ($\theta_0 = 15°$).

The dependence of the cap cross-section $\sigma/\pi a_d^2$ (see Figure 3.20) on the relative maximum depth of the cap is very similar to that of the single cap (see Section 2.4). The main difference is an increase of both cross-sections.

As the parameter θ_0 decreases (hand in hand with h/λ) new phenomena connected with the excitation of two-mirror open resonators appear (see Figures 3.21 and 3.22). The characteristic feature of quasi-optical open resonators is a regular dependence upon ka (or λ). With some correction for diffraction effects, it is clear that the maxima and minima of both $\sigma_T/\pi a_d^2$ and $\sigma/\pi a_d^2$ occur when $2a = (2n+1)\lambda/2$ and $2a = n\lambda$ ($n = 1, 2, \ldots$), respectively.

These features become more pronounced as the depth h diminishes. cross-sections of the structure with $\theta_0 = 5°$ are examined in Figures 3.23–3.25. It is striking that the sonar cross-section $\sigma/\pi a_d^2$ nearly vanishes when the distance between caps is an even number of wavelengths. In this case the spherical two-mirror structure is essentially the well-known Fabri-Perot resonator. From a physical point of view, the open resonator is not excited and releases acoustic energy, if $2a = (2n+1)\lambda/2$; on the other hand, if $2a = n\lambda$, the acoustic energy is locked between the spherical mirrors thus causing deep minima in the frequency dependence of the sonar cross-section.

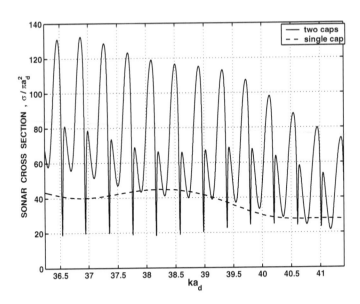

FIGURE 3.22. Comparison of the sonar cross-sections of one and two caps ($\theta_0 = 15°$).

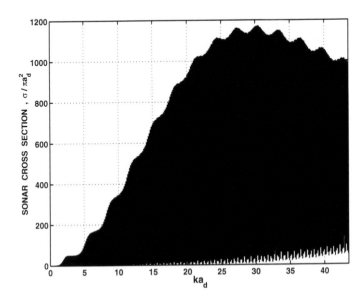

FIGURE 3.23. Sonar cross-sections of two hard spherical caps ($\theta_0 = 5°$.)

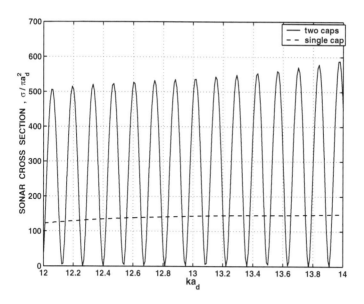

FIGURE 3.24. Comparison of sonar cross-sections of one and two caps ($\theta_0 = 5°$).

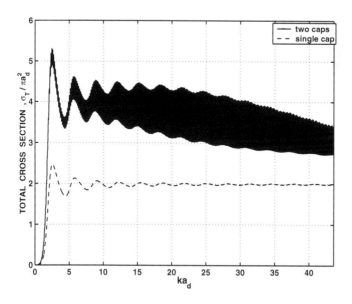

FIGURE 3.25. Comparison of total cross-sections of one and two caps ($\theta_0 = 5°$.)

4

Electromagnetic Diffraction from a Metallic Spherical Cavity.

This chapter examines electromagnetic diffraction from a circular hole in a perfectly conducting spherical shell. Several aspects of this problem are interesting. First it is the simplest three-dimensional conducting structure (of finite extent) that incorporates a cavity and edges. On the other hand, it is the archetypical open resonator with a high Q-factor strongly dependent on aperture size. Finally, any solution that is uniform with respect to aperture size (or angle measure of the spherical cap) also provides the basis for a comprehensive analysis of spherical reflector antennas. The full-wave solution developed in this chapter enables us to analyse accurately a large spherical reflector antenna, of aperture size some hundreds of wavelengths.

In contradistinction to two-dimensional structures, the electromagnetic scattering problem for open spherical shells is intrinsically more complex than the acoustic scattering problem for the same structures, and cannot be reduced to an equivalent scalar (acoustic) problem. Even the simplest problem, described in Section 4.1, of excitation of an open spherical cavity by vertical (electric or magnetic) dipole differs from any conceivable acoustical analogy. In Section 4.2, it will be shown that TE and TM type waves scattered by open shells are intrinsically coupled. However the Mie series solution [13] for the *closed* conducting sphere demonstrates that TE and TM waves are independently scattered.

As will be seen in Sections 4.2 and 4.3, the coupling of TE and TM waves is due to the coupling of boundary conditions for the associated potentials (U, V) and their normal derivatives. In Section 4.3 the dependence of total cross-section and radar cross-section on frequency and aperture size is numerically studied, with special attention to resonant phenomena. The limiting case of a circular disc is also considered in Section 4.3.

The remaining sections (4.4–4.5) are devoted to an accurate analysis of the spherical reflector antenna. With certain parameter choices, these structures have very nearly the same features as the widely deployed parabolic reflector antennas.

Before commencing the first section, some comments on the published literature are perhaps helpful. The very first paper on these topics, published in 1973 by V.P. Shestopalov and A.M. Radin [73], proposed a treatment of electromagnetic scattering from spherical shells. This paper has serious errors; their solution does not belong to the correct solution class, and, furthermore, does not take into account the mutual coupling of TE and TM waves. However the authors did demonstrate the value and utility of dual series equations techniques, based on integral representations of Mehler-Dirichlet type for associated Legendre functions.

A correct solution of the electromagnetic plane wave diffraction problem for ideally conducting spherical shell with circular aperture appeared in 1977 [99]. It followed the publication [98] of a simpler problem, of vertical electric dipole excitation of the same structure, one year earlier (1976).

Further publications mostly dealt with applications ([111], [112], [114]). These Russian papers were unknown in the West for a long time, until Ziolkowski, Johnson and co-workers published some scattering studies ([128], [129]) using dual series techniques.

4.1   Electric or Magnetic Dipole Excitation.

Consider the open spherical shell of radius a with a circular aperture subtending an angle θ_0 at the origin (see Figure 4.1). Locate a dipole, of electric or magnetic type, at an arbitrary point on the z-axis a distance d ($< a$) in the positive direction from the origin; the z-axis is the axis of symmetry and the dipole moment is aligned with the axis.

We may formulate the diffraction problem in terms of a single field component, H_ϕ or E_ϕ, corresponding to the choice of electric or magnetic dipole, respectively (see Section 1.4). The incident field is described by Formula (1.213) and relations (1.89), (1.90).

A correct formulation of the diffraction problem providing a unique solution is as follows. First the solution must satisfy the differential equation (1.91). Secondly, it must be continuous across surface $r = a$, and satisfy appropriate mixed boundary conditions; for electrical dipole excitation (TM

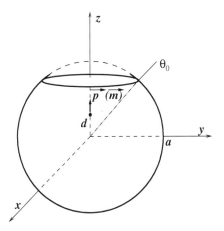

FIGURE 4.1. The spherical cavity excited by an axially located vertical electric or magnetic dipole.

case), the conditions are

$$H_\phi (a - 0, \theta) = H_\phi (a + 0, \theta), \qquad \theta \in (0, \theta_0) \qquad (4.1)$$
$$E_\theta (a - 0, \theta) = E_\theta (a + 0, \theta) = 0, \qquad \theta \in (\theta_0, \pi), \qquad (4.2)$$

and for magnetic dipole excitation (TE case), the conditions are

$$H_\theta (a - 0, \theta) = H_\theta (a + 0, \theta), \qquad \theta \in (0, \theta_0) \qquad (4.3)$$
$$E_\phi (a - 0, \theta) = E_\phi (a + 0, \theta) = 0, \qquad \theta \in (\theta_0, \pi). \qquad (4.4)$$

Third, the scattered field must represent an outgoing spherical wave at infinity, i.e., there exist functions (scattering patterns) $A(\theta)$ and $B(\theta)$, independent of φ, such that

$$H_\phi^s \sim A(\theta) e^{ikr}/r \quad \text{and} \quad E_\phi^s \sim B(\theta) e^{ikr}/r \qquad (4.5)$$

as $r \to \infty$. Finally, the energy of the scattered field in any arbitrary finite region V of space, including the edges, must be finite:

$$W = \frac{1}{2} \iiint_V \left\{ \left| \vec{E}^s \right|^2 + \left| \vec{H}^s \right|^2 \right\} dV < \infty. \qquad (4.6)$$

Let us consider the vertical electric dipole (VED) and vertical magnetic dipole (VMD) separately; certain similarities will allow us to treat the VMD more briefly.

4.1.1   The Vertical Electric Dipole (TM Case).

Taking into account the continuity condition and the far-field behaviour of the scattered field, we decompose the total field as the sum

$$H_\phi = H_\phi^{(0)} + H_\phi^s \tag{4.7}$$

and seek the scattered field in the form

$$H_\phi^s = \frac{pk^2}{r} \sum_{n=1}^{\infty} \frac{2n+1}{n(n+1)} x_n P_n^1 (\cos\theta) \begin{cases} \zeta_n' (ka)\, \psi_n (kr), & r < a \\ \psi_n' (ka)\, \zeta_n (kr), & r > a, \end{cases} \tag{4.8}$$

where the unknown coefficients x_n are to be found. The finite energy condition (4.6) effectively defines the solution class for the coefficient sequence $\{x_n\}_{n=1}^{\infty}$. Taking the integration region in (4.6) to be the sphere of radius a, it is easily deduced that

$$W = W_{TM} = 2\pi p^2 k^4 a \times$$

$$\sum_{n=1}^{\infty} \frac{2n+1}{n(n+1)} |x_n|^2 |\zeta_n' (ka)|^2 \left\{ \psi_n'^2 (ka) - \left[\frac{n(n+1)}{(ka)^2} - 1 \right] \psi_n^2 (ka) \right\} \tag{4.9}$$

must be finite. Using the asymptotic behaviour of the spherical Bessel functions and their derivatives,

$$\psi_n (x) = \frac{2^n n!}{(2n+1)!} x^{n+1} \left\{ 1 - \frac{x^2}{2(2n+3)} + O(n^{-2}) \right\},$$

$$\psi_n' (x) = \frac{2^n (n+1)!}{(2n+1)!} x^n \left\{ 1 - \frac{n+3}{2(n+1)(2n+3)} x^2 + O(n^{-2}) \right\},$$

$$\zeta_n (x) = -i \frac{(2n)!}{2^n n!} x^{-n} \left\{ 1 + \frac{x^2}{2(2n-1)} + O(n^{-2}) \right\},$$

$$\zeta_n' (x) = \frac{i(2n)!}{2^n (n-1)!} x^{-n-1} \left\{ 1 + \frac{n-2}{2n(2n-1)} x^2 + O(n^{-2}) \right\}, \tag{4.10}$$

as $n \to \infty$, it can be readily shown that

$$W_{TM} \leq C_1 (ka) \sum_{n=1}^{\infty} |x_n|^2 \tag{4.11}$$

where $C_1 (ka)$ is a function of ka alone, with finite value. Thus the coefficient sequence $\{x_n\}_{n=1}^{\infty}$ is square summable and belongs to the functional space l_2.

Now enforce the boundary conditions (4.1)–(4.2), bearing in mind the explicit form of the scattered field (4.8), the incident field (1.213) and the relations between the basic field component H_φ and the other field components $(1.89)-(1.90)$. As a result, one obtains the following dual series equations.

$$\sum_{n=1}^{\infty} \frac{2n+1}{n\,(n+1)} x_n P_n^1 (\cos\theta) = 0, \qquad \theta \in (0, \theta_0) \qquad (4.12)$$

$$\sum_{n=1}^{\infty} \frac{2n+1}{n\,(n+1)} x_n \psi_n' (ka)\, \zeta_n' (ka)\, P_n^1 (\cos\theta) =$$

$$- (kd)^{-2} \sum_{n=1}^{\infty} (2n+1)\, \psi_n (kd)\, \zeta_n' (ka)\, P_n^1 (\cos\theta)\,, \qquad \theta \in (\theta_0, \pi)\,. \quad (4.13)$$

The first step in the regularization process is to identify a suitable asymptotically small parameter. Thus define

$$\varepsilon_n = 1 + ika \frac{2n+1}{n\,(n+1)} \psi_n' (ka)\, \zeta_n' (ka)\,. \qquad (4.14)$$

It is readily verified from the asymptotics (4.10) that $\varepsilon_n = O\left((ka/n)^2\right)$ as $n \to \infty$. The dual series equations (4.12) and (4.13) may be rearranged as

$$\sum_{n=1}^{\infty} \frac{2n+1}{n\,(n+1)} x_n P_n^1 (\cos\theta) = 0, \qquad \theta \in (0, \theta_0) \qquad (4.15)$$

$$\sum_{n=1}^{\infty} (1 - \varepsilon_n)\, x_n P_n^1 (\cos\theta) = \sum_{n=1}^{\infty} \alpha_n P_n^1 (\cos\theta)\,, \quad \theta \in (\theta_0, \pi)\,, \qquad (4.16)$$

where

$$\alpha_n = i \frac{ka}{(kd)^2} (2n+1)\, \psi_n (kd)\, \zeta_n' (ka)\,. \qquad (4.17)$$

Because the unknown coefficient sequence is square summable, it can be readily estimated that the general term of series (4.15) and (4.16) decays at rate $O\left(n^{-\frac{3}{2}}\right)$ and $O\left(n^{-\frac{1}{2}}\right)$, respectively, as $n \to \infty$. Thus the first series (4.15) converges uniformly, whilst the second (4.16) is nonuniformly converging. The Abel integral transform may not be directly applied to (4.16); the formal interchange of the order of integration and summation leads to a divergent series. To ameliorate this difficulty, recognise that

$$P_n^1 (\cos\theta) = -\frac{d}{d\theta} P_n (\cos\theta)$$

and now term-by-term integration of the series (4.16) is permissible since $\{x_n\}_{n=1}^{\infty} \in l_2$ (see Appendix D of Volume I). The result is the following uniformly converging series on which Abel integral transforms may be validly applied.

$$\sum_{n=1}^{\infty} (1 - \varepsilon_n)\, x_n\, P_n\,(\cos\theta) = C_1 + \sum_{n=1}^{\infty} \alpha_n\, P_n\,(\cos\theta)\,, \quad \theta \in (\theta_0, \pi) \quad (4.18)$$

where C_1 is a constant of integration.

We will employ the readily deduced integral representations of Mehler-Dirichlet type

$$P_n^1\,(\cos\theta) = \frac{1}{\sin\theta}\,\frac{n\,(n+1)}{2n+1}\,\{P_{n-1}\,(\cos\theta) - P_{n+1}\,(\cos\theta)\}$$

$$= \frac{2\sqrt{2}}{\pi}\,\frac{1}{\sin\theta}\,\frac{n\,(n+1)}{2n+1}\int_0^\theta \frac{\sin\left(n + \frac{1}{2}\right)\phi \sin\phi}{\sqrt{\sin\phi - \cos\theta}}\,d\phi$$

$$= -\frac{2\sqrt{2}}{\pi}\,\frac{1}{\sin\theta}\,\frac{n\,(n+1)}{2n+1}\int_\theta^\pi \frac{\cos\left(n + \frac{1}{2}\right)\phi \sin\phi}{\sqrt{\cos\theta - \cos\phi}}\,d\phi \quad (4.19)$$

to transform the original equations (4.12)–(4.13) to the equivalent preregularised form

$$\sum_{n=1}^{\infty} x_n \sin\left(n + \frac{1}{2}\right)\theta =$$

$$\begin{cases} 0, & \theta \in (0, \theta_0) \\ C_1 \sin\tfrac{1}{2}\theta + \sum_{n=1}^{\infty} (\alpha_n + \varepsilon_n x_n)\sin\left(n + \tfrac{1}{2}\right)\theta, & \theta \in (\theta_0, \pi)\,. \end{cases} \quad (4.20)$$

It is worth observing that the constant C_1 has a more profound meaning than the purely formal role of a constant of integration. The original dual series equations are series expansions in the complete orthogonal family of functions $P_n^1\,(\cos\theta)$ $(n = 1, 2, ...)$. Likewise, the transformed equations are series expansions in another complete orthogonal family, the trigonometric functions $\left\{\sin\left(n + \frac{1}{2}\right)\theta\right\}_{n=0}^{\infty}$; the unknown coefficients x_n are associated with an incomplete subset of this family (indexed by $n = 1, 2, \ldots$), and the constant C_1 is associated with the zero index element $(\sin\frac{1}{2}\theta)$, the addition of which makes the trigonometric family complete.

Making use of the orthogonality, the equations (4.20) are easily transformed to the i.s.l.a.e. of the second kind,

$$(1 - \varepsilon_m)\, x_m + \sum_{n=1}^{\infty} x_n \varepsilon_n\, R_{nm}^{(1)}\,(\theta_0) = \alpha_m - \sum_{n=1}^{\infty} \alpha_n\, R_{nm}^{(1)}\,(\theta_0)\,, \quad (4.21)$$

where $m = 1, 2, \ldots,$

$$R_{nm}^{(1)}(\theta_0) = R_{nm}(\theta_0) - \frac{R_{n0}(\theta_0)}{1 - R_{00}(\theta_0)} R_{0m}(\theta_0),$$

and R_{nm} is the incomplete scalar product

$$R_{nm}(\theta_0) = \widehat{Q}_{nm}^{(\frac{1}{2}, -\frac{1}{2})}(\cos\theta_0)$$

defined in Volume I (Appendix B.6); when $n \neq m$,

$$R_{nm}(\theta_0) = \frac{1}{\pi}\left\{ \frac{\sin(n-m)\theta_0}{n-m} - \frac{\sin(n+m+1)\theta_0}{n+m+1} \right\}, \qquad (4.22)$$

and when $n = m$, the first term on the right hand side of (4.22) is replaced by θ_0.

An alternative form of (4.21) is obtained by introducing the angle $\theta_1 = \pi - \theta_0$ subtended by the spherical shell at its centre. Let Q_{nm} be the incomplete scalar product

$$Q_{nm}(\theta_0) = \widehat{Q}_{nm}^{(-\frac{1}{2}, \frac{1}{2})}(\cos\theta_0)$$

also discussed in Volume I (Appendix B.6); when $n \neq m$,

$$Q_{nm}(\theta_0) = \frac{1}{\pi}\left\{ \frac{\sin(n-m)\theta_0}{n-m} + \frac{\sin(n+m+1)\theta_0}{n+m+1} \right\}, \qquad (4.23)$$

and as before, when $n = m$, the first term on the right hand side of (4.23) is replaced by θ_0. Now define

$$Q_{nm}^{(1)}(\theta_0) = Q_{nm}(\theta_0) - \frac{Q_{n0}(\theta_0)}{Q_{00}(\theta_0)} Q_{0m}(\theta_0).$$

It is quite obvious that

$$Q_{nm}^{(1)}(\theta_0) = \delta_{nm} - (-1)^{n-m} R_{nm}^{(1)}(\theta_1),$$
$$R_{nm}^{(1)}(\theta_0) = \delta_{nm} - (-1)^{n-m} Q_{nm}^{(1)}(\theta_1),$$

and the i.s.l.a.e. (4.21) may be transformed into the more compact form

$$X_m - \sum_{n=1}^{\infty} X_n \varepsilon_n Q_{nm}^{(1)}(\theta_1) = \sum_{n=1}^{\infty} A_n Q_{nm}^{(1)}(\theta_1), \qquad (4.24)$$

where $m = 1, 2, \ldots$ and

$$X_m = (-1)^m x_m, \quad A_m = (-1)^m \alpha_m. \qquad (4.25)$$

Equations (4.21) or (4.24) remain valid when $d > a$ and the dipole is located in the upper half-space $z > 0$, provided the coefficients α_n are replaced by

$$\alpha_n = i \frac{ka}{(kd)^2} (2n + 1) \zeta_n (kd) \psi'_n (ka).$$ (4.26)

If the dipole is located in the lower half-space $z < 0$ then the value of α_n, defined by (4.17) or (4.26), is simply multiplied by the factor $(-1)^{n-1}$.

As in previous chapters the i.s.l.a.e. (4.21) or (4.24) provides the basis for an exhaustive examination of the scattered field. Uniqueness of solutions to these systems, and of those encountered in susbsequent sections in this chapter, is established by adapting the uniqueness argument given in Section 2.1; the truncation number for numerical studies is chosen according to the criterion set out in that section.

Omitting their deduction we briefly state some near-field and far-field characteristics. The *surface current density* has one component and is equal to the jump in the magnetic field intensity across the shell,

$$j_\theta = H_\phi (a - 0, \theta) - H_\phi (a + 0, \theta) = i \frac{pk^2}{a} \sum_{n=1}^{\infty} \frac{2n + 1}{n(n+1)} x_n P_n^1 (\cos \theta).$$

(4.27)

The *radiation pattern* S_1, defined by $E_\theta, H_\phi \backsim S_1 (\theta) e^{ikr}/r$ as $r \to \infty$, equals

$$S_1 (\theta) = pk^2 \sum_{n=1}^{\infty} (2n + 1) \left\{ \frac{\psi'_n (ka)}{n(n+1)} x_n + \frac{\psi_n (kd)}{(kd)^2} \right\} (-1)^{n+1} P_n^1 (\cos \theta).$$

(4.28)

The *radiation resistance* R is defined to be the ratio of the power flux across a sphere S_r of radius r of the total electromagnetic field to that of the VED in free space, as $r \to \infty$. The power flux P may be computed from the Poynting vector via

$$P = \frac{1}{2} \text{Re} \iint_{S_r} \vec{E} \times \vec{H^*} \, dS.$$

As $r \to \infty$ the dominant contribution to the Poynting vector is radially directed,

$$\vec{E} \times \vec{H^*} = E_\theta H_\phi^* \vec{i_r} + O(r^{-3}),$$

and so the radiation resistance equals

$$R = \frac{3}{2} \sum_{n=1}^{\infty} n(n+1)(2n+1) \left| \frac{\psi_n (kd)}{(kd)^2} + \frac{\psi'_n (ka)}{n(n+1)} x_n \right|^2.$$ (4.29)

Amongst near-field characteristics the stored or accumulated energy is of particular interest. We distinguish between the energy W^i accumulated inside $(r \leq a)$ the open spherical shell from that energy $W^{(e)}$ accumulated in the exterior $(r > a)$.

How should we calculate the energy in the infinite domain $(a \leq r \leq \infty)$? The energy of the field is divergent; so, we take into account the part of scattered energy that *adheres* to the open shell, i.e., the reactive part of the energy. From a physical viewpoint the difference between the total energy of the scattered field and the energy carried by the outgoing travelling wave is exactly the desired *stored external energy* $W^{(e)}$. With this understanding

$$W_{TM}^{(e)} = 2\pi p^2 k^4 a \sum_{n=1}^{\infty} \frac{2n+1}{n(n+1)} |x_n|^2 \psi_n'^2(ka) \times$$

$$\left\{ 2 + \left[\frac{n(n+1)}{(ka)^2} - 1 \right] |\zeta_n(ka)|^2 - |\zeta_n'(ka)|^2 \right\}. \quad (4.30)$$

The total stored energy is then defined as the sum of scattered energy that is stored in the internal $(r < a)$ and external $(r > a)$ regions,

$$W_{TM} = W_{TM}^i + W_{TM}^{(e)} \quad (4.31)$$

where values of W_{TM}^i are defined by formula (4.9).

In analysing the high Q-factor oscillations that develop in this open spherical resonator it is useful to decompose

$$W_{TM} = \sum_{n=1}^{\infty} W_{TM}^{(n)} \quad (4.32)$$

into a sum of terms representing the normalised energy contribution of each harmonic, and to examine the ratios

$$q_{TM}^{(n)} = \frac{W_{TM}^{(n)}}{W_{TM}} \quad (4.33)$$

measuring energy distribution amongst each harmonic.

Before presenting numerical results let us consider analytical features of the solution obtained above. First, a careful look at the expression for surface current density (4.27) reveals that the general term of the series decays at the rate $O\left(n^{-\frac{3}{2}}\right)$ as $n \to \infty$, i.e., is relatively slowly converging. Although this rate of decay is not so bad for numerical calculations, the rate of decay of the terms in corresponding surface current expression for the structure excited by the VMD is rather slower $(O\left(n^{-\frac{1}{2}}\right))$ and it is imperative to accelerate the convergence of the series. In the VED case, this is achieved by rearranging the i.s.l.a.e. (4.24) in the form

$$X_n = \sum_{s=1}^{\infty} \left(X_s \varepsilon_s + A_s \right) Q_{sn}^{(1)} \left(\theta_1 \right). \tag{4.34}$$

Setting $X_n = (-1)^n x_n$ into formula (4.27) and interchanging the order of summation produces the computational formula for surface current density

$$j_\theta = -i \frac{pk^2}{a} \sum_{s=1}^{\infty} \left(X_s \varepsilon_s + A_s \right) \Pi_s \left(\vartheta, \theta_1 \right), \tag{4.35}$$

where $\vartheta = \pi - \theta$ and

$$\Pi_s \left(\vartheta, \theta_1 \right) = \sum_{n=1}^{\infty} \frac{2n+1}{n(n+1)} Q_{sn}^{(1)} \left(\theta_1 \right) P_n^1 \left(\cos \vartheta \right). \tag{4.36}$$

This function vanishes when $\vartheta \in (\theta_1, \pi)$; when $\vartheta \in (0, \theta_1)$ it may be expressed in the form

$$\Pi_s \left(\vartheta, \theta_1 \right) = \mathrm{cosec}\, \vartheta \times$$
$$\left\{ \left[G_{s-1} \left(\vartheta, \theta_1 \right) - G_{s+1} \left(\vartheta, \theta_1 \right) \right] + \frac{Q_{s0} \left(\theta_1 \right)}{Q_{00} \left(\theta_1 \right)} \left[G_1 \left(\vartheta, \theta_1 \right) + G_0 \left(\vartheta, \theta_1 \right) \right] \right\},$$
$$\tag{4.37}$$

where the function $G_s \left(\vartheta, \theta_1 \right)$ is defined by (2.178); we recall that it may be easily computed from the formulae (2.180)–(2.181).

The convergence rate of the modified series (4.35) is $O\left(s^{-\frac{7}{2}} \right)$ as $s \to \infty$. Importantly, the vanishing of $\Pi_s \left(\vartheta, \theta_1 \right)$ for $\vartheta \in (\theta_1, \pi)$ shows that the boundary condition (4.1) is satisfied term by term. Moreover, using (2.180)–(2.181), analysis of the form (4.37) of $\Pi_s \left(\vartheta, \theta_1 \right)$ in the interval $\vartheta \in (0, \theta_1)$ reveals that the surface current j_θ has the physically expected behaviour as $\theta \to \theta_0$ (or $\vartheta \to \theta_1$): each term in (4.35) is proportional to $(\cos \vartheta - \cos \theta_1)^{\frac{1}{2}}$ in the vicinity of the sharp edge.

Thus accurate calculation of the surface current density distribution is facilitated by this transformation to a much more rapidly convergent series. This analytical transformation will be used later in the analysis of the spherical reflector antenna, where such calculations are of practical interest.

Now let us consider the quasi-eigenoscillations that develop in the open spherical cavity with a small aperture $(\theta_0 \ll 1)$, when the excitation frequency coincides with one of the quasi-eigenvalues of the spectrum $\varkappa_{sl}$.

	$s = 1$	2	3	4	5	6	7
$l = 1$	2.744	3.870	4.973	6.062	7.140	8.211	9.275
2	6.117	7.443	8.722	9.968	11.189	12.391	13.579
3	9.317	10.713	12.064	13.380	14.670	15.939	17.190

TABLE 4.1. TM$_{ls0}$-oscillations, spectral values $\varkappa_{sl}^{(0)}$

The spectrum of eigenvalues of quasi-eigenoscillations in the *closed* cavity (of TM$_{ls0}$-type) is determined by the roots $\varkappa_{sl}^{(0)}$ $(l = 1, 2, \ldots)$ of the characteristic equation

$$\psi_s'\left(\varkappa_{sl}^{(0)}\right) = 0, \qquad (4.38)$$

where $s = 1, 2, \ldots$; for each s, the roots are indexed in increasing order by l. The indices s and l designate the number of field variations in the radial and angular variables (r and θ), respectively. Some lower order roots are shown in Table 4.1. Following the same idea used in Chapters 2 and 3, the perturbation to the spectral value $\varkappa_{sl}^{(0)}$ caused by a small aperture (with $\theta_0 \ll 1$) may be analytically calculated:

$$\varkappa_{sl}/\varkappa_{sl}^{(0)} = 1 + \frac{s(s+1)}{2s+1}\frac{R_{ss}^{(1)}(\theta_0)}{\varkappa_{sl}^{(0)2} - s(s+1)} -$$

$$i\frac{s(s+1)}{2s+1}\frac{1}{\varkappa_{sl}^{(0)}\left[\varkappa_{sl}^{(0)2} - s(s+1)\right]}\sum_{\substack{n=1 \\ n \neq s}}^{\infty}\frac{n(n+1)}{2n+1}\frac{R_{ns}^{(1)}(\theta_0)R_{sn}^{(1)}(\theta_0)}{\left|\zeta_n'\left(\varkappa_{sl}^{(0)}\right)\right|^2}. \qquad (4.39)$$

Thus the real part of the quasi-eigenvalue $\varkappa_{sl}$ slightly exceeds $\varkappa_{sl}^{(0)}$, and $\mathrm{Im}(\varkappa_{sl}) < 0$, indicating radiative losses; the associated Q-factor

$$Q_{ls0}^{TM} = -2\frac{\mathrm{Re}(\varkappa_{sl})}{\mathrm{Im}(\varkappa_{sl})} \qquad (4.40)$$

is of finite value.

Provided that $(2n+1)\theta_0 \ll 1$ and $(2s+1)\theta_0 \ll 1$,

$$R_{ns}^{(1)}(\theta_0) = R_{sn}^{(1)}(\theta_0) =$$

$$\frac{(2n+1)(2s+1)}{6\pi}\theta_0^3\left\{1 + O\left((2n+1)^2\theta_0^2, (2s+1)^2\theta_0^2\right)\right\} \qquad (4.41)$$

and

$$Q_{ns}^{(1)}(\theta_0) = Q_{sn}^{(1)}(\theta_0) = \frac{2}{45\pi}n(n+1)s(s+1)\theta_0^5 + O\left(n^3s^3\theta_0^7\right). \qquad (4.42)$$

Thus, for small apertures $(\theta_0 \ll 1)$, the following series in (4.39) may be approximated as

$$\sum_{\substack{n=1 \\ n \neq s}}^{\infty} \frac{n(n+1)}{2n+1} \frac{R_{ns}^{(1)}(\theta_0) R_{sn}^{(1)}(\theta_0)}{\left| \zeta_n' \left(\varkappa_{sl}^{(0)} \right) \right|^2} \simeq \frac{(2s+1)^2}{36\pi^2} \theta_0^6 \sum_{\substack{n=1 \\ n \neq s}}^{\infty} \frac{n(n+1)(2n+1)}{\left| \zeta_n' \left(\varkappa_{sl}^{(0)} \right) \right|^2},$$

$$(4.43)$$

and hence

$$Q_{ls0}^{TM} \sim \theta_0^{-6}. \tag{4.44}$$

This result is of restricted interest. To obtain accurate values of the Q-factor for larger apertures, it is necessary to find the complex roots of the full characteristic equation, that may by approximated by the matrix equation of order N (N must be chosen sufficiently large),

$$\det \{I_N - E + A\} = 0 \tag{4.45}$$

where I_N is the identity matrix of order N, E is the diagonal matrix diag $(\varepsilon_1, \varepsilon_2, \varepsilon_3, \ldots, \varepsilon_N)$, and A is the square matrix

$$A = \begin{pmatrix} \varepsilon_1 R_{11}^{(1)} & \varepsilon_2 R_{21}^{(1)} & \varepsilon_3 R_{31}^{(1)} & \cdots & \varepsilon_N R_{N1}^{(1)} \\ \varepsilon_1 R_{12}^{(1)} & \varepsilon_2 R_{22}^{(1)} & \varepsilon_3 R_{32}^{(1)} & \cdots & \varepsilon_N R_{N2}^{(1)} \\ \varepsilon_1 R_{13}^{(1)} & \varepsilon_2 R_{23}^{(1)} & \varepsilon_3 R_{33}^{(1)} & \cdots & \varepsilon_N R_{N3}^{(1)} \\ \cdots & \cdots & \cdots & \cdots & \cdots \\ \varepsilon_1 R_{1N}^{(1)} & \varepsilon_2 R_{2N}^{(1)} & \varepsilon_3 R_{3N}^{(1)} & \cdots & \varepsilon_N R_{NN}^{(1)} \end{pmatrix}. \tag{4.46}$$

Formula (4.40) defines the so-called *unloaded* Q-factor of an open spherical resonator. The *loaded* Q-factor is defined as follows. Consider an oscillatory system that is excited by an external source; let W be the energy stored by the system, and δW be the radiated energy that is irrevocably lost by the oscillating system. Consider the value

$$Q = 2\pi \frac{W}{\delta W} \tag{4.47}$$

as a function of all variables that describe the oscillating system. At resonance, this value is the Q-factor of this resonant oscillation. At the resonance point it has a (locally) maximal value, but in the vicinity its values follow a bell-shaped form. It is not difficult to justify that

$$\delta W = P_0 \cdot R \cdot T \tag{4.48}$$

where P_0 is the power radiated by the source (possibly a dipole) in free space, R the relative radiation resistance (see [4.29]) and T the oscillation

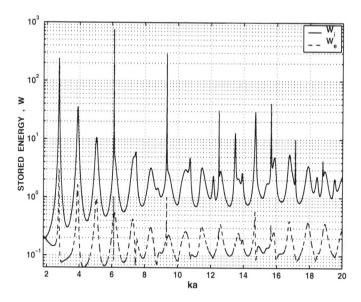

FIGURE 4.2. Internal (solid) and external (dashed) stored energy for cavity $(\theta = 30°)$ excited by a electric dipole located in $z < 0$ with $q = d/a = 0.9$.

period. By elementary algebra the loaded Q-factor of the spherical open resonator is given by

$$Q_{ls0}^{TM} = \frac{3}{2}ka \cdot W_{TM}/R. \qquad (4.49)$$

This equation may be used for systematic studies of resonant phenomena arising from the excitation of metallic thin-walled spherical shells with circular apertures.

By way of illustration, the frequency dependence of the stored internal and external energies ($W_{TM}^{(i)}$ and $W_{TM}^{(e)}$), respectively, is presented in Figure 4.2 for the open spherical resonator with parameter $\theta_0 = 30°$ excited by a vertical electric dipole located in the lower half-space ($z < 0$) with $q = d/a = 0.9$. The energy stored outside the resonator is very small compared to that stored internally. The value of $q_{TM}^{(s)}$, that measures the energy fraction associated with the s-th harmonic, is displayed in Figure 4.3 for $s = 1, 2$. It shows that at least for the first three resonances (solid line) the energy contribution of the first harmonic exceeds 95%.

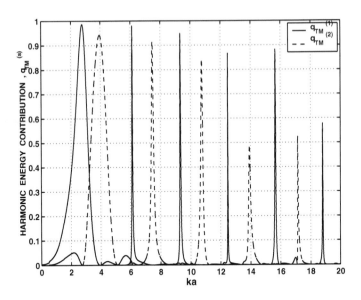

FIGURE 4.3. Excitation of the cavity ($\theta = 30°$) by a electric dipole located in $z < 0$ with $q = d/a = 0.9$. First (solid) and second (dashed) harmonic terms.

4.1.2   The Vertical Magnetic Dip ole (TE Case)

Let us now consider the vertical magnetic dipole replacing the vertical electric dipole described at the beginning of Section 4.1. Taking into account the continuity condition and the far-field behaviour of the scattered field, we decompose the total field as the sum

$$E_\phi = E_\phi^{(0)} + E_\phi^s,$$

and seek the scattered field in the form

$$E_\phi^s = -\frac{mk^2}{r} \sum_{n=1}^{\infty} y_n P_n^1 (\cos \theta) \begin{cases} \zeta_n (ka) \psi_n (kr), & r < a \\ \psi_n (ka) \zeta_n (kr), & r > a \end{cases} \qquad (4.50)$$

where the unknown coefficients y_n are to be found. The finite energy condition (4.6) effectively defined the solution class for the coefficient sequence $\{y_n\}_{n=1}^{\infty}$. Taking the integration region in (4.6) to be the sphere of radius a, it is easily deduced that

$$W = W_{TE} = 2\pi m^2 k^4 a \times$$

$$\sum_{n=1}^{\infty} \frac{n(n+1)}{2n+1} |y_n|^2 |\zeta_n (ka)|^2 \left\{ \psi_n'^2 (ka) - \left[\frac{n(n+1)}{(ka)^2} - 1 \right] \psi_n^2 (ka) \right\}$$

$$(4.51)$$

must be finite. Using the asymptotic behaviour of the spherical Bessel functions and their derivatives (see equations [4.10]) it can be readily shown that

$$W_{TE} \leq C_2 (ka) \sum_{n=1}^{\infty} |y_n|^2 \qquad (4.52)$$

where $C_2 (ka)$ is a function of ka alone, with finite value. Thus the coefficient sequence $\{y_n\}_{n=1}^{\infty}$ is square summable and belongs to the functional space l_2.

Now enforce the boundary conditions (4.3)–(4.4), bearing in mind the explicit form of the scattered field (4.50), the incident field (1.213) and the relations between the basic field component E_φ and the other field components (1.90). As a result, one obtains the following dual series equations.

$$\sum_{n=1}^{\infty} y_n P_n^1 (\cos \theta) = 0, \qquad \theta \in (0, \theta_0) \qquad (4.53)$$

$$\sum_{n=1}^{\infty} y_n \psi_n (ka) \zeta_n (ka) P_n^1 (\cos \theta) =$$

$$- (kd)^{-2} \sum_{n=1}^{\infty} (2n + 1) \psi_n (kd) \zeta_n (ka) P_n^1 (\cos \theta) , \ \theta \in (\theta_0, \pi) . \quad (4.54)$$

The first step is to identify a suitable asymptotically small parameter. Thus define

$$\mu_n = 1 - \frac{i}{ka} \frac{4n (n + 1)}{2n + 1} \psi_n (ka) \zeta_n (ka) . \qquad (4.55)$$

It is readily verified from the asymptotics (4.10) that $\mu_n = O\left(n^{-2}\right)$, as $n \to \infty$. The dual series equations (4.53)–(4.54) may be rearranged as

$$\sum_{n=1}^{\infty} y_n P_n^1 (\cos \theta) = 0, \ \theta \in (0, \theta_0) \qquad (4.56)$$

$$\sum_{n=1}^{\infty} \frac{2n + 1}{n (n + 1)} (1 - \mu_n) y_n P_n^1 (\cos \theta)$$

$$= \sum_{n=1}^{\infty} \frac{2n + 1}{n (n + 1)} \beta_n P_n^1 (\cos \theta) , \ \theta \in (\theta_0, \pi) \quad (4.57)$$

where

$$\beta_n = -i \frac{4}{ka\,(kd)^2} n\,(n+1)\,\psi_n\,(kd)\,\zeta_n\,(ka)\,. \tag{4.58}$$

Because the unknown coefficient sequence is square summable, it can be readily estimated that the general term of series (4.56) and (4.57) decay at rates $O\left(n^{-\frac{1}{2}}\right)$ and $O\left(n^{-\frac{3}{2}}\right)$, respectively, as $n \to \infty$. Thus the first series converges uniformly, whilst the second converges nonuniformly. The difficulty is circumvented in the same way as for the VED case. Set

$$P_n^1\,(\cos\theta) = -\frac{d}{d\theta} P_n\,(\cos\theta)$$

and then validly integrate term by term (because $\{y_n\}_{n=1}^{\infty} \in l_2$) to obtain the following uniformly converging series to which Abel integral transforms may be applied,

$$\sum_{n=1}^{\infty} y_n\,P_n\,(\cos\theta) = C_2,\ \theta \in (0,\theta_0) \tag{4.59}$$

where C_2 is a constant of integration.

We now use the integral representations (4.19) of Mehler-Dirichlet kind to obtain the equivalent pre-regularised form

$$\sum_{n=1}^{\infty} y_n \cos\left(n+\frac{1}{2}\right)\theta = \begin{cases} C_2 \cos\frac{1}{2}\theta, & \theta \in (0,\theta_0) \\ \sum_{n=1}^{\infty} (\beta_n + \mu_n y_n) \cos\left(n+\frac{1}{2}\right)\theta, & \theta \in (\theta_0,\pi)\,. \end{cases} \tag{4.60}$$

The constant C_2 formally arising as the result of integration has an interpretation that parallels that given to the constant C_1 in the previous section. The original dual series equations were series expansions in the complete orthogonal family of functions $P_n^1\,(\cos\theta)$ $(n = 1, 2, ...)$; the transformed equations are series expansions in another complete orthogonal family of functions, the trigonometric functions $\left\{\cos\left(n+\frac{1}{2}\right)\theta\right\}_{n=0}^{\infty}$; the unknown coefficients y_n are associated with an incomplete subset of this family (indexed by $n = 1, 2, ...$); and the constant C_2 is associated with the zero index element $(\cos\frac{1}{2}\theta)$ that makes this family of functions complete.

Making use of orthogonality, the equations (4.60) are easily transformed to the i.s.l.a.e. of the second kind

$$(1 - \mu_m) y_m + \sum_{n=1}^{\infty} y_n \mu_n Q_{nm}^{(1)}\,(\theta_0) = \beta_m - \sum_{n=1}^{\infty} \beta_n Q_{nm}^{(1)}\,(\theta_0)\,, \tag{4.61}$$

where $m = 1, 2, \ldots$ and $Q_{nm}^{(1)}(\theta_0)$ was defined in the previous section.

Introducing the angle $\theta_1 = \pi - \theta_0$ leads to the alternative and more compact form (recall [4.24])

$$Y_m - \sum_{n=1}^{\infty} Y_n \mu_n R_{nm}^{(1)}(\theta_1) = \sum_{n=1}^{\infty} B_n R_{nm}^{(1)}(\theta_1), \qquad (4.62)$$

where $m = 1, 2, \ldots$ and

$$Y_m = (-1)^m y_m, \ B_m = (-1)^m \beta_m. \qquad (4.63)$$

Equations (4.61) or (4.62) remain valid when $d < a$ and the dipole is located in upper half-space $z > 0$, provided the coefficients β_n are replaced by

$$\beta_n = -i \frac{4}{ka(kd)^2} n(n+1) \zeta_n(kd) \psi_n(ka). \qquad (4.64)$$

If the dipole is located in lower half-space ($z < 0$) then the values of β_n, defined by (4.58) or (4.64), are modified by a simple multiplication with the factor $(-1)^{n-1}$.

Let us collect the near-field and far-field characteristics. The *surface current density* has one component and is equal to the jump in the magnetic field intensity across the shell,

$$j_\phi = -\{H_\theta(a-0,\theta) - H_\theta(a+0,\theta)\} = -\frac{mk^2}{a} \sum_{n=1}^{\infty} y_n P_n^1(\cos\theta). \qquad (4.65)$$

The *radiation pattern* S_2 defined by $E_\theta, H_\phi \backsim S_2(\theta) e^{ikr}/r$ as $r \to \infty$ equals

$$S_2(\theta) = mk^2 \sum_{n=1}^{\infty} (2n+1) \left\{ \frac{\psi_n(kd)}{(kd)^2} + \frac{\psi_n(ka)}{2n+1} y_n \right\} (-1)^{n-1} P_n^1(\cos\theta). \qquad (4.66)$$

The *radiation resistance* equals

$$R = \frac{3}{2} \sum_{n=1}^{\infty} n(n+1)(2n+1) \left| \frac{\psi_n(kd)}{(kd)^2} + \frac{\psi_n(ka)}{2n+1} y_n \right|^2. \qquad (4.67)$$

Finally the stored external energy (see the discussion in the previous section) is

$$
W_{TE}^{(e)} = 2\pi m^2 k^4 a \sum_{n=1}^{\infty} \frac{n(n+1)}{2n+1} |y_n|^2 \, \psi_n^2 (ka) \times
$$

$$
\left\{ 2 + \left[\frac{n(n+1)}{(ka)^2} - 1 \right] |\zeta_n (ka)|^2 - |\zeta_n' (ka)|^2 \right\}.
$$

The total stored energy is the sum of scattered energy that is stored in the internal $(r < a)$ and external $(r > a)$ regions

$$
W_{TE} = W_{TE}^i + W_{TE}^{(e)}, \tag{4.68}
$$

where the value of W_{TE}^i is defined by formula (4.51).

In analysing high Q-factor oscillations that develop in this open spherical resonator it we may decompose, as in the TM case,

$$
W_{TE} = \sum_{n=1}^{\infty} W_{TE}^{(n)} \tag{4.69}
$$

into a sum of terms representing the normalised energy contribution of each harmonic, and to examine the ratios

$$
q_{TE}^{(n)} = \frac{W_{TE}^{(n)}}{W_{TE}}. \tag{4.70}
$$

Before presenting numerical results let us consider analytical features of the solution obtained above. First, a careful look at the expressions for surface current density (4.65) reveals that general terms of the series decay at the rate $O\left(n^{-\frac{1}{2}}\right)$ as $n \to \infty$, i.e., the series is slowly converging, and at a rate that is rather slower than the corresponding series for the VED.

Acceleration of convergence is achieved by rearrangement of the i.s.l.a.e. (4.62) in the form

$$
Y_n = \sum_{s=1}^{\infty} (Y_s \mu_s + B_s) R_{sn}^{(1)} (\theta_1). \tag{4.71}
$$

Setting $Y_n = (-1)^n y_n$ into formula (4.65) and interchanging the order of summation produces the computational formula for surface current density

$$
j_\varphi = \frac{mk^2}{a} \sum_{s=1}^{\infty} (Y_s \mu_s + B_s) L_s (\vartheta, \theta_1), \tag{4.72}
$$

where $\vartheta = \pi - \theta$ and

$$L_s \left(\vartheta, \theta_1 \right) = \sum_{n=1}^{\infty} R_{sn}^{(1)} \left(\theta_1 \right) P_n^1 \left(\cos \vartheta \right). \tag{4.73}$$

This function may be expressed in the form

$$L_s \left(\vartheta, \theta_1 \right) = \begin{cases} -F_s \left(\vartheta, \theta_1 \right) - \frac{R_{s0}(\theta_1)}{1 - R_{00}(\theta_1)} F_0 \left(\vartheta, \theta_1 \right), \vartheta \in (0, \theta_1) \\ 0, \qquad\qquad\qquad\qquad\qquad\qquad \vartheta \in (\theta_1, \pi) \end{cases} \tag{4.74}$$

where

$$F_s \left(\vartheta, \theta_1 \right) = \frac{\partial}{\partial \vartheta} G_s \left(\vartheta, \theta_1 \right) \tag{4.75}$$

and G_s was defined by (2.178). The function $F_s \left(\vartheta, \theta_1 \right)$ may be easily computed from the recurrence formula

$$F_s \left(\vartheta, \theta_1 \right) = \frac{2}{\pi s} \sin \vartheta \frac{\cos \left(s - \frac{1}{2} \right) \theta_1}{\sqrt{2 \left(\cos \vartheta - \cos \theta_1 \right)}} - \frac{2s - 1}{s} \sin \vartheta G_{s-1} \left(\vartheta, \theta_1 \right)$$

$$+ \frac{2s - 1}{s} \cos \vartheta F_{s-1} \left(\nu, \theta_1 \right) - \frac{s - 1}{s} F_{s-2} \left(\vartheta, \theta_1 \right), \tag{4.76}$$

valid for $s = 1, 2, \ldots$, initialised by the elementary expression for F_0

$$F_0 \left(\vartheta, \theta_1 \right) = -\frac{2}{\pi} \frac{\tan \left(\frac{1}{2} \vartheta \right)}{\sqrt{2 \left(\cos \vartheta - \cos \theta_1 \right)}} \times$$

$$\frac{\sin \theta_1 \left\{ \sin \frac{1}{2} \theta_1 + \sqrt{2 \left(\cos \vartheta - \cos \theta_1 \right)} \right\} + \cos \frac{1}{2} \theta_1 \left(\cos \vartheta - \cos \theta_1 \right)}{1 + \cos \vartheta - 2 \cos \theta_1 + 2 \sin \frac{1}{2} \theta_1 \sqrt{2 \left(\cos \vartheta - \cos \theta_1 \right)}}. \tag{4.77}$$

The convergence rate of the modified series (4.72) is $O \left(s^{-\frac{5}{2}} \right)$ as $s \to \infty$. More importantly, the vanishing of $L_s \left(\vartheta, \theta_1 \right)$ when $\vartheta \in (\theta_1, \pi)$ shows that the boundary condition (4.3) is satisfied term by term; using (4.76) and (4.77), analysis of the form (4.74) of $L_s \left(\vartheta, \theta_1 \right)$ in the interval $\vartheta \in (0, \theta_1)$ reveals that the surface current j_φ has the expected singular behaviour as $\theta \to \theta_0$ (or $\vartheta \to \theta_1$): each term in (4.72) has the correct behaviour (a singularity of form $(\cos \vartheta - \cos \theta_1)^{-\frac{1}{2}}$) in the vicinity of the sharp edge. Thus accurate calculation of surface current density distribution is facilitated by this transformation to a much more rapidly convergent series.

Now consider the quasi-eigenoscillations that develop in the open spherical cavity with a small aperture ($\theta_0 \ll 1$) when the excitation frequency coincides with one of the quasi-eigenvalues of spectrum ν_{sl}. The spectrum

$s = 1$	2	3	4	5	6	7
$l = 1$ 4.493	5.763	6.988	8.183	9.356	10.513	11.657
2 7.725	9.095	10.417	11.705	12.967	14.207	15.431
3 10.90	12.323	13.698	15.040	16.355	17.648	18.923

TABLE 4.2. TE$_{ls0}$-oscillations, spectral values $\nu_{sl}^{(0)}$

of eigenvalues of quasi-eigenoscillations in the *closed* cavity (of TE$_{ls0}$-type) is determined by the roots $\nu_{sl}^{(0)}$ $(l = 1, 2, \ldots)$ of the characteristic equation

$$\psi_s\left(\nu_{sl}^{(0)}\right) = 0, \tag{4.78}$$

where $s = 1, 2, \ldots$; for each s, the roots are indexed in increasing order by l. The indices s and l designate the number of field variations in the radial and angular variables (r and θ), respectively. Some lower order roots are shown in Table 4.2.

Following the same idea used in Chapters 2 and 3, the perturbation to the spectral value $\nu_{sl}^{(0)}$ caused by a small aperture (with $\theta_0 \ll 1$) may be analytically calculated:

$$\nu_{sl}/\nu_{sl}^{(0)} = 1 - \frac{2s+1}{4s\,(s+1)} Q_{ss}^{(1)}\,(\theta_0) -$$

$$i\frac{2s+1}{16s\,(s+1)}\nu_{sl}^{(0)} \sum_{\substack{n=1 \\ n \neq s}}^{\infty} \frac{2n+1}{n\,(n+1)} \frac{Q_{ns}^{(1)}\,(\theta_0)\,Q_{sn}^{(1)}\,(\theta_0)}{\left|\zeta_n\left(\nu_{sl}^{(0)}\right)\right|^2}. \tag{4.79}$$

Thus $\mathrm{Re}\,(\nu_{sl}) < \nu_{sl}^{(0)}$ and $\mathrm{Im}\,(\nu_{sl}) < 0$, indicating radiative losses with Q-factor

$$Q_{ls0}^{TE} = -2\frac{\mathrm{Re}\,(\nu_{sl})}{\mathrm{Im}\,(\nu_{sl})} \tag{4.80}$$

of finite value.

Providing that $(2n + 1)\,\theta_0 \ll 1$ and $(2s + 1)\,\theta_0 \ll 1$, the angle functions $R_{ns}^{(1)}\,(\theta_0)$ and $Q_{ns}^{(1)}\,(\theta_0)$ may be approximated by formulae (4.41)–(4.41), and so for small apertures ($\theta_0 \ll 1$), the following series in (4.79) may be approximated

$$\sum_{\substack{n=1 \\ n \neq s}}^{\infty} \frac{2n+1}{n\,(n+1)} \frac{Q_{ns}^{(1)}\,(\theta_0)\,Q_{sn}^{(1)}\,(\theta_0)}{\left|\zeta_n\left(\nu_{sl}^{(0)}\right)\right|^2} \simeq \frac{4s^2\,(s+1)^2}{2025\pi^2} \sum_{\substack{n=1 \\ n \neq s}}^{\infty} \frac{n\,(n+1)\,(2n+1)}{\left|\zeta_n\left(\nu_{sl}^{(0)}\right)\right|^2} \theta_0^{10}$$

$$\tag{4.81}$$

and hence

$$Q_{ls0}^{TE} \sim \theta_0^{-10}. \tag{4.82}$$

The above formula is of restricted interest. For accurate values of the Q-factor for larger apertures, it is necessary to find the complex roots of the full characteristic equation, that may by approximated by the matrix equation of order N (N must be chosen sufficiently large),

$$\det \{I_N - M + B\} = 0 \tag{4.83}$$

where I_N is the identity matrix of order N, M is the diagonal matrix $\mathrm{diag}\,(\mu_1, \mu_2, \mu_3, \dots, \mu_N)$, and B is the square matrix

$$B = \begin{pmatrix} \mu_1 Q_{11}^{(1)} & \mu_2 Q_{21}^{(1)} & \mu_3 Q_{31}^{(1)} & \cdots & \mu_N Q_{N1}^{(1)} \\ \mu_1 Q_{12}^{(1)} & \mu_2 Q_{22}^{(1)} & \mu_3 Q_{32}^{(1)} & \cdots & \mu_N Q_{N2}^{(1)} \\ \mu_1 Q_{13}^{(1)} & \mu_2 Q_{23}^{(1)} & \mu_3 Q_{33}^{(1)} & \cdots & \mu_N Q_{N3}^{(1)} \\ \cdots & \cdots & \cdots & \cdots & \cdots \\ \mu_1 Q_{1N}^{(1)} & \mu_2 Q_{2N}^{(1)} & \mu_3 Q_{3N}^{(1)} & \cdots & \mu_N Q_{NN}^{(1)} \end{pmatrix}. \tag{4.84}$$

Finally, in parallel with equation (4.49), the loaded Q-factor of a spherical open resonator is given by

$$Q_{ls0}^{TE} = \frac{3}{2}ka \cdot W_{TE}/R. \tag{4.85}$$

By way of illustration, the frequency dependence of the stored internal and external energies ($W_{TE}^{(i)}$ and $W_{TE}^{(e)}$), respectively, is presented in Figure 4.4 for the open spherical resonator with parameter $\theta_0 = 30°$ excited by a vertical magnetic dipole located in the lower half-space ($z < 0$) with $q = d/a = 0.9$. As we have already seen in the electric dipole case, the energy stored outside the resonator is very small compared to that stored internally. The value of $q_{TE}^{(s)}$, ($s = 1, 2$), is displayed in Figure 4.5, showing that at least for the first three resonances (solid line) the energy contribution of the first harmonics exceeds 95%.

4.2   Plane Wave Diffraction from a Circular Hole in a Thin Metallic Sphere.

The previous section considered vertical dipole excitation of the spherical cavity with a circular hole. If the dipole is located along the vertical axis of symmetry, only TM or TE waves are excited depending upon whether

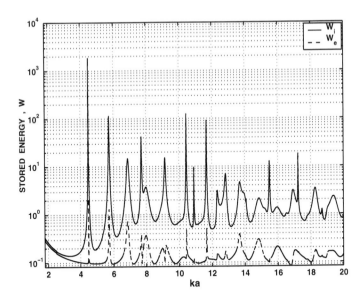

FIGURE 4.4. Internal (solid) and external (dashed) stored energy for cavity ($\theta = 30°$) excited by a magnetic dipole located in $z < 0$ with $q = d/a = 0.9$.

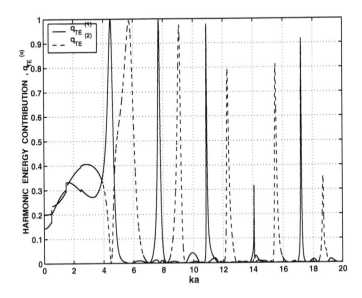

FIGURE 4.5. Harmonic energy fractions for dipole excited cavity of Figure 4.4.

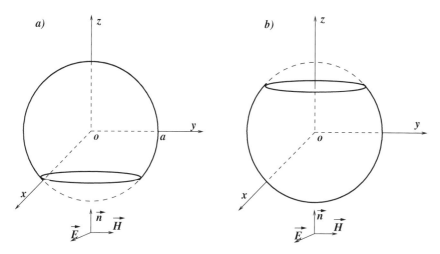

FIGURE 4.6. The spherical cavity illuminated by a plane wave a) at normal incidence $\alpha = 0$ and b) with $\alpha = \pi$.

the dipole is of electric or magnetic type. However, when the cavity is illuminated by an electromagnetic plane wave the TM and TE waves are coupled even if the wave propagates normally to the aperture plane. This more complex cavity scattering scenario is addressed in this section (see also [99], [111]).

Consider then a thin metallic sphere with a circular hole irradiated by a plane electromagnetic wave, as shown in Figure 4.6a. The z-axis is normal to the aperture plane and the direction of propagation coincides with positive z-axis. The incident field is described by

$$E_x^0 = -H_y^0 = \exp(ikz) = \exp(ikr\cos\theta) \tag{4.86}$$

where we have employed the symmetrised form of Maxwell's equations (1.58)–(1.59).

According to (1.243) the associated electric (U^0) and magnetic (V^0) Debye potentials are

$$\left\{ \begin{array}{c} U^0 \\ V^0 \end{array} \right\} = \left\{ \begin{array}{c} \cos\varphi \\ \sin\varphi \end{array} \right\} \frac{1}{ik^2 r} \sum_{n=1}^{\infty} A_n \psi_n(kr) P_n^1(\cos\theta), \tag{4.87}$$

where

$$A_n = i^n \frac{2n+1}{n(n+1)}.$$

Debye potentials (U^s, V^s) associated with the scattered field $\overrightarrow{E^s}, \overrightarrow{H^s}$ must satisfy the Helmholtz equation. Furthermore, the total field must satisfy appropriate boundary conditions (see Chapter 1), and the scattered field must satisfy the radiation conditions

$$E_\theta^s, H_\varphi^s \backsim \frac{e^{ikr}}{r} S_1(\theta, \phi), \quad H_\theta^s, E_\varphi^s \backsim \frac{e^{ikr}}{r} S_2(\theta, \phi) \qquad (4.88)$$

as $kr \to \infty$, where the radiation patterns $S_1(\theta, \phi) = S_1(\theta) \cos \phi$ and $S_2(\theta, \phi) = S_2(\theta) \sin \phi$ take a rather simple form.

Finally the energy W of the scattered field in any finite but arbitrary volume of space incorporating the edges of the scatterer must be bounded (see 1.287):

$$W = \frac{1}{2} \iiint_V \left\{ \left| \overrightarrow{E}^s \right|^2 + \left| \overrightarrow{H}^s \right|^2 \right\} dV < \infty, \qquad (4.89)$$

where $dV = r^2 \sin \theta dr d\theta d\phi$ is the volume element in spherical coordinates.

In the interior region $(r < a)$, we seek Debye potentials for the interior electromagnetic field $\overrightarrow{E}^i, \overrightarrow{H}^i$ in the form

$$U^i = \frac{\cos \phi}{ik^2 r} \sum_{n=1}^{\infty} A_n x_n^i \psi_n(kr) P_n^1(\cos \theta) \qquad (4.90)$$

$$V^i = \frac{\sin \phi}{ik^2 r} \sum_{n=1}^{\infty} A_n y_n^{(i)} \psi_n(kr) P_n^1(\cos \theta) \qquad (4.91)$$

where $x_n^{(i)}, y_n^{(i)}$ are unknown coefficients to be determined, whilst in the exterior region $(r > a)$, we seek Debye potentials for the exterior electromagnetic field $\overrightarrow{E}^e, \overrightarrow{H}^e$ in the form

$$U^e = U^0 + \frac{\cos \phi}{ik^2 r} \sum_{n=1}^{\infty} A_n x_n^{(e)} \zeta_n(kr) P_n^1(\cos \theta) \qquad (4.92)$$

$$V^e = V^0 + \frac{\sin \phi}{ik^2 r} \sum_{n=1}^{\infty} A_n y_n^{(e)} \zeta_n(kr) P_n^1(\cos \theta) \qquad (4.93)$$

where $x_n^{(e)}, y_n^{(e)}$ are unknown coefficients to be determined.

The mixed boundary conditions on the surface $r = a$ for the tangential components of the field require vanishing of the tangential electric field on the metal,

$$\overrightarrow{i_r} \times \overrightarrow{E^i} = \overrightarrow{i_r} \times \overrightarrow{E^e} = \overrightarrow{0}, \qquad (4.94)$$

for $\theta \in (0, \theta_0)$ and $\varphi \in (0, 2\pi)$, and continuity of the tangential magnetic field across the aperture

$$\overrightarrow{i_r} \cdot \overrightarrow{H^i} = \overrightarrow{i_r} \cdot \overrightarrow{H^e} \qquad (4.95)$$

for $\theta \in (0, \theta_0)$ and $\varphi \in (0, 2\pi)$. In fact, the tangential electric field is continuous over the complete spherical surface $r = a$,

$$\overrightarrow{i_r} \times \overrightarrow{E^i} = \overrightarrow{i_r} \times \overrightarrow{E^e},$$

and it is readily deduced that

$$x_n^{(e)} = \frac{\psi_n'(ka)}{\zeta_n'(ka)} \left(x_n^{(i)} - 1 \right), \qquad (4.96)$$

$$y_n^{(e)} = \frac{\psi_n(ka)}{\zeta_n(ka)} \left(y_n^{(i)} - 1 \right). \qquad (4.97)$$

The mixed boundary conditions (4.94) and (4.95) impose equivalent conditions on the Debye potentials and their normal derivatives (on the surface $r = a$), and their enforcement produces the following coupled sets of dual series equations for the unknowns $x_n^{(i)}, y_n^{(i)}$:

$$\sum_{n=1}^{\infty} A_n x_n^{(i)} \psi_n'(ka) P_n^1(\cos\theta) = B_1 \tan\frac{\theta}{2}, \quad \theta \in (0, \theta_0) \qquad (4.98)$$

$$\sum_{n=1}^{\infty} A_n \left(x_n^{(i)} - 1 \right) \frac{P_n^1(\cos\theta)}{\zeta_n'(ka)} = B_3 \cot\frac{\theta}{2}, \quad \theta \in (\theta_0, \pi) \qquad (4.99)$$

$$\sum_{n=1}^{\infty} A_n y_n^{(i)} \psi_n(ka) P_n^1(\cos\theta) = iB_1 \tan\frac{\theta}{2}, \quad \theta \in (0, \theta_0) \qquad (4.100)$$

$$\sum_{n=1}^{\infty} A_n \left(y_n^{(i)} - 1 \right) \frac{P_n^1(\cos\theta)}{\zeta_n(ka)} = -iB_3 \cot\frac{\theta}{2}, \quad \theta \in (\theta_0, \pi) \quad (4.101)$$

where B_1, B_3 are the *polarisation coupling constants*.

In contrast to the approach adopted in the previous section we carry out a regularisation process that reduces to the exact analytical solution of the corresponding static problems (when $ka = 0$). The definition of the parameter μ_n given by (4.55) is replaced by

$$\mu_n = 1 - \frac{i}{ka}(2n+1)\psi_n(ka)\zeta_n(ka), \qquad (4.102)$$

but the definition of ε_n (4.14) is retained. From the asymptotics (4.10),

$$\lim_{ka \to 0} \mu_n = \lim_{ka \to 0} \varepsilon_n = 0 \qquad (4.103)$$

for all n. With these definitions, and the rescaling

$$x_n^{(i)} = (-i)^n \zeta_n'(ka) x_n, \quad y_n^{(i)} = (-i)^n \zeta_n(ka) y_n, \qquad (4.104)$$

the equations (4.19) may be written

$$\sum_{n=1}^{\infty} (1 - \varepsilon_n) x_n P_n^1 (\cos \theta) = -ikaB_1 \tan \frac{\theta}{2}, \quad \theta \in (0, \theta_0) \qquad (4.105)$$

$$\sum_{n=1}^{\infty} \frac{2n+1}{n(n+1)} \left(x_n - \frac{i^n}{\zeta_n'(ka)} \right) P_n^1 (\cos \theta) = B_3 \cot \frac{\theta}{2}, \quad \theta \in (\theta_0, \pi) \quad (4.106)$$

$$\sum_{n=1}^{\infty} \frac{y_n}{n(n+1)} (1 - \mu_n) P_n^1 (\cos \theta) = -\frac{1}{ka} B_1 \tan \frac{\theta}{2}, \quad \theta \in (0, \theta_0) \qquad (4.107)$$

$$\sum_{n=1}^{\infty} \frac{2n+1}{n(n+1)} \left(y_n - \frac{i^n}{\zeta_n(ka)} \right) P_n^1 (\cos \theta) = -iB_3 \cot \frac{\theta}{2}, \quad \theta \in (\theta_0, \pi).$$
$$(4.108)$$

The next step is to identify the solution class of the coefficient sequences $\{x_n\}_{n=1}^{\infty}, \{y_n\}_{n=1}^{\infty}$ imposed by the finite energy condition. Evaluation of the integral (4.89) produces

$$W = \frac{\pi}{k^3} ka \sum_{n=1}^{\infty} (2n+1) \left\{ |x_n|^2 |\zeta_n'(ka)|^2 + |y_n|^2 |\zeta_n(ka)|^2 \right\} \times$$

$$\times \left\{ \psi_n'^2 (ka) - \left[\frac{n(n+1)}{(ka)^2} - 1 \right] \psi_n^2 (ka) \right\} \qquad (4.109)$$

so that

$$\sum_{n=1}^{\infty} (2n+1) |x_n|^2 |\zeta_n'(ka)|^2 \left\{ \psi_n'^2 (ka) - \left[\frac{n(n+1)}{(ka)^2} - 1 \right] \psi_n^2 (ka) \right\} < \infty$$
$$(4.110)$$

and

$$\sum_{n=1}^{\infty} (2n+1) |y_n|^2 |\zeta_n(ka)|^2 \left\{ \psi_n'^2 (ka) - \left[\frac{n(n+1)}{(ka)^2} - 1 \right] \psi_n^2 (ka) \right\} < \infty.$$
$$(4.111)$$

The asymptotics (4.10) now show that

$$\{x_n\}_{n=1}^{\infty} \in l_2 (2), \quad \{y_n\}_{n=1}^{\infty} \in l_2 (0) \equiv l_2. \qquad (4.112)$$

The standard process of regularisation may now be applied. Suppressing the details (see also [111]), one obtains the coupled i.s.l.a.e. of the second kind for the unknowns x_n, y_n,

$$x_m - \sum_{n=1}^{\infty} x_n \varepsilon_n c_{nm} - i \sum_{n=1}^{\infty} y_n \mu_n d_{nm} = \sum_{n=1}^{\infty} \alpha_n (\delta_{nm} - c_{nm}) - i \sum_{n=1}^{\infty} \beta_n d_{nm},$$

(4.113)

$$y_m - \sum_{n=1}^{\infty} y_n \mu_n t_{nm} - i \sum_{n=1}^{\infty} x_n \varepsilon_n p_{nm} = \sum_{n=1}^{\infty} \beta_n (\delta_{nm} - t_{nm}) - i \sum_{n=1}^{\infty} \alpha_n p_{nm},$$

(4.114)

where $m = 1, 2, \ldots$, and

$$\alpha_n = \frac{i^n}{\zeta_n' (ka)},$$

$$\beta_n = \frac{i^n}{\zeta_n (ka)},$$

$$c_{nm} = Q_{nm}^* (\theta_0) - ka \frac{1 - \theta_{00} (\theta_0)}{q (ka, \theta_0)} Q_{n0}^* (\theta_0) V_m (\theta_0),$$

$$d_{nm} = ka \frac{V_m (\theta_0)}{q (ka, \theta_0)} Q_{n0} (\theta_0),$$

$$p_{nm} = \frac{Q_{0m} (\theta_0)}{ka \cdot q (ka, \theta_0)} Q_{n0}^* (\theta_0),$$

$$t_{nm} = Q_{nm} (\theta_0) + \frac{1 + (ka)^2 V_0 (\theta_0)}{ka \cdot q (ka, \theta_0)} Q_{n0} (\theta_0) Q_{0m} (\theta_0),$$

$$V_m = \theta_0 \tan \frac{\theta_0}{2} Q_{0m} (\theta_0) - W_m (\theta_0),$$

$$q (ka, \theta_0) = ka [1 - Q_{00} (\theta_0)] V_0 (\theta_0) - \frac{1}{ka} Q_{00} (\theta_0).$$

(4.115)

In addition,

$$Q_{nm}^* (\theta_0) = Q_{nm} (\theta_0) - \frac{\cos \left(n + \frac{1}{2}\right) \theta_0}{\cos \frac{1}{2} \theta_0} Q_{0m} (\theta_0),$$

$$Q_{nm} (\theta_0) \equiv \hat{Q}_{nm}^{\left(-\frac{1}{2}, \frac{1}{2}\right)} (\cos \theta_0) = \frac{1}{\pi} \left[\frac{\sin (n - m) \theta_0}{n - m} + \frac{\sin (n + m + 1) \theta_0}{n + m + 1} \right]$$

and

$$W_m \left(\theta_0\right) = \frac{\theta_0}{\pi} \left[\frac{\cos m\theta_0}{m} - \frac{\sin \left(m+1\right)\theta_0}{m+1}\right]$$
$$- \frac{1}{\pi}\left[\frac{\sin m\theta_0}{m^2} - \frac{\sin \left(m+1\right)\theta_0}{\left(m+1\right)^2}\right]. \quad (4.116)$$

If the plane wave is incident from the opposite direction (so that the aperture is in shadow, see Figure 4.6b) the i.s.l.a.e. is replaced by

$$x_m^* - \sum_{n=1}^{\infty} x_n^* \varepsilon_n c_{nm} + i \sum_{n=1}^{\infty} y_n^* \mu_n d_{nm} = \sum_{n=1}^{\infty} \alpha_n^* \left(\delta_{nm} - c_{nm}\right) + i \sum_{n=1}^{\infty} \beta_n^* d_{nm},$$
$$(4.117)$$

$$y_m^* - \sum_{n=1}^{\infty} y_n^* \mu_n t_{nm} + i \sum_{n=1}^{\infty} x_n^* \varepsilon_n p_{nm} = \sum_{n=1}^{\infty} \beta_n^* \left(\delta_{nm} - t_{nm}\right) + i \sum_{n=1}^{\infty} \alpha_n^* p_{nm},$$
$$(4.118)$$

where $m = 1, 2, \ldots$, and

$$\{x_m^*, y_m^*, \alpha_m^*, \beta_m^*\} = \left(-1\right)^m \{x_m, y_m, \alpha_m, \beta_m\}. \quad (4.119)$$

The i.s.l.a.e. of the second kind (4.113)–(4.114) and (4.117)–(4.118) are very convenient for deducing approximate analytical expressions in the Rayleigh scattering regime ($ka \ll 1$, i.e., $\lambda \gg a$) that have no restrictions on aperture dimension (i.e., is uniform in the parameter $\theta_0 \in (0, \pi)$). Suppressing the details, the method of successive approximations shows that

$$x_n - \alpha_n = \mp a_{1n} \left(ka\right)^2 - i a_{2n} \left(ka\right)^3 \mp a_{3n} \left(ka\right)^4 + O\left(\left(ka\right)^5\right) \quad (4.120)$$

$$y_n - \beta_n = \mp b_{1n} ka + i b_{2n} \left(ka\right)^2 \mp b_{3n} \left(ka\right)^3 + O\left(\left(ka\right)^4\right) \quad (4.121)$$

where the upper and lower signs refer to the cases shown in Figure 4.6a and Figure 4.6b, respectively; in addition

$$a_{1n} = Q_{1n}^* \left(\theta_0\right), \quad a_{2n} = \frac{1}{6}Q_{2n}^* \left(\theta_0\right) + \frac{Q_{10}\left(\theta_0\right)}{Q_{00}\left(\theta_0\right)} V_n \left(\theta_0\right),$$

$$a_{3n} = \frac{3}{20}Q_{1n}^* \left(\theta_0\right) -$$
$$\frac{\left[1 - Q_{00}\left(\theta_0\right)\right] Q_{10}^* \left(\theta_0\right) - \frac{1}{3}Q_{20}\left(\theta_0\right)}{Q_{00}\left(\theta_0\right)} V_n \left(\theta_0\right) + F_n \left(\theta_0\right),$$

$$b_{1n} = Q_{1n}^{(1)}(\theta_0), \quad b_{2n} = \frac{1}{3}Q_{2n}^{(1)}(\theta_0) + \frac{Q_{10}^*(\theta_0)}{Q_{00}(\theta_0)}Q_{0n}(\theta_0),$$

$$b_{3n} = \frac{1}{10}Q_{1n}^{(1)}(\theta_0) +$$

$$\frac{Q_{10}(\theta_0)V_0(\theta_0) + \frac{1}{6}Q_{20}^*(\theta_0)Q_{00}(\theta_0)}{Q_{00}^2(\theta_0)}Q_{0n}(\theta_0) + G_n(\theta_0),$$

and

$$F_m(\theta_0) = \frac{1}{2}\sum_{n=1}^{\infty}\frac{(2n^2 + 2n + 3)Q_{1n}^*(\theta_0)}{n(n+1)(2n-1)(2n+3)}Q_{nm}^*(\theta_0),$$

$$G_m(\theta_0) = 2\sum_{n=1}^{\infty}\frac{Q_{1n}^{(1)}(\theta_0)Q_{nm}(\theta_0)}{(2n-1)(2n+3)}; \qquad (4.122)$$

the expressions $Q_{nm}^{(1)}(\theta_0)$ are defined by (4.22).

The approximate solution (with $ka \ll 1$) given by formulae (4.120) and (4.121) collapses to the *exact* solution of the corresponding electrostatic and magnetostatic problems at $ka = 0$. The electrostatic U_e and magnetostatic U_m potentials are related to the electrostatic $\overrightarrow{E}$ and magnetostatic $\overrightarrow{H}$ field by

$$\overrightarrow{E} = -\operatorname{grad} U_e, \quad \overrightarrow{H} = -\operatorname{grad} U_m$$

and for the spherical cavity have the expressions

$$U_e = -r\sin\theta\cos\phi +$$

$$a\cos\phi\sum_{n=1}^{\infty}Q_{1n}^*(\theta_0)P_n^1(\cos\theta)\left\{\begin{matrix}(r/a)^n, & r < a \\ (r/a)^{-n-1}, & r > a\end{matrix}\right\}, \qquad (4.123)$$

$$U_m = -r\sin\theta\sin\phi +$$

$$a\sin\phi\sum_{n=1}^{\infty}Q_{1n}^{(1)}(\theta_0)P_n^1(\cos\theta)\left\{\begin{matrix}n^{-1}(r/a)^n, & r < a \\ -(n+1)^{-1}(r/a)^{-n-1}, & r > a\end{matrix}\right\}.$$

$$(4.124)$$

In deducing (4.123) and (4.124) we used the relationship between the static potentials U_e and U_m and the Debye potentials U and V,

$$U_e = -\frac{\partial}{\partial r}(rU), U_m = -\frac{\partial}{\partial r}(rV). \qquad (4.125)$$

Now we show how to avoid the comparatively complicated solution form of (4.113)–(4.114) for the particular case of Rayleigh scattering. For the rest of the frequency band, stretching from low frequencies $(\lambda \gg a)$, through the diffraction (or resonance) regime, up to deep quasi-optics $(\lambda \ll a)$, it is more effective to use a relatively simple algorithm based on these equations, even for calculations at short wavelengths.

To realize this aim, it is convenient to represent the total Debye potentials (4.90)–(4.91) in the alternative form

$$U = U^0 +$$

$$\frac{\cos \phi}{ik^2 r} \sum_{n=1}^{\infty} A_n \left\{ \begin{array}{l} \psi_n' \left(ka\right) \zeta_n \left(kr\right), r > a \\ \zeta_n' \left(ka\right) \psi_n \left(kr\right), r < a \end{array} \right\} x_n P_n^1 \left(\cos \theta\right), \quad (4.126)$$

$$V = V^0 +$$

$$\frac{\sin \phi}{ik^2 r} \sum_{n=1}^{\infty} A_n \left\{ \begin{array}{l} \psi_n \left(ka\right) \zeta_n \left(kr\right), r > a \\ \zeta_n \left(ka\right) \psi_n \left(kr\right), r < a \end{array} \right\} y_n P_n^1 \left(\cos \theta\right). \quad (4.127)$$

By virtue of condition (4.89), both sequences of unknowns $\{x_n\}_{n=1}^{\infty}$ and $\{y_n\}_{n=1}^{\infty}$ belong to the same solution classes as before ($l_2 \left(2\right)$ and l_2, respectively).

In an analogous way enforcement of the mixed boundary conditions produces a pair of coupled dual series equations relatively for the rescaled coefficients

$$X_n = i^n \left(2n + 1\right) x_n, \quad Y_n = i^n y_n, \quad (4.128)$$

in the form

$$\sum_{n=1}^{\infty} \frac{2n + 1}{n \left(n + 1\right)} \left\{ \left(1 - \varepsilon_n\right) X_n - \alpha_n \right\} P_n^1 \left(\cos \theta\right)$$

$$= -4ka C_1 \tan \frac{\theta}{2}, \quad \theta \in \left(0, \theta_0\right), \quad (4.129)$$

$$\sum_{n=1}^{\infty} \frac{X_n}{n \left(n + 1\right)} P_n^1 \left(\cos \theta\right) = C_2 \cot \frac{\theta}{2}, \quad \theta \in \left(\theta_0, \pi\right), \quad (4.130)$$

$$\sum_{n=1}^{\infty} \frac{1}{n \left(n + 1\right)} \left\{ \left(1 - \mu_n\right) Y_n - \beta_n \right\} P_n^1 \left(\cos \theta\right)$$

$$= -\frac{1}{ka} C_1 \tan \frac{\theta}{2}, \quad \theta \in \left(0, \theta_0\right), \quad (4.131)$$

$$\sum_{n=1}^{\infty} \frac{2n+1}{n(n+1)} Y_n P_n^1(\cos\theta) = -iC_2 \cot\frac{\theta}{2}, \quad \theta \in (\theta_0, \pi). \qquad (4.132)$$

Here we have introduced the notation

$$\alpha_n = i^{n+1} 4ka \psi_n'(ka), \quad \beta_n = -i^{n+1} \frac{1}{ka}(2n+1)\psi_n(ka), \qquad (4.133)$$

and C_1 and C_2 are so-called *polarisation constants* (see Section 1.5). Notice that $\{X_n\}_{n=1}^{\infty}$ and $\{Y_n\}_{n=1}^{\infty}$ lie in l_2.

We make use of Abel integral transforms based on the representations (4.19) and on the following representations for the functions $\tan\frac{1}{2}\theta$ and $\cot\frac{1}{2}\theta$ to transform equations (4.129)–(4.132) to regularised form,

$$\tan\frac{1}{2}\theta = \frac{1}{\sin\theta}(1-\cos\theta) = \frac{1}{\sin\theta}\{P_0(\cos\theta) - P_1(\cos\theta)\}$$

$$= \frac{2\sqrt{2}}{\pi}\csc^{-1}\theta \int_0^\theta \frac{\sin\frac{1}{2}\phi\sin\phi}{\sqrt{\cos\phi - \cos\theta}}d\phi, \qquad (4.134)$$

$$\cot\frac{1}{2}\theta = \frac{1}{\sin\theta}(1+\cos\theta) = \frac{1}{\sin\theta}\{P_0(\cos\theta) + P_1(\cos\theta)\}$$

$$= \frac{2\sqrt{2}}{\pi}\csc^{-1}\theta \int_\theta^\pi \frac{\cos\frac{1}{2}\phi\sin\phi}{\sqrt{\cos\theta - \cos\phi}}d\phi. \qquad (4.135)$$

We obtain

$$\sum_{n=1}^{\infty} X_n \sin\left(n+\frac{1}{2}\right)\theta =$$

$$\begin{cases} -i4kaC_1 \sin\frac{1}{2}\theta + \sum_{n=1}^{\infty}(X_n\varepsilon_n + \alpha_n)\sin\left(n+\frac{1}{2}\right)\theta, & \theta \in (0,\theta_0) \\ -C_2 \sin\frac{1}{2}\theta, & \theta \in (\theta_0,\pi) \end{cases} \qquad (4.136)$$

$$\sum_{n=1}^{\infty} Y_n \cos\left(n+\frac{1}{2}\right)\theta =$$

$$\begin{cases} -\frac{1}{ka}C_1 \cos\frac{1}{2}\theta + \sum_{n=1}^{\infty}(Y_n\mu_n + \beta_n)\cos\left(n+\frac{1}{2}\right)\theta, & \theta \in (0,\theta_0) \\ iC_2 \cos\frac{1}{2}\theta, & \theta \in (\theta_0,\pi). \end{cases}$$
$$(4.137)$$

It is worth noticing that the transform from equations (4.129)–(4.132) to (4.136)–(4.137) replaces a series expansion in one complete set of orthogonal functions (namely $\left\{P_n^1\left(\cos\theta\right)\right\}_{n=1}^{\infty}$) by another complete set of orthogonal functions ($\left\{\sin\left(n+\frac{1}{2}\right)\theta\right\}_{n=0}^{\infty}$ or $\left\{\cos\left(n+\frac{1}{2}\right)\theta\right\}_{n=0}^{\infty}$).

The polarisation constants make the change of basis invertible; without their presence the transform process described would induce a map of l_2 into a proper subspace of itself. Both C_1 and C_2 may be taken as zero index Fourier coefficients ($n=0$).

Making use of the orthogonality of the trigonometric functions leads to the coupled i.s.l.a.e. of the second kind

$$X_m - \sum_{n=1}^{\infty} X_n \varepsilon_n R_{nm}\left(\theta_0\right) + i4ka R_{0m}\left(\theta_0\right) C_1 - R_{0m}\left(\theta_0\right) C_2$$

$$= \sum_{n=1}^{\infty} \alpha_n R_{nm}\left(\theta_0\right), \quad (4.138)$$

where $m = 1, 2, \ldots$, and

$$-\sum_{n=1}^{\infty} X_n \varepsilon_n R_{n0}\left(\theta_0\right) + i4ka R_{00}\left(\theta_0\right) C_1 + \left[1 - R_{00}\left(\theta_0\right)\right] C_2$$

$$= \sum_{n=1}^{\infty} \alpha_n R_{n0}\left(\theta_0\right), \quad (4.139)$$

and

$$Y_m - \sum_{n=1}^{\infty} Y_n \mu_n Q_{nm}\left(\theta_0\right) + \frac{1}{ka} Q_{0m}\left(\theta_0\right) C_1 + i Q_{0m}\left(\theta_0\right) C_2$$

$$= \sum_{n=1}^{\infty} \beta_n Q_{nm}\left(\theta_0\right), \quad (4.140)$$

where $m = 1, 2, \ldots$, and

$$-\sum_{n=1}^{\infty} Y_n \mu_n Q_{n0}\left(\theta_0\right) + \frac{1}{ka} Q_{00}\left(\theta_0\right) C_1 - i \left[1 - Q_{00}\left(\theta_0\right)\right] C_2$$

$$= \sum_{n=1}^{\infty} \beta_n Q_{n0}\left(\theta_0\right); \quad (4.141)$$

here $R_{nm}\left(\theta_0\right)$ and $Q_{nm}\left(\theta_0\right)$ are the well-known functions given by (4.22) and (4.23). The constants C_1 and C_2 may be eliminated from this system, by use of equations (4.139) and (4.141), but we prefer not to do so; this solution form is simpler for numerical investigation.

Before presenting numerical results let us use the approximate analytical solution (4.120)–(4.121) when $ka \ll 1$ to analyse the total and radar cross-sections. The *radar cross-section* (RCS) σ_B is defined (see, for example, [77]) by

$$\sigma_B = \lim_{r \to \infty} 4\pi r^2 \left(|E_\theta^s|^2 / |E_\theta^0|^2 \right) \qquad (4.142)$$

where scattered field E_θ^s is that observed in the direction opposite to the propagating incident wave ($\theta = \pi$ in this particular case).

The radiation patterns (see [4.88]) take the form

$$S_1 (\theta, \phi) = \frac{\cos \phi}{ik} \sum_{n=1}^{\infty} \frac{2n+1}{n(n+1)} \left\{ x_n^{(e)} \tau_n (\cos \theta) + y_n^{(e)} \pi_n (\cos \theta) \right\}, \quad (4.143)$$

$$S_2 (\theta, \phi) = \frac{\sin \phi}{ik} \sum_{n=1}^{\infty} \frac{2n+1}{n(n+1)} \left\{ x_n^{(e)} \pi_n (\cos \theta) + y_n^{(e)} \tau_n (\cos \theta) \right\}, \quad (4.144)$$

where

$$\tau_n (\cos \theta) = \frac{d}{d\theta} P_n^1 (\cos \theta), \quad \pi_n (\cos \theta) = \frac{1}{\sin \theta} P_n^1 (\cos \theta). \qquad (4.145)$$

Since

$$\tau_n (-1) = -\pi_n (-1) = (-1)^n \frac{n(n+1)}{2}, \qquad (4.146)$$

$$|S_1 (\pi)| = |S_2 (\pi)| = \frac{1}{2k} \left| \sum_{n=1}^{\infty} (-1)^n (2n+1) \left(x_n^{(e)} - y_n^{(e)} \right) \right|, \qquad (4.147)$$

and the normalised RCS equals

$$\frac{\sigma_B}{\pi a^2} = (ka)^{-2} \left| \sum_{n=1}^{\infty} (-1)^n (2n+1) \left(x_n^{(e)} - y_n^{(e)} \right) \right|^2. \qquad (4.148)$$

Making use of the complex power theorem with respect to the electromagnetic field for the spherical layer incorporating the open spherical shell, it is possible to deduce the energy relation (using Equation [1.19])

$$\frac{2\pi}{k^2} \sum_{n=1}^{\infty} (2n+1) \left\{ \mathrm{Re}\, x_n^{(e)} + \mathrm{Re}\, y_n^{(e)} + \left| x_n^{(e)} \right|^2 + \left| y_n^{(e)} \right|^2 \right\} = 0, \qquad (4.149)$$

that is a conservation law for scattered energy. Associated with (4.149) is the total cross-section (TCS) σ; in its normalised form, it is defined analogously to Formula (2.110) and equals

$$\frac{\sigma}{\pi a^2} = \frac{2}{(ka)^2} \sum_{n=1}^{\infty} (2n+1) \left\{ \left| x_n^{(e)} \right|^2 + \left| y_n^{(e)} \right|^2 \right\}. \qquad (4.150)$$

In the Rayleigh scattering regime $(\lambda \gg a)$, the cross-sections are

$$\frac{\sigma_B}{\pi a^2} = (2a_{11} + b_{11})^2 (ka)^4 \left(2a_{21} + b_{21} + a_{12} + \frac{1}{3} b_{12} \right)^2 -$$

$$2 (2a_{11} + b_{11}) \left(\frac{2a_{11}}{5} - 2a_{31} + \frac{b_{11}}{10} + b_{31} + a_{22} + \frac{b_{22}}{3} + \frac{4a_{13}}{15} + \frac{b_{13}}{15} \right) (ka)^6$$

$$+ O \left((ka)^8 \right) \qquad (4.151)$$

and

$$\frac{\sigma}{\pi a^2} = \frac{2}{3} \{ (4a_{11}^2 + b_{11}^2) (ka)^4 +$$

$$[4 (a_{11}^2 + 2a_{11}a_{31}) + b_{21}^2 - 2b_{11}b_{31} - \frac{8}{5} a_{11}^2 - \frac{1}{5} b_{11}^2 + \frac{3}{5} a_{12}^2 + \frac{1}{15} b_{12}^2] (ka)^6$$

$$+ O \left((ka)^8 \right) \}; \qquad (4.152)$$

to this order neither RCS nor TCS depend upon the aperture orientation.

For a closed sphere $(\theta_0 = \pi)$, it follows from (4.122) that $a_{11} = b_{11} = 1$, and

$$\frac{\sigma_B}{\pi a^2} = 9 (ka)^4 + O \left((ka)^6 \right), \qquad (4.153)$$

$$\frac{\sigma}{\pi a^2} = \frac{10}{3} (ka)^4 + O \left((ka)^6 \right). \qquad (4.154)$$

At the other extreme, when $\theta_0 \ll 1$ and the structure is very nearly a circular disc of radius $a_d = a \sin \theta_0 \simeq a\theta_0$, we have

$$a_{11} = \frac{4}{3\pi} \theta_0^3 + O (\theta_0^5), \quad b_{11} = O (\theta_0^5)$$

and the cross-sections are

$$\frac{\sigma_B}{\pi a_d^2} = \frac{64}{9\pi^2} (ka_d)^4 + O \left((ka_d)^6 \right), \qquad (4.155)$$

$$\frac{\sigma}{\pi a_d^2} = \frac{128}{27\pi^2} (ka_d)^4 + O\left((ka_d)^6\right).$$ (4.156)

These results for the ideally conducting sphere (4.153)–(4.154) and circular disc (4.155)–(4.156) are in complete agreement with well-known classical results [49]. Although it is possible in principle to extend the number of terms in formulae (4.151) and (4.152), it is more productive to use the i.s.l.a.e. as a basis for numerical algorithms to investigate the reflectivity of the cavity across various frequency bands; what is more important is that it is possible to devise suitable numerical algorithms to carry out all desired calculations with guaranteed accuracy.

4.3   Reflectivity of Open Spherical Shells.

In this section we make use of the system (4.138)–(4.141) to effect extensive radar cross-section computations for open spherical shells. It is convenient to rearrange the formula (4.148) in terms of the coefficients X_n and Y_n. We distinguish two cases, the *spherical disc* or cap (with $\theta_0 < 90°$) and the spherical cavity with an aperture (with $\theta_0 > 90°$ or $\theta_1 = 180° - \theta_0 < 90°$). The hemisphere demarcates the boundary between these classes; its RCS (to be discussed below) is shown in Figure 4.7. Set

$$W = \sum_{n=1}^{\infty} i^n \left\{ \psi_n'(ka) X_n - (2n+1) Y_n \psi_n(ka) \right\};$$ (4.157)

we then normalise the RCS for the spherical disc in the form

$$\frac{\sigma_B}{\pi a_d^2} = \frac{1}{(ka_d)^2} |W|^2,$$ (4.158)

where $a_d = a \sin \theta_0$, and in the form

$$\frac{\sigma_B}{\pi a^2} = \frac{1}{(ka)^2} |W|^2$$ (4.159)

for the spherical cavity.

Acoustic and electromagnetic scattering mechanisms for *open* spherical shells have much in common, at least for reflectivity. The crucial distinctive feature of electromagnetic scattering lies in the more complicated structure of the quasi-eigenoscillations that develop inside the cavity, due to the mutual coupling of TM and TE modes, formally recognized in the emergence of constants such as B_1 and B_3 (see [4.98]–[4.101]). Pure TM or TE oscillations do not develop for modes with azimuthal indices $m \geq 1$; such

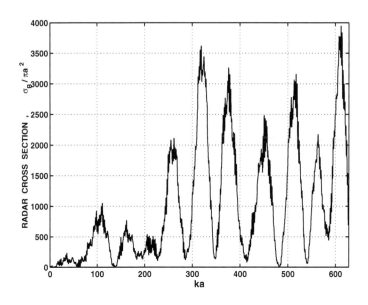

FIGURE 4.7. Normalised radar cross-section for the hemisphere.

isolated modes are only possible in the axially symmetric situation (with index $m = 0$) considered in Section 4.1. This coupling completely vanishes as the aperture closes $(\theta_1 \to 0)$, so if the aperture is small $(\theta_1 \ll 1)$ oscillations that are *dominantly* of TM or TE type may arise, according as $|E_r| \gg |H_r|$ or $|H_r| \gg |E_r|$.

We begin our calculations of reflectivity with the most powerful reflector in the class of open spherical shells, the hemispherical shell (with $\theta_0 = \theta_1 = 90°$); its radar cross-section is shown in Figure 4.7 over the range $0 < ka \leq 628$ $(0 < a/\lambda \leq 100)$. The RCS is extremely difficult, if not impossible, to predict by high frequency methods such as geometric optics because of multiple reflections of rays inside the hemispherical cavity. Similar behaviour is observed for deep spherical dishes; for example the RCS for spherical dishes with $\theta_0 = 60°$ and $\theta_0 = 45°$ is shown in the next two figures (4.8 and 4.9). There is an overall decrease in peak RCS values in the same frequency band.

As we have already seen in the acoustic context (Chapter 2), the radar cross-section of shallower dishes is easier to predict via geometric optics because there are no multiple reflections and just two types of rays contribute dominantly to the reflected field, one from the edge and the other from the central point of the dish. The constructive or destructive interference of these rays gives rise to the relatively simple dependence of RCS on frequency. We recall that two nondimensional parameters, ka_d and h/λ,

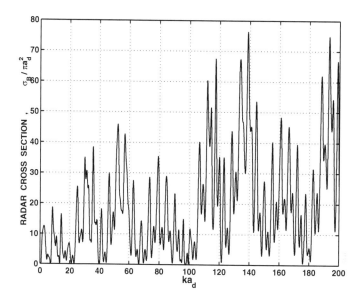

FIGURE 4.8. Normalised radar cross-section for the open spherical shell ($\theta_0 = 60°$).

where $h = a\left(1 - \cos\theta_0\right)$ is dish depth, were useful. The radar cross-section results displayed in Figures 4.10 and 4.11 for spherical dishes with $\theta_0 = 30°$ and $20°$ confirm the simple and regular dependence on these parameters.

We complete this first set of calculations by comparing three shallow dishes with $\theta_0 = 20°, 10°$ and $5°$. The RCS (see Figure 4.12) is very close to that for the circular disc provided that $h/\lambda < 1/4$.

We now consider the RCS of spherical cavities with apertures of $\theta_1 < 90°$. The scattered response now is dominated by resonant features that occur at frequencies lying close to quasi-eigenvalues of quasi-eigenoscillations of *hybrid* type, when the coupling of TM oscillations and TE oscillations is so strong that $|E_r|$ and $|H_r|$ are commensurate. This phenomenon occurs for larger apertures with $30° < \theta_1 < 90°$.

High Q-factor quasi-eigenoscillations that are dominantly of TM or TE type develop if the apertures are small enough ($\theta_1 < 30°$). The radar cross-section of the closed metallic sphere is readily differentiated from that of the high Q-factor system comprising the sphere with a small aperture. Effectively the RCS of the cavity structure is that of the closed sphere superimposed with resonance doublets in the vicinity of the quasi-eigenvalues of quasi-eigenoscillations. Radar cross-sections of the closed sphere and the cavity with an aperture (θ_1 was taken to be $10°$) are plotted in Figures 4.13 and 4.14, respectively. In the range on $0 < ka \le 6$ the curves practically

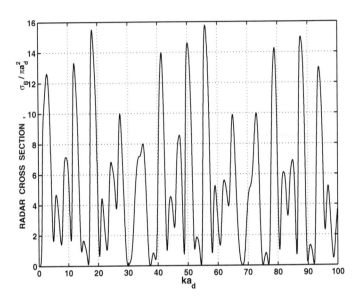

FIGURE 4.9. Normalised radar cross-section for the open spherical shell $(\theta_0 = 45°)$.

coincide, except in those narrow frequency intervals where high Q-factor oscillations are excited. The reasons for the existence of these resonance doublets that characterise the rapid variation of the radar cross-section in the vicinity of quasi-eigenvalues are entirely similar to those discussed in the acoustic context in Chapter 2, where the same phenomenon in the sonar cross-section of the structure excited by an acoustic plane wave was observed.

As might be anticipated, this striking feature disappears as the aperture widens, because the Q-factor diminishes, resonant regions overlap, and there is increasingly strong coupling between TM and TE oscillations. This change is illustrated in Figures 4.15 and 4.16 for the cavity structures with θ_1 taking values 30° and 60°, respectively.

In conclusion we note that the radar cross-section values calculated for the cavity with $\theta_1 = 60°$ are in excellent agreement with the experimental results reported in [43].

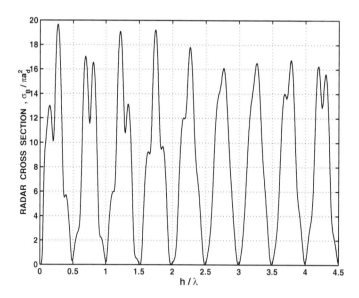

FIGURE 4.10. Normalised radar cross-section for the open spherical shell $(\theta_0 = 30°)$.

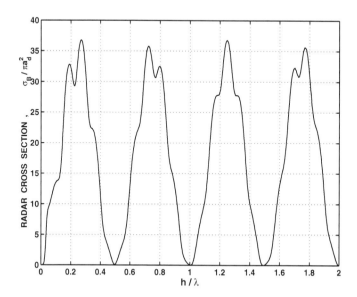

FIGURE 4.11. Radar cross-section for the spherical dish $(\theta_0 = 20°)$.

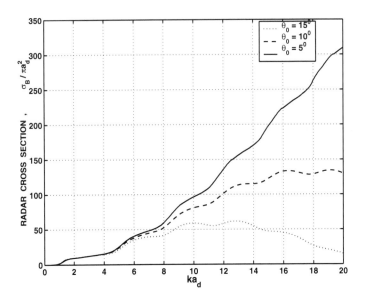

FIGURE 4.12. Radar cross-sections for the spherical dish, various θ_0.

4.4   The Focal Region of a Receiving Spherical Reflector Antenna.

The regularised system obtained in Section 4.2 allows us, perhaps for the first time, to place a theoretical analysis of the spherical reflector antenna (SRA) on a completely sound footing, and to obtain numerical results of justifiable and guaranteed accuracy. It is important to note that it is uniformly valid and applicable to any size SRA, small or large. The electrical size of an SRA is given by the ratio D/λ, where D is aperture diameter and λ wavelength; in the notation of Section 4.2 $D = 2a \sin \theta_0$. We shall call an SRA *small, moderate* or *large* according as $1 \le D/\lambda \le 20$, $20 < D/\lambda \le 100$ or $D/\lambda \ge 100$.

A study of the focal region is essential for the proper design of any focusing system. Theoretically, the best shape of a reflector is parabolic. From a geometric optics perspective, any bundle of parallel rays reflected from the concave surface of a parabolic mirror, collects at the same point of the optical axis, the *focus*. However, in practice, instead of focusing to a single point, the rays concentrate in a region known as the *focal spot*. In better designs, the extent of this spot is about a wavelength. However the centre of the spot is usually shifted some distance from the focal point predicted by GO, even for the supposedly ideal parabolic dish (albeit of finite extent).

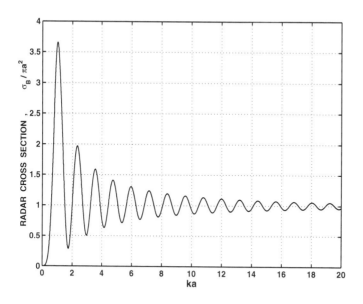

FIGURE 4.13. Radar cross-section of the closed sphere.

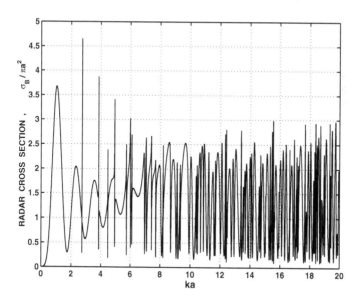

FIGURE 4.14. Normalised radar cross-section for the open spherical cavity $(\theta_1 = 10°)$.

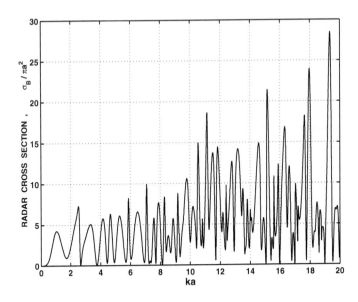

FIGURE 4.15. Normalised radar cross-section for the open spherical cavity ($\theta_1 = 30°$).

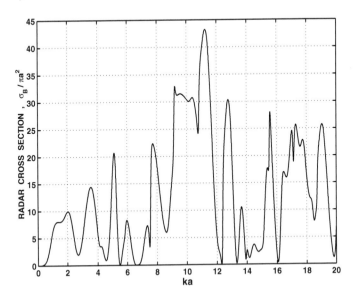

FIGURE 4.16. Normalised radar cross-section for the open spherical cavity ($\theta_1 = 60°$).

This deviation is due to the small but nonzero size of a wavelength (neglected in GO) and the impact of edge rays diffracted by the sharp rim of a parabolic reflector. The situation is worse for spherical reflectors. Spherical aberration (see [87], [33]) plays an important part in focal spot formation. Suppression of such distortion requires a knowledge of the electromagnetic field distribution in the neighbourhood of the focal spot. No approximate high-frequency technique works totally satisfactorily in this region: although the location of the focal spot may be found, fine detail of its structure is invariably absent.

Although this is not a handbook on antenna theory, we wish to make a preliminary study of the basic features of SRA, especially for those readers who may wish to pursue this interest. Thus, in this section we concentrate on an accurate study within the SRA near-field zone of the normalised electromagnetic energy density

$$W = \frac{1}{2}\left\{ \left|\vec{E}\right|^2 + \left|\vec{H}\right|^2 \right\} \qquad (4.160)$$

measured in decibels ($10\log_{10}(W)$). Note for comparative purposes that the plane electromagnetic wave propagating in free space has unit energy density (0 dB).

We first the energy density distribution along the optical axis of the SRA (z-axis). The diameter of a focal spot along a line is defined as distance between two points where the value of the energy density W falls to $1/e$ of its maximal value $W_{\max}$. The focal region is that volume in which W exceeds $W_{\max}/e$.

We consider two spherical dishes. The first has angular size $\theta_0 = 15°$ and $f/D = 0.966$, where $f = a/2$ is the focal distance according to geometric optics. The second employs the parameters of the world's largest SRA, the Arecibo Observatory [48], with aperture dimension $D = 305m$ and radius of curvature $a = 265m$. Thus its angular size is $\theta_0 = 35.13°$ and $f/D \simeq 0.434$. Clearly spherical aberration is significant for a dish of this depth.

According to a well-known "rule of thumb" in antenna design, the spherical reflector is expected to behave as a parabolic reflector provided that the maximal deviation from the ideal parabolic shape does not exceed $\lambda/16$. This limits the electrical size D/λ of the SRA with parameter θ_0 equal to 15° and 35.13° to 53.5 and 4, respectively, the latter value being much smaller than any likely operating wavelength.

Axial distributions of energy density for the shallower SRA ($\theta_0 = 15°$) are displayed in Figure 4.17, with D/λ equalling $10, 50$ and 200. The reflector focuses the electromagnetic energy into one simple spot. Furthermore, the main peak is well separated from the next local maximum that is at least ~ 11.5 dB lower (this value occurs when $D/\lambda = 200$). This SRA may be excited effectively by a single point source, for instance, a Huygens source.

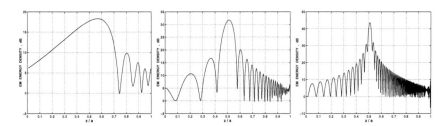

FIGURE 4.17. Axial distributions of energy density for SRA ($\theta_0 = 15°$) with D/λ equal to 10 (left), 50 (centre) and 200 (right).

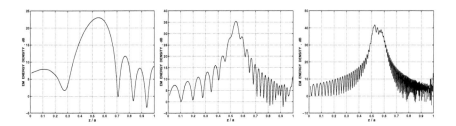

FIGURE 4.18. Axial distributions of energy density for SRA ($\theta_0 = 35.13°$) with D/λ equal to 10 (left), 50 (centre) and 200 (right).

From the reciprocity theorem [28], location of the primary source at that point of the optical axis where the maximum energy is focused results in the formation of a well-collimated main beam.

Axial distributions of energy density for the deeper SRA ($\theta_0 = 35.13°$) are displayed in Figures 4.18 and 4.19, with D/λ equalling $10, 50, 200$ and 437, the last value corresponding accurately to the line source frequency used at Arecibo [48]. Only when the reflector is electrically small is there formation of a focal point. Only in the range ($5 \leq D/\lambda \leq 20$) do we observe formation of a *bright* spot well separated (in level of energy) from other local maxima (*darker* spots, typically 10–10.5 dB lower). In transmission then, single point source is effective only in this range.

At larger electrical sizes, spherical aberration tends to level local energy maxima. The difference between the main and next local maximum for $D/\lambda = 50$ (see Figure 4.18) is about 7.5 dB; for larger sizes ($D/\lambda = 200, 437$) the difference is within 1–1.5 dB (see Figure 4.19). The focal region breaks up into separate spots in the range $50 \leq D/\lambda \leq 100$; the energy level of the nearest local maximum to the main peak exceeds the cut-off level $W_{\max}/e$. In such circumstances excitation of SRA by a single source is of limited effectiveness, and more effective excitation is possible through

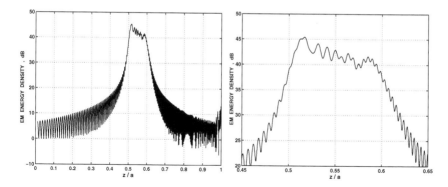

FIGURE 4.19. Axial distributions of energy density for SRA ($\theta_0 = 35.13°$) with $D/\lambda = 437$ (left), and expanded view (right).

the use of *line sources* or *distributed sources* that approximate the multiple focal spot distribution.

Let us investigate the shape of the focal spot for both shallow and deeper reflectors. Calculations in both the E-plane ($\phi = 0°$) and H-plane ($\phi = 90°$) were carried out with a rectangular mesh of spacing $\lambda/20$. From Figure 4.20 it can be seen that the shallower reflector focuses the energy into a single spot of shape similar to that of a prolate spheroid with semi-minor and semi-major axes a and b, respectively, that vary from $(a, b) = (0.35, 1.67)\lambda$ to $(a, b) = (0.33, 2.1)\lambda$ as D/λ varies from 10 to 100.

The deeper reflector performs differently. Within the range $10 \le D/\lambda \le 50$ it focuses the energy into a single spot of less elongated prolate spheroidal shape (Figure 4.21) with (a, b) equalling $(0.4, 1)\lambda$ and $(0.42, 1.08)\lambda$ when D/λ equals 10 and 50, respectively. As D/λ increases beyond 50, the focal spot deforms noticeably and then completely breaks up with several secondary focal spots (Figure 4.22).

4.5   The Transmitting Spherical Reflector Antenna.

In this section we continue to use the regularised solution developed in Section 4.5 to analyse the SRA, with some attention to extremely large reflectors. As in Chapter 2, instead of the traditional primary radiation pattern of form $\cos^n \psi$ $(n = 0, 1, ...)$ we use a *complex-point Huygens source* (CPHS) for more realistic models of actual antenna feeds, being close to a beam-like distribution of Gaussian type. Beamwidth of the primary beam and the illumination levels at the reflector edges (the *tapering*) are easily specified

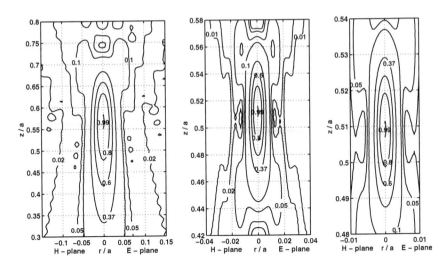

FIGURE 4.20. Energy distribution in the focal region of SRA ($\theta_0 = 15°$) with D/λ equal to 10 (left), 50 (centre) and 100 (right).

for the CPHS, and only minor amendments to the regularised system previously developed for a stable and accurate computational algorithm are needed.

A comprehensive analysis of the SRA is outside the scope of this book, and belongs more to handbooks of antenna design. However we wish to give a flavour of how the techniques introduced in this book may be adapted to this purpose with substantive success. Let us first describe the simulation of a beam-like radiation pattern of a primary source or antenna feed.

4.5.1   The Complex-Point Huygens Source: Debye Potentials.

The term *Huygens source*, or *real Huygens source*, is usually applied to the field generated by a pair of perpendicularly crossed electric and magnetic dipoles with dipole moments $\vec{p}$ and $\vec{m}$, respectively, of equal magnitude ($|\vec{p}| = |\vec{m}|$, see Figure 1.2[right]). We may describe the electromagnetic field of such a source, located at $r_0 = (x_0, y_0, z_0)$, by the electric $\vec{\Pi}_0$ and magnetic $\vec{\Pi}_0^{(m)}$ Hertz vectors

$$\vec{\Pi}_0 = \vec{i}_x \Pi_0, \ \vec{\Pi}_0^{(m)} = \vec{i}_y \Pi_0 \qquad (4.161)$$

where

$$\Pi_0 = \exp(ikR)/R, \qquad (4.162)$$

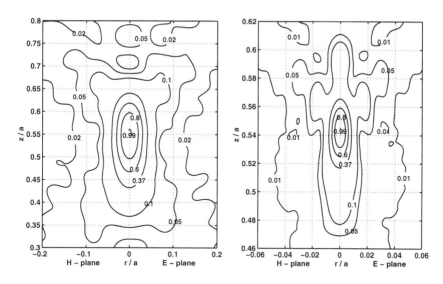

FIGURE 4.21. Energy distribution in the focal region of SRA ($\theta_0 = 35.13°$) with D/λ equal to 10 (left) and 50 (right).

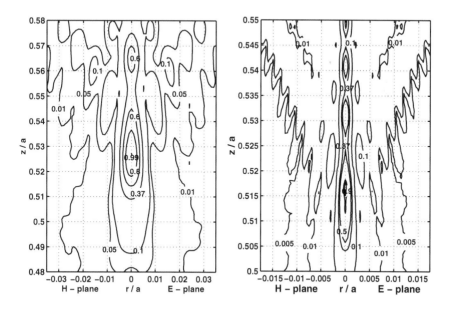

FIGURE 4.22. Energy distribution in the focal region of SRA ($\theta_0 = 35.13°$) with D/λ equal to 100 (left) and 437 (right).

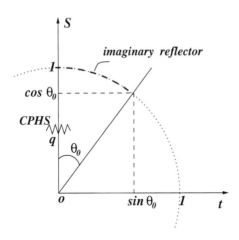

FIGURE 4.23. Reflector illumination by a real or complex point Huygens source.

with $R = \sqrt{(x - x_0)^2 + (y - y_0)^2 + (z - z_0)^2}$. A complex-point Huygens source is obtained by formally allowing the source location to be complex-valued; we will set

$$z_0 = d + ib \qquad (4.163)$$

with a suitable choice of branch cut for R defined as in Section 2.7.

The three-dimensional free-space Green function $G\left(\overrightarrow{r}, \overrightarrow{r_0}\right)$ is equal to $-(4\pi)^{-1}\Pi_0\left(\overrightarrow{r}, \overrightarrow{r_0}\right)$ and satisfies the inhomogeneous wave equation

$$\Delta G\left(\overrightarrow{r}, \overrightarrow{r_0}\right) + k^2 G\left(\overrightarrow{r}, \overrightarrow{r_0}\right) = -\delta\left(\overrightarrow{r} - \overrightarrow{r_0}\right) \qquad (4.164)$$

even if $\overrightarrow{r_0}$ is complex-valued.

To determine the associated Debye potentials U^0, V^0 we use results of Chapter 1. Making use of the readily verified relations

$$\frac{\partial \Pi_0}{\partial r}\cos\theta - \frac{\sin\theta}{r}\frac{\partial \Pi_0}{\partial\theta} = -\frac{\partial \Pi_0}{\partial r_s}$$

$$\frac{\partial \Pi_0}{\partial r} = \frac{\sin\theta}{r_s}\frac{\partial \Pi_0}{\partial\theta} - \cos\theta\frac{\partial \Pi_0}{\partial r_s},$$

where $r_s = d + ib$, the relation between the radial field components (E_r and H_r) and the function Π_0 is

$$\left\{\begin{matrix} E_r^0 \\ H_r^0 \end{matrix}\right\} = \frac{1}{r}\left[ik\frac{\partial \Pi_0}{\partial\theta} + \frac{1}{r_s}\frac{\partial}{\partial r_s}\left(r_s\frac{\partial \Pi_0}{\partial\theta}\right)\right]\left\{\begin{matrix}\cos\phi \\ \sin\phi\end{matrix}\right\}. \qquad (4.165)$$

The expansion of Π_0 in spherical wave harmonics is

$$\Pi_0 = \frac{i}{krr_s} \sum_{n=0}^{\infty} (2n+1) P_n (\cos\theta) \begin{cases} \zeta_n (kr_s) \psi_n (kr), r < |r_s| \\ \psi_n (kr_s) \zeta_n (kr), r > |r_s| \end{cases}. \quad (4.166)$$

Substitution of (4.166) into (4.165) produces spherical wave harmonic expansions for the radial components E_r^0, H_r^0.

A comparison of these results with relations between U^0, V^0 and the electromagnetic field components (see formulae [1.49], [1.50]) yields

$$\begin{Bmatrix} U^0 \\ V^0 \end{Bmatrix} = \frac{k}{r} \begin{Bmatrix} \cos\phi \\ \sin\phi \end{Bmatrix} \sum_{n=1}^{\infty} \frac{2n+1}{n(n+1)} P_n^1 (\cos\theta) \begin{Bmatrix} t_n (kr_s) \zeta_n (kr), r > |r_s| \\ q_n (kr_s) \psi_n (kr), r < |r_s| \end{Bmatrix}$$

$$(4.167)$$

where

$$t_n = i \frac{\psi_n' (kr_s)}{kr_s} + \frac{\psi_n (kr_s)}{kr_s}, \quad (4.168)$$

$$q_n = i \frac{\zeta_n' (kr_s)}{kr_s} + \frac{\zeta_n (kr_s)}{kr_s}. \quad (4.169)$$

The representation (4.167) of U^0 and V^0 makes it possible to treat the transmission problem in a similar manner to the problem of plane wave scattering from a spherical cap (see Section 4.2).

What value of kb provides the desired level of illumination (or *taper*) at the edge of the reflector? From (4.167) and (1.49)–(1.50) one can calculate the spatial field distribution generated by the CPHS located in a free space. Let us now imagine that a perfectly transparent spherical reflector occupies some region of space, i.e., the transparent structure has no effect on the radiated field of the CPHS (see Figure 4.23), and locate the source at an arbitrary point along the optical axis. Calculate the electromagnetic field energy densities, w_t and w_e, at the centre point and edges of the transparent spherical reflector. The ratio w_e/w_t is a direct measure of the tapering. Given a desired level of tapering L (specified in decibels, as is usual in antenna design, so that $L < 0$), we can then solve the equation

$$10 \log_{10} (w_e/w_t) - L = 0, \quad (4.170)$$

for the unknown kb (all the other SRA parameters being fixed).

This direct approach is convenient for small to moderate values of D/λ. Although valid at larger electrical sizes, it requires substantive calculation of spherical Bessel functions, and it is more convenient to use elementary functions and base calculations on equations (4.161) and (1.38). Thus consider normalised coordinates $s = z/a$ and $t = \rho/a$, shown in Figure 4.23;

a is the radius of the transparent reflector. In addition, let $q = d/a$ and $p = b/a$ be the normalised parameter to be found. Equation (4.170) takes the form

$$10 \log_{10} \left[\frac{w \left(\cos \theta_0, \sin \theta_0 \right)}{w \left(1, 0 \right)} \right] - L = 0, \qquad (4.171)$$

where we regard $w = w \left(s, t \right)$ as a function of two variables s and t. The complex distance r_s in (s, t)-coordinates is $r_s = a \left(q + ip \right)$.

Suppressing an elementary but lengthy deduction, the analytical expression for w is

$$w = \frac{1}{2 \left(ks \right)^2} \left(A + B + C \right) e^{ika(R - R^*)}, \qquad (4.172)$$

where the star $(*)$ denotes the complex conjugate, $R^2 = t^2 + \left(s - q - ip \right)^2$ with a cut at $t = 0$ given by

$$R \left(s, 0 \right) = \begin{cases} s - q + ip, & s > q \\ q - s + ip, & s < q \end{cases},$$

and the values of A, B and C are

$$A = |R|^{-6} |A_1|^2, \qquad (4.173)$$

where

$$A_1 = - \left(s - q - ip \right) \left(s - q - ip + R \right) +$$
$$i \left[\frac{2t^2 - \left(s - q - ip \right) \left(s - q - ip + R \right)}{ka \cdot R} \right] + \left[\frac{-2t^2 - \left(s - q - ip \right)^2}{\left(ka \cdot R \right)^2} \right], \qquad (4.174)$$

$$B = |R|^{-2} \left| 1 + \frac{i + ka \left(s - q - ip \right)}{ka \cdot R} + \frac{-1 + ika \left(s - q - ip \right)}{\left(ka \cdot R \right)^2} \right|^2, \qquad (4.175)$$

and

$$C = |R|^{-6} t^2 \left| \left(s - q - ip \right) + R + i \frac{3}{ka} \frac{s - q - ip}{R} - \frac{i}{ka} + \frac{3}{\left(ka \right)^2} \frac{s - q - ip}{R^2} \right|^2. \qquad (4.176)$$

Illustrative values of kb, obtained by solving (4.171) for a CPHS placed at $q = 0.5$, are shown in Table 4.3 for various values of θ_0 and D/λ; the taper L is fixed at -10.8 dB.

θ_0	Electrical Size D/λ				
	10	20	50	100	200
15°	9.106	8.917	8.867	8.860	–
30°	1.975	1.959	1.955	1.954	1.954
35.13°	1.311	1.302	1.299	1.299	1.299

TABLE 4.3. Imaginary part kb of the complex source point location.

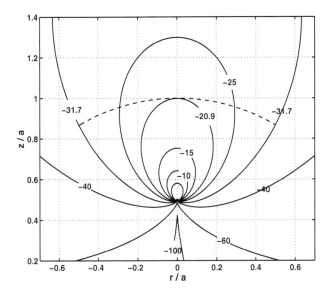

FIGURE 4.24. Energy distribution of the CPHS with $D/\lambda = 50$, $\theta_0 = 30°$, $q = 0.5$, $kb = 1.955$. The countour levels are in dB.

It is instructive to compare the energy distribution for the CPHS with $kb = 1.955$ with that of the real Huygens source ($kb = 0$), shown in Figures 4.24 and 4.25, respectively. The first is *Gaussian beam-like* whilst the second has the well-known *cardioid-like* distribution. The location of the SRA is shown as a dashed line. The value $kb = 1.955$ corresponds to that value of L, which provides tapering of an imaginary spherical reflector with parameters $\theta_0 = 30°$, $D/\lambda = 50$. The next plot (Figure 4.26) shows the relative distribution of energy density at the surface of the transparent reflector, as a function of angle θ; the angles $\theta = 0°$ and $\theta = 30°$ indicate the centre-point and rims, respectively. The cardioid-like distribution of the real Huygens source provides an intrinsic tapering (-3.85 dB for $\theta_0 = 30°$, $q = 0.5$).

Accepting that the concept of CPHS is useful for the simulation of primary feeds, we may employ the solution scheme of Section 4.2 with one change, the replacement of plane wave excitation by CPHS excitation.

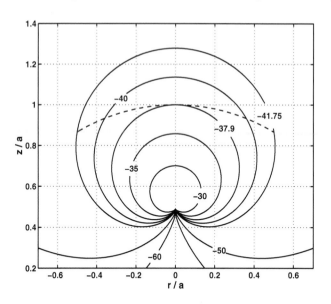

FIGURE 4.25. Energy distribution of the real Huygens source with $D/\lambda = 50$, $\theta_0 = 30°$, $q = 0.5$, $kb = 0$. The countour levels are in dB.

4.5.2   Excitation of the Reflector by a CPHS.

We seek the total solution for Debye potentials U and V in the form

$$U = U^0 + U^s, V = V^0 + V^s \tag{4.177}$$

where U^0 and V^0 are defined by (4.167), and the scattered Debye potentials are

$$U^s = \frac{k}{r} \cos \phi \sum_{n=1}^{\infty} \frac{1}{n(n+1)} X_n P_n^1 (\cos \theta) \left\{ \begin{array}{l} \psi_n' (ka) \zeta_n (kr), r > a \\ \zeta_n' (ka) \psi_n (kr), r < a \end{array} \right\}, \tag{4.178}$$

$$V^s = \frac{k}{r} \sin \phi \sum_{n=1}^{\infty} \frac{2n+1}{n(n+1)} Y_n P_n^1 (\cos \theta) \left\{ \begin{array}{l} \psi_n (ka) \zeta_n (kr), r > a \\ \zeta_n (ka) \psi_n (kr), r < a \end{array} \right\}, \tag{4.179}$$

where $\{X_n\}_{n=1}^{\infty}, \{Y_n\}_{n=1}^{\infty}$ are unknown coefficients to be found. The finite energy condition (4.89) requires that $\{X_n\}_{n=1}^{\infty}$ and $\{Y_n\}_{n=1}^{\infty}$ belong to l_2.

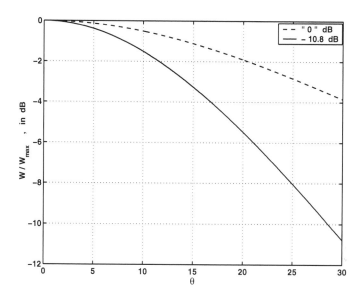

FIGURE 4.26. Normalised incident energy (dB scale) on the reflector surface for a real (dashed line, $kb = 0$) and complex (solid line, $kb = 1.955$) Huygens source. $D/\lambda = 50$, $\theta_0 = 30°$, $q = 0.5$.

The scheme developed in Section 4.2 produces the following i.s.l.a.e. of the second kind

$$i4kaR_{0m}C_1 - R_{0m}C_2 + X_m - \sum_{n=1}^{\infty} X_n \varepsilon_n R_{nm} = \sum_{n=1}^{\infty} \alpha_n R_{nm}, \qquad (4.180)$$

$$\frac{1}{ka}Q_{0m}C_1 + iQ_{0m}C_2 + Y_m - \sum_{n=1}^{\infty} Y_n \mu_n Q_{nm} = \sum_{n=1}^{\infty} \beta_n Q_{nm}, \qquad (4.181)$$

where $m = 1, 2, \ldots$, with the accompanying equations

$$-i4kaR_{00}C_1 - (1 - R_{00})C_2 + \sum_{n=1}^{\infty} X_n \varepsilon_n R_{n0} = -\sum_{n=1}^{\infty} \alpha_n R_{n0}, \qquad (4.182)$$

$$-\frac{1}{ka}Q_{00}C_1 + i(1 - Q_{00})C_2 + \sum_{n=1}^{\infty} Y_n \mu_n Q_{n0} = -\sum_{n=1}^{\infty} \beta_n Q_{n0}; \qquad (4.183)$$

here C_1 and C_2 are so-called *polarisation constants*, $R_{nm} = R_{nm}(\theta_0)$ and $Q_{nm} = Q_{nm}(\theta_0)$ are defined by (4.22), and

$$\varepsilon_n = 1 + i4ka\frac{\psi'_n(ka)\,\zeta'_n(ka)}{2n+1}, \tag{4.184}$$

$$\mu_n = 1 - \frac{i}{ka}(2n+1)\,\psi_n(ka)\,\zeta_n(ka) \tag{4.185}$$

are the asymptotically small parameters, satisfying

$$\varepsilon_n = O\left(n^{-2}\right), \quad \mu_n = O\left((ka/n)^2\right)$$

as $n \to \infty$. In addition, α_n and β_n are known coefficients given by

$$\alpha_n = i4kat_n(kr_s)\,\zeta'_n(ka),$$

$$\beta_n = -\frac{i}{ka}(2n+1)\,t_n(kr_s)\,\zeta_n(ka). \tag{4.186}$$

Numerical computation of the coefficients X_n, Y_n employs the standard truncation method. After computation of a finite set of coefficients $\{X_n\}_{n=1}^{N_{tr}}$ $\{Y_n\}_{n=1}^{N_{tr}}$, where $N_{tr} > ka$ is the truncation number, the antenna characteristics may be computed. (Clear definitions of basic antenna parameters are given by Kraus [52].)

A complete description of the field intensity in far-field zone in every direction θ, ϕ is given by the so-called *principal plane patterns* and *maximum cross-polarisation pattern*. At large distances ($r \to \infty$), both radial components E_r and H_r decay as $O\left(r^{-2}\right)$, tangential components dominate, and

$$E_\theta = H_\phi + O\left(r^{-2}\right), E_\phi = -H_\theta + O\left(r^{-2}\right) \tag{4.187}$$

as $r \to \infty$. In far-field zone ($r \to \infty$, $kr \gg ka$, $r \gg D$ [aperture size]),

$$E_\theta = \frac{e^{ikr}}{r}S_1(\theta)\cos\phi + O\left(r^{-2}\right), \tag{4.188}$$

$$E_\phi = \frac{e^{ikr}}{r}S_2(\theta)\sin\phi + O\left(r^{-2}\right), \tag{4.189}$$

where

$$S_1(\theta) = ik^2 \sum_{n=1}^{\infty}(-i)^n A_n\left[a_n\tau_n(\cos\theta) + b_n\pi_n(\cos\theta)\right], \tag{4.190}$$

$$S_2(\theta) = -ik^2 \sum_{n=1}^{\infty}(-i)^{n+1} A_n\left[a_n\pi_n(\cos\theta) + b_n\tau_n(\cos\theta)\right], \tag{4.191}$$

and $A_n = (2n + 1) / n (n + 1)$,

$$
\begin{aligned}
a_n &= t_n (kr_s) + (2n + 1)^{-1} \psi_n' (ka) X_n, && (4.192) \\
b_n &= t_n (kr_s) + \psi_n Y_n, && (4.193)
\end{aligned}
$$

and the angle functions $\pi_n (\cos \theta)$ and $\tau_n (\cos \theta)$ are defined by (4.145).

When θ is fixed, the maxima of E_θ and E_ϕ occur at $\phi = 0$ and $\phi = \frac{\pi}{2}$, respectively. The planes $\phi = 0$ and $\phi = \frac{\pi}{2}$ are called the *principal planes*, and $S_1 (\theta)$ and $S_2 (\theta)$ the *principal plane patterns*. Denoting their absolute maximum values by $S_1^{\max}$ and $S_2^{\max}$, respectively, it is convenient to calculate the so-called normalised principal plane patterns (in dB) via

$$
\begin{aligned}
S_1^{norm} (\theta) &= 20 \log_{10} (|S_1 (\theta)| / S_1^{\max}) \\
S_2^{norm} (\theta) &= 20 \log_{10} (|S_2 (\theta)| / S_2^{\max}) . && (4.194)
\end{aligned}
$$

It is easy to see that in our particular case $S_1^{\max} = S_1 (\pi)$ and $S_2^{\max} = S_2 (\pi)$.

It is obvious that total pattern

$$
|S (\theta)|^2 = |S_1 (\theta)|^2 \cos^2 \phi + |S_2 (\theta)|^2 \sin^2 \phi
$$

does not possess rotational symmetry, as $|S_1 (\theta)| \neq |S_2 (\theta)|$. The *cross-polar* pattern measures the maximum difference between these patterns via

$$
S_3^{norm} (\theta) = 20 \log_{10} \left(\frac{|S_1 (\theta) + S_2 (\theta)|}{2 S_1^{\max}} \right) . \tag{4.195}
$$

The directivity D of an antenna [52] is the ratio of the maximum value of the radiation intensity (power per unit solid angle) $U_{\max}$ to its average U_{av} (averaged over all spherical directions). In the far-field of the antenna the directivity may be expressed as the ratio of the maximum value $P_{\max}$ of the Poynting vector to its average P_{av}. Thus

$$
D = U_{\max}/U_{av} = P_{\max}/P_{av}. \tag{4.196}
$$

The beam area Ω_A is related to directivity by $\Omega_A = 4\pi/D$. We recall that the Poynting vector is defined by $\vec{P} = \frac{1}{2} \operatorname{Re} \left\{ \vec{E} \times \vec{H} \right\}$, so that as $r \to \infty$ it is essentially radial,

$$
\vec{P} = P \vec{i}_r + O (r^{-3}) \tag{4.197}
$$

where $P = \frac{1}{2} \left(|E_\theta|^2 + |E_\phi|^2 \right) r^{-2}$. Using (4.194), (4.195) we deduce

$$
P_{\max} = P (\pi, \phi) = \frac{k^4}{8r^2} \left| \sum_{n=1}^{\infty} i^{n+1} (2n + 1) (a_n - b_n) \right|^2 \tag{4.198}
$$

and

$$P_{av} = \frac{k^4}{4r^2} \sum_{n=1}^{\infty} (2n+1) \left(|a_n|^2 + |b_n|^2 \right), \qquad (4.199)$$

and directivity is computed from (4.196).

4.5.3   Numerical Results

Our examination of the SRA aims to find the location of the antenna feed (modelled by a CPHS) that is optimal according to one of the following criteria, (a) the location giving the maximum directivity D, (b) the location giving the lowest level of sidelobes (without significant widening of the main beam), or (c) a location providing some compromise between the first two criteria, where the directivity D is close to its maximum but the sidelobe levels are reduced below those occurring at that location where directivity is maximal. It is generally accepted that the optimal taper for an SRA is about -11 dB. Using the rigorous mathematical method developed above, we verify this assertion of antenna theory.

First let us compute the dependence of directivity D on normalised source location z/a for small to moderate SRA ($5 \lesssim D/\lambda \lesssim 50$) with a real Huygens source ($kb = 0$) and a CPHS of taper -10.8 dB. The calculated maximal value $D_{\max}$ of directivity is compared with the ideal directivity, given by $D_0 = (\pi D/\lambda)^2$, via the aperture efficiency

$$\varepsilon_{ap} = D_{\max}/D_0. \qquad (4.200)$$

In Figures 4.27–4.28 the solid curves depict the dependence of directivity D on the CPHS normalised source location z/a with taper -10.8 dB for the SRA with angular extent $\theta_0 = 15°$; the dashed curves depict the directivity for the real Huygens source.

As electrical size increases from 5 to 50, the optimal location $q_{\max}$ of complex point source decreases from 0.542 to 0.508, and aperture efficiency ε_{ap} decreases very slightly from the values 81.5% to 80.1%; the same is true of the real Huygens source. The results confirm, as asserted in the previous section, that, when $D/\lambda < 53.5$, the shallow SRA (with parameter $\theta_0 = 15°$) can be regarded as a parabolic reflector from an engineering point of view. The effectiveness of the shallower SRA ($\theta_0 = 15°$) is retained at higher frequencies, for $D/\lambda \le 200$.

The optimal position of the CPHS providing maximal directivity also provides the lowest level of the first sidelobes, i.e., criteria (a) and (b) are satisfied simultaneously. The principal plane patterns for optimal complex source location ($q_{\max} = 0.508$) are displayed in Figure 4.29. The next figure

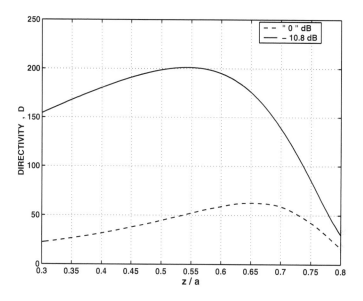

FIGURE 4.27. Dependence of directivity on CPHS location for the SRA with $\theta_0 = 15°$ $(D/\lambda = 5)$.

(4.30) shows a fragment of principal plane patterns close to the main beam aligned at $\theta = 0°$; the pattern for two source locations, $q = q_{max} = 0.508$ (solid line) and $q = 0.500$ (dashed line), demonstrate the optimal source location simultaneously satisfies criteria (a) and (b); maximal directivity and the lowest levels of the first sidelobe (about -25.4 dB) are achieved when $q = q_{max} = 0.508$.

The directivity calculations for the deeper SRA (with $\theta_0 = 35.13°$) are depicted in Figures 4.31 and 4.32. Spherical aberration is much more significant: as its electrical size increases from 5 to 50, aperture efficiency ε_{ap} decreases dramatically, from 78.1% to 53.7%. The optimal location of the CPHS over this range is nearly the same, about $q = 0.540$.

Principal plane patterns for the SRA with parameters $\theta_0 = 35.13°$ and $D/\lambda = 50$, and two complex source locations are shown in Figure 4.33. It is impossible to satisfy the directivity criterion (a) and the sidelobe level criterion (b) simultaneously; the first is achieved at the point $q = 0.540$, but the second at the point $q = 0.530$. Also reduction of the sidelobe level (from -14.2 dB for $q = 0.540$ to -16.0 dB for $q = 0.525$) is achieved with significant widening (lower directivity) of the main beam (see Figure 4.34). As the electrical size of the SRA increases, the directivity reduces dramatically.

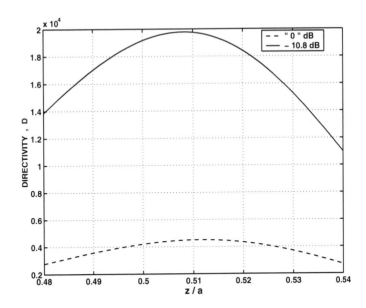

FIGURE 4.28. Dependence of directivity on CPHS location for the SRA with $\theta_0 = 15°$ $(D/\lambda = 50)$.

We may conclude that a single point source is a very ineffective feed, with dramatically reduced aperture efficiency over the larger range of electrical sizes examined in this section.

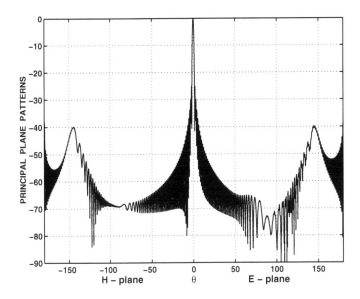

FIGURE 4.29. Principal plane patterns of the SRA ($\theta_0 = 15°$) with optimal source location.

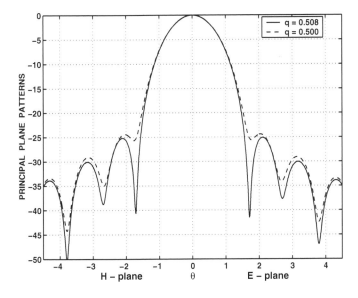

FIGURE 4.30. Principal plane patterns of the SRA ($\theta_0 = 15°$) with various source locations.

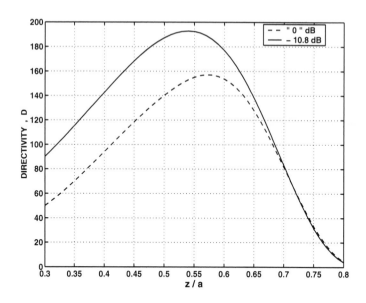

FIGURE 4.31. Dependence of directivity on CPHS location for the SRA with $\theta_0 = 35.13°$ $(D/\lambda = 5)$.

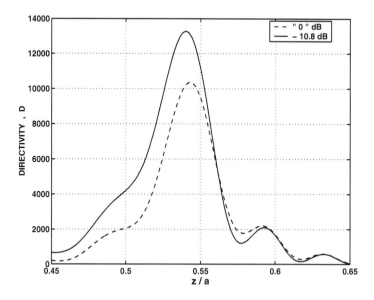

FIGURE 4.32. Dependence of directivity on CPHS location for the SRA with $\theta_0 = 35.13°$ $(D/\lambda = 50)$.

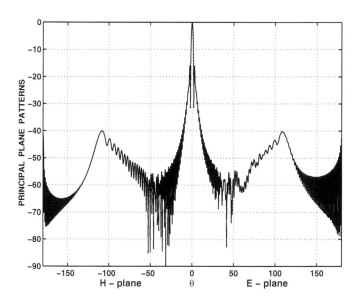

FIGURE 4.33. Principal plane patterns of the SRA ($\theta_0 = 35.13°$) with various source locations.

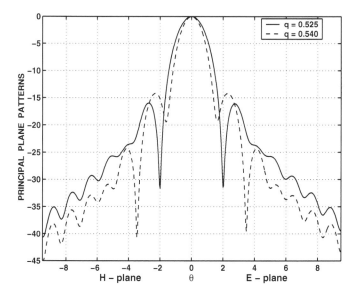

FIGURE 4.34. Principal plane patterns of the SRA ($\theta_0 = 35.13°$) with various source locations.

5

Electromagnetic Diffraction from Various Spherical Cavities.

In this chapter we consider electromagnetic scattering from spherical cavity structures that are more complicated than those encountered in the previous chapter. The first group contains structures consisting of pairs of open spherical concentric shells, of unequal radii. The special case of two symmetrical spherical caps (of equal radii) is treated separately. The second group consists of doubly-connected shells, the so-called *spherical barrels*.

First we consider electromagnetic plane wave scattering from two concentric spherical open shells (Section 5.1). Whilst this is a *two-body problem*, the complexity of the solution method, from a mathematical viewpoint, is only marginally greater than that of the single spherical shell with a circular aperture, though of course there is an increase in the number of dual series equations to be treated.

If the two spherical caps lie on a common spherical surface of specified radius, the diffraction problem is intrinsically more complex and is closely related to the diffraction problem for *spherical barrels*. For these structures we develop a solution based upon the analysis of some appropriate triple series equations.

Vertical dipole excitation of the slotted spherical shell and of the spherical barrel are considered in Sections 5.2 and 5.3, respectively. Plane wave excitation of the same structures requires more complicated analysis because of a coupling of TM and TE modes that is absent in vertical dipole excitation. The reasons for this increase of complexity, manifesting itself through the appearance of polarisation constants, are very similar to those encountered for the single aperture spherical cavity. Plane wave scattering from the perfectly conducting slotted spherical shell is analysed in Section 5.4.

The analysis of horizontal dipole excitation of the slotted shell or barrel has a similar level of complexity because of the coupling of TM and TE modes, and is discussed in Section 5.5.

Finally in Section 5.6, a resonant structure comprising a spherical cap and a circular disc is considered. Apart from its intrinsic physical interest, it is interesting from the analytical point of view because it gives rise to integro-series equations that are amenable to analytical regularisation methods.

5.1   EM Plane Wave Scattering by Two Concentric Spherical Shells.

In this section we consider electromagnetic plane wave scattering by two concentric spherical shells. The radii of the inner and outer shells are a and b, respectively $(b > a)$; circular apertures in the inner and outer shells subtend angles θ_1 and θ_0, respectively. The structure is axisymmetric; let the axis of symmetry be the z-axis. The problem geometry is shown in Figure 5.1. An incident electromagnetic plane wave propagates in the positive direction of the z-axis, with electric field $\vec{E}_0$ and magnetic field $\vec{H}_0$ aligned parallel to the x-axis and y-axis, respectively. As usual, the Debye potentials U^0, V^0 of the incident plane wave are given by formulae (4.87).

Subdivide space into three regions, labelled 1, 2 and 3, being the interior of the smaller shell $(0 \leq r < a)$, the spherical annular region between the smaller and larger shell $(a < r < b)$, and the exterior of the whole structure $(r > b)$, respectively.

The total electromagnetic field in each region i $(i = 1, 2, 3)$ is derived from Debye potentials U_i^t and V_i^t that may be decomposed as a sum of incident and scattered potentials via

$$U_i^t = U^0 + U_i^s, \quad V_i^t = V^0 + V_i^s, \tag{5.1}$$

where the scattered potentials U_i^s and V_i^s may be expanded in spherical wave harmonics as

$$U_i^s = \frac{\cos \phi}{ik^2 r} \sum_{n=1}^{\infty} A_n \left\{ a_n^{(i)} \psi_n(kr) + b_n^{(i)} \zeta_n(kr) \right\} P_n^1(\cos \theta), \tag{5.2}$$

$$V_i^s = \frac{\sin \phi}{ik^2 r} \sum_{n=1}^{\infty} A_n \left\{ \tilde{a}_n^{(i)} \psi_n(kr) + \tilde{b}_n^{(i)} \zeta_n(kr) \right\} P_n^1(\cos \theta), \tag{5.3}$$

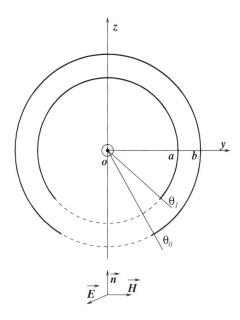

FIGURE 5.1. Concentric spherical shell cavity.

where $A_n = i^n (2n + 1) / n (n + 1)$ and the coefficients $a_n^{(i)}$, $b_n^{(i)}$, $\tilde{a}_n^{(i)}$ and $\tilde{b}_n^{(i)}$ are to be determined. The boundedness of the field at the origin implies that $b_n^{(1)} = \tilde{b}_n^{(1)} = 0$; the radiation conditions require that $a_n^{(3)} = \tilde{a}_n^{(3)} = 0$.

Continuity of the electric field on the surfaces $r = a$ and $r = b$ allows us to eliminate four of the unknowns $(a_n^{(2)}, \tilde{a}_n^{(2)}, b_n^{(2)}$ and $\tilde{b}_n^{(2)})$ in favour of the remaining four $(a_n^{(1)}, \tilde{a}_n^{(1)}, b_n^{(3)}$ and $\tilde{b}_n^{(3)})$, via

$$
\begin{aligned}
a_n^{(2)} &= \Delta_n^{-1} \Delta_{a_n^{(2)}}, & b_n^{(2)} &= \Delta_n^{-1} \Delta_{b_n^{(2)}}, \\
\tilde{a}_n^{(2)} &= \tilde{\Delta}_n^{-1} \Delta_{\tilde{a}_n^{(2)}}, & \tilde{b}_n^{(2)} &= \tilde{\Delta}_n^{-1} \Delta_{\tilde{b}_n^{(2)}},
\end{aligned} \tag{5.4}
$$

where

$$
\begin{aligned}
\Delta_n &= \psi_n'(ka) \zeta_n'(kb) - \psi_n'(kb) \zeta_n'(ka), & (5.5) \\
\tilde{\Delta}_n &= \psi_n(ka) \zeta_n(kb) - \psi_n(kb) \zeta_n(ka), & (5.6) \\
\Delta_{a_n^{(2)}} &= \left\{ \psi_n'(ka) a_n^{(1)} - \zeta_n'(ka) b_n^{(3)} \right\} \zeta_n'(kb), & (5.7) \\
\Delta_{b_n^{(2)}} &= \left\{ \zeta_n'(kb) b_n^{(3)} - \psi_n'(kb) a_n^{(1)} \right\} \psi_n'(ka), & (5.8) \\
\Delta_{\tilde{a}_n^{(2)}} &= \left\{ \psi_n(ka) \tilde{a}_n^{(1)} - \zeta_n(ka) \tilde{b}_n^{(3)} \right\} \zeta_n(kb), & (5.9) \\
\Delta_{\tilde{b}_n^{(2)}} &= \left\{ \zeta_n(kb) \tilde{b}_n^{(3)} - \psi_n(kb) \tilde{a}_n^{(1)} \right\} \psi_n(ka). & (5.10)
\end{aligned}
$$

The deduction of suitable series equations for the four unknowns ($a_n^{(1)}$, $\tilde{a}_n^{(1)}$, $b_n^{(3)}$ and $\tilde{b}_n^{(3)}$) is quite similar to that described in Section 4.2. Suppressing the obvious but lengthy details, the functional equations describing the problem are

$$\sum_{n=1}^{\infty} A_n \left(1 + a_n^{(1)}\right) \psi_n' (ka) P_n^1 (\cos\theta) = C_1 \tan \frac{\theta}{2}, \quad \theta \in (0, \theta_1) \qquad (5.11)$$

$$\sum_{n=1}^{\infty} A_n \Delta_n^{-1} \left\{ \psi_n' (kb) a_n^{(1)} - \zeta_n' (kb) b_n^{(3)} \right\} P_n^1 (\cos\theta) = -C_2 \cot \frac{\theta}{2},$$
$$\theta \in (\theta_1, \pi) \quad (5.12)$$

$$\sum_{n=1}^{\infty} A_n \left(1 + \tilde{a}_n^{(1)}\right) \psi_n (ka) P_n^1 (\cos\theta) = iC_1 \tan \frac{\theta}{2}, \quad \theta \in (0, \theta_1) \qquad (5.13)$$

$$\sum_{n=1}^{\infty} A_n \tilde{\Delta}_n^{-1} \left\{ \psi_n (kb) \tilde{a}_n^{(1)} - \zeta_n (kb) \tilde{b}_n^{(3)} \right\} P_n^1 (\cos\theta) = -C_2 \cot \frac{\theta}{2},$$
$$\theta \in (\theta_1, \pi) \quad (5.14)$$

and

$$\sum_{n=1}^{\infty} A_n \left\{ \psi_n' (kb) + \zeta_n' (kb) b_n^{(3)} \right\} P_n^1 (\cos\theta) = C_3 \tan \frac{\theta}{2}, \quad \theta \in (0, \theta_0) \quad (5.15)$$

$$\sum_{n=1}^{\infty} A_n \Delta_n^{-1} \left\{ \zeta_n' (ka) b_n^{(3)} - \psi_n' (ka) a_n^{(1)} \right\} P_n^1 (\cos\theta) = -iC_4 \cot \frac{\theta}{2},$$
$$\theta \in (\theta_0, \pi) \quad (5.16)$$

$$\sum_{n=1}^{\infty} A_n \left\{ \psi_n (kb) + \zeta_n (kb) \tilde{b}_n^{(3)} \right\} P_n^1 (\cos\theta) = iC_3 \tan \frac{\theta}{2}, \quad \theta \in (0, \theta_0)$$
$$(5.17)$$

$$\sum_{n=1}^{\infty} A_n \tilde{\Delta}_n^{-1} \left\{ \zeta_n (kb) \tilde{b}_n^{(3)} - \psi_n (ka) \tilde{a}_n^{(1)} \right\} P_n^1 (\cos\theta) = -C_4 \cot \frac{\theta}{2},$$
$$\theta \in (\theta_0, \pi), \quad (5.18)$$

where C_i $(i = 1, 2, 3, 4)$ are constants to be determined (as well as the unknowns $a_n^{(1)}$, $b_n^{(3)}$, $\tilde{a}_n^{(1)}$ and $\tilde{b}_n^{(3)}$).

To effect the regularisation process introduce rescaled unknowns by

$$a_n^{(1)} = \frac{(-i)^n}{2n+1} \frac{\Delta_n}{\psi'_n(kb)} x_n, \quad \tilde{a}_n^{(1)} = (-i)^n \frac{\tilde{\Delta}_n}{\psi_n(kb)} z_n, \tag{5.19}$$

$$b_n^{(3)} = \frac{(-i)^n}{2n+1} \frac{\Delta_n}{\zeta'_n(ka)} y_n, \quad \tilde{b}_n^{(3)} = (-i)^n \frac{\tilde{\Delta}_n}{\zeta_n(ka)} v_n. \tag{5.20}$$

The asymptotically small parameters are defined by

$$\varepsilon_n^{(1)} = 1 + i4ka(2n+1)^{-1} \psi'_n(ka) \zeta'_n(ka) - \\ i4ka(2n+1)^{-1} \psi'^2_n(ka) \zeta'_n(kb) / \psi'_n(kb), \tag{5.21}$$

$$\varepsilon_n^{(2)} = 1 + i4kb(2n+1)^{-1} \psi'_n(kb) \zeta'_n(kb) - \\ i4kb(2n+1)^{-1} \psi'_n(ka) \zeta'^2_n(kb) / \zeta'_n(ka), \tag{5.22}$$

$$\mu_n^{(1)} = 1 - \frac{i}{ka}(2n+1) \psi_n(ka) \zeta_n(ka) + \\ \frac{i}{ka}(2n+1) \psi^2_n(ka) \zeta_n(kb) / \psi_n(kb), \tag{5.23}$$

$$\mu_n^{(2)} = 1 - \frac{i}{ka} \psi_n(kb) \zeta_n(kb) (2n+1) + \\ \frac{i}{kb}(2n+1) \psi_n(ka) \zeta^2_n(kb) / \zeta_n(ka). \tag{5.24}$$

We shall see that they are closely related to the asymptotically small parameters already encountered in Sections 4.1 and 4.2.

It is convenient to introduce the notations

$$q_n^{(1)} = \zeta'_n(kb) / \zeta'_n(ka), \quad q_n^{(2)} = \psi'_n(ka) / \psi'_n(kb), \tag{5.25}$$

$$t_n^{(1)} = \zeta_n(kb) / \zeta_n(ka), \quad t_n^{(2)} = \psi_n(ka) / \psi_n(kb), \tag{5.26}$$

and

$$\alpha_n^{(1)} = -4i^{n+1} ka \psi'_n(ka), \quad \alpha_n^{(2)} = -4i^{n+1} kb \psi'_n(kb), \tag{5.27}$$

$$\beta_n^{(1)} = \frac{1}{ka} i^{n+1} (2n+1) \psi_n (ka), \quad \beta_n^{(2)} = \frac{1}{kb} i^{n+1} (2n+1) \psi_n (kb). \quad (5.28)$$

With this notation, the functional equations (5.11)–(5.18) may be written

$$\sum_{n=1}^{\infty} \frac{2n+1}{n(n+1)} \left\{ \left(1 - \varepsilon_n^{(1)}\right) x_n - \alpha_n^{(1)} \right\} P_n^1 (\cos\theta) = i4kaC_1 \tan\frac{\theta}{2}, \quad \theta \in (0, \theta_1)$$
$$(5.29)$$

$$\sum_{n=1}^{\infty} \frac{1}{n(n+1)} \left\{ x_n - q_n^{(1)} y_n \right\} P_n^1 (\cos\theta) = -iC_2 \cot\frac{\theta}{2}, \quad \theta \in (\theta_1, \pi) \quad (5.30)$$

$$\sum_{n=1}^{\infty} \frac{1}{n(n+1)} \left\{ \left(1 - \mu_n^{(1)}\right) z_n - \beta_n^{(1)} \right\} P_n^1 (\cos\theta) = \frac{1}{ka} C_1 \tan\frac{\theta}{2}, \quad \theta \in (0, \theta_1)$$
$$(5.31)$$

$$\sum_{n=1}^{\infty} \frac{2n+1}{n(n+1)} \left\{ z_n - t_n^{(1)} v_n \right\} P_n^1 (\cos\theta) = -iC_2 \cot\frac{\theta}{2}, \quad \theta \in (\theta_1, \pi) \quad (5.32)$$

and

$$\sum_{n=1}^{\infty} \frac{2n+1}{n(n+1)} \left\{ \left(1 - \varepsilon_n^{(2)}\right) y_n - \alpha_n^{(2)} \right\} P_n^1 (\cos\theta) = i4kbC_3 \tan\frac{\theta}{2}, \quad \theta \in (0, \theta_0)$$
$$(5.33)$$

$$\sum_{n=1}^{\infty} \frac{1}{n(n+1)} \left\{ y_n - q_n^{(2)} x_n \right\} P_n^1 (\cos\theta) = -iC_4 \cot\frac{\theta}{2}, \quad \theta \in (\theta_0, \pi) \quad (5.34)$$

$$\sum_{n=1}^{\infty} \frac{1}{n(n+1)} \left\{ \left(1 - \mu_n^{(2)}\right) v_n - \beta_n^{(2)} \right\} P_n^1 (\cos\theta) = \frac{1}{kb} C_3 \tan\frac{\theta}{2}, \quad \theta \in (0, \theta_0)$$
$$(5.35)$$

$$\sum_{n=1}^{\infty} \frac{2n+1}{n(n+1)} \left\{ v_n - t_n^{(2)} z_n \right\} P_n^1 (\cos\theta) = -iC_4 \cot\frac{\theta}{2}, \quad \theta \in (\theta_0, \pi). \quad (5.36)$$

The asymptotically small parameters $\varepsilon_n^{(i)}$, $\mu_n^{(i)}$ $(i = 1, 2)$ are closely related to the parameters ε_n, μ_n introduced in Sections 4.1 and 4.2, but incorporate additional terms accounting for the interaction between the two open spheres. Making use of very first terms in the asymptotic expansions (4.10) one can readily show that the additional terms depend upon the parameter $q = a/b$ and, in fact,

$$\varepsilon_n^{(1)} - \varepsilon_n(ka) = q^{2n+1} + O(q^{2n+2}), \tag{5.37}$$
$$\mu_n^{(1)} - \mu_n(ka) = q^{2n+1} + O(q^{2n+2}), \tag{5.38}$$
$$\varepsilon_n^{(2)} - \varepsilon_n(kb) = -q^{2n+1} + O(q^{2n+2}), \tag{5.39}$$
$$\mu_n^{(2)} - \mu_n(kb) = q^{2n+1} + O(q^{2n+2}), \tag{5.40}$$

as $n \to \infty$. Moreover,

$$q_n^{(1)} = q^{n+1} + O(q^{n+2}), \quad q_n^{(2)} = q^n + O(q^{n+1}), \tag{5.41}$$
$$t_n^{(1)} = q^n + O(q^{n+1}), \quad t_n^{(2)} = q^{n+1} + O(q^{n+2}), \tag{5.42}$$

as $n \to \infty$, so that the coupling between the sequences $\{x_n\}_{n=1}^\infty$ and $\{y_n\}_{n=1}^\infty$ is asymptotically weak, as is that between $\{z_n\}_{n=1}^\infty$ and $\{v_n\}_{n=1}^\infty$.

The regularisation procedure expounded in Section 4.2 and elsewhere may be adapted to transform equations (5.29)–(5.32) and (5.33)–(5.36) to the following coupled i.s.l.a.e. for the unknown coefficients $\{x_n\}_{n=1}^\infty$, $\{y_n\}_{n=1}^\infty$, $\{z_n\}_{n=1}^\infty$ and $\{v_n\}_{n=1}^\infty$ that all belong to the class l_2:

$$x_m - \sum_{n=1}^\infty x_n \varepsilon_n^{(1)} R_{nm}(\theta_1) - \sum_{n=1}^\infty y_n q_n^{(1)} [\delta_{nm} - R_{nm}(\theta_1)]$$

$$- i4ka R_{0m}(\theta_1) C_1 + i R_{0m}(\theta_1) C_2 = \sum_{n=1}^\infty a_n^{(1)} R_{nm}(\theta_1), \quad (5.43)$$

where $m = 1, 2, \ldots,$

$$- \sum_{n=1}^\infty x_n \varepsilon_n^{(1)} R_{n0}(\theta_1) + \sum_{n=1}^\infty y_n q_n^{(1)} R_{n0}(\theta_1)$$

$$- i4ka R_{00}(\theta_1) C_1 - i[1 - R_{00}(\theta_1)] C_2 = \sum_{n=1}^\infty a_n^{(1)} R_{n0}(\theta_1), \quad (5.44)$$

$$z_m - \sum_{n=1}^\infty z_n \mu_n^{(1)} Q_{nm}(\theta_1) - \sum_{n=1}^\infty v_n t_n^{(1)} [\delta_{nm} - Q_{nm}(\theta_1)]$$

$$- \frac{1}{ka} Q_{0m}(\theta_1) C_1 + Q_{0m}(\theta_1) C_2 = \sum_{n=1}^\infty \beta_n^{(1)} Q_{nm}(\theta_1), \quad (5.45)$$

where $m = 1, 2, \ldots,$

$$-\sum_{n=1}^{\infty} z_n \mu_n^{(1)} Q_{n0} (\theta_1) + \sum_{n=1}^{\infty} v_n t_n^{(1)} Q_{n0} (\theta_1)$$

$$-\frac{1}{ka} Q_{00} (\theta_1) C_1 - [1 - Q_{00} (\theta_1)] C_2 = \sum_{n=1}^{\infty} \beta_n^{(1)} Q_{n0} (\theta_1) , \quad (5.46)$$

$$y_m - \sum_{n=1}^{\infty} y_n \varepsilon_n^{(2)} R_{nm} (\theta_0) - \sum_{n=1}^{\infty} x_n q_n^{(2)} [\delta_{nm} - R_{nm} (\theta_0)]$$

$$- i4kb R_{0m} (\theta_0) C_3 + i R_{0m} (\theta_0) C_4 = \sum_{n=1}^{\infty} \alpha_n^{(2)} R_{nm} (\theta_0) , \quad (5.47)$$

where $m = 1, 2, \ldots,$

$$-\sum_{n=1}^{\infty} y_n \varepsilon_n^{(2)} R_{n0} (\theta_0) + \sum_{n=1}^{\infty} x_n q_n^{(2)} R_{n0} (\theta_0)$$

$$- i4kb R_{00} (\theta_0) C_3 - i [1 - R_{00} (\theta_0)] C_4 = \sum_{n=1}^{\infty} \alpha_n^{(2)} R_{n0} (\theta_0) , \quad (5.48)$$

$$v_m - \sum_{n=1}^{\infty} v_n \mu_n^{(2)} Q_{nm} (\theta_0) - \sum_{n=1}^{\infty} z_n t_n^{(2)} [\delta_{nm} - Q_{nm} (\theta_0)]$$

$$- \frac{1}{kb} Q_{0m} (\theta_0) C_3 + Q_{0m} (\theta_0) C_4 = \sum_{n=1}^{\infty} \beta_n^{(2)} Q_{nm} (\theta_0) , \quad (5.49)$$

where $m = 1, 2, \ldots,$ and

$$-\sum_{n=1}^{\infty} v_n \mu_n^{(2)} Q_{n0} (\theta_0) + \sum_{n=1}^{\infty} z_n t_n^{(2)} Q_{n0} (\theta_0)$$

$$- \frac{1}{kb} Q_{00} (\theta_0) C_3 - [1 - Q_{00} (\theta_0)] C_4 = \sum_{n=1}^{\infty} \beta_n^{(2)} Q_{n0} (\theta_0) . \quad (5.50)$$

This multi-parametric scattering system is characterised by the independent parameters θ_0, θ_1, kb and $q = a/b$. If θ_0 and θ_1 exceed $90°$ the structure may usefully be regarded as two coupled cavities, the interior spherical cavity $(r < a)$ and the spherical annular layer $(a < r < b)$. Our previous studies indicate that the scattered response is dominated by the quasi-eigenoscillations that develop when the excitation frequency is close to one of the quasi-eigenvalues of these oscillations.

For the complete range of the four parameters in this diffraction problem, a complete examination of the radar or total cross-section is too lengthy and we restrict ourselves to one or two limiting cases in which the aperture of the inner cavity closes (so $\theta_1 = \pi$). For any q ($q < 1$), the quasi-eigenvalues are, when θ_1 is small ($\theta_1 \ll 1$), perturbations of the eigenvalue spectrum of the interior of the annular region obtained when both apertures close; this is obtained as the roots of the expressions Δ_n and $\tilde{\Delta}_n$ (defined by [5.5]–[5.6]).

In fact, the characteristic equation for the TM oscillations takes the form, for each $n = 1, 2, \ldots$,

$$\psi_n' \left(q\varkappa_{nl}^{(0)} \right) \zeta_n' \left(\varkappa_{nl}^{(0)} \right) - \psi_n' \left(\varkappa_{nl}^{(0)} \right) \zeta_n' \left(q\varkappa_{nl}^{(0)} \right) = 0, \qquad (5.51)$$

where the roots are ordered in ascending order ($\varkappa_{n1}^{(0)} < \varkappa_{n2}^{(0)} < \ldots$), whilst the characteristic equation for the TE oscillations takes the form, for each $n = 1, 2, \ldots$,

$$\psi_n \left(q\nu_{nl}^{(0)} \right) \zeta_n \left(\nu_{nl}^{(0)} \right) - \psi_n \left(\nu_{nl}^{(0)} \right) \zeta_n \left(q\nu_{nl}^{(0)} \right) = 0, \qquad (5.52)$$

where the roots are also ordered in ascending order ($\nu_{n1}^{(0)} < \nu_{n2}^{(0)} < \ldots$). If $a \to 0$, the roots coincide with the roots of the characteristic equations obtained for the empty spherical cavity (see equations [4.38] and [4.78]).

If $q \ll 1$, the inner sphere may be considered as a small embedding in the outer open spherical cavity of radius b, and one may simplify both equations, using formula (4.10). An application of Newton's method shows that the presence of the small metallic sphere perturbs the eigenvalues $\varkappa_{nl}^{(0)}$ and $\nu_{nl}^{(0)}$ for closed spherical cavity, for TM and TE types of oscillations, respectively, according to

$$\varkappa_{nl}/\varkappa_{nl}^{(0)} = 1 -$$

$$\frac{(2^n n!)^2}{(2n)!\,(2n+1)!} \frac{n+1}{n} \frac{\left\{ \left(\varkappa_{nl}^{(0)} \right)^{n+1} \operatorname{Im} \zeta_n' \left(\varkappa_{nl}^{(0)} \right) \right\}^2}{\varkappa_{nl}^{(0)2} - n\,(n+1)} q^{2n+1} + O\left(q^{2n+3} \right)$$

$$(5.53)$$

$$\nu_{nl}/\nu_{nl}^{(0)} = 1 + \frac{(2^n n!)^2}{(2n)!\,(2n+1)!} \left[\left(\nu_{nl}^{(0)} \right)^n \operatorname{Im} \zeta_n \left(\nu_{nl}^{(0)} \right) \right]^2 q^{2n+1} + O\left(q^{2n+3} \right).$$

$$(5.54)$$

Thus the frequencies of oscillations of TM type in the spherical cavity with an embedding are reduced by an amount proportional to q^{2n+1}, whilst the frequencies for TE type oscillations are increased by an amount also proportional to q^{2n+1}. For lowest order TM or TE oscillations ($n = 1$), the shift (q^3) is proportional to the volume occupied by the small metallic sphere.

Another extreme is the narrow spherical gap (or concentric layer) characterised by $p = 1 - q = (b - a)/a \ll 1$. Under this condition, one may justify that no eigenoscillations of TE type occur. Only TM oscillations develop at frequencies close to the spectrum of eigenvalues defined by

$$\varkappa_n = \sqrt{n\,(n+1)}\left\{1 + \frac{1}{2}\left(p - p^2 + p^3\right) + O\left(p^4\right)\right\}. \qquad (5.55)$$

These oscillations play an important part in studies of the natural annular resonator formed by the Earth and ionosphere, and are known as *Schumann resonances*. The spectrum of doubly indexed TM eigenvalues $\varkappa_{nl}^{(0)}$ obtained when $q = 0$ degenerates into a singly indexed sequence $\varkappa_n$ when $p \ll 1$, due to the collapse in radial variation previously indexed by l.

This preliminary quite simple information is useful in sorting out which resonances occur in the more general situation. The total and radar cross-sections for a hemispherical shell ($\theta_0 = 90°$) enclosing a smaller closed sphere ($\theta_1 = 180°$) are shown in Figures 5.2 and 5.3, respectively, for values of q equal to 0, 0.1 and 0.5. The plots corresponding to the values $q = 0$ and $q = 0.1$ are practically indistinguishable over the range $ka \leq 20$.

In Figures 5.4 and 5.5 are plotted the total and radar cross-sections for the narrow concentric hemispherical cavity ($q = 0.98$).

Assuming that both Earth and ionosphere are perfectly conducting, a crude model of the Earth–ionosphere resonator is provided by the fully closed spherical annular region described above, and we may use (5.55) to calculate low-frequency resonances that develop inside this resonator, being

$$f_n = \frac{c}{2\pi R}\sqrt{n\,(n+1)}\left\{1 + \frac{1}{2}\left(p - p^2 + p^3\right)\right\} \qquad (5.56)$$

(in Hz), where R is the radius of Earth, $p = h/R$ and h is ionosphere altitude. Using the values $R = 6.409 \cdot 10^6$ and $h = 10^5$ m predicts the lowest resonance frequencies to be $f_1 = 10.45Hz$, $f_2 = 18.10Hz$, $f_3 = 25.6Hz$, $f_4 = 33.06Hz, \ldots$. In fact, the values of the observed spectrum are somewhat less than this, for example, $f_1 \simeq 8Hz$.

Recognising the diurnal variation of the ionosphere, a slightly less crude approach is to replace the outer spherical shell of this model with a perfectly conducting hemispherical shell. This case corresponds to the data plotted in Figure 5.4 and 5.5. However the values of f_n ($n = 1, 2, 3, 4$) are nearly

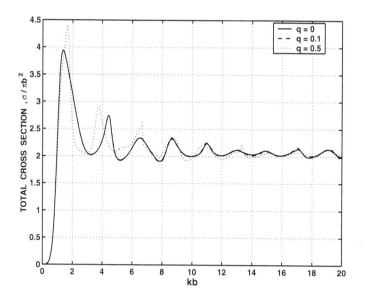

FIGURE 5.2. Total cross-section of the concentric shells with $\theta_1 = 180°$ and $\theta_0 = 90°$, for various values of q.

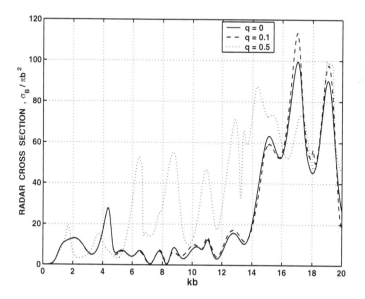

FIGURE 5.3. Radar cross-section for the concentric shells with $\theta_1 = 180°$ and $\theta_0 = 90°$, for various values of q.

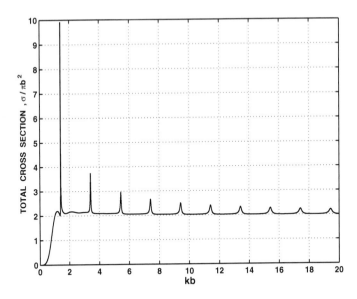

FIGURE 5.4. Total cross-section for the concentric shells with $\theta_1 = 180°$ and $\theta_0 = 90°$.

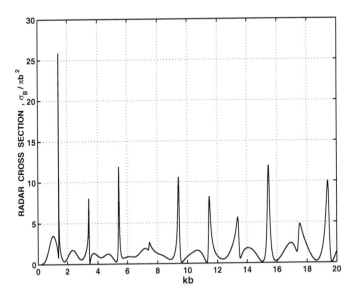

FIGURE 5.5. Radar cross-section for the concentric shells with $\theta_1 = 180°$ and $\theta_0 = 90°$.

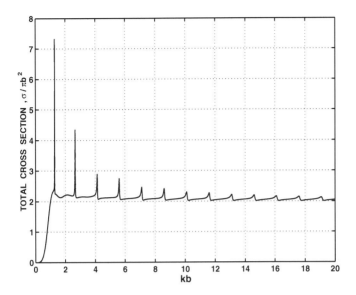

FIGURE 5.6. Total cross-section for the concentric shells with $\theta_1 = 180°$ and $\theta_0 = 120°$.

unchanged from the previous values, and we are forced to conclude that the model is too simplistic for accurate prediction of the Schumann resonances.

As the aperture becomes narrower, increasingly large Q-factor oscillations are excited. This phenomenon is illustrated in the total cross-sections plotted in Figures 5.6 and 5.7 for aperture angles θ_0 equal to 120° and 150°, respectively and in the corresponding radar cross-sections plotted in Figures 5.8 and 5.9.

5.2   Dipole Excitation of a Slotted Sphere.

Cavity backed slot antennae are of widespread importance in antenna theory and design [54]. A canonical problem of relevance is the spherical shell with a thin equatorial slot, where the cavity is excited by a vertical (electric or magnetic) dipole located on the axis of symmetry. Consider then a thin-walled hollow metallic sphere of radius a in which an equatorial slot is symmetrically removed so that the two spherical shells each subtend an angle θ_0 at the origin; set $\theta_1 = \frac{\pi}{2} - \theta_0$, so that the slot subtends an angle $2\theta_0$ at the origin. See Figure 5.10.

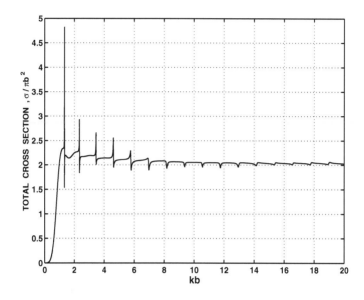

FIGURE 5.7. Total cross-section for the concentric shells with $\theta_1 = 180°$ and $\theta_0 = 150°$.

Let the z-axis coincide with the axis of symmetry, on which is located a vertical electric or magnetic dipole, in the cavity interior; without loss of generality, it may be located in the upper half space $z \geq 0$.

The problem may be formulated in a manner very similar to that studied in Section 4.1 (dipole excitation of a spherical cavity with a single circular aperture). We retain the notations employed in that section, and *mutatis mutandis*, we may write down at once symmetrical triple series equations to be solved for the unknown coefficients x_n and y_n.

5.2.1   The Vertical Electric Dipole.

Deferring the vertical magnetic dipole (the TE case), we consider first the vertical electric dipole (the TM case) for which the functional equations

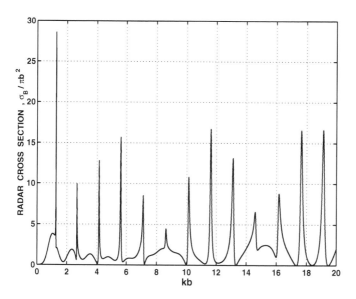

FIGURE 5.8. Radar cross-section for the concentric shells with $\theta_1 = 180°$ and $\theta_0 = 120°$.

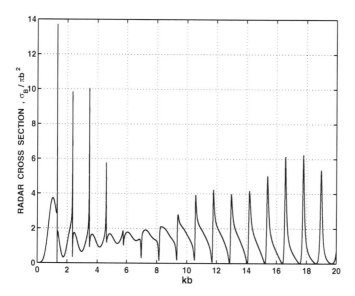

FIGURE 5.9. Radar cross-section for the concentric shells with $\theta_1 = 180°$ and $\theta_0 = 150°$.

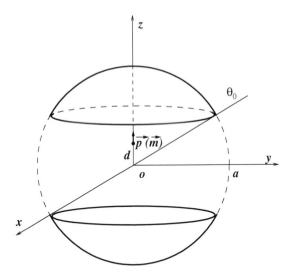

FIGURE 5.10. Dipole excitation of a slotted spherical cavity.

are

$$\sum_{n=1}^{\infty} (1 - \varepsilon_n)\, x_n P_n^1 (\cos \theta) = \sum_{n=1}^{\infty} \alpha_n P_n^1 (\cos \theta)\,, \quad \theta \in (0, \theta_0) \qquad (5.57)$$

$$\sum_{n=1}^{\infty} \frac{2n + 1}{n(n + 1)} x_n P_n^1 (\cos \theta) = 0,\quad \theta \in (\theta_0, \pi - \theta_0) \qquad (5.58)$$

$$\sum_{n=1}^{\infty} (1 - \varepsilon_n)\, x_n P_n^1 (\cos \theta) = \sum_{n=1}^{\infty} \alpha_n P_n^1 (\cos \theta)\,, \quad \theta \in (\pi - \theta_0, \pi)\,. \qquad (5.59)$$

As usual we may transform the triple series equations (5.57)–(5.59) to a decoupled pair of dual series equations for the even and odd index coefficients and treat each separately. Setting

$$u = 2z^2 - 1 = \cos(2\theta_0)\,, \qquad u_0 = 2z_0^2 - 1 = \cos(2\theta_0)\,, \qquad (5.60)$$

and using the relationships

$$P_{2n}^1 (z) \;=\; (1 - u^2)^{\frac{1}{2}} \left(n + \frac{1}{2} \right) P_{n-1}^{(1, \frac{1}{2})} (u)\,, \qquad (5.61)$$

$$P_{2n+1}^1 (z) \;=\; \sqrt{2}\,(1 - u)^{\frac{1}{2}} \left(n + \frac{1}{2} \right) P_n^{(1, -\frac{1}{2})} (u)\,, \qquad (5.62)$$

the system for the even index coefficients is

$$\sum_{n=1}^{\infty} \frac{n + \frac{1}{4}}{n} x_{2n} P_{n-1}^{(1,\frac{1}{2})}(u) = 0, \qquad u \in (-1, u_0), \qquad (5.63)$$

$$\sum_{n=0}^{\infty} \left(n + \frac{1}{2}\right) (1 - \varepsilon_{2n}) x_{2n} P_{n-1}^{(1,\frac{1}{2})}(u)$$

$$= \sum_{n=1}^{\infty} \left(n + \frac{1}{2}\right) \alpha_{2n} P_{n-1}^{(1,\frac{1}{2})}, \qquad u \in (u_0, 1), \quad (5.64)$$

and that for the odd index coefficients is

$$\sum_{n=1}^{\infty} \frac{n + \frac{3}{4}}{n + 1} x_{2n+1} P_n^{(1,-\frac{1}{2})}(u) = 0, \qquad u \in (u_0, -1), \qquad (5.65)$$

$$\sum_{n=0}^{\infty} \left(n + \frac{1}{2}\right) (1 - \varepsilon_{2n+1}) x_{2n+1} P_n^{(1,-\frac{1}{2})}(u)$$

$$= \sum_{n=0}^{\infty} \left(n + \frac{1}{2}\right) \alpha_{2n+1} P_n^{(1,-\frac{1}{2})}(u), \qquad u \in (u_0, 1). \quad (5.66)$$

A direct use of the Abel integral transform method identifies the asymptotically small parameter

$$p_n = 1 - \left(n + \frac{1}{4}\right) \left[\frac{\Gamma\left(n + \frac{1}{2}\right)}{\Gamma(n + 1)}\right]^2 \qquad (5.67)$$

that satisfies $p_n = O\left(n^{-2}\right)$, as $n \to \infty$, and transforms equations (5.63)–(5.64) to their pre-regularised form

$$\sum_{n=1}^{\infty} X_{2n} P_{n-1}^{(\frac{3}{2},0)}(u) =$$

$$\begin{cases} \sum_{n=1}^{\infty} X_{2n} p_n P_{n-1}^{(\frac{3}{2},0)}(u), & u \in (-1, u_0) \\ \sum_{n=1}^{\infty} (A_{2n} + \varepsilon_{2n} X_{2n}) P_{n-1}^{(\frac{3}{2},0)}(u), & u \in (u_0, 1), \end{cases} \qquad (5.68)$$

where

$$\{A_{2n}, X_{2n}\} = \frac{\Gamma\left(n + \frac{1}{2}\right)}{\Gamma(n + 1)} \left\{h_{n-1}^{(\frac{3}{2},0)}\right\}^{-\frac{1}{2}} \{\alpha_{2n}, x_{2n}\}. \qquad (5.69)$$

From (5.68) we deduce immediately the i.s.l.a.e. of the second kind for the even index coefficients X_{2n},

$$(1 - p_m) X_{2m} - \sum_{n=1}^{\infty} X_{2n} (\varepsilon_{2n} - p_n) Q_{n-1,m-1}^{(\frac{3}{2},0)} (u_0)$$

$$= \sum_{n=1}^{\infty} A_{2n} Q_{n-1,m-1}^{(\frac{3}{2},0)} (u_0), \quad (5.70)$$

where $m = 1, 2, 3, \dots$.

In contrast to (5.63) and (5.64), equations (5.65) and (5.66) are not directly solvable by the Abel integral transform method because the second index β of the relevant Jacobi polynomials $P_{n-1}^{(\alpha,\beta)}$ takes the value $-\frac{1}{2}$, and the methods of Chapter 2 of Volume I may not be validly applied. However as noted at the end of Section 2.2 of Volume I [1] the standard device to surmount this difficulty is to integrate both equations (5.65) and (5.66) making the use of formula (see formula [1.174] of Volume I)

$$\int (1 + u)^{-\frac{1}{2}} P_n^{(1,-\frac{1}{2})} (u) \, du = \frac{(1 + u)^{\frac{1}{2}}}{n + \frac{1}{2}} P_n^{(0,\frac{1}{2})} (u) . \quad (5.71)$$

Term-by-term integration of these expansions in orthogonal Jacobi polynomials may be justified as in Appendix D.2 of Volume I. Thus

$$\sum_{n=0}^{\infty} \frac{n + \frac{3}{4}}{(n + \frac{1}{2}) (n + 1)} x_{2n+1} P_n^{(0,\frac{1}{2})} (u) = 0, \quad u \in (-1, u_0), \quad (5.72)$$

$$\sum_{n=0}^{\infty} (1 - \varepsilon_{2n+1}) x_{2n+1} P_n^{(0,\frac{1}{2})} (u) =$$

$$(1 + u)^{-\frac{1}{2}} C + \sum_{n=0}^{\infty} \alpha_{2n+1} P_n^{(0,\frac{1}{2})} (u), \quad u \in (u_0, 1), \quad (5.73)$$

where C is a constant of integration to be determined subsequently by the requirement that the solution $\{x_{2n+1}\}_{n=0}^{\infty}$ belongs to the functional class l_2.

The integrated equations (5.72) and (5.73) are now amenable to treatment by the standard Abel integral transform method, and one obtains

$$(1 - u)^{\frac{1}{2}} \sum_{n=0}^{\infty} X_{2n+1} \hat{P}_n^{(\frac{1}{2},0)} (u) = \begin{cases} F_1 (u), & u \in (-1, u_0) \\ F_2 (u), & u \in (u_0, 1), \end{cases} \quad (5.74)$$

where

$$F_1(u) = (1 - u)^{\frac{1}{2}} \sum_{n=0}^{\infty} X_{2n+1} q_n \hat{P}_n^{(\frac{1}{2},0)}(u),$$

$$F_2(u) = (1 - u)^{\frac{1}{2}} \sum_{n=0}^{\infty} \{A_{2n+1} + X_{2n+1}\varepsilon_{2n+1}\} \hat{P}_n^{(\frac{1}{2},0)}(u)$$

$$+ \frac{2}{\sqrt{\pi}} C \log \left[\frac{\sqrt{2} + \sqrt{1 - u}}{\sqrt{1 + u}} \right],$$

$$\{a_{2n+1}, x_{2n+1}\} = \frac{\Gamma\left(n + \frac{3}{2}\right)}{\Gamma\left(n + 1\right)} \left\{ h_n^{(\frac{1}{2},0)} \right\}^{-\frac{1}{2}} \{A_{2n+1}, X_{2n+1}\} \qquad (5.75)$$

and

$$q_n = 1 - \frac{n + \frac{3}{4}}{\left(n + \frac{1}{2}\right)(n + 1)} \left\{ \frac{\Gamma\left(n + \frac{3}{2}\right)}{\Gamma\left(n + 1\right)} \right\}^2 \qquad (5.76)$$

is the asymptotically small parameter satisfying $q_n = O\left(n^{-2}\right)$ as $n \to \infty$.

Since $\{x_{2n+1}\}_{n=0}^{\infty}$ lies in l_2, it follows that $\{X_{2n+1}\}_{n=0}^{\infty}$ lies in $l_2(2)$. This constrains the value of C, for if it could be chosen arbitrarily we could only make the weaker assertion that $\{X_{2n+1}\}_{n=0}^{\infty}$ lies in $l_2(0) \equiv l_2$. The constant C is thus determined by recognising that the series on the left hand side of (5.74) is uniformly convergent to a function that is continuous on the interval $[-1, 1]$, and is therefore continuous at the point $u = u_0$ (also see Chapter 2 of Volume I [1]). Thus

$$C = -\frac{\sqrt{\pi}}{2} \frac{(1 - u_0)^{\frac{1}{2}}}{\log\left[\left(\sqrt{2} + \sqrt{1 - u_0}\right)/\sqrt{1 + u_0}\right]} \times$$

$$\sum_{n=0}^{\infty} \{A_{2n+1} + (\varepsilon_{2n+1} - q_n) X_{2n+1}\} \hat{P}_n^{(\frac{1}{2},0)}(u_0). \qquad (5.77)$$

Now multiply both sides of (5.74) by $\hat{P}_m^{(\frac{1}{2},0)}(u)$ and integrate over $(-1,1)$. Taking into account the readily evaluated interval

$$I_m = \int_{u_0}^1 \log\left[\frac{\sqrt{2}+\sqrt{1-u}}{\sqrt{1+u}}\right] \hat{P}_m^{(\frac{1}{2},0)}(u)\, du =$$

$$\frac{1+u_0}{\sqrt{\left(m+\frac{1}{2}\right)(m+1)}} \log\left[\frac{\sqrt{2}+\sqrt{1-u_0}}{\sqrt{1+u_0}}\right] \hat{P}_m^{(-\frac{1}{2},1)}(u_0)$$

$$+ \frac{1}{\sqrt{2}} \frac{(1-u_0)^{\frac{1}{2}}}{\left(m+\frac{1}{2}\right)(m+1)} \hat{P}_m^{(\frac{1}{2},0)}(u_0), \quad (5.78)$$

and the identity (see Appendix B.6, Volume I)

$$\hat{Q}_{nm}^{(\frac{1}{2},0)}(u_0) = -\frac{(1+u_0)^{\frac{1}{2}}(1+u_0)}{\sqrt{\left(m+\frac{1}{2}\right)(m+1)}} \hat{P}_m^{(-\frac{1}{2},1)}(u_0)\, \hat{P}_n^{(\frac{1}{2},0)}(u_0)$$

$$+ \sqrt{\frac{\left(n+\frac{1}{2}\right)(n+1)}{\left(m+\frac{1}{2}\right)(m+1)}} \hat{Q}_{nm}^{(-\frac{1}{2},1)}(u_0), \quad (5.79)$$

and the value (5.77) for C we obtain the i.s.l.a.e. of the second kind

$$(1-q_m) X_{2m+1} - \sum_{n=0}^{\infty} X_{2n+1}\left(\varepsilon_{2n+1} q_n\right) \hat{Q}_{nm}^{(\frac{1}{2},0)}(u_0)$$

$$= \sum_{n=1}^{\infty} A_{2n+1} \hat{Q}_{nm}^{(\frac{1}{2},0)}(u_0) + \frac{2}{\sqrt{\pi}} C \cdot I_m, \quad (5.80)$$

where $m = 0,1,2,\ldots$.

One may verify that the choice of the constant C eliminates all terms of order $O\left(m^{-1}\right)$ and the solution thus lies in $l_2(2)$. Introducing the coefficients

$$\left\{X_{2m+1}^*, A_{2m+1}^*\right\} = \sqrt{\left(m+\frac{1}{2}\right)(m+1)}\left\{X_{2m+1}, A_{2m+1}\right\}, \quad (5.81)$$

so that $\left\{X_{2m+1}^*\right\}_{m=0}^{\infty} \in l_2$, the final form of the i.s.l.a.e. is

$$(1-q_m) X_{2m+1}^* - \sum_{n=0}^{\infty} X_{2m+1}^*\left(\varepsilon_{2n+1} q_n\right) S_{nm}(u_0) = \sum_{n=0}^{\infty} A_n^* S_{nm}(u_0),$$

$$\qquad(5.82)$$

where $m = 0, 1, 2, \ldots$, and

$$S_{nm}(u_0) = \hat{Q}_{nm}^{(-\frac{1}{2},1)}(u_0) - Q_n(u_0) Q_m(u_0) / \log\left[\frac{\sqrt{2} + \sqrt{1 - u_0}}{\sqrt{1 + u_0}}\right],$$

$$(5.83)$$

with

$$Q_s(u_0) = 2^{1/4} \left(\frac{1 - u_0}{2}\right)^{\frac{1}{2}} \frac{\hat{P}_s^{(\frac{1}{2},0)}(u_0)}{\sqrt{\left(s + \frac{1}{2}\right)(s + 1)}}.$$

Thus the problem of slotted sphere excitation by an electric vertical dipole produces a pair of i.s.l.a.e. (5.70) and (5.82) of the second kind for the even index coefficients $\{X_{2n}\}_{n=1}^{\infty}$ and odd coefficients $\{X_{2n+1}^*\}_{n=0}^{\infty}$, respectively. Uniqueness of solutions to these systems, and of those encountered in other sections in this chapter, is established by adapting the uniqueness argument given in Section 2.1; the truncation number for numerical studies is chosen according to the criterion set out in that section.

5.2.2   The Vertical Magnetic Dipole

We now consider vertical magnetic dipole excitation of the slotted spherical cavity. The relevant functional equations for this excitation (the TE case) corresponding to (5.57) and (5.59) are

$$\sum_{n=1}^{\infty} \frac{2n + 1}{n(n + 1)} (1 - \mu_n) y_n P_n^1(\cos\theta) = \sum_{n=1}^{\infty} \frac{2n + 1}{n(n + 1)} \beta_n P_n^1(\cos\theta),$$

$$\theta \in (0, \theta_0) \quad (5.84)$$

$$\sum_{n=1}^{\infty} y_n P_n^1(\cos\theta) = 0, \qquad \theta \in (\theta_0, \pi - \theta_0) \qquad (5.85)$$

$$\sum_{n=1}^{\infty} \frac{2n + 1}{n(n + 1)} (1 - \mu_n) y_n P_n^1(\cos\theta) = \sum_{n=1}^{\infty} \frac{2n + 1}{n(n + 1)} \beta_n P_n^1(\cos\theta),$$

$$\theta \in (\pi - \theta_0, \pi). \quad (5.86)$$

As for the electric dipole case, we may transform the triple series equations (5.84)–(5.86) to a decoupled pair of dual series equations for the even and odd index coefficients and treat them separately. Recalling the relationships (5.60), (5.61) and (5.62), we obtain

$$\sum_{n=1}^{\infty} \left(n + \frac{1}{2} \right) y_{2n} P_{n-1}^{(1,\frac{1}{2})} (u) = 0, \qquad u \in (-1, u_0) \qquad (5.87)$$

$$\sum_{n=1}^{\infty} \frac{n + \frac{1}{4}}{n} (1 - \mu_{2n}) y_{2n} P_{n-1}^{(1,\frac{1}{2})} (u) = \sum_{n=1}^{\infty} \frac{n + \frac{1}{4}}{n} \beta_{2n} P_{n-1}^{(1,\frac{1}{2})} (u) \,,$$

$$u \in (u_0, 1) \quad (5.88)$$

$$\sum_{n=0}^{\infty} \left(n + \frac{1}{2} \right) y_{2n+1} P_n^{(1,-\frac{1}{2})} (u) = 0, \qquad u \in (-1, u_0) \qquad (5.89)$$

$$\sum_{n=0}^{\infty} \frac{n + \frac{3}{4}}{n + 1} (1 - \mu_{2n+1}) y_{2n+1} P_n^{(1,-\frac{1}{2})} (u) = \sum_{n=0}^{\infty} \frac{n + \frac{3}{4}}{n + 1} \beta_{2n+1} P_n^{(1,-\frac{1}{2})} (u) \,,$$

$$u \in (u_0, 1) . \quad (5.90)$$

In contrast to the dual series equations obtained in the electric dipole excitation case, both the dual series equations (5.87)–(5.88) and (5.89)–(5.90) are directly amenable to treatment by the Abel integral transform method, and may be transformed reduced to their pre-regularised form

$$\sum_{n=1}^{\infty} Y_{2n} \hat{P}_{n-1}^{(\frac{1}{2},1)} (u) =$$

$$\begin{cases} \sum_{n=1}^{\infty} t_n Y_{2n} \hat{P}_{n-1}^{(\frac{1}{2},1)} (u) \,, & u \in (-1, u_0) \\ \sum_{n=1}^{\infty} \{ B_{2n+1} + \mu_{2n} Y_{2n} \} \hat{P}_{n-1}^{(\frac{1}{2},1)} (u) \,, & u \in (u_0, 1) \end{cases} \quad (5.91)$$

$$\sum_{n=0}^{\infty} Y_{2n+1} \hat{P}_n^{(\frac{1}{2},0)} (u) =$$

$$\begin{cases} \sum_{n=1}^{\infty} r_n Y_{2n+1} \hat{P}_n^{(\frac{1}{2},0)} (u) \,, & u \in (-1, u_0) \\ \sum_{n=1}^{\infty} \{ B_{2n+1} + \mu_{2n+1} Y_{2n+1} \} \hat{P}_n^{(\frac{1}{2},0)} (u) \,, & u \in (u_0, 1) \end{cases} \quad (5.92)$$

and subsequently to the well-conditioned i.s.l.a.e. (second kind Fredholm matrix equations)

$$(1 - t_m) Y_{2m} - \sum_{n=1}^{\infty} (\mu_{2n} - t_n) \hat{Q}_{n-1,m-1}^{(\frac{1}{2},1)} (u_0)$$

$$= \sum_{n=1}^{\infty} B_{2n} \hat{Q}_{n-1,m-1}^{(\frac{1}{2},1)} (u_0) , \quad (5.93)$$

where $m = 1, 2, \ldots$, and

$$(1 - r_m) Y_{2m+1} - \sum_{n=0}^{\infty} (\mu_{2n+1} - r_n) \hat{Q}_{nm}^{(\frac{1}{2},0)} (u_0)$$

$$= \sum_{n=0}^{\infty} B_{2n+1} \hat{Q}_{nm}^{(\frac{1}{2},0)} (u_0) , \quad (5.94)$$

where $m = 0, 1, 2, \ldots$; the variables and coefficients have been rescaled so that

$$\{Y_{2n}, B_{2n}\} = \left(n + \frac{1}{4}\right) \frac{\Gamma(n)}{\Gamma\left(n + \frac{1}{2}\right)} \left\{h_{n-1}^{(\frac{1}{2},1)}\right\}^{\frac{1}{2}} \{y_{2n}, \beta_{2n}\} , \quad (5.95)$$

$$\{Y_{2n+1}, B_{2n+1}\} = \left(n + \frac{3}{4}\right) \frac{\Gamma\left(n + \frac{1}{2}\right)}{\Gamma\left(n + \frac{3}{2}\right)} \left\{h_n^{(\frac{1}{2},0)}\right\}^{\frac{1}{2}} \{y_{2n+1}, \beta_{2n+1}\} , \quad (5.96)$$

and the parameters

$$t_n = 1 - \frac{n\left(n + \frac{1}{2}\right)}{n + 1/4} \left\{\frac{\Gamma\left(n + \frac{1}{2}\right)}{\Gamma(n + 1)}\right\}^2 , \quad (5.97)$$

$$r_n = 1 - \left(n + \frac{3}{4}\right)^{-1} \left\{\frac{\Gamma\left(n + \frac{3}{2}\right)}{\Gamma(n + 1)}\right\}^2 , \quad (5.98)$$

are asymptotically small: both t_n and r_n are $O\left(n^{-2}\right)$, as $n \to \infty$.

This completes our brief discussion of magnetic dipole excitation of the cavity. Before leaving this section, we note that the slotted spherical cavity is symmetrical about the plane $z = 0$. If the lower cap (in the half space $z < 0$) is discarded and replaced by the perfectly conducting infinite plane $z = 0$, we may apply image theory to deduce that the excitation of the structure consisting of spherical cap and infinite plane by a vertical electric dipole (axially located) is described by equation (5.82), and all the even index coefficients X_{2m} vanish. Magnetic dipole excitation of the same structure is described by equation (5.93) and all the odd index coefficients Y_{2m} vanish. In both cases the dipole strength should be doubled, as should all field components.

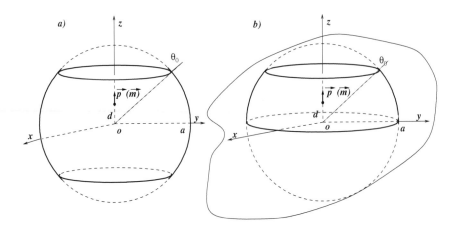

FIGURE 5.11. Dipole excitation of a) a spherical cavity with two equal circular apertures , and b) a semispherical cavity with a circular aperture, mounted on a perfectly conducting infinite ground plane.

5.3   Dipole Excitation of Doubly-Connected Spherical Shells.

The dipole excitation of a spherical cavity with two equal circular apertures symmetrically located is closely related to the problems studied in the previous section. The geometry is shown in Figure 5.11a.

In fact, from a mathematical poin t of view there is a single change. The equations to be solved are given by (5.63)–(5.66) and (5.87)–(5.90), except for the definition of subintervals on which they hold. Precisely, θ_0 now denotes the angle subtended by the aperture at the origin (instead of the angular measure of a spherical cap); also $u_0 = \cos 2\theta_0$.

For vertical electric dipole excitation, the series equations to be solved are

$$\sum_{n=1}^{\infty} \left(n + \frac{1}{2} \right) (1 - \varepsilon_{2n}) \, x_{2n} P_{n-1}^{(1,\frac{1}{2})} (u) =$$

$$\sum_{n=1}^{\infty} \left(n + \frac{1}{2} \right) \alpha_{2n} P_{n-1}^{(1,\frac{1}{2})} (u), \qquad u \in (-1, u_0) \quad (5.99)$$

$$\sum_{n=1}^{\infty} \frac{n + \frac{1}{4}}{n} x_{2n} P_{n-1}^{(1,\frac{1}{2})} (u) = 0, \qquad u \in (u_0, 1), \qquad (5.100)$$

and

$$\sum_{n=0}^{\infty} \left(n + \frac{1}{2} \right) (1 - \varepsilon_{2n+1}) \, x_{2n+1} P_n^{(1, -\frac{1}{2})} (u) =$$

$$\sum_{n=0}^{\infty} \left(n + \frac{1}{2} \right) \alpha_{2n+1} P_n^{(1, -\frac{1}{2})} (u) , \qquad u \in (-1, u_0) \quad (5.101)$$

$$\sum_{n=1}^{\infty} \frac{n + \frac{3}{4}}{n + 1} x_{2n+1} P_n^{(1, -\frac{1}{2})} (u) = 0, \qquad u \in (u_0, 1) . \quad (5.102)$$

For vertical magnetic dipole excitation, the series equations to be solved are

$$\sum_{n=1}^{\infty} \frac{n + \frac{1}{4}}{n} (1 - \mu_{2n}) \, y_{2n} P_{n-1}^{(1, \frac{1}{2})} (u) =$$

$$\sum_{n=1}^{\infty} \frac{n + \frac{1}{4}}{n} \beta_{2n} P_{n-1}^{(1, \frac{1}{2})} (u) , \qquad u \in (-1, u_0) \quad (5.103)$$

$$\sum_{n=1}^{\infty} \left(n + \frac{1}{2} \right) y_{2n} P_{n-1}^{(1, \frac{1}{2})} (u) = 0, \qquad u \in (u_0, 1) \quad (5.104)$$

and

$$\sum_{n=0}^{\infty} \frac{n + \frac{3}{4}}{n + 1} (1 - \mu_{2n+1}) \, y_{2n+1} P_n^{(1, -\frac{1}{2})} (u) =$$

$$\sum_{n=0}^{\infty} \frac{n + \frac{3}{4}}{n + 1} \beta_{2n+1} P_n^{(1, -\frac{1}{2})} (u) , \qquad u \in (-1, u_0) \quad (5.105)$$

$$\sum_{n=1}^{\infty} \left(n + \frac{1}{2} \right) y_{2n+1} P_n^{(1, -\frac{1}{2})} (u) = 0, \qquad u \in (u_0, 1) . \quad (5.106)$$

The pair of equations (5.99)–(5.100) is formally very similar to (5.87)–(5.88); likewise the pair of equations (5.101)–(5.102) is formally very similar to (5.89)-(5.90); the same similarity exists between the pairs (5.103)–(5.104) and (5.63)–(5.64); likewise the same similarity exists between the pairs (5.105)–(5.106) and (5.65)-(5.66). For the sake of brevity, we omit details of the regularisation of equations (5.99)–(5.102) and equations (5.103)–(5.106) and simply state the i.s.l.a.e. obtained in the standard way.

For the vertical electric dipole excitation, the i.s.l.a.e. for the even index coefficients is

$$(1 - \varepsilon_{2m}) X_{2m} + \sum_{n=1}^{\infty} X_{2n} \left(\varepsilon_{2n} - f_n \right) \hat{Q}_{n-1,m-1}^{(\frac{1}{2},1)} (u_0)$$

$$= \sum_{n=1}^{\infty} A_{2n} \left\{ \delta_{nm} - \hat{Q}_{n-1,m-1}^{(\frac{1}{2},1)} (u_0) \right\}, \quad (5.107)$$

where $m = 1, 2, \ldots,$

$$\{X_{2n}, A_{2n}\} = \frac{\Gamma \left(n + \frac{2}{3} \right)}{\Gamma \left(n + 1 \right)} \left\{ h_{n-1}^{(\frac{1}{2},1)} \right\}^{\frac{1}{2}} \{x_{2n}, \alpha_{2n}\},$$

and the asymptotically small parameter

$$f_n = 1 - \frac{n + \frac{1}{4}}{n \left(n + \frac{1}{2} \right)} \left\{ \frac{\Gamma \left(n + 1 \right)}{\Gamma \left(n + \frac{1}{2} \right)} \right\}^2$$

satisfies $f_n = O \left(n^{-2} \right)$, as $n \to \infty$; the i.s.l.a.e. for the odd index coefficients is

$$(1 - \varepsilon_{2m+1}) X_{2m+1} + \sum_{n=0}^{\infty} X_{2n+1} \left(\varepsilon_{2n+1} - g_n \right) \hat{Q}_{n,m}^{(\frac{1}{2},0)} (u_0)$$

$$= \sum_{n=0}^{\infty} A_{2n+1} \left\{ \delta_{nm} - \hat{Q}_{n,m}^{(\frac{1}{2},1)} (u_0) \right\}, \quad (5.108)$$

where $m = 1, 2, \ldots,$

$$\{X_{2n+1}, A_{2n+1}\} = \frac{\Gamma \left(n + \frac{2}{3} \right)}{\Gamma \left(n + 1 \right)} \left\{ h_n^{(\frac{1}{2},0)} \right\}^{\frac{1}{2}} \{x_{2n+1}, \alpha_{2n+1}\},$$

and the asymptotically small parameter

$$g_n = 1 - \left(n + \frac{3}{4} \right) \left\{ \frac{\Gamma \left(n + 1 \right)}{\Gamma \left(n + \frac{3}{2} \right)} \right\}^2$$

satisfies $g_n = O \left(n^{-2} \right)$, as $n \to \infty$.

For the vertical magnetic dipole excitation, the i.s.l.a.e. for the even index coefficients is

$$(1 - \mu_{2m}) Y_{2m} + \sum_{n=1}^{\infty} Y_{2n} \left(\mu_{2n} - h_n \right) \hat{Q}_{n-1,m-1}^{(\frac{3}{2},1)} (u_0)$$

$$= \sum_{n=1}^{\infty} B_{2n} \left\{ \delta_{nm} - \hat{Q}_{n-1,m-1}^{(\frac{3}{2},1)} (u_0) \right\}, \quad (5.109)$$

where $m = 1, 2, \ldots ,$

$$\{Y_{2n}, B_{2n}\} = \left(n + \frac{1}{4}\right) \frac{\Gamma\left(n + \frac{1}{2}\right)}{\Gamma(n + 1)} \left\{h_{n-1}^{\left(\frac{3}{2}, 0\right)}\right\}^{\frac{1}{2}} \{y_{2n}, \beta_{2n}\} ,$$

and the asymptotically small parameter

$$h_n = 1 - \left(n + \frac{1}{4}\right)^{-1} \left\{\frac{\Gamma(n + 1)}{\Gamma\left(n + \frac{1}{2}\right)}\right\}^2$$

satisfies $h_n = O\left(n^{-2}\right)$, as $n \to \infty$; the i.s.l.a.e. for the odd index coefficients is

$$\left(1 - \mu_{2m+1}\right) Y_{2m+1}^* + \sum_{n=0}^{\infty} \left(\varepsilon_{2n+1} - g_n\right) Y_{2n+1}^* S_{nm}(u_0)$$

$$= \sum_{n=0}^{\infty} B_{2n+1}^* \left\{\delta_{nm} - S_{nm}(u_0)\right\}, \tag{5.110}$$

where $m = 0, 1, 2, \ldots ,$

$$\left\{B_{2m+1}^*, Y_{2m+1}^*\right\} = \left(m + \frac{1}{2}\right)^{\frac{1}{2}} (m + 1)^{\frac{1}{2}} \{B_{2m+1}, Y_{2m+1}\}$$

and the asymptotically small parameter

$$e_n = 1 - \frac{\left(n + \frac{1}{2}\right)(n + 1)}{n + \frac{3}{4}} \left\{\frac{\Gamma(n + 1)}{\Gamma\left(n + \frac{3}{2}\right)}\right\}^2$$

satisfies $e_n = O\left(n^{-2}\right)$, as $n \to \infty$. Furthermore, the functions $S_{nm}(u_0)$ are defined by (5.83).

This completes our discussion of dipole excitation of the barrel. Before leaving this section, we note that the structure is symmetrical about the plane $z = 0$. If the lower half (in the half space $z < 0$) is removed and the remaining upper part is mounted, as shown in Figure 5.11b, on the perfectly conducting infinite plane $z = 0$, we may apply image theory in a very similar fashion to that described at the end of the previous section to arrive at a regularised i.s.l.a.e. in which half the desired solution coefficients vanish.

5.4   Plane Wave Diffraction: Perfectly Conducting Slotted Spherical Shell.

It was shown in Section 4.2 that when a perfectly conducting spherical shell with circular aperture is illuminated by an electromagnetic plane wave, the

TM and TE modes couple and the final solution form is more complicated than the situation of axisymmetric excitation by a vertical electric or magnetic dipole. This may be recognized by comparing the solutions. The excitation of purely axisymmetric TM modes and TE modes is described by the systems (4.24) and (4.62), respectively. Diffraction of normally incident plane wave is described by the i.s.l.a.e. of the second kind (4.117)–(4.118), or equivalently (4.138)–(4.141).

The next level of complexity is found in the spherical shell with two equal circular holes (the *spherical barrel*) or with a symmetrical equatorial slot. Where the structure is excited by a vertical electric or magnetic dipole axially located, axisymmetry is retained, and an effective solution of some suitable triple series equations may be developed, based on the techniques of Volume I. One obtains two decoupled i.s.l.a.e. of the second kind for the even and odd index Fourier coefficients (see Sections 5.2 and 5.3).

Plane wave diffraction from a perfectly conducting slotted spherical shell presents an intrinsically more complex problem. The TM and TE modes of the scattered electromagnetic field are inextricably coupled through the appearance of so-called *polarisation constants*. Because of its complexity and importance, we will give rather fuller details of the derivation of the final solution.

Let us formulate the problem briefly. An incident plane electromagnetic wave propagating along the positive z-axis is scattered by a spherical shell with an equatorial slot (see Figure 5.12); the z-axis is the axis of symmetry. The equatorial slot subtends an angle $2\theta_1$ at the origin; each spherical cap subtends an angle $\theta_0 = \pi - \theta_1$ at the origin.

The statement of the problem is quite analogous to that given in Section 4.2. The only inessential difference is the form of the solution. Thus, formulating the problem in terms of Debye potentials, we seek the solution as a sum of incident and scattered potentials in the form

$$
U = \frac{\cos \phi}{ik^2 r} \sum_{n=1}^{\infty} \frac{2n+1}{n(n+1)} i^n \psi_n (kr) P_n^1 (\cos \theta) +
$$

$$
\frac{\cos \phi}{ik^2 r} \sum_{n=1}^{\infty} \frac{2n+1}{n(n+1)} \frac{x_n}{2n+1} P_n^1 (\cos \theta) \left\{ \begin{array}{ll} \zeta_n' (ka) \psi_n (kr), & r < a \\ \psi_n' (ka) \zeta_n (kr), & r > a \end{array} \right. \qquad (5.111)
$$

and

$$
V = \frac{\sin \phi}{ik^2 r} \sum_{n=1}^{\infty} \frac{2n+1}{n(n+1)} i^n \psi_n (kr) P_n^1 (\cos \theta) +
$$

$$
\frac{\sin \phi}{ik^2 r} \sum_{n=1}^{\infty} \frac{2n+1}{n(n+1)} y_n P_n^1 (\cos \theta) \left\{ \begin{array}{ll} \zeta_n (ka) \psi_n (kr), & r < a \\ \psi_n (ka) \zeta_n (kr), & r > a, \end{array} \right. \qquad (5.112)
$$

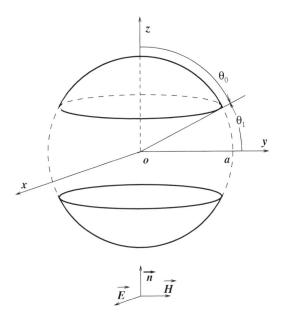

FIGURE 5.12. Plane wave excitation of a slotted spherical cavity.

where $\{x_n\}_{n=1}^{\infty}$ and $\{y_n\}_{n=1}^{\infty}$ are the unknown sets of Fourier coefficients to be determined. As in Section 4.2, both $\{x_n\}_{n=1}^{\infty}$ and $\{y_n\}_{n=1}^{\infty}$ lie in l_2. The tangential electric field components vanish at the conducting surface, and both electric and magnetic field components are continuous across the aperture. Thus the tangential electric field is continuous on the spherical surface $r = a$,

$$\vec{i}_r \times \vec{E}\Big|_{r=a-0} = \vec{i}_r \times \vec{E}\Big|_{r=a+0} \tag{5.113}$$

for all θ and ϕ. Moreover the mixed boundary conditions to be enforced on the spherical surface $r = a$ are

$$\vec{i}_r \times \vec{E}\Big|_{r=a-0} = \vec{i}_r \times \vec{E}\Big|_{r=a+0} = 0, \ \theta \in (0, \theta_0) \cup (\pi - \theta_0, \pi), \tag{5.114}$$

$$\vec{i}_r \times \vec{H}\Big|_{r=a-0} = \vec{i}_r \times \vec{H}\Big|_{r=a+0}, \qquad \theta \in (\theta_0, \pi - \theta_0), \tag{5.115}$$

for all ϕ. The continuity condition is satisfied if Debye potentials obey

$$\frac{\partial (rU)}{\partial r}\Big|_{r=a-0} = \frac{\partial (rU)}{\partial r}\Big|_{r=a+0}, \qquad V|_{r=a-0} = V|_{r=a+0}, \tag{5.116}$$

for all θ and ϕ on the spherical surface $r = a$.

Imposing these conditions leads to the coupled systems of triple series equations

$$\sum_{n=1}^{\infty} \frac{2n+1}{n(n+1)} \left\{ (1 - \varepsilon_n) x_n - \alpha_n \right\} P_n^1(z) =$$
$$- i4kaC_4 \sqrt{\frac{1+z}{1-z}}, \quad z \in (-1, -z_0) \quad (5.117)$$

$$\sum_{n=1}^{\infty} \frac{x_n}{n(n+1)} P_n^1(z) = C_2 \sqrt{\frac{1-z}{1+z}} + C_3 \sqrt{\frac{1+z}{1-z}}, \quad z \in (-z_0, z_0) \quad (5.118)$$

$$\sum_{n=1}^{\infty} \frac{2n+1}{n(n+1)} \left\{ (1 - \varepsilon_n) x_n - \alpha_n \right\} P_n^1(z) =$$
$$- i4kaC_1 \sqrt{\frac{1-z}{1+z}}, \quad z \in (z_0, 1) \quad (5.119)$$

and

$$\sum_{n=1}^{\infty} \frac{1}{n(n+1)} \left\{ (1 - \mu_n) y_n - \beta_n \right\} P_n^1(z) =$$
$$\frac{1}{ka} C_4 \sqrt{\frac{1+z}{1-z}}, \quad z \in (-1, -z_0) \quad (5.120)$$

$$\sum_{n=1}^{\infty} \frac{2n+1}{n(n+1)} y_n P_n^1(z) = iC_2 \sqrt{\frac{1-z}{1+z}} - iC_3 \sqrt{\frac{1+z}{1-z}}, \quad z \in (-z_0, z_0)$$
$$(5.121)$$

$$\sum_{n=1}^{\infty} \frac{1}{n(n+1)} \left\{ (1 - \mu_n) y_n - \beta_n \right\} P_n^1(z) =$$
$$- \frac{1}{ka} C_1 \sqrt{\frac{1-z}{1+z}}, \quad z \in (z_0, 1) \quad (5.122)$$

where $z = \cos(\theta)$, $z_0 = \cos(\theta_0)$,

$$\alpha_n = 4i^{n+1} ka\psi_n'(ka), \qquad \beta_n = -\frac{i^{n+1}}{ka}(2n+1)\psi_n(ka), \qquad (5.123)$$

and the asymptotically small parameters

$$\varepsilon_n = 1 + i4ka\frac{\psi'_n(ka)\,\zeta'_n(ka)}{2n+1}, \tag{5.124}$$

$$\mu_n = 1 - \frac{i}{ka}(2n+1)\,\psi_n(ka)\,\zeta_n(ka) \tag{5.125}$$

satisfy $\varepsilon_n = O\left((n^{-2})\right)$, $\mu_n = O\left((ka/n)^2\right)$ as $n \to \infty$. In addition, the constants C_i ($i = 1, 2, 3, 4$) are *polarisation constants* to be determined subsequently.

Making use of the symmetry

$$P_n^1(-z) = (-1)^{n+1}P_n^1(z), \tag{5.126}$$

we may split the triple series equations (5.117)–(5.119) and (5.120)–(5.122) into two *independent groups*,

$$\sum_{n=1}^{\infty}\frac{x_{2n}}{n\left(n+\frac{1}{2}\right)}P_{2n}^1(z) = \tilde{A}_1\frac{z}{\sqrt{1-z^2}}, \quad z \in (0, z_0) \tag{5.127}$$

$$\sum_{n=1}^{\infty}\frac{n+\frac{1}{4}}{n\left(n+\frac{1}{2}\right)}\{(1-\varepsilon_{2n})x_{2n} - \alpha_{2n}\}P_{2n}^1(z) =$$
$$-\frac{i}{2}ka\tilde{A}_2\sqrt{\frac{1-z}{1+z}}, \quad z \in (z_0, 1) \tag{5.128}$$

$$\sum_{n=0}^{\infty}\frac{n+\frac{3}{4}}{\left(n+\frac{1}{2}\right)(n+1)}y_{2n+1}P_{2n+1}^1(z) = -\frac{i}{4}\frac{\tilde{A}_1}{\sqrt{1-z^2}}, \quad z \in (0, z_0) \tag{5.129}$$

$$\sum_{n=0}^{\infty}\frac{1}{\left(n+\frac{1}{2}\right)(n+1)}\{(1-\mu_{2n+1})y_{2n+1} - \beta_{2n+1}\}P_{2n+1}^1(z) =$$
$$-\frac{1}{2ka}\tilde{A}_2\sqrt{\frac{1-z}{1+z}}, \quad z \in (z_0, 1) \tag{5.130}$$

and

$$\sum_{n=0}^{\infty}\frac{x_{2n+1}}{\left(n+\frac{1}{2}\right)(n+1)}P_{2n+1}^1(z) = \frac{\tilde{A}_3}{\sqrt{1-z^2}}, \quad z \in (0, z_0) \tag{5.131}$$

$$\sum_{n=0}^{\infty} \frac{n + \frac{3}{4}}{\left(n + \frac{1}{2}\right)(n+1)} \left\{(1 - \varepsilon_{2n+1})\, x_{2n+1} - \alpha_{2n+1}\right\} P_{2n+1}^{1}(z) =$$

$$-\frac{i}{2} k a \tilde{A}_4 \sqrt{\frac{1-z}{1+z}}, \quad z \in (z_0, 1) \quad (5.132)$$

$$\sum_{n=1}^{\infty} \frac{n + \frac{1}{4}}{n\left(n + \frac{1}{2}\right)} y_{2n} P_{2n}^{1}(z) = -\frac{i}{4} \tilde{A}_3 \frac{z}{\sqrt{1-z^2}}, \quad z \in (0, z_0) \quad (5.133)$$

$$\sum_{n=1}^{\infty} \frac{1}{n\left(n + \frac{1}{2}\right)} \left\{(1 - \mu_{2n})\, y_{2n} - \beta_{2n}\right\} P_{2n}^{1}(z) =$$

$$-\frac{1}{2ka} \tilde{A}_4 \sqrt{\frac{1-z}{1+z}}, \quad z \in (z_0, 1), \quad (5.134)$$

where the polarisation constants $C_i (i = 1, 2, 3, 4)$ are replaced by

$$\tilde{A}_1 = 4(C_3 - C_2), \quad \tilde{A}_2 = 4(C_1 - C_4),$$
$$\tilde{A}_3 = 4(C_2 + C_3), \quad \tilde{A}_4 = 4(C_1 + C_4). \quad (5.135)$$

The symmetry of the situation dictates that the systems (5.127)–(5.130) and (5.131)–(5.134) are decoupled, so that the spherical harmonics in the first group scatter independently of the second. The first group (5.127)–(5.130) consists of even spherical harmonics of TM type and odd spherical harmonics of TE type that are coupled solely through the presence of the modified polarisation constants $\tilde{A}_1$ and $\tilde{A}_2$. Likewise, the presence of the modified polarisation constants $\tilde{A}_3$ and $\tilde{A}_4$ is responsible for the coupling of the odd spherical harmonics of TM type and even spherical harmonics of TE type in the second group (5.131)–(5.134).

We now make use of the relation between the associated Legendre functions P_n^1 and the Jacobi polynomials $P_n^{(1, \pm \frac{1}{2})}$,

$$P_{2n}^{1}(z) = \left(n + \frac{1}{2}\right)(1 - u^2)^{\frac{1}{2}} P_{n-1}^{(1, \frac{1}{2})}(u), \quad (5.136)$$

$$P_{2n+1}^{1}(z) = \sqrt{2}\left(n + \frac{1}{2}\right)(1 - u)^{\frac{1}{2}} P_{n}^{(1, -\frac{1}{2})}(u), \quad (5.137)$$

where $u = 2z^2 - 1$, to transform the groups (5.127)–(5.130) and (5.131)–(5.134) to functional equations in the variable $u \in (-1, 1)$. Setting $u_0 = 2z_0^2 - 1$, we obtain

$$(1 - u) \sum_{n=1}^{\infty} \frac{x_{2n}}{n} P_{n-1}^{(1, \frac{1}{2})}(u) = \tilde{A}_1, \quad u \in (-1, u_0) \quad (5.138)$$

$$(1-u)\sum_{n=1}^{\infty}\frac{n+\frac{1}{4}}{n}\left\{(1-\varepsilon_{2n})\,x_{2n}-\alpha_{2n}\right\}P_{n-1}^{(1,\frac{1}{2})}\left(u\right)=$$

$$\frac{i}{2}ka\tilde{A}_2\left\{1-\left(\frac{1+u}{2}\right)^{-\frac{1}{2}}\right\},\quad u\in(u_0,1)\quad(5.139)$$

$$(1-u)\sum_{n=0}^{\infty}\frac{n+\frac{3}{4}}{n+1}y_{2n+1}P_n^{(1,-\frac{1}{2})}\left(u\right)=-\frac{i}{4}\tilde{A}_1,\quad u\in(-1,u_0)\qquad(5.140)$$

$$(1-u)\sum_{n=0}^{\infty}\frac{1}{n+1}\left\{(1-\mu_{2n+1})\,y_{2n+1}-\beta_{2n+1}\right\}P_n^{(1,-\frac{1}{2})}\left(u\right)$$

$$=-\frac{1}{2ka}\tilde{A}_2\left\{1-\left(\frac{1+u}{2}\right)^{\frac{1}{2}}\right\},u\in(u_0,1)\quad(5.141)$$

and

$$(1-u)\sum_{n=0}^{\infty}\frac{x_{2n+1}}{n+1}P_n^{(1,-\frac{1}{2})}\left(u\right)=\tilde{A}_3,\quad u\in(-1,u_0)\qquad(5.142)$$

$$(1-u)\sum_{n=0}^{\infty}\frac{n+\frac{3}{4}}{n+1}\left\{(1-\varepsilon_{2n+1})\,x_{2n+1}-\alpha_{2n+1}\right\}P_n^{(1,-\frac{1}{2})}\left(u\right)$$

$$=-\frac{i}{2}ka\tilde{A}_4\left\{1-\left(\frac{1+u}{2}\right)^{\frac{1}{2}}\right\},\quad u\in(u_0,1)\quad(5.143)$$

$$(1-u)\sum_{n=1}^{\infty}\frac{n+\frac{1}{4}}{n}y_{2n}P_{n-1}^{(1,\frac{1}{2})}\left(u\right)=-\frac{i}{4}\tilde{A}_3,\quad u\in(-1,u_0)\qquad(5.144)$$

$$(1-u)\sum_{n=1}^{\infty}\frac{1}{n}\left\{(1-\mu_{2n})\,y_{2n}-\beta_{2n}\right\}P_{n-1}^{(1,\frac{1}{2})}\left(u\right)$$

$$=\frac{1}{2ka}\tilde{A}_4\left\{1-\left(\frac{1+u}{2}\right)^{-\frac{1}{2}}\right\},\quad u\in(u_0,1)\,.\quad(5.145)$$

The series (5.138)–(5.145) have relatively good convergence rates. The terms of the series in (5.138), (5.141), (5.142) and (5.145) are $O\left(n^{-\frac{5}{2}}\right)$ as

$n \to \infty$; the terms in the remaining are $O\left(n^{-\frac{3}{2}}\right)$ as $n \to \infty$. Thus all the series converge uniformly, and term-by-term integration is justified; term-by-term differentiation is justified for the more rapidly convergent series. The treatment of the pairs (5.142)–(5.143) and (5.144)–(5.145) is quite similar to that of (5.138)–(5.139), but the dual series equations (5.140)–(5.141) warrant special consideration.

We first consider the pair (5.138)–(5.139), differentiating both sides of the first equation according to the result

$$\frac{d}{du}\left[(1-u)\,P_{n-1}^{(1,\frac{1}{2})}(u)\right] = -nP_{n-1}^{(0,\frac{3}{2})}(u)\,, \qquad (5.146)$$

to obtain

$$\sum_{n=1}^{\infty} x_{2n}P_{n-1}^{(0,\frac{3}{2})}(u) = 0,\, u \in (-1, u_0)\,. \qquad (5.147)$$

Equation (5.147) is equivalent to (5.138) provided we retain information about the constant A_1. This is most conveniently done by evaluating the uniformly convergent series (5.138) at the point $u = u_0$ to obtain

$$\tilde{A}_1 = (1 - u_0) \sum_{n=1}^{\infty} \frac{x_{2n}}{n} P_{n-1}^{(1,\frac{1}{2})}(u_0)\,. \qquad (5.148)$$

The Abel integral transform technique may now be applied in the usual way to (5.147) and (5.139). The relevant integral representations for the kernels of both equations

$$P_{n-1}^{(0,\frac{3}{2})}(u) = (1+u)^{-\frac{3}{2}} \frac{\Gamma\left(n+\frac{3}{2}\right)}{\Gamma\left(\frac{1}{2}\right)\Gamma\left(n+1\right)} \int_{-1}^{u} \frac{(1+x)\,P_{n-1}^{(\frac{1}{2},1)}(x)}{(u-x)^{\frac{1}{2}}} dx \qquad (5.149)$$

and

$$P_{n-1}^{(1,\frac{1}{2})}(u) = (1-u)^{-1} \frac{\Gamma\left(n+1\right)}{\Gamma\left(\frac{1}{2}\right)\Gamma\left(n+\frac{1}{2}\right)} \int_{u}^{1} \frac{(1-x)^{\frac{1}{2}}\,P_{n-1}^{(\frac{1}{2},1)}(x)}{(x-u)^{\frac{1}{2}}} dx \qquad (5.150)$$

are substituted into (5.147) and (5.139), respectively. The order of summation and integration may be interchanged (uniform convergence of these series being ensured by the decay rate $O\left(n^{-\frac{3}{2}}\right)$ of the terms as $n \to \infty$). The resulting Abel integral equations are readily inverted (by the standard inversion formulae) to obtain the following pre-regularised system employing the rescaled coefficients

$$\{X_{2n}, A_{2n}\} = \frac{n+\frac{1}{4}}{n} \frac{\Gamma\left(n+1\right)}{\Gamma\left(n+\frac{1}{2}\right)} \left\{h_{n-1}^{(\frac{1}{2},1)}\right\}^{\frac{1}{2}} \{x_{2n}, \alpha_{2n}\}\,, \qquad (5.151)$$

in the form

$$F(u) = \begin{cases} F_1(u), & u \in (-1, u_0) \\ F_2(u), & u \in (u_0, 1), \end{cases} \tag{5.152}$$

where F is the continuous function (on $(-1, 1)$)

$$F(u) = (1-u)^{\frac{1}{2}} (1+u) \sum_{n=1}^{\infty} X_{2n} \hat{P}_{n-1}^{(\frac{1}{2},1)}(u), \tag{5.153}$$

$$F_1(u) = (1-u)^{\frac{1}{2}} (1+u) \sum_{n=1}^{\infty} X_{2n} t_n \hat{P}_{n-1}^{(\frac{1}{2},1)}(u), \tag{5.154}$$

$$F_2(u) = -\frac{i}{2\sqrt{\pi}} ka \tilde{A}_2 (1-u)^{\frac{1}{2}} +$$

$$(1-u)^{\frac{1}{2}} (1+u) \sum_{n=1}^{\infty} (X_{2n} \varepsilon_{2n} + A_{2n}) \hat{P}_{n-1}^{(\frac{1}{2},1)}(u), \tag{5.155}$$

and the asymptotically small parameter

$$t_n = 1 - \frac{n\left(n+\frac{1}{2}\right)}{n+\frac{1}{4}} \left\{ \frac{\Gamma\left(n+\frac{1}{2}\right)}{\Gamma(n+1)} \right\}^2 \tag{5.156}$$

satisfies $t_n = O(n^{-2})$, as $n \to \infty$.

The family $\left\{ \hat{P}_{n-1}^{(\frac{1}{2},1)} \right\}_{n=1}^{\infty}$ is orthonormal, with weight $(1-x)^{\frac{1}{2}} (1+x)$, on $(-1, 1)$:

$$\int_{-1}^{1} (1-x)^{\frac{1}{2}} (1+x) \hat{P}_n^{(\frac{1}{2},1)}(x) \hat{P}_m^{(\frac{1}{2},1)}(x) \, dx = \delta_{nm}.$$

Exploiting orthonormality transforms functional equation (5.152) into the second kind i.s.l.a.e. for the unknowns X_{2n},

$$(1 - t_m) X_{2m} + \sum_{n=1}^{\infty} X_{2n} (t_n - \varepsilon_{2n}) \hat{Q}_{n-1,m-1}^{(\frac{1}{2},1)}(u_0)$$

$$= \sum_{n=1}^{\infty} A_{2n} \hat{Q}_{n-1,m-1}^{(\frac{1}{2},1)}(u_0) - ika \left(\frac{2}{\pi}\right)^{\frac{1}{2}} \tilde{A}_2 S_m(u_0), \tag{5.157}$$

where $m = 1, 2, \ldots$ and

$$S_m (u_0) = \left(\frac{1 - u_0}{2}\right)^{\frac{3}{2}} \frac{\hat{P}_{m-1}^{\left(\frac{3}{2}, 0\right)} (u_0)}{\sqrt{m \left(m + \frac{1}{2}\right)}}. \tag{5.158}$$

In terms of unknowns X_{2n}, the constant $\tilde{A}_1$ (see [5.148]) takes the form

$$\tilde{A}_1 = (1 - u_0) \sum_{n=1}^{\infty} \frac{X_{2n}}{n + \frac{1}{4}} \frac{\Gamma \left(n + \frac{1}{2}\right)}{\Gamma (n + 1)} \hat{P}_{n-1}^{\left(1, \frac{1}{2}\right)} (u_0). \tag{5.159}$$

Because the dual series equations (5.138)–(5.139) and (5.140)–(5.141) are coupled through the constants $\tilde{A}_1$ and $\tilde{A}_2$, the system (5.157) together with condition (5.159) is not soluble without taking account of the pair (5.140)–(5.141). We thus consider the regularisation of this pair.

Rearrange (5.140) as

$$(1 + u)^{-\frac{1}{2}} \sum_{n=0}^{\infty} \frac{n + \frac{3}{4}}{n + 1} y_{2n+1} P_n^{\left(1, -\frac{1}{2}\right)} (u) = -\frac{i}{4} \frac{\tilde{A}_1}{(1 - u) (1 + u)^{\frac{1}{2}}},$$
$$u \in (-1, u_0). \tag{5.160}$$

Apply fractional integration both sides of this equation using the formulae (see formula [1.172] of Volume I [1])

$$\int_{-1}^{u} \frac{(1 + x)^{-\frac{1}{2}} P_n^{\left(1, -\frac{1}{2}\right)} (x)}{(u - x)^{\frac{1}{2}}} = \frac{\Gamma \left(\frac{1}{2}\right) \Gamma \left(n + \frac{1}{2}\right)}{\Gamma (n + 1)} P_n^{\left(\frac{1}{2}, 0\right)} (u) \tag{5.161}$$

and

$$\int_{-1}^{u} \frac{dx}{(1 - x) \sqrt{(u - x) (1 + x)}} = \frac{\pi}{\sqrt{2}} (1 - u)^{-\frac{1}{2}} \tag{5.162}$$

to obtain

$$(1 - u)^{\frac{1}{2}} \sum_{n=0}^{\infty} \left(n + \frac{3}{4}\right) \frac{\Gamma \left(n + \frac{1}{2}\right)}{\Gamma (n + 1)} y_{2n+1} P_n^{\left(\frac{1}{2}, 0\right)} (u) = -\frac{i}{4} \left(\frac{2}{\pi}\right)^{\frac{1}{2}} A_1,$$
$$u \in (-1, u_0). \tag{5.163}$$

Further transformation of (5.163) and (5.141) follows the same steps used to obtain equation (5.152) and produces

$$G (u) = \begin{cases} G_1 (u), & u \in (-1, u_0) \\ G_2 (u), & u \in (u_0, 1) \end{cases} \tag{5.164}$$

where the rescaled variables

$$\{Y_{2n+1}, B_{2n+1}\} = \frac{\Gamma(n+1)}{\Gamma\left(n+\frac{3}{2}\right)} \left\{h_n^{\left(\frac{1}{2},0\right)}\right\}^{\frac{1}{2}} \{y_{2n+1}, \beta_{2n+1}\} \tag{5.165}$$

are employed,

$$G(u) = (1-u)^{\frac{1}{2}} \sum_{n=0}^{\infty} Y_{2n+1} \hat{P}_n^{\left(\frac{1}{2},0\right)}(u), \tag{5.166}$$

$$G_1(u) = -\frac{i}{4}\left(\frac{2}{\pi}\right)^{\frac{1}{2}} \tilde{A}_1 + (1-u)^{\frac{1}{2}} \sum_{n=0}^{\infty} Y_{2n+1} q_n \hat{P}_n^{\left(\frac{1}{2},0\right)}(u), \tag{5.167}$$

$$G_2(u) = -\frac{\tilde{A}_2}{2ka}(2\pi)^{-\frac{1}{2}} \log\left[\frac{\sqrt{2}+\sqrt{1-u}}{\sqrt{1+u}}\right] +$$

$$(1-u)^{\frac{1}{2}} \sum_{n=0}^{\infty} \left(Y_{2n+1}\mu_{2n+1} + B_{2n+1}\right) \hat{P}_n^{\left(\frac{1}{2},0\right)}(u), \tag{5.168}$$

and the asymptotically small parameter

$$q_n = 1 - \frac{n+\frac{3}{4}}{\left(n+\frac{1}{2}\right)(n+1)} \left\{\frac{\Gamma\left(n+\frac{3}{2}\right)}{\Gamma(n+1)}\right\}^2 \tag{5.169}$$

satisfies $q_n = O\left(n^{-2}\right)$, as $n \to \infty$.

Because the sequence $\{y_{2n+1}\}_{n=0}^{\infty}$ lies in l_2, the relation (5.165) implies that $\{Y_{2n+1}\}_{n=0}^{\infty}$ lies in $l_2(2)$. Uniform convergence of the series (5.167) defining the function G implies its continuity at the point $u = u_0$, i.e.,

$$G_1(u_0) = G_2(u_0). \tag{5.170}$$

Thus, the equations (5.140)–(5.141) are transformed to the second kind i.s.l.a.e.

$$(1-q_m) Y_{2m+1} + \sum_{n=0}^{\infty} Y_{2n+1} \left(q_n - \mu_{2n+1}\right) \hat{Q}_{nm}^{\left(\frac{1}{2},0\right)}(u_0) =$$

$$\sum_{n=0}^{\infty} B_{2n+1} \hat{Q}_{nm}^{\left(\frac{1}{2},0\right)}(u_0) - \frac{i}{2}\left(\frac{\pi}{2}\right)^{\frac{1}{2}} V_m(u_0) \tilde{A}_1 +$$

$$\frac{\tilde{A}_2}{2ka}(2\pi)^{-\frac{1}{2}} \left\{2V_m(u_0) \log\left[\frac{\sqrt{2}+\sqrt{1-u_0}}{\sqrt{1+u_0}}\right] - \frac{W_m(u_0)}{\sqrt{\left(m+\frac{1}{2}\right)(m+1)}}\right\}, \tag{5.171}$$

where $m = 0, 1, 2, \ldots$, and

$$V_m(u_0) = \frac{1 + u_0}{2} \frac{\hat{P}_m^{(-\frac{1}{2},1)}(u_0)}{\sqrt{(m + \frac{1}{2})(m + 1)}}, \tag{5.172}$$

$$W_m(u_0) = \left(\frac{1 - u_0}{2}\right)^{\frac{1}{2}} \frac{\hat{P}_m^{(\frac{1}{2},0)}(u_0)}{\sqrt{(m + \frac{1}{2})(m + 1)}}. \tag{5.173}$$

Furthermore, the system (5.171) must be supplemented by the condition imposed by the continuity of G (see (5.170)),

$$\frac{i}{4}\left(\frac{\pi}{2}\right)^{\frac{1}{2}} \tilde{A}_1 - \frac{(2\pi)^{-\frac{1}{2}}}{2ka} \log\left[\frac{\sqrt{2} + \sqrt{1 - u_0}}{\sqrt{1 + u_0}}\right] A_2 =$$

$$\sqrt{2} \sum_{n=0}^{\infty} \sqrt{\left(n + \frac{1}{2}\right)(n + 1)} \left\{Y_{2n+1}\left(q_n - \mu_{2n+1}\right) - B_{2n+1}\right\} W_n(u_0)$$

$$\tag{5.174}$$

The solution of the system comprising equations (5.157) and (5.171) combined with equations (5.159) and (5.174) completely specifies the scattering by a slotted sphere of even harmonics of TM type coupled to odd harmonics of TE type.

From the computational point of view some additional analytic manipulation of the obtained equations is desirable. Although the unknowns $\{Y_{2n+1}\}_{n=0}^{\infty}$ should lie in the functional class $l_2(2)$, how well their numerical approximations obey the same requirement depends strongly on how accurately the relation (5.174) is fulfilled numerically.

Let us make an analytic rearrangement so that the solution automatically lies in $l_2(2)$. Use (5.174) to express $\tilde{A}_2$ in terms of $\tilde{A}_1$ and $\{Y_{2n+1}\}_{n=1}^{\infty}$, and substitute into (5.171). Use the property of the incomplete scalar product $\hat{Q}_{nm}^{(\alpha,\beta)}$ (see Appendix B.6 of Volume I [1]),

$$\hat{Q}_{nm}^{(\frac{1}{2},0)}(u_0) = -\frac{(1 - u_0)^{\frac{1}{2}}(1 + u_0)}{\sqrt{(m + \frac{1}{2})(m + 1)}} \hat{P}_m^{(-\frac{1}{2},1)}(u_0) \hat{P}_n^{(\frac{1}{2},0)}(u_0)$$

$$+ \sqrt{\frac{(n + \frac{1}{2})(n + 1)}{(m + \frac{1}{2})(m + 1)}} \hat{Q}_{nm}^{(-\frac{1}{2},1)}(u_0) \tag{5.175}$$

that may also be expressed (see [5.172] and [5.173]) as

$$\hat{Q}_{nm}^{(\frac{1}{2},0)}(u_0) = -2\sqrt{2}\sqrt{\left(n + \frac{1}{2}\right)(n+1)} W_n(u_0) V_m(u_0)$$

$$+ \sqrt{\frac{\left(n + \frac{1}{2}\right)(n+1)}{\left(m + \frac{1}{2}\right)(m+1)}} \hat{Q}_{nm}^{(-\frac{1}{2},1)}(u_0). \quad (5.176)$$

Introducing the rescaled unknowns

$$\left\{\tilde{Y}_{2n+1}, \tilde{B}_{2n+1}\right\} = \sqrt{\left(n + \frac{1}{2}\right)(n+1)} \left\{Y_{2n+1}, B_{2n+1}\right\} \quad (5.177)$$

we may state the final form of the second kind i.s.l.a.e. as

$$(1 - q_m)\tilde{Y}_{2m+1} + \sum_{n=0}^{\infty} \tilde{Y}_{2n+1}\left(q_n - \mu_{2n+1}\right) R_{nm}(u_0)$$

$$= -i\tilde{A}_1 \bar{\Phi}_m(u_0) + \sum_{n=0}^{\infty} \tilde{B}_{2n+1} R_{nm}(u_0), \quad (5.178)$$

where $m = 0, 1, \ldots$ and

$$R_{nm}(u_0) = \hat{Q}_{nm}^{(-\frac{1}{2},1)}(u_0) - \sqrt{2}\frac{W_n(u_0) W_m(u_0)}{\log\left[\left(\sqrt{2} + \sqrt{1 - u_0}\right)/\sqrt{1 + u_0}\right]}, \quad (5.179)$$

$$\Phi_m(u_0) = \left(\frac{\pi}{2}\right)^{\frac{1}{2}} \frac{W_m(u_0)}{4\log\left[\left(\sqrt{2} + \sqrt{1 - u_0}\right)/\sqrt{1 + u_0}\right]}. \quad (5.180)$$

In the rearrangement (5.177) it is now obvious that $\left\{\tilde{Y}_{2m+1}\right\}_{m=0}^{\infty}$ belongs to the functional class l_2, and so $\{Y_{2n+1}\}_{n=0}^{\infty} \in l_2(2)$. Thus, the final solution now consists of equations (5.178), (5.157) and relations (5.159) and (5.174). It is convenient to represent the very last relation (5.174) in terms of $\tilde{Y}_{2n+1}, \tilde{B}_{2n+1}$ as

$$\frac{i}{4}\left(\frac{\pi}{2}\right)^{\frac{1}{2}}\tilde{A}_1 - \frac{(2\pi)^{-\frac{1}{2}}}{2ka}\log\left[\frac{\sqrt{2} + \sqrt{1 - u_0}}{\sqrt{1 + u_0}}\right]\tilde{A}_2$$

$$= \sqrt{2}\sum_{n=0}^{\infty}\left\{\tilde{Y}_{2n+1}\left(q_n - \mu_{2n+1}\right) - \tilde{B}_{2n+1}\right\} W_n(u_0). \quad (5.181)$$

Let us now turn to the equations (5.142)–(5.145) describing the scattering of odd harmonics of TM type that are coupled to even harmonics of TE type. They are solved in a manner similar to that which was used to obtain the i.s.l.a.e. (5.157) with the additional relation (5.159), and so we supply an abridged description of their solution. According to this scheme equations (5.142) and (5.143) are first transformed to their pre-regularised form

$$
(1-u)^{\frac{1}{2}} \sum_{n=0}^{\infty} X_{2n+1} \hat{P}_n^{(\frac{1}{2},0)}(u) = \begin{cases} H_1(u), & u \in (-1, u_0) \\ H_2(u), & u \in (u_0, 1) \end{cases} \tag{5.182}
$$

where

$$
H_1(u) = (1-u)^{\frac{1}{2}} \sum_{n=0}^{\infty} X_{2n+1} p_n \hat{P}_n^{(\frac{1}{2},0)}(u)
$$

and

$$
H_2(u) = -\frac{i}{2(2\pi)^{\frac{1}{2}}} ka \log\left[\frac{\sqrt{2}+\sqrt{1-u}}{\sqrt{1+u}}\right] \tilde{A}_4 +
$$

$$
(1-u)^{\frac{1}{2}} \sum_{n=0}^{\infty} (X_{2n+1}\varepsilon_{2n+1} + A_{2n+1}) \hat{P}_n^{(\frac{1}{2},0)}(u),
$$

with the additional relation

$$
\tilde{A}_3 = (1-u_0) \sum_{n=0}^{\infty} \frac{x_{2n+1}}{n+1} \hat{P}_n^{(1,-\frac{1}{2})}(u_0). \tag{5.183}
$$

In terms of the rescaled variables

$$
\{X_{2n+1}, A_{2n+1}\} = \left(n+\frac{3}{4}\right) \frac{\Gamma(n+1)}{\Gamma(n+\frac{3}{2})} \left\{h_n^{(\frac{1}{2},0)}\right\}^{\frac{1}{2}} \{x_{2n+1}, \alpha_{2n+1}\}
$$

the second kind i.s.l.a.e. is

$$
(1-p_m) X_{2m+1} + \sum_{n=0}^{\infty} X_{2n+1} (q_n - \varepsilon_{2n+1}) \hat{Q}_{nm}^{(\frac{1}{2},0)}(u_0)
$$

$$
= -\frac{ika}{2(2\pi)^{\frac{1}{2}}} K_m(u_0) \tilde{A}_4 + \sum_{n=0}^{\infty} A_{2n+1} \hat{Q}_{nm}^{(\frac{1}{2},0)}(u_0), \tag{5.184}
$$

where $m = 0, 1, 2, \ldots$,

$$
K_m(u_0) = -(1+u_0) \frac{\hat{P}_m^{(-\frac{1}{2},1)}(u_0)}{\sqrt{(m+\frac{1}{2})(m+1)}} \log\left[\frac{\sqrt{2}+\sqrt{1-u_0}}{\sqrt{1+u_0}}\right]
$$

$$
+ \left(\frac{1-u_0}{2}\right)^{\frac{1}{2}} \frac{\hat{P}_m^{(\frac{1}{2},0)}(u_0)}{(m+\frac{1}{2})(m+1)}, \tag{5.185}
$$

and the asymptotically small parameter

$$p_n = 1 - \left(n + \frac{3}{4}\right)^{-1} \left\{ \frac{\Gamma\left(n + \frac{3}{2}\right)}{\Gamma\left(n + 1\right)} \right\}^2 \tag{5.186}$$

obeys $p_n = O\left(n^{-2}\right)$, as $n \to \infty$.

Again, in a similar way, equations (5.144)–(5.145) may be transformed to the functional equations

$$(1 - u)^{\frac{3}{2}} \sum_{n=1}^{\infty} Y_{2n} \hat{P}_{n-1}^{\left(\frac{3}{2},0\right)}(u) = \left\{ \begin{array}{ll} K_1\left(u\right), & u \in \left(-1, u_0\right) \\ K_2\left(u\right), & u \in \left(u_0, 1\right) \end{array} \right. \tag{5.187}$$

where

$$K_1\left(u\right) = (1 - u)^{\frac{3}{2}} \sum_{n=1}^{\infty} Y_{2n} r_n \hat{P}_{n-1}^{\left(\frac{3}{2},0\right)}(u)$$

and

$$K_2\left(u\right) = -\frac{i}{2ka} \frac{\tilde{A}_4}{\sqrt{\pi}} \left\{ (1 - u)^{\frac{1}{2}} - \sqrt{2} \log \left[\frac{\sqrt{2} + \sqrt{1 - u}}{\sqrt{1 + u}} \right] \right\}$$

$$+ (1 - u)^{\frac{3}{2}} \sum_{n=1}^{\infty} \left(Y_{2n} \mu_{2n} + B_{2n}\right) \hat{P}_{n-1}^{\left(\frac{3}{2},0\right)}(u)$$

with the additional relation

$$\sum_{n=1}^{\infty} \frac{1}{n} \left\{ (1 - \mu_{2n}) y_{2n} - \beta_{2n} \right\} \hat{P}_{n-1}^{\left(1,\frac{1}{2}\right)}(1) = -\frac{1}{8ka} \tilde{A}_4. \tag{5.188}$$

Then (5.187) is transformed in the usual way, with the rescaling of variables

$$\{Y_{2n}, B_{2n}\} = \frac{n - \frac{1}{2}}{n} \frac{\Gamma\left(n + 2\right)}{\Gamma\left(n + \frac{3}{2}\right)} \left\{ h_{n-1}^{\left(\frac{3}{2},0\right)} \right\}^{\frac{1}{2}} \{y_{2n}, \beta_{2n}\}$$

into the second kind i.s.l.a.e.

$$(1 - r_m) Y_{2m} + \sum_{n=1}^{\infty} Y_{2n} \left(r_n - \mu_{2n}\right) \hat{Q}_{n-1,m-1}^{\left(\frac{3}{2},0\right)}(u_0) =$$

$$\frac{1}{2ka} \left(\frac{2}{\pi}\right)^{\frac{1}{2}} \hat{\Phi}_m\left(u_0\right) \tilde{A}_4 + \sum_{n=1}^{\infty} B_{2n} \hat{Q}_{n-1,m-1}^{\left(\frac{3}{2},0\right)}(u_0), \tag{5.189}$$

where $m = 1, 2, \ldots$,

$$\bar{\Phi}_m (u_0) = \frac{(1 + u_0) \, \hat{P}_{m-1}^{(\frac{1}{2},1)} (u_0)}{\sqrt{m \left(m + \frac{1}{2}\right)}} \left\{ \left(\frac{1 - u_0}{2}\right)^{\frac{1}{2}} - \log \left[\frac{\sqrt{2} + \sqrt{1 - u_0}}{\sqrt{1 + u_0}} \right] \right\}$$
$$+ \left(\frac{1 - u_0}{2}\right)^{\frac{3}{2}} \frac{\hat{P}_{m-1}^{(\frac{3}{2},0)} (u_0)}{m \left(m + \frac{1}{2}\right)}, \quad (5.190)$$

and the asymptotically small parameter

$$r_n = 1 - \left(n + \frac{1}{4}\right) \frac{n}{n - \frac{1}{2}} \frac{\Gamma \left(n + \frac{1}{2}\right) \Gamma \left(n + \frac{3}{2}\right)}{\Gamma \left(n + 1\right) \Gamma \left(n + 2\right)} \quad (5.191)$$

satisfies $r_n = O\left(n^{-2}\right)$, as $n \to \infty$.

This concludes our analysis of plane wave diffraction from the slotted spherical shell. These mathematical tools are equally applicable to plane wave scattering from the *spherical barrel* (the cavity with two symmetrically located circular apertures) and many other related problems. The apparent complexity of the solution presented is rewarded by a system of well-conditioned equations for which a stable numerical scheme of well-defined accuracy over a wide frequency range is easily implemented.

5.5 Magnetic Dipole Excitation of an Open Spherical Resonator.

In this section we consider a pair of related electromagnetic problems, horizontal magnetic dipole excitation of an open spherical resonator, and horizontal electric dipole excitation of a grounded *barrel-type* open spherical shell. The geometry is shown in Figure 5.13. In contrast to the problems studied in Sections 5.2 and 5.3, there is mutual coupling of TM and TE modes in the scattered electromagnetic field. As the figure suggests, the solution to both problems relies on the image principle.

Since the solutions to both problems have many similarities, we restrict attention to the first and consider only the excitation by a magnetic horizontal dipole of an open resonator formed by an infinite metallic plane and spherical cap (or mirror). As we shall see, the standard formulation of this problem leads to a system of equations that is very similar to that encountered in the previous section, so that the regularised solution format may immediately be deduced.

Let the ground plane be $z = 0$, and suppose that the structure is axisymmetrical about the z-axis. Suppose the cap has radius of curvature a

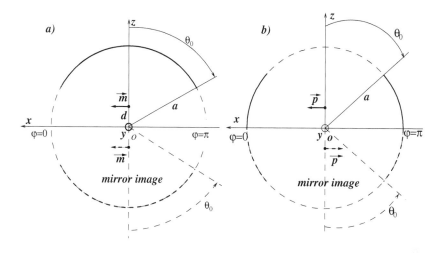

FIGURE 5.13. Dipole excitation of various spherical cavities.

and subtends an angle θ_0 at the origin. The dipole is located in the upper half-space on the axis of symmetry, and we may suppose that its moment $\vec{p}$ is aligned with the positive x-axis. We shall employ spherical coordinates, centred at the origin. The real physical structure is located in the upper half-space. It is equivalent, after removal of the infinite metallic plane, to the structure obtained by the addition of the mirror image of the cap in the plane $z = 0$, and of a dipole of moment $-\vec{p}$ located at the image point in the lower half space. Thus the initially posed problem is equivalent to the excitation, by a pair of horizontal magnetic dipoles, of a spherical cavity in which an equatorial slot has been removed as shown on Figure 5.13a.

Let us formulate the problem in terms of electric U and magnetic V Debye potentials. The potentials associated with the horizontal magnetic dipole and its image may be expressed as

$$U^0 = U^+ + U^-, \quad V^0 = V^+ + V^- \tag{5.192}$$

where

$$V^\pm = \frac{im}{rd} \cos\phi \times$$

$$\sum_{n=1}^{\infty} (\pm 1)^{n+1} A_n P_n^1 (\cos\theta) \begin{cases} \zeta_n'(kd) \psi_n(kr), & r < d \\ \psi_n'(kd) \zeta_n(kr), & r > d, \end{cases} \tag{5.193}$$

$$U^{\pm} = -\frac{m}{rd}\sin\phi \times$$

$$\sum_{n=1}^{\infty} (\pm 1)^{n+1} A_n P_n^1 (\cos\theta) \begin{cases} \zeta_n(kd)\,\psi_n(kr)\,, & r < d \\ \psi_n(kd)\,\zeta_n(kr)\,, & r > d, \end{cases} \qquad (5.194)$$

and the plus and minus signs refer to the potentials generated by the dipoles in the upper and lower half space, respectively; as usual $A_n = (2n+1)/n(n+1)$. Thus

$$V^0 = 2\frac{im}{rd}\cos\phi \times$$

$$\sum_{n=0}^{\infty} \frac{n + \frac{3}{4}}{\left(n + \frac{1}{2}\right)(n+1)} P_{2n+1}^1 (\cos\theta) \begin{cases} \zeta'_{2n+1}(kd)\,\psi_{2n+1}(kr)\,, & r < d \\ \psi'_{2n+1}(kd)\,\zeta_{2n+1}(kr)\,, & r > d \end{cases}$$

$$(5.195)$$

and

$$U^0 = 2\frac{m}{rd}\sin\phi \times$$

$$\sum_{n=1}^{\infty} \frac{n + \frac{1}{4}}{n\left(n + \frac{1}{2}\right)} P_{2n}^1 (\cos\theta) \begin{cases} \zeta_{2n}(kd)\,\psi_{2n}(kr)\,, & r < d \\ \psi_{2n}(kd)\,\zeta_{2n}(kr)\,, & r > d. \end{cases} \qquad (5.196)$$

In the limit when $d \to 0$, we obtain a model of a magnetic slot in the plane and the representations (5.195)–(5.196) simplify to

$$V^0 = 2\frac{imk}{r}\zeta_1(kr)\sin\theta\cos\phi, \qquad U^0 = 0. \qquad (5.197)$$

One may justify that the tangential electric field components vanish on the plane $z = 0$,

$$E_r\left(r, \frac{\pi}{2}, \phi\right) = E_\phi\left(r, \frac{\pi}{2}, \phi\right) = 0. \qquad (5.198)$$

It is quite clear from the problem symmetry that the scattered electromagnetic field comprises only odd oscillations of TE type and even oscillations of TM type. Thus we seek the scattered potentials in the form

$$V^s = 2\frac{imk}{r}\cos\phi \times$$

$$\sum_{n=0}^{\infty} \frac{n + \frac{3}{4}}{\left(n + \frac{1}{2}\right)(n+1)} b_{2n+1} P_{2n+1}^1 (\cos\theta) \begin{cases} \psi_{2n+1}(kr)\,, & r < a \\ \frac{\psi_{2n+1}(ka)}{\zeta_{2n+1}(ka)}\zeta_{2n+1}(kr)\,, & r > a \end{cases}$$

$$(5.199)$$

$$U^s = 2\frac{mk}{r} \sin\phi \times$$

$$\sum_{n=1}^{\infty} \frac{n+\frac{1}{4}}{n\left(n+\frac{1}{2}\right)} a_{2n} P_{2n}^1 \left(\cos\theta\right) \begin{cases} \psi_{2n}\left(kr\right), & r < a \\ \frac{\psi'_{2n}(ka)}{\zeta'_{2n}(ka)}\zeta_{2n}\left(kr\right), & r > a \end{cases} \quad (5.200)$$

Imposition of the boundary conditions on surface $r = a$ leads to the following two systems of dual series equations, coupled through *polarisation* constants B_1 and B_2, to be solved for rescaled coefficients $\{x_{2n}\}_{n=1}^{\infty}$ and $\{y_{2n+1}\}_{n=0}^{\infty}$,

$$\sum_{n=1}^{\infty} \frac{x_{2n}}{n\left(n+\frac{1}{2}\right)} P_{2n}^1 \left(z\right) = 4B_2 \frac{z}{\sqrt{1-z^2}}, \quad z \in (0, z_0) \quad (5.201)$$

$$\sum_{n=1}^{\infty} \frac{n+\frac{1}{4}}{n\left(n+\frac{1}{2}\right)} \{x_{2n}\left(1-\varepsilon_{2n}\right) - \alpha_{2n}\} P_{2n}^1 \left(z\right)$$

$$= -i4kaB_1 \left(\frac{1-z}{1+z}\right)^{\frac{1}{2}}, \quad z \in (z_0, 1) \quad (5.202)$$

and

$$\sum_{n=0}^{\infty} \frac{n+\frac{3}{4}}{\left(n+\frac{1}{2}\right)\left(n+1\right)} y_{2n+1} P_{2n+1}^1 \left(z\right) = -\frac{B_2}{\sqrt{1-z^2}}, \quad z \in (0, z_0) \quad (5.203)$$

$$\sum_{n=0}^{\infty} \{y_{2n+1}\left(1-\mu_{2n+1}\right) - \beta_{2n+1}\} \frac{P_{2n+1}^1 \left(z\right)}{\left(n+\frac{1}{2}\right)\left(n+1\right)}$$

$$= i\frac{4}{ka}B_1 \left(\frac{1-z}{1+z}\right)^{\frac{1}{2}}, \quad z \in (z_0, 1) \quad (5.204)$$

where $z_0 = \cos\theta_0$,

$$a_{2n} = \frac{\zeta'_{2n}(ka)}{4n+1} x_{2n}, \quad b_{2n+1} = \zeta_{2n+1}(ka) y_{2n+1} \quad (5.205)$$

and the asymptotically small parameters are defined by

$$\varepsilon_n = 1 + i4ka\frac{\psi'_n(ka)\zeta'_n(ka)}{2n+1},$$

$$\mu_n = 1 - \frac{i}{ka}(2n+1)\psi_n(ka)\zeta_n(ka). \quad (5.206)$$

Furthermore,

$$\alpha_{2n} = i4ka\frac{\psi_{2n}(kd)}{kd}\zeta'_{2n}(ka),$$

$$\beta_{2n+1} = -\frac{i}{ka}(4n+3)\frac{\psi'_{2n+1}(kd)}{kd}\zeta_{2n+1}(ka). \qquad (5.207)$$

The regularisation of this system is very similar to that considered in the previous section, and so the details are omitted.

5.6   Open Resonators Composed of Spherical and Disc Mirrors.

In the previous section, an open spherical resonator with a plane mirror of infinite extent was considered; in this section the mirror is replaced by a finite disc. There are extensive studies of this type of resonator and a comprehensive treatment is beyond the scope of this book. The aim of this section is to illustrate how the rigorous approach developed in this book may be applied to such problems, and we restrict ourselves to the simplest configuration shown in Figure 5.14.

The upper mirror is composed of part of a spherical surface with radius of curvature b and angular extent θ_0, so that its location is described in spherical polar coordinates by $\{(r,\theta,\phi) \mid r = b, 0 \leq \theta < \theta_0, 0 \leq \phi \leq 2\pi\}$; the lower mirror is a thin metallic circular disc of radius a, lying in the half-plane $z = h$, where $h > 0$.

For the sake of simplicity we excite the open resonator by a vertical electric dipole (VED), of moment $\overrightarrow{p} = p\overrightarrow{k}$, located a distance d, from the origin O along the axis of symmetry (the z-axis). The VED produces an axisymmetric electromagnetic field of TM type; by virtue of the axial symmetry of the resonator, the scattered electromagnetic field is of the same type, i.e., it is axially symmetric.

We will need to consider spherical and cylindrical coordinates with a common origin O aligned so that the positive z-axis corresponds to the half-line $\theta = 0$. In spherical coordinates three components (H_r, H_θ, E_ϕ) of an axisymmetric electromagnetic field of TM type vanish (see Chapter 1), and the field is described by the remaining three (E_r, E_θ, H_ϕ); in cylindrical coordinates the three vanishing components are (H_ρ, H_z, E_ϕ) and the field is described by the remaining three (E_ρ, H_ϕ, E_z). In both cases the electric components (E_r, E_θ) are simply expressed in terms of the magnetic component H_ϕ, and it is sufficient to solve any TM scattering problem in terms of this component alone.

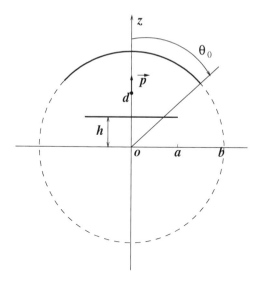

FIGURE 5.14. An open resonator composed of spherical and disc mirrors.

In cylindrical coordinates, it can be readily shown that the electromagnetic field of the VED is described by the z-component of an electric Hertz vector $\overrightarrow{\Pi^0} = \Pi_z^0 \overrightarrow{k}$ where

$$\Pi_z^0 = p \int_0^\infty \frac{e^{-\sqrt{\nu^2 - k^2}|z-d|}}{\sqrt{\nu^2 - k^2}} \nu J_0(\nu\rho)\, d\nu, \qquad (5.208)$$

where $\mathrm{Im}\left(\sqrt{\nu^2 - k^2}\right) < 0$ when $v < k$. Due to the well-known relation between $\overrightarrow{\Pi^0}$ and the electromagnetic field,

$$\overrightarrow{H} = -ik\, \mathrm{curl}\, \overrightarrow{\Pi^0}, \quad \overrightarrow{E} = \mathrm{curl}\, \mathrm{curl}\, \overrightarrow{\Pi^0}, \qquad (5.209)$$

the nonvanishing components are

$$H_\phi^0 = ik\frac{\partial \Pi_z^0}{\partial \rho} = -ikp \int_0^\infty \frac{e^{-\sqrt{\nu^2 - k^2}|z-d|}}{\sqrt{\nu^2 - k^2}} \nu^2 J_1(\nu\rho)\, d\nu \qquad (5.210)$$

$$E_\rho^0 = \frac{\partial^2 \Pi_z^0}{\partial z \partial \rho} = sign\,(z-d) \int_0^\infty e^{-\sqrt{\nu^2 - k^2}|z-d|} \nu^2 J_1(\nu\rho)\, d\nu \qquad (5.211)$$

$$E_z^0 = -\frac{1}{\rho}\frac{\partial}{\partial \rho}\left(\rho \frac{\partial \Pi_z^0}{\partial \rho}\right) = p \int_0^\infty \frac{e^{-\sqrt{\nu^2 - k^2}|z-d|}}{\sqrt{\nu^2 - k^2}} \nu^3 J_0(\nu\rho)\, d\nu. \qquad (5.212)$$

In spherical coordinates the nonvanishing components of the dipolar electromagnetic field are

$$H_\phi^0 = \frac{pk^3}{kr\,(kd)^2} \sum_{n=1}^{\infty} (2n+1)\, P_n^1 (\cos\theta) \left\{ \begin{array}{l} \psi_n\,(kd)\,\zeta_n\,(kr)\,, r > d \\ \zeta_n\,(kd)\,\psi_n\,(kr)\,, r < d \end{array} \right. , \quad (5.213)$$

$$E_\theta^0 = \frac{-ipk^3}{(kr)\,(kd)^2} \sum_{n=1}^{\infty} (2n+1)\, P_n^1 (\cos\theta) \left\{ \begin{array}{l} \psi_n\,(kd)\,\zeta_n'\,(kr)\,, r > d \\ \zeta_n\,(kd)\,\psi_n'\,(kr)\,, r < d \end{array} \right. , \\ \hspace{9cm} (5.214)$$

$$E_r^0 = \frac{ipk^3}{(kr)^2\,(kd)^2} \times$$
$$\sum_{n=1}^{\infty} n\,(n+1)\,(2n+1)\, P_n^1 (\cos\theta) \left\{ \begin{array}{l} \psi_n\,(kd)\,\zeta_n\,(kr)\,, r > d \\ \zeta_n\,(kd)\,\psi_n\,(kr)\,, r < d. \end{array} \right. \quad (5.215)$$

It is convenient to represent the component H_ϕ of the total field as a superposition of the incident component H_ϕ^0 and contributions from the spherical mirror H_ϕ^s and disc H_ϕ^d in the form

$$H_\phi = H_\phi^0 + H_\phi^s + H_\phi^d, \quad (5.216)$$

where H_ϕ^0 is defined in (5.210)–(5.212) and (5.213)–(5.215), the spherical mirror contribution is sought in the form

$$H_\phi^s = \frac{pk^3}{kr} \sum_{n=1}^{\infty} (2n+1)\, a_n P_n^1 (\cos\theta) \left\{ \begin{array}{ll} \psi_n\,(kr)\,, & r < b \\ \psi_n'\,(kb)\,\zeta_n\,(kr)\,/\zeta_n'\,(kb)\,, & r > b, \end{array} \right\} \\ \hspace{9cm} (5.217)$$

and the disc contribution is sought in the form

$$H_\phi^d = -ik^2 sign\,(z-h) \int_0^\infty g\,(\nu)\, J_1\,(\nu\rho)\, e^{-\sqrt{\nu^2-k^2}|z-h|} d\nu. \quad (5.218)$$

The unknown modal coefficients $\{a_n\}_{n=1}^{\infty}$ and unknown spectral function g are to be determined subsequently by the requirements that the solution form satisfies Maxwell's equations, radiation conditions at infinity, and continuity conditions for tangential electric field components on the boundary surfaces $r = b$ and $z = h$. The edge condition (see Section 1.5) determines the class in which the solution coefficients a_n and function g lie.

The pertinent mixed boundary conditions are

$$
\begin{align}
E_\theta \left(b + 0, \theta \right) &= E_\theta \left(b - 0, \theta \right) = 0, \quad \theta \in \left(0, \theta_0 \right) \tag{5.219} \\
H_\phi \left(b + 0, \theta \right) &= H_\phi \left(b - 0, \theta \right), \quad \theta \in \left(\theta_0, \pi \right) \tag{5.220} \\
E_\rho \left(\rho, h + 0 \right) &= E_\rho \left(\rho, h - 0 \right) = 0, \quad \rho < a \tag{5.221} \\
H_\phi \left(\rho, h + 0 \right) &= H_\phi \left(\rho, h - 0 \right), \quad \rho > a. \tag{5.222}
\end{align}
$$

In order to effect these boundary conditions the basic solutions in spherical coordinates for Maxwell's equations must be re-expressed in cylindrical coordinate format and vice versa. The general principle was enunciated in Section 2.9 of Volume I [1], and the integro-series equations that arise in the electrostatic analogue of the above structure were as analysed by these methods in Section 8.5 of Volume I.

We first examine the dual integral equations that arise from the enforcement of the mixed boundary conditions (5.221) and (5.222),

$$
\int_0^\infty g\left(\nu \right) \sqrt{\nu^2 - k^2} J_1 \left(\nu\rho \right) d\nu = \frac{1}{k} \int_0^\infty e^{-\sqrt{\nu^2 - k^2}\left(d - h \right)} \nu^2 J_1 \left(\nu\rho \right) d\nu +
$$

$$
ik \sum_{n=1}^\infty \left(2n + 1 \right) a_n \frac{d}{dz} \left[j_n \left(k\sqrt{\rho^2 + z^2} \right) P_n^1 \left(\frac{z}{\sqrt{\rho^2 + z^2}} \right) \right]_{z=h}, \quad \rho < a
$$

$$
\tag{5.223}
$$

$$
\int_0^\infty g\left(\nu \right) J_1 \left(\nu\rho \right) d\nu = 0, \quad\quad \rho > a. \tag{5.224}
$$

To re-expand the spherical wave harmonics on the right-hand side of (5.223) as cylindrical wave functions or, more accurately, as a Fourier-Bessel integral, we make use of the two results (see [24])

$$
\left(\frac{\pi}{2} \right)^{\frac{1}{2}} \frac{\left(-1 \right)^n y^{\mu + \frac{1}{2}}}{\left(a^2 + y^2 \right)^{\frac{\mu}{2} + \frac{1}{4}}} C_{2n+1}^{\left(\mu + \frac{1}{2} \right)} \left(\frac{\alpha}{\left(y^2 + a^2 \right)^{\frac{1}{2}}} \right) J_{\mu + \frac{3}{2} + 2n} \left(\left(a^2 + y^2 \right)^{\frac{1}{2}} \right)
$$

$$
= \int_0^1 \frac{x^{\mu + \frac{1}{2}}}{\left(1 - x^2 \right)^{\frac{1}{2}}} \sin \left(\alpha \left(1 - x^2 \right)^{\frac{1}{2}} \right) C_{2n+1}^{\left(\mu + \frac{1}{2} \right)} \left(\left(1 - x^2 \right)^{\frac{1}{2}} \right) J_\mu \left(xy \right) \left(xy \right)^{\frac{1}{2}} dx
$$

$$
\tag{5.225}
$$

and

$$
\left(\frac{\pi}{2} \right)^{\frac{1}{2}} \frac{\left(-1 \right)^n y^{\mu + \frac{1}{2}}}{\left(a^2 + y^2 \right)^{\frac{\mu}{2} + \frac{1}{4}}} C_{2n}^{\left(\mu + \frac{3}{2} \right)} \left(\frac{\alpha}{\left(y^2 + a^2 \right)^{\frac{1}{2}}} \right) J_{\mu + \frac{1}{2} + 2n} \left(\left(a^2 + y^2 \right)^{\frac{1}{2}} \right)
$$

$$
= \int_0^1 \frac{x^{\mu + \frac{1}{2}}}{\left(1 - x^2 \right)^{\frac{1}{2}}} \cos \left(\alpha \left(1 - x^2 \right)^{\frac{1}{2}} \right) C_{2n}^{\left(\mu + \frac{1}{2} \right)} \left(\left(1 - x^2 \right)^{\frac{1}{2}} \right) J_\mu \left(xy \right) \left(xy \right)^{\frac{1}{2}} dx,
$$

$$
\tag{5.226}
$$

where ultraspherical polynomials $C_k^{(\gamma)}$ are related to the associated Legendre functions $P_{k+1}^{(\gamma-\frac{1}{2})}$ by

$$C_{2n}^{(\frac{3}{2})}\left(\frac{\alpha}{(y^2+\alpha^2)^{\frac{1}{2}}}\right) = \frac{(y^2+\alpha^2)^{\frac{1}{2}}}{y}P_{2n+1}^1\left(\frac{\alpha}{(y^2+\alpha^2)^{\frac{1}{2}}}\right),$$

$$C_{2n-1}^{(\frac{3}{2})}\left(\frac{\alpha}{(y^2+\alpha^2)^{\frac{1}{2}}}\right) = \frac{(y^2+\alpha^2)^{\frac{1}{2}}}{y}P_{2n}^1\left(\frac{\alpha}{(y^2+\alpha^2)^{\frac{1}{2}}}\right). \quad (5.227)$$

Thus the spherical wave harmonics on the right-hand side of (5.223) may be expanded in the surprisingly simple form

$$\frac{d}{dz}\left[j_{2n+l}\left(k\sqrt{\rho^2+z^2}\right)P_{2n+l}^1\left(\frac{z}{\sqrt{\rho^2+z^2}}\right)\right]_{z=h} =$$

$$\frac{(-1)^{n-1}}{k}\int_0^k d\nu\cdot\nu P_{2n+l}^1\left(\left(1-\nu^2/k^2\right)^{\frac{1}{2}}\right)J_1\left(\nu\rho\right)\times$$

$$\begin{cases} \cos\left(\left(1-\nu^2/k^2\right)^{\frac{1}{2}}kh\right), & l=0 \\ \sin\left(\left(1-\nu^2/k^2\right)^{\frac{1}{2}}kh\right), & l=1. \end{cases} \quad (5.228)$$

Thus the dual integral equations (5.223)–(5.224) may be rearranged in the compact and elegant form

$$\int_0^\infty \nu g(\nu)\left[1-\varepsilon(\nu,k)\right]J_1(\nu\rho)\,d\nu = \frac{1}{k}\int_0^\infty e^{-\sqrt{\nu^2-k^2}(d-h)}\nu^2 J_1(\nu\rho)\,d\nu$$

$$+ i\int_0^k\left\{\sum_{n=1}^\infty (2n+1)a_n\sigma_n(\nu,k;h)\right\}\nu J_1(\nu\rho)\,d\nu, \quad \rho < a, \quad (5.229)$$

$$\int_0^\infty g(\nu)J_1(\nu\rho)\,d\nu = 0, \qquad \rho > a, \quad (5.230)$$

where

$$\sigma_{2n} = (-1)^{n-1}\cos\left(\sqrt{k^2-\nu^2}h\right)P_{2n}^1\left(\frac{\sqrt{k^2-\nu^2}}{k}\right), \quad (5.231)$$

$$\sigma_{2n+1} = (-1)^{n-1}\sin\left(\sqrt{k^2-\nu^2}h\right)P_{2n+1}^1\left(\frac{\sqrt{k^2-\nu^2}}{k}\right) \quad (5.232)$$

Now as usual in the regularisation method, we identify the parameter

$$\varepsilon(\nu,k) = \frac{\nu-\sqrt{\nu^2-k^2}}{\nu} \quad (5.233)$$

that is asymptotically small: $\varepsilon\left(\nu, k\right) = O\left(k^2/\nu^2\right)$, as $\nu \to \infty$.

The edge condition demands that g belongs to $L_2\left(0, \infty\right)$ and justifies the validity of subsequent transformations on dual integral equations in this section. Employing the Abel integral transform method (as explained in Section 2.6 of Volume I [1]) leads to the pair of functional equations

$$\int_0^\infty \nu^{\frac{1}{2}} g\left(\nu\right) J_{\frac{3}{2}}\left(\nu\rho\right) d\nu = \int_0^\infty \nu^{\frac{1}{2}} g\left(\nu\right) \varepsilon\left(\nu, k\right) J_{\frac{3}{2}}\left(\nu\rho\right) d\nu$$

$$+ \frac{1}{k}\int_0^\infty e^{-\sqrt{\nu^2 - k^2}(d-h)} \nu^{\frac{1}{2}} J_{\frac{3}{2}}\left(\nu\rho\right) d\nu$$

$$+ \frac{1}{k}\int_0^\infty \left\{\sum_{n=1}^\infty (2n+1) a_n \sigma_n\right\} \nu^{\frac{1}{2}} J_{\frac{3}{2}}\left(\nu\rho\right) d\nu, \quad \rho < a, \quad (5.234)$$

$$\int_0^\infty \nu^{\frac{1}{2}} g\left(\nu\right) J_{\frac{3}{2}}\left(\nu\rho\right) d\nu = 0, \quad \rho > a. \quad (5.235)$$

Further transformation of (5.234)–(5.235) uses the Fourier-Bessel transform and the representation of the unknown function g in a Neumann series

$$g\left(\mu\right) = \left(\mu a\right)^{-\frac{1}{2}} \sum_{s=0}^\infty (4s+5)^{\frac{1}{2}} y_s J_{2s+\frac{5}{2}}\left(\mu a\right), \quad (5.236)$$

where $\{y_s\}_{s=0}^\infty \in l_2$. We use also the discontinuous Weber-Schafheitlin integral (see [23])

$$\int_0^\infty J_{\frac{3}{2}}\left(\mu\rho\right) J_{2l+\frac{5}{2}}\left(\mu a\right) d\mu = \begin{cases} a^{-1}\left(\rho/a\right)^{\frac{3}{2}} P_l^{\left(\frac{3}{2},0\right)}\left(1 - 2\rho^2/a^2\right), & \rho < a \\ 0, & \rho > a \end{cases}$$

$$(5.237)$$

that implies the important result

$$\int_0^a \rho J_{\frac{3}{2}}\left(\nu\rho\right)\left\{\int_0^\infty J_{\frac{3}{2}}\left(\mu\rho\right) J_{2l+\frac{5}{2}}\left(\mu a\right) d\mu\right\} d\rho = \nu^{-1} J_{2l+\frac{5}{2}}\left(\nu a\right). \quad (5.238)$$

Finally the dual integral equations are transformed into the second kind i.s.l.a.e.

$$y_l - \sum_{s=0}^\infty \alpha_{sl} y_s - \sum_{n=1}^\infty \beta_{2nl} a_n = \gamma_l, \quad (5.239)$$

where $l = 0, 1, 2, \ldots$,

$$\alpha_{sl} = \left[(4s+5)(4l+5)\right]^{\frac{1}{2}} \int_0^\infty \frac{\varepsilon\left(\nu, k\right)}{\nu} J_{2s+\frac{5}{2}}\left(\nu a\right) J_{2l+\frac{5}{2}}\left(\nu a\right) d\nu, \quad (5.240)$$

$$\beta_{nl} = ia^{\frac{1}{2}} (4l+5)^{\frac{1}{2}} (2n+1) \int_0^k \sigma_n (\nu, k, h) \nu^{-\frac{1}{2}} J_{2l+\frac{5}{2}} (\nu a) \, d\nu, \qquad (5.241)$$

and

$$\gamma_l = \frac{a^{\frac{1}{2}}}{k} (4l+5)^{\frac{1}{2}} \int_0^k e^{-\sqrt{\nu^2 - k^2}(d-h)} \nu^{\frac{1}{2}} J_{2l+\frac{5}{2}} (\nu a) \, d\nu. \qquad (5.242)$$

Enforcing the mixed boundary conditions (5.219), (5.220) leads to the following dual series equations that hold providing that the disc does not intersect the spherical mirror (so $b \cos \theta_0 > h$),

$$\sum_{n=1}^{\infty} (2n+1) a_n \psi_n' (kb) P_n^1 (\cos \theta) =$$

$$\frac{i}{k} \int_0^{\infty} g(\nu) e^{\sqrt{\nu^2 - k^2} h} \left\{ \frac{d}{dr} \left[r J_1 (\nu r \sin \theta) e^{-\sqrt{\nu^2 - k^2} r \cos \theta} \right] \right\} \Bigg|_{r=b} d\nu$$

$$- \sum_{n=1}^{\infty} (2n+1) \frac{\psi_n (kd)}{(kd)^2} \zeta_n' (kb) P_n^1 (\cos \theta), \quad \theta \in (0, \theta_0) \quad (5.243)$$

$$\sum_{n=1}^{\infty} (2n+1) a_n \frac{P_n^1 (\cos \theta)}{\zeta_n' (kb)} = 0, \qquad \theta \in (\theta_0, \pi). \qquad (5.244)$$

We exclude the case in which the disc intersects the spherical mirror $(b \cos \theta_0 < h)$ from further consideration. Start the formal re-expansion via

$$F (\nu, k; \theta) =$$

$$\begin{cases} e^{\sqrt{\nu^2 - k^2} h} \frac{d}{dr} \left[r J_1 (\nu r \sin \theta) e^{-\sqrt{\nu^2 - k^2} r \cos \theta} \right]_{r=b}, & b \cos \theta_0 > h \\ e^{-\sqrt{\nu^2 - k^2} h} \frac{d}{dr} \left[r J_1 (\nu r \sin \theta) e^{\sqrt{\nu^2 - k^2} r \cos \theta} \right]_{r=b}, & b \cos \theta_0 < h \end{cases}$$

$$= \frac{1}{2} \sum_{n=1}^{\infty} \frac{2n+1}{n(n+1)} z_n (\nu, k) P_n^1 (\cos \theta) \qquad (5.245)$$

where

$$z_n (\nu, k) = \int_0^{\pi} F (\nu, k; \theta) P_n^1 (\cos \theta) \sin \theta d\theta. \qquad (5.246)$$

Then transform (5.243)–(5.244) to

$$\sum_{n=1}^{\infty} x_n (1 - \varepsilon_n) P_n^1 (\cos \theta) = ikb \sum_{n=1}^{\infty} (2n+1) \frac{\psi_n (kd)}{(kd)^2} \zeta_n' (kb) P_n^1 (\cos \theta)$$

$$- ikb \sum_{n=1}^{\infty} \frac{2n+1}{n(n+1)} \left\{ \sum_{s=0}^{\infty} y_s A_{sn} \right\} P_n^1 (\cos \theta), \quad \theta \in (0, \theta_0), \qquad (5.247)$$

$$\sum_{n=1}^{\infty} \frac{2n+1}{n(n+1)} x_n P_n^1 (\cos\theta), \qquad \theta \in (\theta_0, \pi), \qquad (5.248)$$

where

$$a_n = \frac{\zeta_n'(kb)}{n(n+1)} x_n \qquad (5.249)$$

$$A_{sn} = \frac{(4s+5)^{\frac{1}{2}}}{a^{\frac{1}{2}}} \int_0^{\infty} \nu^{-\frac{1}{2}} J_{2s+\frac{5}{2}}(\nu a) z_n(\nu, k) \, d\nu \qquad (5.250)$$

and the asymptotically small parameter

$$\varepsilon_n = 1 + ikb \frac{2n+1}{n(n+1)} \psi_n'(kb) \zeta_n'(kb) \qquad (5.251)$$

satisfies $\varepsilon_n = O\left((kb/n)^2\right)$ as $n \to \infty$.

In the customary way we may transform equations (5.247)–(5.248) to the second kind i.s.l.a.e. (noting that $\{x_m\}_{n=1}^{\infty} \in l_2$),

$$x_m - \sum_{n=1}^{\infty} x_n \varepsilon_n Q_{nm}^{(1)}(\theta_0) + \sum_{n=0}^{\infty} y_n t_{nm} = \sum_{n=1}^{\infty} R_n Q_{nm}^{(1)}(\theta_0), \qquad (5.252)$$

where $m = 1, 2, \ldots$, and

$$Q_{nm}^{(1)}(\theta_0) = Q_{nm}(\theta_0) - \frac{Q_{n0}(\theta_0)}{Q_{00}(\theta_0)} Q_{0m}(\theta_0), \qquad (5.253)$$

$$t_{nm} = ikb \sum_{s=1}^{\infty} \frac{2s+1}{s(s+1)} A_{ns} Q_{sm}^{(1)}(\theta_0), \qquad (5.254)$$

$$R_n = ikb(2n+1) \frac{\psi_n(kd)}{(kd)^2} \zeta_n'(kb). \qquad (5.255)$$

Taking into account (5.249) we may rearrange (5.239) in the form

$$y_m - \sum_{n=0}^{\infty} \alpha_{nm} y_n - \sum_{n=1}^{\infty} \beta_{nm}^* x_n = \gamma_m, \qquad (5.256)$$

where $m = 0, 1, 2, \ldots$, and

$$\beta_{nm}^* = \frac{\zeta_n'(kb)}{n(n+1)} \beta_{nm}. \qquad (5.257)$$

Thus, the originally stated problem is reduced to a coupled pair of second kind i.s.l.a.e. for the unknowns x_m and y_m that are amenable to standard numerical solution techniques employing truncation methods.

6

Spherical Cavities with Spherical Dielectric Inclusions.

In this chapter we consider electromagnetic diffraction from cavity structures that contain dielectric obstacles. In common with previous chapters, a rigorous treatment based upon regularisation methods is possible in the context of spherical geometry, so that we shall consider spherical shell cavities with spherically symmetric dielectric inclusions. Our choice of problems is constrained to a few but characteristic situations that arise frequently in practical applications.

The first, examined in Section 6.1, concerns the resonant heating of bodies with absorbing properties; thus, instead of the scattered field, the quantity of greatest interest is the electromagnetic energy coupled to the cavity region.

The second concerns the reflectivity of dielectric spheres partially screened by a metallic spherical shell. In Sections 6.2 and 6.3, we consider homogeneous dielectric spheres and multi-layer structures consisting of homogeneous dielectric shells. If several suitably graded dielectric layers are employed, a model of the Luneberg lens is obtained, and so a rigorous theory of the so-called *Luneberg lens reflector* is obtained.

Whilst this selection of problems is not exhaustive, it indicates the scope and applicability of regularisation methods for various configurations of dielectric loaded metallic structures.

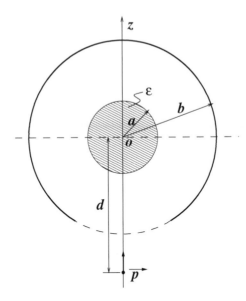

FIGURE 6.1. Spherical cavity containing a concentrically located small lossy dielectric sphere, excited by a dipole source.

6.1   Resonant Cavity Heating of a Small Lossy Dielectric Sphere.

Consider a perfectly conducting spherical cavity of radius b containing a small lossy dielectric sphere, concentrically located and of radius a. In this section we examine the coupling of electromagnetic waves to the cavity region and calculate the energy absorbed by the dielectric sphere. The absorbed energy manifests itself as heat, and it is of great practical interest to design structures that are effective for heating purposes, particularly in the microwave regime. We shall excite the cavity by a vertical electric dipole (VED) located outside the cavity at a distance d, along the axis of rotational symmetry. The geometry is illustrated in Figure 6.1.

Suppose the sphere is composed of homogeneous material of relative permittivity ε, relative permeability μ and conductivity σ. The angle subtended at the centre of the cavity by the aperture is $\theta_1 = \pi - \theta_0$. The structure is excited by the spherical wave radiated by a VED of moment p located along the z-axis at the point given by $r = d, \theta = \pi$.

The electromagnetic field of the VED is described by the Debye potential (see Chapter 1)

$$U^0(r, \theta) = \frac{ipk_0^3}{k_0^2 r (k_0 d)^2} \sum_{n=0}^{\infty} (-1)^n (2n+1) P_n (\cos \theta) \times$$

$$\begin{cases} \psi_n (k_0 r) \zeta_n (k_0 d), & r < d \\ \zeta_n (k_0 r) \psi_n (k_0 d), & r > d \end{cases} \quad (6.1)$$

where $k_0 = \omega/c$. Here and henceforth the time harmonic dependence factor $\exp(-i\omega t)$ is suppressed.

Subdivide the space into three regions, where regions I, II and III comprise the interior of the dielectric sphere ($r < a$), the interior of the cavity excluding the dielectric sphere ($a < r < b$), and the exterior of the cavity ($r > b$), respectively. In each region we seek the total field in the form

$$U^I(r, \theta) = \frac{ip}{kd^2 r} \sum_{n=0}^{\infty} (-1)^n (2n+1) x_n \psi_n (kr) P_n (\cos \theta), \quad (6.2)$$

$$U^{II}(r, \theta) =$$
$$\frac{ip}{k_0 d^2 r} \sum_{n=0}^{\infty} (-1)^n (2n+1) \{y_n \psi_n (k_0 r) + z_n \zeta_n (k_0 r)\} P_n (\cos \theta), \quad (6.3)$$

and

$$U^{III}(r, \theta) = U^0(r, \theta) + \frac{ip}{kd^2 r} \sum_{n=0}^{\infty} (-1)^n (2n+1) v_n \zeta_n (kr) P_n (\cos \theta), \quad (6.4)$$

where

$$k^2 = k_0^2 (\mu \varepsilon + i\mu (\sigma/\varepsilon_0 \omega)). \quad (6.5)$$

We suppose henceforth that $\mu = 1$. The enforcement of continuity of electric and magnetic flux densities across the surface of the dielectric sphere requires

$$k^2 U^I(a, \theta) = k_0^2 U^{II}(a, \theta) \quad (6.6)$$

and

$$\left. \frac{\partial}{\partial r} (rU^I) \right|_{r=a} = \left. \frac{\partial}{\partial r} (rU^{II}) \right|_{r=a} \quad (6.7)$$

for all $\theta \in (0, \pi)$. Thus the coefficients y_n and z_n may be expressed in terms of x_n, as

$$y_n = -iq_n x_n, \quad z_n = -is_n x_n \tag{6.8}$$

where

$$
\begin{aligned}
q_n &= \sqrt{\varepsilon}\psi_n\left(ka\right)\zeta_n'\left(k_0a\right) - \psi_n'\left(ka\right)\zeta_n\left(k_0a\right), &(6.9)\\
s_n &= \psi_n\left(k_0a\right)\psi_n'\left(ka\right) - \sqrt{\varepsilon}\psi_n'\left(k_0a\right)\psi_n\left(ka\right). &(6.10)
\end{aligned}
$$

The mixed boundary conditions to be enforced on the surface $r = b$ are

$$E_\theta^{II}(b, \theta) = E_\theta^{III}(b, \theta) = 0, \qquad \theta \in (0, \theta_0) \tag{6.11}$$

and

$$H_\varphi^{II}(b, \theta) = H_\varphi^{III}(b, \theta) = 0, \qquad \theta \in (\theta_0, \pi). \tag{6.12}$$

It follows from the continuity conditions that the tangential electric field component E_θ is continuous on the complete spherical surface $r = b$,

$$E_\theta^{II}(b, \theta) = E_\theta^{III}(b, \theta), \qquad \theta \in (0, \pi),$$

so that

$$v_n = -ix_n\left\{s_n - q_n\frac{\psi_n'\left(k_0b\right)}{\zeta_n'\left(k_0b\right)}\right\} - \frac{\psi_n'\left(k_0b\right)}{\zeta_n'\left(k_0b\right)}\zeta_n\left(k_0d\right). \tag{6.13}$$

Imposition of the boundary conditions (6.11) and (6.12) now leads to the following dual series equations to be solved for the unknowns x_n,

$$\sum_{n=0}^{\infty} (-1)^n (2n+1)\left\{q_n\psi_n'\left(k_0b\right) + s_n\zeta_n'\left(k_0b\right)\right\}x_n P_n^1\left(\cos\theta\right)$$

$$= 0, \quad \theta \in (0, \theta_0) \tag{6.14}$$

$$\sum_{n=0}^{\infty} (-1)^n (2n+1)\frac{q_n}{\zeta_n'\left(k_0b\right)}x_n P_n^1\left(\cos\theta\right)$$

$$= i\sum_{n=0}^{\infty} (-1)^n (2n+1)\frac{\zeta_n\left(k_0d\right)}{\zeta_n'\left(k_0b\right)}P_n^1\left(\cos\theta\right), \quad \theta \in (\theta_0, \pi). \tag{6.15}$$

We now use the method of regularisation described in Chapter 2 of Volume I to transform this set of dual series equations to the i.s.l.a.e. of second kind

$$X_m - \sum_{n=1}^{\infty} X_n \varepsilon_n Q_{nm}^* (\theta_0) =$$

$$i \sum_{n=1}^{\infty} n(n+1) \frac{\zeta_n(k_0 d)}{\zeta_n'(k_0 b)} (-1)^n \{\delta_{nm} - Q_{nm}^*(\theta_0)\} \quad (6.16)$$

where $m = 1, 2, \ldots,$

$$x_n = (-1)^n \frac{X_n}{n(n+1)} \frac{\zeta_n'(k_0 b)}{q_n}, \quad (6.17)$$

$$\varepsilon_n = 1 + ik_0 b \frac{2n+1}{n(n+1)} \left\{ \psi_n'(k_0 b) \zeta_n'(k_0 b) + \frac{s_n}{q_n} \left(\zeta_n'(k_0 b)\right)^2 \right\} \quad (6.18)$$

and

$$Q_{nm}^* (\theta_0) = Q_{nm}(\theta_0) - \frac{Q_{0m}(\theta_0)}{Q_{00}(\theta_0)} Q_{n0}(\theta_0). \quad (6.19)$$

The parameter ε_n is asymptotically small as $n \to \infty$: using asymptotics for the spherical Bessel functions, it is easy to show that when $a < b$,

$$\varepsilon_n = \frac{2n^2 + 2n + 3}{2n(n+1)(2n-1)(2n+3)} (k_0 b)^2 + \frac{\varepsilon - 1}{\varepsilon + 1} \left(\frac{a}{b}\right)^{2n+1}$$

$$+ O\left(\left(\frac{a}{b}\right)^{2n+3}, \left(\frac{k_0 b}{n}\right)^4\right) \quad (6.20)$$

as $n \to \infty$.

The case when $a = b$ warrants separate consideration. In this case the regularisation procedure produces the i.s.l.a.e. given by (6.16) except that the parameter ε_n is replaced by the parameter

$$\widehat{\varepsilon}_n = 1 - \frac{1}{2} k_0 b (\varepsilon + 1) \frac{2n+1}{n(n+1)} \frac{\psi_n'(k_0 b) \zeta_n'(k_0 b)}{q_n} \quad (6.21)$$

that is asymptotically small as $n \to \infty$.

It is convenient to transform (6.16) by introducing the angle $\theta_1 = \pi - \theta_0$ to

$$(1 - \varepsilon_m) X_m^* + \sum_{n=1}^{\infty} X_n^* \varepsilon_n R_{nm}^* (\theta_1) = i \sum_{n=1}^{\infty} n(n+1) \frac{\zeta_n(k_0 d)}{\zeta_n'(k_0 b)} R_{nm}^* (\theta_1)$$

$$(6.22)$$

where $m = 1, 2, \ldots$, $X_m^* = (-1)^m X_m$ and

$$R_{nm}^* (\theta_1) = R_{nm} (\theta_1) + \frac{R_{n0} (\theta_1)}{1 - R_{00} (\theta_1)} R_{0m} (\theta_1).$$

We have suppressed the familiar arguments about the solution class of the coefficients, but simply note that the square summability of the coefficient sequence $\{X_n^*\}_{n=1}^{\infty}$ provides the correct singular behaviour of the field components in the vicinity of the edge (where $\theta \to \theta_0$). Uniqueness of solutions to this systems, and of those encountered in susbsequent sections in this chapter, is established by adapting the uniqueness argument given in Section 2.1; the truncation number for numerical studies is chosen according to the criterion set out in that section.

Let us now discuss the application of this solution to the of a small dielectric sphere located in the open spherical resonator. Laser beams are often employed for heating regions of plasma of diameter around 10^{-2} cm. This corresponds to a multi- mode regime, and its effectiveness is due to the high concentration of energy in a tiny volume. The resonant excitation of an open spherical cavity provides an alternative mechanism for heating small objects. In contrast to the laser, a single mode is generated in the cavity.

A measure of the effectiveness of such heating is the effectiveness factor η that is defined to be the ratio of the power $\overline{P} (\theta_1)$ deposited in the body inside the cavity with aperture half-angle θ_1 to that power deposited in the body $\overline{P} (\pi)$ in free space (which correspond to the "cavity" $\theta_1 = \pi$):

$$\eta = \frac{\overline{P} (\theta_1)}{\overline{P} (\pi)}. \tag{6.23}$$

The average power, per period of oscillation, deposited as heat in a target that occupies a volume V and has dielectric constant $\varepsilon = \varepsilon' + i\varepsilon''$ with imaginary part ε'' is (see [67] and [97])

$$\overline{P} = \frac{1}{2} k_0 c \varepsilon'' \int_V \left| \overrightarrow{E} \right|^2 dV. \tag{6.24}$$

This formula may be used to calculate the heat deposited in the lossy dielectric sphere in the absence of the cavity.

We assume that the dielectric sphere is small enough and satisfies

$$k_0 a \ll 1, \quad |ka| \ll 1 \tag{6.25}$$

so that only the first term (with index $n = 1$) need be retained in the spherical wave expansions (see [6.2]–[6.4]). Then it is readily shown that

$$\overline{P} (\pi) \approx 24\pi P_0 \left(\frac{a}{d} \right)^3 \frac{1 + (k_0 d)^{-2}}{k_0 d} \frac{\varepsilon''}{(\varepsilon' + 2)^2 + (\varepsilon'')^2} \tag{6.26}$$

where $P_0 = p^2 k_0^4 c$ is the power of the source (in watts). When the same object is located in the cavity resonator, a similar calculation shows that

$$\overline{P}(\theta_1) \approx 6\pi P_0 \left(\frac{a}{d}\right)^3 \frac{1 - (k_0 b)^{-2} + (k_0 b)^{-4}}{k_0 d} |X_1^*|^2 . \tag{6.27}$$

Thus the effectiveness factor is

$$\eta = \frac{\overline{P}(\theta_1)}{\overline{P}(\pi)} = \frac{1 - (k_0 b)^{-2} + (k_0 b)^{-4}}{4\left(1 + (k_0 d)^{-2}\right)} |X_1^*|^2 . \tag{6.28}$$

Its value is essentially determined by the value of $|X_1^*|^2$.

We make use of the mathematical tools used in Chapters 2 and 3 to find approximate analytical expressions for the coefficients X_k^* $(k = 1, 2, \ldots)$ when $\theta_1 \ll 1$. The result is

$$X_k^* \approx G_{ks} \frac{\theta_1^3}{k_0 b - y_{ks}^{(1)} + i\Gamma_{ks}} \tag{6.29}$$

where

$$G_{ks} = -\frac{k_0 b}{6\pi} \frac{k(k+1)}{\left(y_{ks}^{(0)}\right)^2 - k(k+1)} \sum_{n=1}^{\infty} (2n+1) g_n, \tag{6.30}$$

and

$$g_n = in(n+1) \frac{\zeta_n(k_0 d)}{\zeta_n'(k_0 b)}. \tag{6.31}$$

Furthermore

$$y_{ks}^{(1)} = y_{ks}^{(0)} \left\{ 1 + \frac{k(k+1)(2k+1)}{\left(y_{ks}^{(0)}\right)^2 - k(k+1)} \frac{\theta_1^3}{6\pi} - y_{ks}^{(0)} \lambda_k^{(2)} \frac{\left[\mathrm{Im}\,\zeta_k'\left(y_{ks}^{(0)}\right)\right]^2}{\left(y_{ks}^{(0)}\right)^2 - k(k+1)} \right\} \tag{6.32}$$

are the real parts of the quasi-eigenoscillations developed in the spherical cavity containing the small lossy dielectric sphere. In this formula, $y_{ks}^{(0)}$ denotes the eigenvalues of eigenoscillations for the closed sphere without any dielectric inclusions. In these formulae, the index k is the number of field variations in the angular coordinate θ, whilst the index s denotes field variations in the radial coordinate r. The term proportional to θ_1^3 in (6.32) describes the perturbation to the value of $y_{ks}^{(0)}$ by a small aperture (with

$\theta_1 \ll 1$). The following term provides the perturbation due to the presence of the small lossy dielectric sphere (recall $k_0 a \ll 1, |ka| \ll 1$) where

$$\lambda_k^{(2)} \approx \frac{2^{2k}(k-1)!(k+1)!}{(2k)!(2k+1)!} \frac{(\varepsilon'-1)(\varepsilon'+\frac{k+1}{k})+(\varepsilon'')^2}{(\varepsilon'+\frac{k+1}{k})^2+(\varepsilon'')^2}(k_0 a)^{2k+1}. \quad (6.33)$$

The heat losses and radiation losses in the range $\left|k_0 b - y_{ks}^{(1)}\right| \ll 1$ are defined by the value

$$\Gamma_{ks} = \Gamma_{ks}^{(1)} + \Gamma_{ks}^{(2)} \quad (6.34)$$

where

$$\Gamma_{ks}^{(1)} = \left(y_{ks}^{(0)}\right)^2 \lambda_k^{(1)} \frac{\left[\operatorname{Im}\zeta_k'\left(y_{ks}^{(0)}\right)\right]^2}{\left(y_{ks}^{(0)}\right)^2 - k(k+1)}, \quad (6.35)$$

$$\lambda_k^{(1)} = \frac{2^{2k}(k-1)!(k-1)!(k+1)!}{[(2k)!]^2} \frac{\varepsilon''}{(\varepsilon'+\frac{k+1}{k})^2+(\varepsilon'')^2}(k_0 a)^{2k+1}, \quad (6.36)$$

$$\Gamma_{ks}^{(2)} = \frac{\theta_1^6}{36\pi^2} \frac{k(k+1)(2k+1)}{\left(y_{ks}^{(0)}\right)^2 - k(k+1)} \sum_{n \neq k} \frac{n(n+1)(2n+1)}{\left|\zeta_n'\left(y_{ks}^{(0)}\right)\right|^2}. \quad (6.37)$$

It is evident that the perturbation of the spectral values $y_{ks}^{(0)}$ is maximal when $k = 1$. The perturbation is proportional to the volume of the small lossy dielectric sphere; furthermore, it is negative. Physically it is quite clear that the presence of the dielectric effectively reduces the resonant volume. The perturbation of the spectral values $y_{ks}^{(0)}$ is much smaller when $k > 1$. This occurs because the maximal field intensity of the higher modes is not located at the origin, as it is for the first mode.

It follows from (6.28) and (6.29) that when the excitation wavenumber $k_0 b$ is not near the quasi-eigenvalue $y_{ks}^{(1)}$, the effectiveness factor η is proportional to θ_1^6. In the vicinity of the quasi-eigenvalue $y_{ks}^{(1)}$, η increases, but its precise value depends upon the balance between heat and radiation losses given by $\Gamma_{1s}^{(1)}$ and $\Gamma_{1s}^{(2)}$. If $\Gamma_{1s}^{(1)} \sim \Gamma_{1s}^{(2)}$ and $k_0 b = y_{ks}^{(1)}$, the value of η is large and proportional to θ_1^{-6}. If $\Gamma_{1s}^{(1)} \gg \Gamma_{1s}^{(2)}$, the value of η is approximately that for a lossy sphere located in free space, i.e., $\eta \sim 1$. The inference is that optimal absorption of electromagnetic energy, and consequent heating of the lossy sphere, is achieved when heat and radiation losses are comparable ($\Gamma_{1s}^{(1)} \sim \Gamma_{1s}^{(2)}$); in this situation the Q-factor is maximal ($Q_{1s} \sim \theta_1^{-6}$).

Let us consider the physical situation. If the parameters of a lossy dielectric sphere $(a, \varepsilon', \varepsilon'')$ are specified, what choice of aperture angle optimises η? To obtain the optimal angle θ_1^{opt}, first set $k_0 b = y_{ks}^{(1)}$ in formula (6.29). The "resonant" values of the coefficients are

$$X_{1s}^* = iG_{1s} \frac{\theta_1^3}{\alpha_s (k_0 a)^3 + \beta_s \theta_1^6} \tag{6.38}$$

where the dimensionless value

$$\alpha_s = \left(y_{1s}^{(0)}\right)^2 \frac{\left[\operatorname{Im} \zeta_1'\left(y_{1s}^{(0)}\right)\right]^2}{\left(y_{ks}^{(0)}\right)^2 - 2} \frac{2\varepsilon''}{(\varepsilon' + 2)^2 + (\varepsilon'')^2} \tag{6.39}$$

measures the loss density in the volume of the sphere, and

$$\beta_s = \frac{1}{6\pi^2} \frac{1}{\left(y_{ks}^{(0)}\right)^2 - 2} \sum_{n \neq 1} \frac{n(n+1)(2n+1)}{\left|\zeta_n'\left(y_{1s}^{(0)}\right)\right|^2} \tag{6.40}$$

describes the radiation losses of the structure at resonance. The minimum value of X_{1s}^* and hence of η occurs when

$$\theta_1 = \theta_1^{opt} = \left(\frac{\alpha_s}{\beta_s}\right)^{\frac{1}{2}} (k_0 a)^{\frac{1}{2}}. \tag{6.41}$$

The deposition power may be represented by the formula

$$\overline{P}(\theta_1) = P_0 C(\varepsilon) F Q_{1s} \tag{6.42}$$

where

$$C(\varepsilon) = \frac{6\pi\varepsilon''}{(\varepsilon' + 2)^2 + (\varepsilon'')^2} \tag{6.43}$$

and

$$F = \left(\frac{a}{d}\right)^3 \frac{1 - (k_0 b)^{-2} + (k_0 b)^{-4}}{k_0 d}. \tag{6.44}$$

The geometrical factor F reflects the various locations in the elements of the resonant system.

Thus deposition power at resonant excitation depends upon the dielectric constant (via the factor $C(\varepsilon)$), the geometrical factor F, the power P_0 of the source and the Q-factor Q_{1s}. This formula is applicable to other configurations as well as the specific spherical structure of present interest.

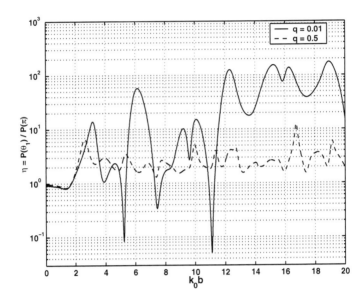

FIGURE 6.2. Effectiveness factor for dielectric spheres ($\epsilon = 2.58 + i0.0215$) in a spherical cavity ($\theta_1 = 90°$).

W e compute the effectiveness factor η for a small ($q = a/b = 0.01$) and a moderately sized ($q = a/b = 0.5$) dielectric inclusion in the spherical cavity, for three aperture sizes ($\theta_1 = 90°, 60°$ and $30°$). Results for the hemispherical shell ($\theta_1 = 90°$) are shown in Figure 6.2. The small inclusion ($q = 0.01$) does not significantly disturb the quasi-eigenoscillation structure and the effectiveness factor η takes large values exceeding 100. By contrast the larger dielectric inclusion ($q = 0.5$) completely destroys the modal structure and η is small.

As the aperture decreases in size, the value of η is large at resonant points in the frequency ranges displayed in Figures 6.3 and 6.4. This is true even for the larger inclusion ($q = 0.5$) at the first resonance point, where η exceeds 10^3 (see Figure 6.4); this is attributable to the extremely high Q-factor for the unloaded cavity.

Finally the dependence of η on the filling fraction q is examined in Figure 6.5; the value of η at resonant points is smaller when q is larger.

This concludes our study of the coupling of electromagnetic energy to a dielectric inclusion in a spherical cavity. There are a number of related questions about optimal configurations for electromagnetic w ave heating, but they lie outside the scope of this book.

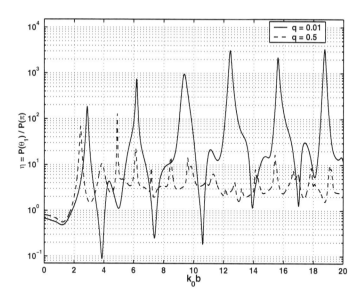

FIGURE 6.3. Factor η for spheres ($\epsilon = 2.58 + i0.0215$) in the cavity ($\theta_1 = 60°$).

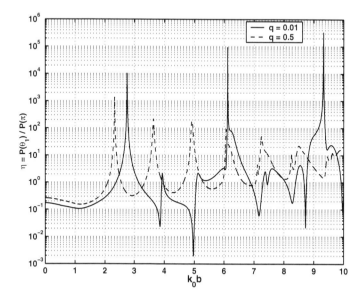

FIGURE 6.4. Factor η for spheres ($\epsilon = 2.58 + i0.0215$) in the cavity ($\theta_1 = 30°$).

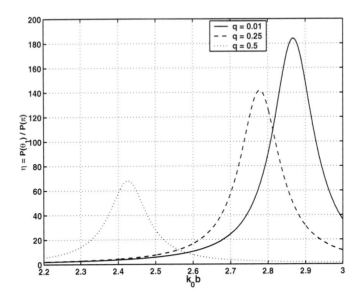

FIGURE 6.5. Factor η for spheres ($\epsilon = 2.58 + i0.0215$) in the cavity ($\theta_1 = 60°$).

6.2   Reflectivity of a Partially Screened Dielectric Sphere.

The mathematical treatment of a spherical cavity filled with dielectric material has much in common with the approach for the empty cavity employed in Section 4.2. The same is true if the aperture angle is increased so much that the cavity shell becomes a spherical cap that partially screens the dielectric sphere when it is present.

Consider therefore a sphere of radius a composed of material with complex dielectric constant $\varepsilon = \varepsilon' + i\varepsilon''$ that is partially enclosed by a perfectly conducting spherical shell (or cap) of the same radius and subtending an angle θ_0 at the origin (see Figure 6.6). The z-axis is normal to the plane of the circular aperture formed in the spherical shell; the structure is illuminated by a plane electromagnetic wave travelling in the positive direction along this axis.

As usual, the electromagnetic field may be prescribed in terms of Debye potentials, and it is convenient to subdivide the whole space into an interior region I ($r < a$) and an exterior region II ($r > a$), with spherical interface $r = a$. Seek the total solution Debye potentials in each region as the sum of the incident Debye potentials U^0, V^0 and scattered Debye potentials

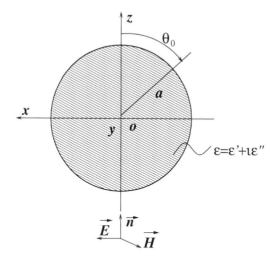

FIGURE 6.6. Dielectric sphere partially screened by a spherical cap of angle θ_0.

U_{sc}^i, V_{sc}^i $(i = I, II)$ in the form

$$U_t^i = U_{sc}^i + U^0, \qquad V_t^i = V_{sc}^i + V^0. \tag{6.45}$$

The incident potentials are defined by equation (4.87), and the scattered potentials have the form

$$U_{sc}^I = \frac{\cos\varphi}{ik_0^2 r} \sum_{n=1}^{\infty} A_n \psi_n(kr) x_n P_n^1(\cos\theta) \tag{6.46}$$

$$V_{sc}^I = \frac{\sin\varphi}{ik_0^2 r} \sum_{n=1}^{\infty} A_n \psi_n(kr) y_n P_n^1(\cos\theta) \tag{6.47}$$

$$U_{sc}^{II} = \frac{\cos\varphi}{ik_0^2 r} \sum_{n=1}^{\infty} A_n \zeta_n(k_0 r) z_n P_n^1(\cos\theta) \tag{6.48}$$

$$V_{sc}^{II} = \frac{\sin\varphi}{ik_0^2 r} \sum_{n=1}^{\infty} A_n \zeta_n(k_0 r) v_n P_n^1(\cos\theta) \tag{6.49}$$

where $k_0 = \omega/c$, $k = \sqrt{\varepsilon}k_0$, $A_n = (2n+1)/n(n+1)$, and x_n, y_n, z_n, v_n are the unknown coefficients to be determined (the branch of the square root being chosen so that it is positive when ε is real and positive).

It may be easily verified that the form of solution specified by (6.45) and (6.46)–(6.49) satisfies Maxwell's equations and the radiation condition at infinity. Keeping in mind the boundedness condition for the scattered energy (cf. [4.89]), the other conditions to be enforced are the continuity conditions

$$E_\theta^I (a, \theta, \varphi) = E_\theta^{II} (a, \theta, \varphi), \qquad (6.50)$$
$$E_\varphi^I (a, \theta, \varphi) = E_\varphi^{II} (a, \theta, \varphi) \qquad (6.51)$$

for all θ and φ, and the mixed boundary conditions

$$E_\theta^I (a, \theta, \varphi) = E_\theta^{II} (a, \theta, \varphi) = 0, \qquad \theta \in (0, \theta_0) \qquad (6.52)$$
$$E_\varphi^I (a, \theta, \varphi) = E_\varphi^{II} (a, \theta, \varphi) = 0, \qquad \theta \in (0, \theta_0) \qquad (6.53)$$
$$H_\varphi^I (a, \theta, \varphi) = H_\varphi^{II} (a, \theta, \varphi), \qquad \theta \in (\theta_0, \pi) \qquad (6.54)$$
$$H_\theta^I (a, \theta, \varphi) = H_\theta^{II} (a, \theta, \varphi), \qquad \theta \in (\theta_0, \pi), \qquad (6.55)$$

that are valid for all φ. It is evident that the continuity conditions (6.50) and (6.51) are equivalent to the conditions

$$\left. \frac{\partial}{\partial r} \left(r U^I \right) \right|_{r=a} = \left. \frac{\partial}{\partial r} \left(r U^{II} \right) \right|_{r=a}, \qquad (6.56)$$
$$\left. V^I \right|_{r=a} = \left. V^{II} \right|_{r=a} \qquad (6.57)$$

for all θ and φ. It follows that the relations between the interior region coefficients (x_n, y_n) and exterior region coefficients (z_n, v_n) are

$$x_n = \frac{\zeta_n' (k_0 a)}{\sqrt{\varepsilon} \psi_n' (ka)} z_n, \qquad (6.58)$$
$$y_n = \frac{\zeta_n (k_0 a)}{\psi_n (ka)} v_n. \qquad (6.59)$$

It should be noted that at this stage there is no coupling between electromagnetic oscillations of E type ($E_r \neq 0$) and H type ($H_r \neq 0$); the relation (6.58) connects coefficients of electric type, whilst the relation (6.59) connects coefficients of magnetic type.

We now enforce the mixed boundary conditions (6.52)–(6.55), and following the argument outlined in Section 4.2, obtain the following coupled pairs of dual series equations containing so-called *polarisation constants* B_1 and B_2.

$$\sum_{n=1}^{\infty} A_n \zeta_n' (k_0 a) z_n P_n^1 (\cos \theta) =$$

$$B_1 \tan \frac{\theta}{2} - \sum_{n=1}^{\infty} A_n \psi_n' (k_0 a) P_n^1 (\cos \theta), \qquad \theta \in (0, \theta_0) \quad (6.60)$$

$$\sum_{n=1}^{\infty} A_n \frac{q_n}{\psi_n'(ka)} z_n P_n^1(\cos\theta) =$$

$$B_2 \cot\frac{\theta}{2} + (\sqrt{\varepsilon} - 1) \sum_{n=1}^{\infty} A_n \psi_n(k_0 a) P_n^1(\cos\theta), \quad \theta \in (\theta_0, \pi) \quad (6.61)$$

$$\sum_{n=1}^{\infty} A_n \zeta_n(k_0 a) v_n P_n^1(\cos\theta) =$$

$$iB_1 \tan\frac{\theta}{2} - \sum_{n=1}^{\infty} A_n \psi_n(k_0 a) P_n^1(\cos\theta), \quad \theta \in (0, \theta_0) \quad (6.62)$$

$$\sum_{n=1}^{\infty} A_n \frac{s_n}{\psi_n(ka)} v_n P_n^1(\cos\theta) = B_2 \cot\frac{\theta}{2}, \quad \theta \in (\theta_0, \pi), \quad (6.63)$$

where

$$q_n = \psi_n'(ka)\zeta_n(k_0 a) - \zeta_n'(k_0 a)\psi_n(ka), \quad (6.64)$$
$$s_n = \psi_n(ka)\zeta_n'(k_0 a) - \sqrt{\varepsilon}\psi_n'(ka)\zeta_n(k_0 a). \quad (6.65)$$

In the absence of the dielectric sphere (i.e., $\varepsilon = 1$), it is readily shown that $s_n = -q_n = i$. Now introduce the rescaled coefficients

$$Z_n = i^n \frac{2n+1}{\psi_n'(ka)} q_n z_n, \quad V_n = i^n \frac{1}{\psi_n(ka)} s_n v_n. \quad (6.66)$$

Since

$$\lim_{n\to\infty} \frac{\zeta_n'(k_0 a)\psi_n'(ka)}{(2n+1)q_n} = -\frac{1}{2k_0 a(1 + \sqrt{\varepsilon})} \quad (6.67)$$

and

$$\lim_{n\to\infty} \frac{(2n+1)\zeta_n(k_0 a)\psi_n(ka)}{s_n} = -k_0 a, \quad (6.68)$$

we are led to define the parameters

$$\varepsilon_n = 1 + 2k_0 a(1 + \sqrt{\varepsilon})\frac{\zeta_n'(k_0 a)\psi_n'(ka)}{(2n+1)q_n} \quad (6.69)$$

and

$$\mu_n = 1 + \frac{(2n+1)}{(k_0 a)s_n}\zeta_n(k_0 a)\psi_n(ka); \quad (6.70)$$

they are asymptotically small as $n \to \infty$.

We observe that when $\varepsilon = 1$ (so that $s_n = -q_n = i$), these parameters simplify to those previously encountered for the empty spherical cavity (or cap), i.e.,

$$\varepsilon_n = 1 + 4ik_0a \frac{\zeta_n'(k_0a)\,\psi_n'(k_0a)}{(2n+1)} \tag{6.71}$$

and

$$\mu_n = 1 - \frac{i(2n+1)}{(k_0a)}\zeta_n(k_0a)\,\psi_n(k_0a)\,; \tag{6.72}$$

as $n \to \infty$, $\varepsilon_n = O(n^{-2})$ and $\mu_n = O((k_0a/n)^2)$.

Let us establish the corresponding asymptotic behaviour when $\varepsilon \neq 1$. Employing the standard asymptotics of the spherical Bessel functions, it is found that the parameter μ_n retains the same rate of decay ($\mu_n = O(n^{-2})$ as $n \to \infty$), but that the companion parameter ε_n decays less rapidly according to

$$\varepsilon_n = \frac{\sqrt{\varepsilon}-1}{2(\sqrt{\varepsilon}+1)}n^{-1} + O(n^{-2})\,. \tag{6.73}$$

In the absence of the dielectric sphere, this formula reproduces the quadratic rate of decay stated above.

At this stage we remark that similar features are encountered for spherical screens with impedance boundary conditions.

Following the argument given in Section 4.2, the standard regularisation process leads to the following functional equations for the unknowns X_n, V_n and the polarisation constants B_1, B_2.

$$\sum_{n=1}^{\infty} Z_n \sin\left(n+\frac{1}{2}\right)\theta =$$
$$\begin{cases} -2k_0a\left(\sqrt{\varepsilon}+1\right)B_1 \sin\frac{1}{2}\theta + \\ \quad \sum_{n=1}^{\infty}\left(\alpha_n^{(1)} + Z_n\varepsilon_n\right)\sin\left(n+\frac{1}{2}\right)\theta, & \theta \in (0,\theta_0) \\ -B_2 \sin\frac{1}{2}\theta + \sum_{n=1}^{\infty}\alpha_n^{(2)}\sin\left(n+\frac{1}{2}\right)\theta, & \theta \in (\theta_0,\pi)\,, \end{cases} \tag{6.74}$$

$$\sum_{n=1}^{\infty} V_n \cos\left(n+\frac{1}{2}\right)\theta =$$
$$\begin{cases} -\frac{i}{k_0a}B_1 \cos\frac{1}{2}\theta + \sum_{n=1}^{\infty}(\beta_n + V_n\mu_n)\cos\left(n+\frac{1}{2}\right)\theta, & \theta \in (0,\theta_0) \\ -B_2 \cos\frac{1}{2}\theta, & \theta \in (\theta_0,\pi)\,, \end{cases} \tag{6.75}$$

where

$$\alpha_n^{(1)} = 2k_0 a \left(\sqrt{\varepsilon} + 1 \right) i^n \psi_n' \left(k_0 a \right), \tag{6.76}$$

$$\alpha_n^{(2)} = \left(\sqrt{\varepsilon} - 1 \right) i^n \left(2n + 1 \right) \psi_n \left(k_0 a \right), \tag{6.77}$$

and

$$\beta_n = \frac{1}{k_0 a} i^n \left(2n + 1 \right) \psi_n \left(k_0 a \right). \tag{6.78}$$

The functional equations (6.74)–(6.75) give rise in the standard way to the following coupled sets of i.s.l.a.e. in which the matrix operator is a completely continuous perturbation of the identity operator in l_2.

$$Z_m - \sum_{n=1}^{\infty} Z_n \varepsilon_n R_{nm} \left(\theta_0 \right) + 2 k_0 a \left(\sqrt{\varepsilon} + 1 \right) B_1 R_{0m} \left(\theta_0 \right) - B_2 R_{0m} \left(\theta_0 \right)$$

$$= \sum_{n=1}^{\infty} \alpha_n^{(1)} R_{nm} \left(\theta_0 \right) + \sum_{n=1}^{\infty} \alpha_n^{(2)} \left[\delta_{nm} - R_{nm} \left(\theta_0 \right) \right], \quad (6.79)$$

where $m = 1, 2, \ldots$, with

$$- \sum_{n=1}^{\infty} Z_n \varepsilon_n R_{n0} \left(\theta_0 \right) + 2 k_0 a \left(\sqrt{\varepsilon} + 1 \right) B_1 R_{00} \left(\theta_0 \right) + B_2 \left[1 - R_{00} \left(\theta_0 \right) \right]$$

$$= \sum_{n=1}^{\infty} \alpha_n^{(1)} R_{n0} \left(\theta_0 \right) - \sum_{n=1}^{\infty} \alpha_n^{(2)} R_{n0} \left(\theta_0 \right), \quad (6.80)$$

and

$$V_m - \sum_{n=1}^{\infty} V_n \mu_n Q_{nm} \left(\theta_0 \right) + \frac{i}{k_0 a} B_1 Q_{0m} \left(\theta_0 \right) - i B_2 Q_{0m} \left(\theta_0 \right)$$

$$= \sum_{n=1}^{\infty} \beta_n Q_{nm} \left(\theta_0 \right), \quad (6.81)$$

where $m = 1, 2, \ldots$, with

$$- \sum_{n=1}^{\infty} V_n \mu_n Q_{n0} \left(\theta_0 \right) + \frac{i}{k_0 a} B_1 Q_{00} \left(\theta_0 \right) + i B_2 \left[1 - Q_{00} \left(\theta_0 \right) \right]$$

$$= \sum_{n=1}^{\infty} \beta_n Q_{n0} \left(\theta_0 \right). \quad (6.82)$$

The expressions $Q_{nm}(\theta_0)$ and $R_{nm}(\theta_0)$ are as previously defined.

The reflectivity of the partially screened dielectric sphere is defined by its radar cross-section (RCS, cf. Section 4.2)

$$\sigma_B = \lim_{r \to \infty} 4\pi r^2 \frac{|E_\theta^{sc}|^2}{|E_\theta^0|^2} = \frac{\pi}{k_0^2} \left| \sum_{n=1}^{\infty} (-1)^n (2n+1) (z_n - v_n) \right|^2 . \qquad (6.83)$$

Before we consider the dependence of the RCS on the electrical size $2a/\lambda$ of the dielectric sphere, it is worth recalling that the homogeneous (unscreened) dielectric sphere acts as a focusing system. In the older literature, such a sphere was known as a "Constant-K Lens" [6]. A simple geometric optics argument shows that a dielectric sphere of permittivity $\varepsilon = 3.5$ will focus a parallel bundle of rays to a point on the surface of the sphere; otherwise the bundle is focused to a small region or spot inside or outside the sphere according as ε is greater than or less than 3.5. This approximate result provides a useful guide to the expected behaviour of these lenses, for antenna design and other purposes. For our purposes, the reflectivity of the partially screened dielectric sphere is obviously strongly influenced by the focusing process of the unscreened lens.

As a preliminary step, it is of interest to determine accurately the distribution of the electromagnetic energy density around dielectric spheres of varying electrical size, with attention to the formation of well defined focal regions or spots. As well as the ideal value of 3.5, we examine dielectric spheres with $\varepsilon' = 2.1$ (Teflon) and 5.6 (Alsimag ceramic). The loss tangent of these materials, $\tan\delta = \varepsilon''/\varepsilon'$ is small, being $2 \cdot 10^{-4}$ and $16 \cdot 10^{-4}$, respectively. Fused quartz and sublimed sulphur have permittivity near the ideal value, with values of ε' equal to 3.78 and 3.69, respectively, and corresponding loss tangents equal to $4 \cdot 10^{-4}$ and $11 \cdot 10^{-4}$, respectively. Since the loss tangent is small, we take it to be zero in all cases.

According to the geometric optics approach, a Teflon sphere would focus radiation outside the sphere, whilst an Alsimag sphere would focus the radiation inside, and the quartz or sulphur sphere would focus the energy close to the surface.

Let us calculate the axial distribution (along the z-axis) of the electromagnetic energy density for a range of electrical sizes $2a/\lambda$ between 50 and 200. The results for a Teflon sphere are displayed in Figure 6.7. As the electrical size increases, the distance f of the focal spot (the point at which energy density is maximal) increases from $1.29a$ to $1.49a$. When $\varepsilon = 3.5$, the results displayed in Figure 6.8 show that f increases from $0.986a$ to $0.995a$ when $2a/\lambda$ increases from 50 and 100; when $2a/\lambda = 200$, two energy density maxima are observed at $f = 0.96a$ and $f = 1.02a$. Finally if $\varepsilon = 5.6$, the focal spot moves from $f \approx 0.64a$ to $f = 0.671a$ as electrical size increases from 50 to 100.

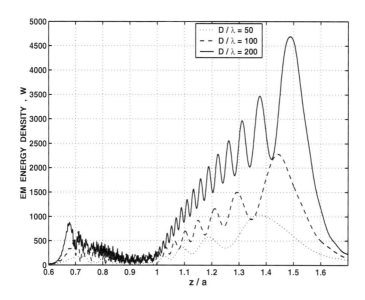

FIGURE 6.7. Axial distribution of electromagnetic energy density for a "constant-K" lens ($\epsilon = 2.1$).

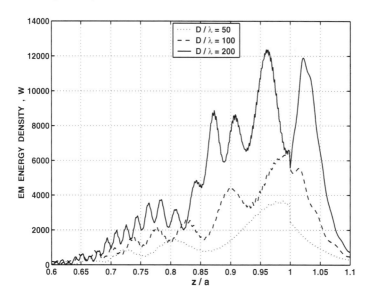

FIGURE 6.8. Axial distribution of electromagnetic energy density for a "constant-K" lens ($\epsilon = 3.5$).

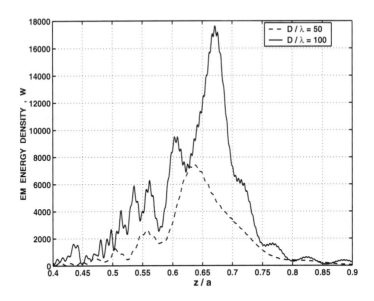

FIGURE 6.9. Axial distribution of electromagnetic energy density for a "constant-K" lens ($\epsilon = 5.6$).

When $\varepsilon = 3.5$, the axial distribution of energy is very useful in predicting powerful reflection effects from the screened sphere because the focal spot is very close to the screen. However it is rather less useful for other dielectric values because the focal spot is located well away from the metallic surface. This assertion is demonstrated by the following results. In Figures 6.10 and 6.11 are plotted the radar cross-sections for spheres of dielectric $\varepsilon = 2.1$ and 5.6 with a moderate aperture size ($\theta_0 = 30°$).

The dependence of RCS on frequency is reminiscent of that for the single spherical cap considered in Chapter 4. The only difference arises from reflection at the dielectric-free space interface and manifests itself as jitter in the plots.

Let us now examine what is potentially the most interesting structure, that with $\varepsilon = 3.5$, for which we have already calculated the axial distribution of energy (in the absence of the cap): see Figure 6.8. It is useful to explore the transverse as well as axial extent of the focal spot because the angular extent of the reflecting metallic cap can then be optimally matched to the extent of the focal region. The precise definition of *spot size* is to some extent conventional; we take it to be the region in which the electromagnetic energy density W exceeds W_{max}/e, where W_{max} denotes the maximal electromagnetic energy density. In Figure 6.12 *normalised* electromagnetic energy density $W_r = W/W_{max}$ is plotted for the dielectric

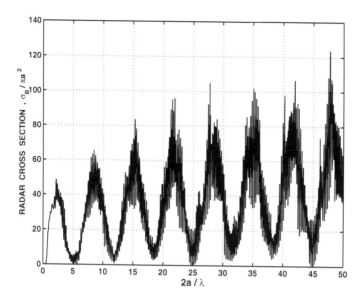

FIGURE 6.10. Radar cross-section of a screened dielectric sphere, $\epsilon = 2.1$ and $\theta_0 = 30°$.

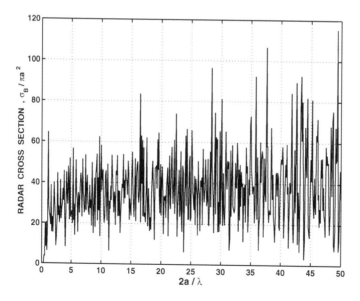

FIGURE 6.11. Radar cross-section of a screened dielectric sphere, $\epsilon = 5.6$ and $\theta_0 = 30°$.

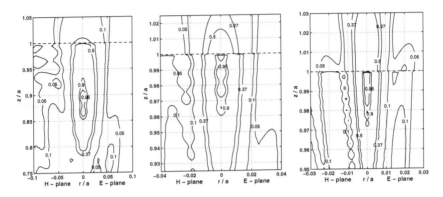

FIGURE 6.12. Normalised electromagnetic energy density W_r for a dielectric sphere ($\epsilon = 3.5$) with D/λ equal to 20 (left), 50 (centre) and 100 (right).

sphere ($\varepsilon = 3.5$) of electric size $2a/\lambda$ ranging from 20 to 100. The dashed line indicates the potential location for a spherical cap ($r/a = 0$; $z/a = 1$). The focal spot is asymmetric with respect to the axis of symmetry (compare contour values for $W_r = 0.1$ in the E-plane [$\varphi = 0°$] and the H-plane [$\varphi = 90°$]). The transverse extent r_t of the focal spot is about $0.05a$, $0.022a$ and $0.024a$ when electric size $2a/\lambda$ equals 20, 50 and 100, respectively.

A suitable choice of cap size for the best reflector would seem to be that comparable to the focal spot size. Thus in the range examined above, the cap angle θ_0 may be taken to be $2.5°$; a cap of larger extent might not be expected to improve the overall reflectivity. Let us examine this assertion by an accurate computation of the RCS. The RCS of the $2.5°$ cap is plotted in Figure 6.13 for $2a/\lambda$ ranging from 0.1 to 100. The RCS is dramatically increased compared to that from the Teflon and Alsimag spheres (Figures 6.10–6.11). The RCS is also plotted as the cap angle is increased from $2.5°$ to $90°$ (Figure 6.14). No dramatic changes in the RCS are observed over this range, and similar dependence is observed for different cap angles.

An interesting phenomenon appears when the cap angle θ_0 exceeds $90°$; of course the structure is more properly viewed as a dielectric filled spherical cavity. It might be supposed that the reflectivity would decrease for θ_0 in this range, but this statement is not completely true. The RCS is displayed in Figure 6.15 for the cavities defined by $\theta_0 = 120°, 135°$ and $150°$. The RCS levels for $\theta_0 = 120°$ are, perhaps surprisingly, comparable to those for spherical caps of various angles (with $\theta_0 \leq 90°$). Actually a geometric optics explanation is available. It is known (see for example [4]) that spherical aberration reduces the apparent aperture of the dielectric sphere (with $\varepsilon = 3.5$) by about 10%, so that metallic coverings of somewhat greater angular extent than $90°$ do not significantly block the lens aperture for

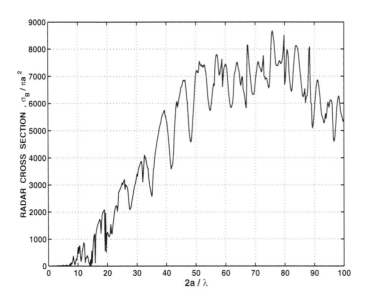

FIGURE 6.13. Radar cross-section for the screened dielectric sphere with $\epsilon = 3.5$ and $\theta_0 = 2.5°$.

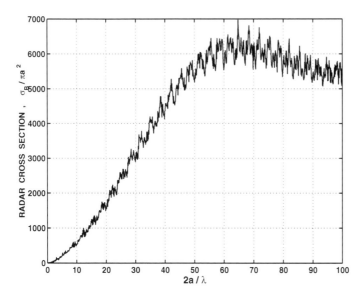

FIGURE 6.14. Radar cross-section for the screened dielectric sphere with $\epsilon = 3.5$ and $\theta_0 = 90°$.

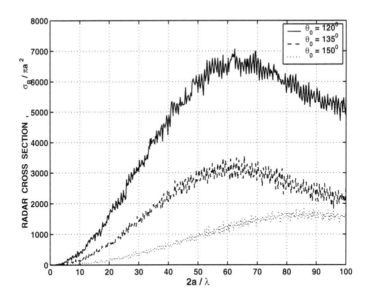

FIGURE 6.15. Radar cross-section for the screened dielectric sphere with $\epsilon = 3.5$ and various values of θ_0.

the reception of a bundle of parallel rays. This 10% reduction means that the angle θ_0 may be increased to around 115° or 120° before a noticeable degradation in reflectivity appears.

In conclusion, this section provides rigorously accurate models of dielectric filled cavities and lens-reflector combinations that have many practical purposes.

6.3   The Luneberg Lens Reflector.

In this section we consider a multi-layered dielectric filling of a metallic spherical shell with a circular aperture; the dielectric filling is composed of several spherical layers, each of homogeneous material so that the overall structure is rotationally symmetric about an axis normal to the plane of the circular aperture. The angle θ_0 subtended by the aperture at the centre of the spherical structure may be greater than or less than 90°; in the latter case the structure is best viewed as a multi-layered dielectric sphere with a metallic cap reflector.

Such structures are of great practical importance in the design and construction of various lens-reflector combinations. A particularly interesting lens is the *Luneberg lens* that has been extensively studied by analytical

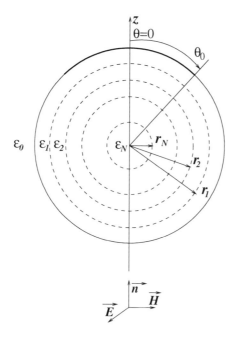

FIGURE 6.16. The multi-layered dielectric sphere partially screened by a spherical cap subtending angle θ_0.

methods [91] and by numerical techniques such as ray tracing [58]. The index n of refraction ($n = \sqrt{\varepsilon}$) of this spherically symmetric lens continuously increases from 1 at the surface to 2 at the centre: at distance r from the centre,

$$n = \sqrt{2 - (r/r_1)^2} \qquad (6.84)$$

where r_1 is the radius of the spherical lens. It has the property that the rays of an incident plane wave are focused to a point on the spherical surface diametrically opposite to that point where the wavefront first impacts the lens. This focusing feature is somewhat idealised, and holds only in the quasi-optical regime where ray tracing descriptions are valid. The addition of a properly designed spherical metallic cap produces a powerful reflector (the *Luneberg lens reflector*) that is very useful in calibrating microwave radar systems.

In practice it is difficult, if not impossible, to construct a lens continuously graded according to (6.84), and a multilayered spherical structure composed of piecewise constant dielectric shells surrounding a dielectric core is usually manufactured. Moreover these lens are often deployed in an intermediate frequency regime where the results of ray tracing techniques are approximate. Thus there is a need for an accurate full-wave solution for the multilayered structure that is valid in the resonance as well as the high frequency regime, and incorporates the addition of a spherical cap reflector. It is asserted in [78] that the Luneberg lens reflector is not "able to be analysed as a classical boundary value problem," but we shall see that this multilayered structure is amenable to a semi-analytic treatment by the method of regularisation, and yields benchmark solutions of guaranteed accuracy against which purely numerical solutions may be compared.

Thus the aim of this section to develop a rigorous solution of a new canonical problem, diffraction of a plane electromagnetic wave by a multi-layered dielectric sphere partially screened by a perfectly conducting and conformal spherical cap. This is an obvious generalisation of the partially screened homogeneous dielectric sphere considered in Section 6.2. The regularisation procedure provides an infinite system of equations that may be truncated to a finite system which is rapidly convergent as the truncation order increases, and produces solutions of guaranteed numerical accuracy.

The problem is multi-parametric, characterised by wavelength, number of layers, choice of dielectric in each layer and cap size and thus provides a whole class of canonical diffraction structures.

Subdivide the interior of a sphere of radius r_1 into N concentric spherical layers, so that the external radius of each layer is r_s ($s = 1, 2, \ldots, N$). This subdivision of the interior of the sphere comprises an innermost spherical region, defined by $0 \leq r < r_N$, and $N - 1$ annular regions given by $r_N < r < r_{N-1}, r_{N-1} < r < r_{N-2}, \ldots$, and $r_2 < r < r_1$; the exterior of this region is defined by $r_1 < r < \infty$. Let $\varepsilon_s = \varepsilon_s' + \varepsilon_s''$ denote the complex relative dielectric constant of layer s ($s = 1, 2, \ldots, N$); the relative permittivity of the free space region (denoted ε_0) has of course unit value. Attached conformally to the surface of the multilayered dielectric sphere is spherical cap (of radius r_1) that subtends an angle $\theta_0 \in (0, \pi)$. A plane electromagnetic wave propagating in the positive direction of the z-axis illuminates the partially screened structure.

As usual we may formulate a boundary value problem for the scattered wave using Maxwell's equations and the associated Debye potentials U, V. The Debye potentials for the incident field (cf. equation [4.87]) are

$$\begin{Bmatrix} U_{inc} \\ V_{inc} \end{Bmatrix} = \begin{Bmatrix} \cos \varphi \\ \sin \varphi \end{Bmatrix} \frac{1}{ik_0 r^2} \sum_{n=1}^{\infty} A_n \psi_n (k_0 r) P_n^1 (\cos \theta) . \qquad (6.85)$$

Within the annular layer s ($s = 1, 2, \ldots, N$), we seek Debye potentials U_s, V_s for the total field in the form

$$U_s = \frac{\cos \varphi}{ik_0 r^2} \sum_{n=1}^{\infty} A_n \left\{ a_n^{(s)} \psi_n (k_s r) + b_n^{(s)} \zeta_n (k_s r) \right\} P_n^1 (\cos \theta) , \qquad (6.86)$$

$$V_s = \frac{\sin\varphi}{ik_0 r^2} \sum_{n=1}^{\infty} A_n \left\{ \tilde{a}_n^{(s)} \psi_n \left(k_s r \right) + \tilde{b}_n^{(s)} \zeta_n \left(k_s r \right) \right\} P_n^1 \left(\cos\theta \right). \qquad (6.87)$$

In the free space region $(s = 0)$ we represent the total potentials in the form

$$U_0 = U_{inc} + \frac{\cos\varphi}{ik_0 r^2} \sum_{n=1}^{\infty} A_n b_n^{(0)} \zeta_n \left(k_0 r \right) P_n^1 \left(\cos\theta \right), \qquad (6.88)$$

$$V_0 = V_{inc} + \frac{\sin\varphi}{ik_0 r^2} \sum_{n=1}^{\infty} A_n \tilde{b}_n^{(0)} \zeta_n \left(k_0 r \right) P_n^1 \left(\cos\theta \right), \qquad (6.89)$$

where $k_0 = \omega/c$, $k_s = \sqrt{\varepsilon_s} k_0$, and the coefficients $a_n^{(s)}, \tilde{a}_n^{(s)}, b_n^{(s)}, \tilde{b}_n^{(s)}$ are to be determined for $s = 0, 1, 2, \ldots, N$; we adopt the convention $a_n^{(0)} = \tilde{a}_n^{(0)} = 0$, and note that the boundedness of the total field at the origin implies that $b_n^{(N)} = \tilde{b}_n^{(N)} = 0$.

This form of potential provides a solution of Maxwell's equations that also satisfies the radiation condition at infinity. At the external interface $r = r_1$, the mixed boundary conditions to be enforced are

$$E_\theta^{(0)} \left(r_1, \theta, \varphi \right) = E_\theta^{(1)} \left(r_1, \theta, \varphi \right) = 0, \ \theta \in (0, \theta_0) \qquad (6.90)$$

$$E_\varphi^{(0)} \left(r_1, \theta, \varphi \right) = E_\varphi^{(1)} \left(r_1, \theta, \varphi \right) = 0, \ \theta \in (0, \theta_0) \qquad (6.91)$$

$$H_\theta^{(0)} \left(r_1, \theta, \varphi \right) = H_\theta^{(1)} \left(r_1, \theta, \varphi \right), \qquad \theta \in (\theta_0, \pi) \qquad (6.92)$$

$$H_\varphi^{(0)} \left(r_1, \theta, \varphi \right) = H_\varphi^{(1)} \left(r_1, \theta, \varphi \right), \qquad \theta \in (\theta_0, \pi) \qquad (6.93)$$

for all φ, together with

$$\frac{\partial}{\partial r} \left(rU_0 \right) \bigg|_{r=r_1} = \frac{\partial}{\partial r} \left(rU_1 \right) \bigg|_{r=r_1} \qquad (6.94)$$

$$V_0 \big|_{r=r_1} = V_1 \big|_{r=r_1} \qquad (6.95)$$

for all θ and φ. At each internal interface $r = r_s$ (where $s = 2, 3, \ldots, N$), the boundary conditions to be imposed are

$$k_{s-1}^2 U_{s-1} = k_s^2 U_s \qquad (6.96)$$

$$\frac{\partial}{\partial r} \left(rU_{s-1} \right) = \frac{\partial}{\partial r} \left(rU_s \right) \qquad (6.97)$$

$$V_{s-1} = V_s \qquad (6.98)$$

$$\frac{\partial}{\partial r} \left(rV_{s-1} \right) = \frac{\partial}{\partial r} \left(rV_s \right) \qquad (6.99)$$

for all θ and φ.

Upon enforcing the boundary conditions (6.90)–(6.93) we obtain the following coupled pair of dual series equations

$$\sum_{n=1}^{\infty} A_n \left\{ \psi_n' (k_0 r_1) + b_n^{(0)} \zeta_n' (k_0 r_1) \right\} P_n^1 (\cos \theta) = C_1 \tan \frac{\theta}{2}, \quad \theta \in (0, \theta_0)$$

$$(6.100)$$

$$\sum_{n=1}^{\infty} A_n \left\{ \psi_n (k_0 r_1) + b_n^{(0)} \zeta_n (k_0 r_1) - \sqrt{\varepsilon_1} \left[a_n^{(1)} \psi_n (k_1 r_1) + b_n^{(1)} \zeta_n (k_1 r_1) \right] \right\} \times$$

$$P_n^1 (\cos \theta) = C_2 \tan \frac{\theta}{2}, \quad \theta \in (\theta_0, \pi) \quad (6.101)$$

$$\sum_{n=1}^{\infty} A_n \left\{ \psi_n (k_0 r_1) + \widetilde{b}_n^{(0)} \zeta_n (k_0 r_1) \right\} P_n^1 (\cos \theta) = iC_1 \tan \frac{\theta}{2}, \quad \theta \in (0, \theta_0)$$

$$(6.102)$$

$$\sum_{n=1}^{\infty} A_n \left\{ \psi_n' (k_0 r_1) + \widetilde{b}_n^{(0)} \zeta_n' (k_0 r_1) - \sqrt{\varepsilon_1} \left[\widetilde{a}_n^{(1)} \psi_n' (k_1 r_1) + \widetilde{b}_n^{(1)} \zeta_n' (k_1 r_1) \right] \right\} \times$$

$$P_n^1 (\cos \theta) = iC_2 \tan \frac{\theta}{2}, \quad \theta \in (\theta_0, \pi) . \quad (6.103)$$

From the continuity conditions (6.94)–(6.95) we obtain the following two relationships between the unknown coefficients occurring in (6.100)–(6.103),

$$\psi_n' (k_0 r_1) + b_n^{(0)} \zeta_n' (k_0 r_1) = \sqrt{\varepsilon_1} \left[a_n^{(1)} \psi_n' (k_1 r_1) + b_n^{(1)} \zeta_n' (k_1 r_1) \right] \quad (6.104)$$

and

$$\psi_n (k_0 r_1) + \widetilde{b}_n^{(0)} \zeta_n (k_0 r_1) = \widetilde{a}_n^{(1)} \psi_n (k_1 r_1) + \widetilde{b}_n^{(1)} \zeta_n (k_1 r_1) . \quad (6.105)$$

Finally imposition of the boundary conditions (6.96)–(6.99) leads to the following relations between the unknown coefficients in adjacent layers,

$$a_n^{(s-1)} \psi_n (k_{s-1} r_s) + b_n^{(s-1)} \zeta_n (k_{s-1} r_s) =$$

$$\frac{\varepsilon_s}{\varepsilon_{s-1}} \left\{ a_n^{(s)} \psi_n (k_s r_s) + b_n^{(s)} \zeta_n (k_s r_s) \right\}, \quad (6.106)$$

$$a_n^{(s-1)} \psi_n' (k_{s-1} r_s) + b_n^{(s-1)} \zeta_n' (k_{s-1} r_s) =$$

$$\sqrt{\frac{\varepsilon_s}{\varepsilon_{s-1}}} \left\{ a_n^{(s)} \psi_n' (k_s r_s) + b_n^{(s)} \zeta_n' (k_s r_s) \right\} \quad (6.107)$$

and

$$\widetilde{a}_n^{(s-1)}\psi_n\left(k_{s-1}r_s\right) + \widetilde{b}_n^{(s-1)}\zeta_n\left(k_{s-1}r_s\right) =$$
$$\widetilde{a}_n^{(s)}\psi_n\left(k_sr_s\right) + \widetilde{b}_n^{(s)}\zeta_n\left(k_sr_s\right), \quad (6.108)$$

$$\widetilde{a}_n^{(s-1)}\psi_n'\left(k_{s-1}r_s\right) + \widetilde{b}_n^{(s-1)}\zeta_n'\left(k_{s-1}r_s\right) =$$
$$\sqrt{\frac{\varepsilon_s}{\varepsilon_{s-1}}}\left\{\widetilde{a}_n^{(s)}\psi_n'\left(k_sr_s\right) + \widetilde{b}_n^{(s)}\zeta_n'\left(k_sr_s\right)\right\}. \quad (6.109)$$

These equations hold for $s = 2, 3, \ldots, N$; we recall that $a_n^{(0)} = \widetilde{a}_n^{(0)} = 0$, and $b_n^{(N)} = \widetilde{b}_n^{(N)} = 0$.

From equations (6.106) and (6.107) we deduce that all the coefficients $a_n^{(s-1)}$ and $b_n^{(s-1)}$ for $s = 2, 3, \ldots, N$ are expressible in terms of the single core coefficient $a_n^{(N)}$; likewise, from equations (6.108) and (6.109) that all the coefficients $\widetilde{a}_n^{(s-1)}$ and $\widetilde{b}_n^{(s-1)}$ for $s = 2, 3, \ldots, N$ are expressible in terms of the single core coefficient $\widetilde{a}_n^{(N)}$. The explicit expressions are

$$a_n^{(s)} = -i\Delta_{a_n^{(s)}}a_n^{(N)}, \quad b_n^{(s)} = -i\Delta_{b_n^{(s)}}a_n^{(N)} \quad (6.110)$$

$$\widetilde{a}_n^{(s)} = -i\Delta_{\widetilde{a}_n^{(s)}}\widetilde{a}_n^{(N)}, \quad \widetilde{b}_n^{(s)} = -i\Delta_{\widetilde{b}_n^{(s)}}\widetilde{a}_n^{(N)} \quad (6.111)$$

where $s = 1, 2, \ldots, N-1$; the values of the determinantal expressions $\Delta_{a_n^{(s)}}$ and $\Delta_{b_n^{(s)}}$ are calculated by the recurrence formulae

$$\Delta_{a_n^{(s-1)}} = -i\sqrt{\frac{\varepsilon_s}{\varepsilon_{s-1}}} \times$$
$$\left\{\left[\sqrt{\frac{\varepsilon_s}{\varepsilon_{s-1}}}\psi_n\left(k_sr_s\right)\zeta_n'\left(k_{s-1}r_s\right) - \psi_n'\left(k_sr_s\right)\zeta_n\left(k_{s-1}r_s\right)\right]\Delta_{a_n^{(s)}}\right.$$
$$\left. + \left[\sqrt{\frac{\varepsilon_s}{\varepsilon_{s-1}}}\zeta_n\left(k_sr_s\right)\zeta_n'\left(k_{s-1}r_s\right) - \zeta_n'\left(k_sr_s\right)\zeta_n\left(k_{s-1}r_s\right)\right]\Delta_{b_n^{(s)}}\right\}, \quad (6.112)$$

$$\Delta_{b_n^{(s-1)}} = -i\sqrt{\frac{\varepsilon_s}{\varepsilon_{s-1}}} \times$$
$$\left\{\left[\psi_n\left(k_{s-1}r_s\right)\psi_n'\left(k_sr_s\right) - \sqrt{\frac{\varepsilon_s}{\varepsilon_{s-1}}}\psi_n'\left(k_{s-1}r_s\right)\psi_n\left(k_sr_s\right)\right]\Delta_{a_n^{(s)}}\right.$$
$$\left. + \left[\psi_n\left(k_{s-1}r_s\right)\zeta_n'\left(k_sr_s\right) - \sqrt{\frac{\varepsilon_s}{\varepsilon_{s-1}}}\psi_n'\left(k_{s-1}r_s\right)\zeta_n\left(k_sr_s\right)\right]\Delta_{b_n^{(s)}}\right\}, \quad (6.113)$$

where $s = 2, 3, \ldots, N$, initialised by the values

$$\Delta_{a_n^{(N)}} = i, \quad \Delta_{b_n^{(N)}} = 0;$$

the values of the determinantal expressions $\Delta_{\tilde{a}_n^{(s)}}$ and $\Delta_{\tilde{b}_n^{(s)}}$ are similarly calculated by the recurrence formulae

$$\Delta_{\tilde{a}_n^{(s-1)}} =$$
$$- i\{[\psi_n\,(k_s r_s)\,\zeta_n'\,(k_{s-1} r_s) - \sqrt{\frac{\varepsilon_s}{\varepsilon_{s-1}}}\psi_n'\,(k_s r_s)\,\zeta_n\,(k_{s-1} r_s)]\Delta_{\tilde{a}_n^{(s)}}$$
$$+ [\zeta_n\,(k_s r_s)\,\zeta_n'\,(k_{s-1} r_s) - \sqrt{\frac{\varepsilon_s}{\varepsilon_{s-1}}}\zeta_n'\,(k_s r_s)\,\zeta_n\,(k_{s-1} r_s)]\Delta_{\tilde{b}_n^{(s)}}\}, \quad (6.114)$$

$$\Delta_{\tilde{b}_n^{(s-1)}} =$$
$$- i\{[\sqrt{\frac{\varepsilon_s}{\varepsilon_{s-1}}}\psi_n\,(k_{s-1} r_s)\,\psi_n'\,(k_s r_s) - \psi_n'\,(k_{s-1} r_s)\,\psi_n\,(k_s r_s)]\Delta_{\tilde{a}_n^{(s)}}$$
$$+ [\sqrt{\frac{\varepsilon_s}{\varepsilon_{s-1}}}\psi_n\,(k_{s-1} r_s)\,\zeta_n'\,(k_s r_s) - \psi_n'\,(k_{s-1} r_s)\,\zeta_n\,(k_s r_s)]\Delta_{\tilde{b}_n^{(s)}}\}, \quad (6.115)$$

where $s = 2, 3, \ldots, N$, again initialised by the values

$$\Delta_{\tilde{a}_n^{(N)}} = i, \quad \Delta_{\tilde{b}_n^{(N)}} = 0.$$

We return to the pair of dual series equations (6.100)–(6.103) and re-formulate them so that the only unknown coefficients are $b_n^{(0)}$ and $\tilde{b}_n^{(0)}$. This is accomplished by setting $s = 1$ in equations (6.110)–(6.111) and then using (6.104) and (6.105) to express the coefficients $a_n^{(N)}$ and $\tilde{a}_n^{(N)}$ in terms of $b_n^{(0)}$ and $\tilde{b}_n^{(0)}$, respectively, obtaining

$$a_n^{(N)} = \frac{i}{\sqrt{\varepsilon_1}}\frac{\psi_n'\,(k_0 r_1) + \zeta_n'\,(k_0 r_1)\,b_n^{(0)}}{\Delta_{a_n^{(1)}}\psi_n'\,(k_1 r_1) + \Delta_{b_n^{(1)}}\zeta_n'\,(k_1 r_1)} \quad (6.116)$$

and

$$\tilde{a}_n^{(N)} = i\frac{\psi_n\,(k_0 r_1) + \zeta_n\,(k_0 r_1)\,\tilde{b}_n^{(0)}}{\Delta_{\tilde{a}_n^{(1)}}\psi_n\,(k_1 r_1) + \Delta_{\tilde{b}_n^{(1)}}\zeta_n\,(k_1 r_1)}. \quad (6.117)$$

This allows us to eliminate the coefficients $a_n^{(1)}$ and $b_n^{(1)}$ from equation (6.101) and the coefficients $\tilde{a}_n^{(1)}$ and $\tilde{b}_n^{(1)}$ from equation (6.103) in favour of

$b_n^{(0)}$ and $\widetilde{b}_n^{(0)}$, respectively. The resulting functional equations, to be paired with equations (6.100) and (6.102) are, respectively,

$$\sum_{n=1}^{\infty} A_n \left\{ \frac{q_n f_n + t_n g_n}{\psi_n'(k_1 r_1) f_n + \zeta_n'(k_1 r_1) g_n} + \frac{\xi_n f_n + \kappa_n g_n}{\psi_n'(k_1 r_1) f_n + \zeta_n'(k_1 r_1) g_n} b_n^{(0)} \right\} \times$$

$$P_n^1(\cos\theta) = C_2 \tan\frac{\theta}{2}, \qquad \theta \in (\theta_0, \pi) \quad (6.118)$$

and

$$\sum_{n=1}^{\infty} A_n \left\{ \frac{\widetilde{q}_n \widetilde{f}_n + \widetilde{t}_n \widetilde{g}_n}{\psi_n(k_1 r_1) \widetilde{f}_n + \zeta_n(k_1 r_1) \widetilde{g}_n} + \frac{\widetilde{\xi}_n \widetilde{f}_n + \widetilde{\kappa}_n \widetilde{g}_n}{\psi_n(k_1 r_1) \widetilde{f}_n + \zeta_n(k_1 r_1) \widetilde{g}_n} \widetilde{b}_n^{(0)} \right\} \times$$

$$P_n^1(\cos\theta) = iC_2 \tan\frac{\theta}{2}, \qquad \theta \in (\theta_0, \pi), \quad (6.119)$$

where we have employed the notation

$$f_n = \Delta_{a_n^{(1)}}, g_n = \Delta_{b_n^{(1)}}, \widetilde{f}_n = \Delta_{\widetilde{a}_n^{(1)}}, \widetilde{g}_n = \Delta_{\widetilde{b}_n^{(1)}}, \qquad (6.120)$$

$$
\begin{aligned}
q_n &= \psi_n(k_0 r_1) \psi_n'(k_1 r_1) - \psi_n'(k_0 r_1) \psi_n(k_1 r_1), & (6.121) \\
t_n &= \psi_n(k_0 r_1) \zeta_n'(k_1 r_1) - \psi_n'(k_0 r_1) \zeta_n(k_1 r_1), & (6.122) \\
\xi_n &= \psi_n'(k_1 r_1) \zeta_n(k_0 r_1) - \psi_n(k_1 r_1) \zeta_n'(k_0 r_1), & (6.123) \\
\kappa_n &= \zeta_n(k_0 r_1) \zeta_n'(k_1 r_1) - \zeta_n'(k_0 r_1) \zeta_n(k_1 r_1), & (6.124)
\end{aligned}
$$

and

$$
\begin{aligned}
\widetilde{q}_n &= \psi_n'(k_0 r_1) \psi_n(k_1 r_1) - \sqrt{\varepsilon_1} \psi_n(k_0 r_1) \psi_n'(k_1 r_1), & (6.125) \\
\widetilde{t}_n &= \psi_n'(k_0 r_1) \zeta_n(k_1 r_1) - \sqrt{\varepsilon_1} \psi_n(k_0 r_1) \zeta_n'(k_1 r_1), & (6.126) \\
\widetilde{\xi}_n &= \psi_n(k_1 r_1) \zeta_n'(k_0 r_1) - \sqrt{\varepsilon_1} \psi_n'(k_1 r_1) \zeta_n(k_0 r_1), & (6.127) \\
\widetilde{\kappa}_n &= \zeta_n(k_1 r_1) \zeta_n'(k_0 r_1) - \sqrt{\varepsilon_1} \zeta_n'(k_1 r_1) \zeta_n(k_0 r_1). & (6.128)
\end{aligned}
$$

We may now apply the regularisation procedure to the sets of dual series equations given by (6.100), (6.118) and (6.102), (6.119), being formulated in terms of the unknown coefficients $b_n^{(0)}$ and $\widetilde{b}_n^{(0)}$, respectively; it is worth noting that the electromagnetic problem has thus been reduced to finding the scattering coefficients associated with the spherical interface.

First we rescale the unknowns so that

$$X_n = i^n (2n+1) \frac{\xi_n f_n + \kappa_n g_n}{\psi_n'(k_1 r_1) f_n + \zeta_n'(k_1 r_1) g_n} b_n^{(0)} \qquad (6.129)$$

and

$$Y_n = i^n \frac{\tilde{\xi}_n \tilde{f}_n + \tilde{\kappa}_n \tilde{g}_n}{\psi_n (k_1 r_1) \tilde{f}_n + \zeta_n (k_1 r_1) \tilde{g}_n} \tilde{b}_n^{(0)}. \tag{6.130}$$

The sets of dual series equations may then be written

$$\sum_{n=1}^{\infty} \frac{2n+1}{n(n+1)} \{i^n \psi_n' (k_0 r_1) + \Phi_n^{(E)} X_n\} P_n^1 (\cos\theta) = C_1 \tan\frac{\theta}{2}, \quad \theta \in (0, \theta_0) \tag{6.131}$$

$$\sum_{n=1}^{\infty} \frac{1}{n(n+1)} \{i^n (2n+1) \frac{q_n f_n + t_n g_n}{\psi_n' (k_1 r_1) f_n + \zeta_n' (k_1 r_1) g_n} + X_n\} P_n^1 (\cos\theta)$$
$$= C_2 \tan\frac{\theta}{2}, \quad \theta \in (\theta_0, \pi) \tag{6.132}$$

$$\sum_{n=1}^{\infty} \frac{1}{n(n+1)} \{i^n (2n+1) \psi_n (k_0 r_1) + \Phi_n^{(H)} Y_n\} P_n^1 (\cos\theta)$$
$$= iC_1 \tan\frac{\theta}{2}, \quad \theta \in (0, \theta_0) \tag{6.133}$$

$$\sum_{n=1}^{\infty} \frac{2n+1}{n(n+1)} \{i^n \frac{\tilde{q}_n \tilde{f}_n + \tilde{t}_n \tilde{g}_n}{\psi_n (k_1 r_1) \tilde{f}_n + \zeta_n (k_1 r_1) \tilde{g}_n} + Y_n\} P_n^1 (\cos\theta)$$
$$= iC_2 \tan\frac{\theta}{2}, \quad \theta \in (\theta_0, \pi), \tag{6.134}$$

where

$$\Phi_n^{(E)} = \frac{\zeta_n' (k_0 r_1)}{2n+1} \frac{\psi_n' (k_1 r_1) f_n + \zeta_n' (k_1 r_1) g_n}{\xi_n f_n + \kappa_n g_n} \tag{6.135}$$

and

$$\Phi_n^{(H)} = (2n+1) \zeta_n (k_0 r_1) \frac{\psi_n (k_1 r_1) \tilde{f}_n + \zeta_n (k_1 r_1) \tilde{g}_n}{\tilde{\xi}_n \tilde{f}_n + \tilde{\kappa}_n \tilde{g}_n}. \tag{6.136}$$

It may be proved that

$$\lim_{n \to \infty} \Phi_n^{(E)} = -\frac{1}{2k_0 r_1 (\sqrt{\varepsilon_1} + 1)} \tag{6.137}$$

and

$$\lim_{n \to \infty} \Phi_n^{(H)} = -k_0 r_1. \tag{6.138}$$

These quantities are completely analogous to those quantities arising in the treatment of the partially screened homogeneous sphere (see formulae [6.67] and [6.68]).

The transformations leading to the regularisation of these dual series equations are similar to those described in Section 6.2. Thus we simply state the two i.s.l.a.e. for the rescaled coefficients X_n and Y_n. The first is

$$X_m - \sum_{n=1}^{\infty} X_n \varepsilon_n R_{nm} (\theta_0) + 2k_0 r_1 (\sqrt{\varepsilon_1} + 1) R_{0m} (\theta_0) C_1 + R_{0m} (\theta_0) C_2$$

$$= \sum_{n=1}^{\infty} \alpha_n R_{nm} (\theta_0) + \sum_{n=1}^{\infty} \beta_n [\delta_{nm} - R_{nm} (\theta_0)] \tag{6.139}$$

for $m = 1, 2, \ldots$, together with

$$- \sum_{n=1}^{\infty} X_n \varepsilon_n R_{n0} (\theta_0) + 2k_0 r_1 (\sqrt{\varepsilon_1} + 1) R_{00} (\theta_0) C_1 - [1 - R_{0m} (\theta_0)] C_2$$

$$= \sum_{n=1}^{\infty} (\alpha_n - \beta_n) R_{nm} (\theta_0) ; \tag{6.140}$$

the second is

$$Y_m - \sum_{n=1}^{\infty} Y_n \mu_n Q_{nm} (\theta_0) + \frac{i}{k_0 r_1} Q_{0m} (\theta_0) C_1 + i Q_{0m} (\theta_0) C_2$$

$$= \sum_{n=1}^{\infty} \tilde{\alpha}_n Q_{nm} (\theta_0) + \sum_{n=1}^{\infty} \tilde{\beta}_n [\delta_{nm} - Q_{nm} (\theta_0)] \tag{6.141}$$

for $m = 1, 2, \ldots$, together with

$$- \sum_{n=1}^{\infty} Y_n \mu_n Q_{n0} (\theta_0) + \frac{i}{k_0 r_1} Q_{00} (\theta_0) C_1 - i [1 - Q_{00} (\theta_0)] C_2$$

$$= \sum_{n=1}^{\infty} \left[\tilde{\alpha}_n - \tilde{\beta}_n \right] Q_{n0} (\theta_0) . \tag{6.142}$$

Here

$$\varepsilon_n = 1 + 2k_0 r_1 (\sqrt{\varepsilon_1} + 1) \Phi_n^{(E)} \tag{6.143}$$

and

$$\mu_n = 1 + \frac{\Phi_n^{(H)}}{k_0 r_1} \tag{6.144}$$

have the asymptotic behaviour $\varepsilon_n = O\left(n^{-1}\right)$ and $\mu_n = O\left(n^{-2}\right)$ as $n \to \infty$; also

$$\alpha_n = 2k_0 r_1 \left(\sqrt{\varepsilon_1} + 1\right) i^n \psi_n' \left(k_0 r_1\right) \tag{6.145}$$

$$\beta_n = -i^n \left(2n + 1\right) \frac{q_n f_n + t_n g_n}{\psi_n' \left(k_1 r_1\right) f_n + \zeta_n' \left(k_1 r_1\right) g_n} \tag{6.146}$$

$$\tilde{\alpha}_n = \frac{1}{k_0 r_1} \left(2n + 1\right) i^n \psi_n \left(k_0 r_1\right) \tag{6.147}$$

$$\tilde{\beta}_n = -i^n \frac{\tilde{q}_n \tilde{f}_n + \tilde{t}_n \tilde{g}_n}{\psi_n \left(k_1 r_1\right) \tilde{f}_n + \zeta_n \left(k_1 r_1\right) \tilde{g}_n}. \tag{6.148}$$

We recall that

$$\left\{\begin{matrix} Q_{nm} \left(\theta_0\right) \\ R_{nm} \left(\theta_0\right) \end{matrix}\right\} = \frac{1}{\pi} \left\{ \frac{\sin \left(n - m\right) \theta_0}{n - m} \pm \frac{\sin \left(n + m + 1\right) \theta_0}{n + m + 1} \right\}.$$

Moreover the boundedness for the scattered energy requires that the coefficient sequences $\{X_n\}_{n=1}^{\infty}$ and $\{Y_n\}_{n=1}^{\infty}$ are square summable, i.e., lie in l_2; this assertion follows from the asymptotic behaviour of $Q_{nm} \left(\theta_0\right)$ and $R_{nm} \left(\theta_0\right)$ as $m \to \infty$. The asymptotic rates of decay for the quantities defined by (6.143) and (6.144) make it possible to establish that the matrix operator associated with i.s.l.a.e. (6.139) is a completely continuous perturbation of the identity operator in l_2. Thus numerical solutions of these equations may be obtained by truncation to a finite number N_{tr} of equations, with guaranteed accuracy depending directly on N_{tr}.

The reflectivity of the Luneberg Lens Reflector considers the radar cross-section (RCS), equal to

$$\sigma_B = \frac{\pi}{k_0^2} \left| \sum_{n=1}^{\infty} (-1)^n \left(2n + 1\right) \left(b_n^{(0)} - \tilde{b}_n^{(0)}\right) \right|^2. \tag{6.149}$$

The ideal Luneberg lens described in the introduction has its index of refraction $n = \sqrt{\varepsilon}$ graded continuously according to equation (6.84) where $R = r_1$. However as already mentioned, practically manufactured structures

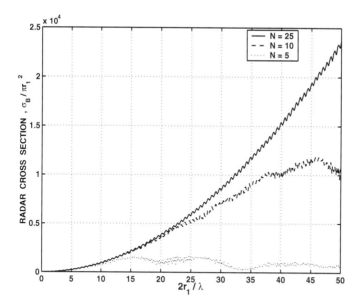

FIGURE 6.17. Radar cross-section of the Luneberg Lens Reflector ($\theta_0 = 90°$).

employ a multilayered dielectric sphere, each layer being homogeneous of constant dielectric. The choice of layer thickness and dielectric constant determines the lens behaviour; in reality the electromagnetic wave will be focused not to a point on the lens surface, but concentrated to a focal region. As in the previous section, the extent of the optimally chosen metallic reflecting cap (to be attached to the surface near this focal region) is determined by the angular spread of the focal region; this diminishes as electrical size increases, so that the angle θ_0 may be as small as 2.5° or less.

Our results examine structures with a metallic cap angle θ_0 taking values 90°, 30° and 5°; the distribution (6.84) is approximated by N layers of equal thickness in each of which the refractive index is constant and equal to $n(r_i)$, where r_i is the radial midpoint of each layer. In our results, N equals 5, 10 or 25.

We first investigate the RCS dependence over the interval $0.1 < 2r_1/\lambda < 50$ upon the number of layers, displayed in Figures 6.17, 6.18 and 6.19. As N increases, the RCS stabilises and changes very little when $N > 20$. Moreover there is only a slight dependence on cap angle θ_0, the results for $\theta_0 = 90°$ being mildly oscillatory about those for $\theta_0 = 30°$.

Also for each fixed θ_0, the RCS plots are identical over that part of the frequency range where the wave thickness does not exceed half a wavelength, whatever the number N of layers is used.

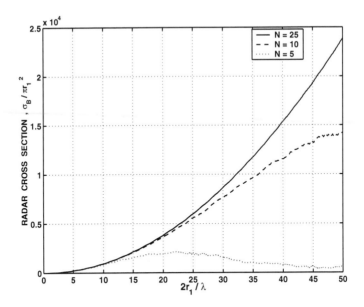

FIGURE 6.18. Radar cross-section of Luneberg Lens Reflector ($\theta_0 = 30°$).

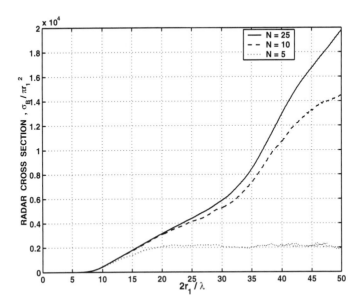

FIGURE 6.19. Radar cross-section of the Luneberg Lens Reflector ($\theta_0 = 5°$).

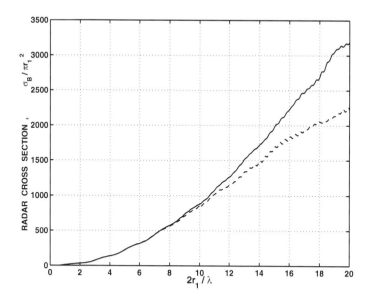

FIGURE 6.20. Radar cross-section of the Luneberg Lens Reflector ($\theta_0 = 30°$),
various dielectric layerings.

At any fixed frequency the maximal value of RCS is obtained by gradu-
ally enlarging the number of layers N. In Figure 6.20, the RCS of a seven
layer dielectric distribution (described in [78]), with spherical layers of unequal thickness (dashed line), is compared to the RCS of a seven layer
dielectric distribution (solid line) with spherical layers of equal thickness
and refractive index uniformly increasing from layer to layer. A similar
comparison of the six layer structure (dashed line) described in [61] with
the comparable uniform six layer structure (solid line) is shown in Figure
6.21. In both cases we may conclude that the uniform dielectric distribution
is preferable because it maximises the RCS.

Finally we observe that, in some circumstances, the partially screened
homogeneous dielectric sphere (PSHDS) has a higher reflectivity than a
poorly designed Luneberg lens reflector. An example is shown in Figure
6.22 where the RCS of a five layer Luneberg lens reflector (dashed line)
is compared to that of a partially screened homogeneous dielectric sphere
(solid line) of relative permittivity $\varepsilon = 3.5$.

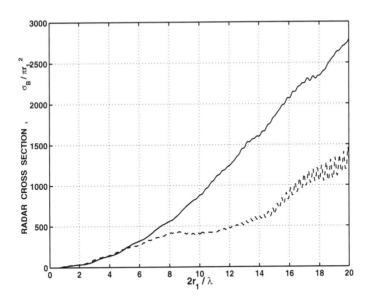

FIGURE 6.21. Radar cross-section of the Luneberg Lens Reflector ($\theta_0 = 30°$), various dielectric layerings.

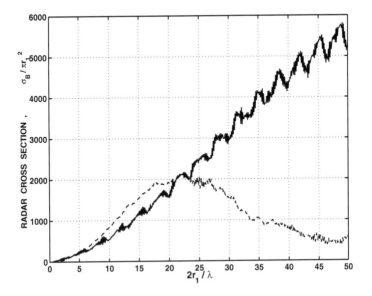

FIGURE 6.22. Radar cross-section of the Luneberg Lens Reflector and PSHDS ($\theta_0 = 30°$).

7
Diffraction from Spheroidal Cavities.

In this chapter we consider dual series equations with kernels that are asymptotically similar to Jacobi polynomials. Such equations arise in the study of diffraction from open spheroidal shells and elliptic cylindrical shells where the relevant kernels are the angular spheroidal functions and the Mathieu functions (a special case of spheroidal functions), respectively. Regularisation techniques for such equations require a significant extension of those described in previous chapters for treating series equations with Jacobi polynomial kernels.

After spherical geometry, spheroidal geometry provides the simplest setting for three-dimensional scattering theory. In this setting a significant extension and generalisation of the spherical shell studies is possible. Depending on its aspect ratio, a closed spheroidal surface takes widely differing shapes ranging from the highly oblate spheroid (the disc is a limiting form) to the sphere to the elongated prolate spheroid, that in a limiting form is a thin cylinder of finite length (or needle-shaped structure). All these structures are interesting for physical and engineering applications.

According as the spheroidal shell (or cavity) is prolate or oblate, various axially symmetric diffraction problems give rise to series equations with kernels that are prolate or oblate angular spheroidal functions, respectively. Dual or triple series equations arise by enforcement of mixed boundary conditions on spheroidal harmonic expansions on the conducting surface or the aperture as appropriate. We shall concentrate on prolate spheroidal shells in this chapter; however the techniques and conclusions described are equally valid for oblate spheroidal geometry and can be applied in the same way.

The solution of the full-wave electromagnetic scattering problem for spheroids is not without mathematical difficulties. This derives from the fact that in spheroidal coordinates (either prolate or oblate), Maxwell's equations with the applied boundary conditions are not separable, in the following sense: in addition to the usual reducibility of the original partial differential equations to a set of linear ordinary differential equations, separability requires that the separated solutions satisfy the boundary conditions term by term. As a result, a mathematically rigorous formulation of plane electromagnetic wave diffraction from a spheroid leads to an infinite system of coupled equations for the Fourier coefficients.

However in *axisymmetric* situations the separation of Maxwell's equations can be achieved if the scattered field does not depend on the azimuthal coordinate ϕ; in this case the electromagnetic problem reduces to two scalar problems (see Section 1.2). In this chapter we consider such axially symmetric situations. The scalar wave equation (the Helmholtz equation) is also separable in both prolate and oblate spheroidal coordinate systems (see Section 1.2).

Many papers have been devoted to the analysis of either perfectly conducting or dielectric closed spheroids that are excited by an axial electric or magnetic dipole field (see, for example [13], [21], [95]) or by an annular slot [122]. Also various full wave diffraction problems for perfectly conducting or dielectric spheroids and their groups (see [5], [25], [26] and [66]) have been solved rigorously.

By contrast diffraction problems for open spheroidal shells – shells with apertures or slots – have not received much rigorous analysis to date. In this chapter we address this deficiency and develop a rigorous approach to diffraction from open prolate spheroidal shells. We consider some structures of this type that, although geometrically simple, nevertheless exhibit considerable complexity of scattering mechanisms. Some of these results have appeared in [116], [120], [82], [117], [118], [83] and [119].

The form of regularisation employed in previous chapters was well suited to spherical geometry; however it must be significantly modified in order to be applicable in a spheroidal coordinate setting. The modifications depend upon the expansions described in Section 7.1.

In Section 7.2 the scalar wave problem for the prolate spheroidal shell with one circular hole is considered; the infinitely thin, acoustically hard spheroidal screen is excited by plane acoustic wave normally incident on the aperture plane. The problem is described by a Neumann mixed boundary value problem for the Helmholtz equation and results in dual series equations in prolate angle spheroidal functions. In Section 7.3 the regularised system is analysed to develop a rigorous theory of the spheroidal Helmholtz resonator.

The metallic prolate spheroidal cavity with one circular hole excited by an axial electric dipole field is considered in Section 7.4. A boundary value problem for the azimuthal component of magnetic field H_ϕ arises from Maxwell's equations in this axially symmetric situation. The regularised solution is used to calculate the far-field radiation pattern and radiation resistance of the open prolate spheroidal antenna in the resonant frequency range. High Q-factor responses due to cavity resonances overlay the spectrum of low Q-factor responses characteristic of closed body oscillations. Section 7.5 examines axially symmetric magnetic dipole excitation of the same cavity that gives rise to a boundary value problem for the azimuthal component of electric field E_ϕ .

Triple series equations with p.a.s.f. kernels are encountered in Section 7.6 where the axial electric dipole excitation of a prolate spheroidal cavity with two symmetrically located circular holes (the prolate spheroidal barrel) is considered. The effect of aperture size and dipole location (inside or outside the cavity) on radiation features is examined.

Accurate assessment of the impact of impedance loading on metallic structures is a central aim of radar cross-section studies and antenna performance studies. Thus in Section 7.7 we consider impedance boundary conditions on the spheroidal barrel examined in the previous section, and analyse the effect of this loading on the resonant properties of the spheroidal shell.

Finally Section 7.8 deals with a more complicated structure, the closed prolate spheroid embedded in a spheroidal cavity. The solution is employed to analyse the resonant properties of the shielded dipole antenna.

We use the notation and normalisation of spheroidal functions employed in the classical monograph of C. Flammer [29], and use the algorithms described therein for their computation. Recently developed numerical algorithms [21] for the computation of spheroidal functions over a wide range of frequency parameter c enable us to calculate the wideband radiation characteristics of various spheroidal cavities.

Thus this chapter describes a set of exact analytical tools that is now available for diffraction studies of a rich class of geometric structures, whole families of spheroidal shells of widely varying aspect ratio and widely varying aperture size, with interior and exterior surface impedances that may be arbitrarily specified.

7.1   Regularisation in Spheroidal Coordinates.

We recall that the prolate spheroidal coordinate system (ξ, η, ϕ) is related to rectangular coordinates by [29]

$$
\begin{aligned}
x &= \frac{d}{2}\left[\left(1-\eta^2\right)\left(\xi^2-1\right)\right]^{\frac{1}{2}} \cos \phi \\
y &= \frac{d}{2}\left[\left(1-\eta^2\right)\left(\xi^2-1\right)\right]^{\frac{1}{2}} \sin \phi \\
z &= \frac{d}{2}\eta\xi
\end{aligned}
\tag{7.1}
$$

where d is the interfocal distance and $-1 \leq \eta \leq 1, 1 \leq \xi < \infty, 0 \leq \phi \leq 2\pi$; ξ plays the role of a radial parameter, η that of angular parameter ($\eta = \cos\theta$, $\theta \in [0, \pi]$), ϕ is the azimuthal coordinate.

If $\xi_0 > 1$ is a constant, the coordinate surface $\xi = \xi_0$ is a prolate spheroid (or an elongated ellipsoid of revolution), with major semi-axis of length $b = \frac{d}{2}\xi_0$, and minor semi-axis of length $a = \frac{d}{2}\left(\xi_0^2 - 1\right)^{\frac{1}{2}}$. The degenerate surface $\xi = 1$ is the straight line segment $|z| \leq \frac{d}{2}$ along the z-axis. The coordinate surfaces $|\eta| = $ constant and $\phi = $ constant are, respectively, a hyperboloid of revolution of two sheets, and a half-plane originating in the z-axis. In the limit when the interfocal distance d approaches zero and ξ tends to infinity, the prolate spheroidal system (ξ, η, ϕ) reduces to the spherical system $(r, \theta, \phi_{sphere})$ with the identification

$$
\frac{d}{2}\xi = r, \ \eta = \cos\theta, \ \phi \equiv \phi_{sphere}.
\tag{7.2}
$$

The form of regularisation employed in previous chapters was well suited to spherical geometry; however, as noted in the introduction, it must be significantly modified in order to be applicable in a spheroidal coordinate setting. The reason lies in the angular eigenfunctions of the Laplace and Helmholtz operators. In the spherical coordinate system (r, θ, ϕ) the eigenfunctions are the same, namely,

$$
P_n^m\left(\cos\theta\right)\begin{Bmatrix} \cos m\phi \\ \sin m\phi \end{Bmatrix},
$$

where P_n^m denotes the associated Legendre functions (and $m = 0, 1, 2, \ldots$, $n = m, m+1, \ldots$); in spheroidal coordinates (ξ, η, ϕ), the eigenfunctions are different, being respectively,

$$
P_n^m\left(\eta\right)\begin{Bmatrix} \cos m\phi \\ \sin m\phi \end{Bmatrix} \text{ and } S_{mn}\left(c, \eta\right)\begin{Bmatrix} \cos m\phi \\ \sin m\phi \end{Bmatrix}
$$

(with the same choice of integers m, n), where $c = k\frac{d}{2}$ denotes the *wave parameter* (depending upon the wavenumber k and the interfocal distance d), and $S_{mn}(c, \eta)$ denotes the prolate angular spheroidal functions (p.a.s.f.).

In contradistinction to the associated Legendre functions, the angular spheroidal functions are *not* of hypergeometric type, nor do they possess Abel type integral representations. Furthermore the p.a.s.f. depends both on the wave parameter c and the angular coordinate η. So, splitting an operator with p.a.s.f. kernels into singular and regular terms (formally, this means introducing an asymptotically small parameter) must be done in both the *radial* and *angular* parts of the equations. So, the choice of the parameter to be extracted is considerably more complicated than in the spherical geometric setting.

The key idea that makes it possible to treat a large class of diffraction problems for open spheroidal shells successfully is to convert the dual series equations with prolate angle spheroidal function kernels to dual series equations with the associated Legendre function kernels; thus converted the equations are amenable to regularisation techniques employing the Abel integral transform. This relies on the expansion of the p.a.s.f. as a series in associated Legendre functions [29], [50], [122]

$$S_{ml}(c, \eta) = \sum_{r=0,1}^{\infty} {}'d_r^{ml}(c) \, P_{r+m}^m(\eta) \tag{7.3}$$

where the expansion coefficients $d_r^{ml}(c)$ depend *only on the wave parameter* c; in this summation the prime indicates that it is taken over only even or odd values of r according as $l - m$ is even or odd, respectively ($r \equiv l - m$ mod 2). The expansion (7.3) converges absolutely for all η.

The theory of spheroidal functions indicates that the expansion coefficients rapidly decrease, provided $r > l$, and the ratio $d_r^{ml}(c) / d_{r-2}^{ml}(c) = O(c^2/4r^2)$ as $r \to \infty$. For given l and m, and provided $l > c$, the dominant coefficient is $d_{l-m}^{ml}(c)$ [29], [50]. The diffraction problems of this chapter all employ the p.a.s.f. with azimuthal index $m = 0$ or 1. In Figures 7.1 and 7.2 the rapid decay of the ratio $d_r^{ml}(c) / d_{l-m}^{ml}(c)$ as $r \to \infty$ is illustrated for these two values of the index m.

Thus we may separate the *dominant* term in the expansion (7.3) and rewrite it in the form

$$S_{ml}(c, \eta) = d_{l-m}^{ml}(c) \left\{ P_l^m(\eta) + \sum_{\substack{r=0,1 \\ r \neq l-m}}^{\infty} {}' \frac{d_r^{ml}(c)}{d_{l-m}^{ml}(c)} P_{r+m}^m(\eta) \right\}. \tag{7.4}$$

The asymptotic expansion of the angular spheroidal function is (see [25])

$$S_{ml}(c, \eta) = P_l^m(\eta) \left[1 + O\left(l^{-1} \right) \right] \tag{7.5}$$

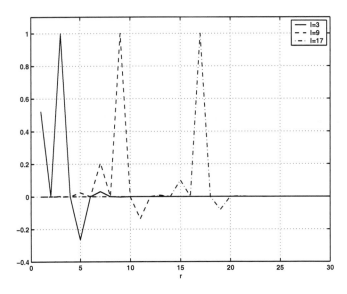

FIGURE 7.1. The coefficient ratio $d_r^{0l}(5)/d_l^{0l}(5)$ as a function of r (various l). The ratio decays rapidly when $r > l$.

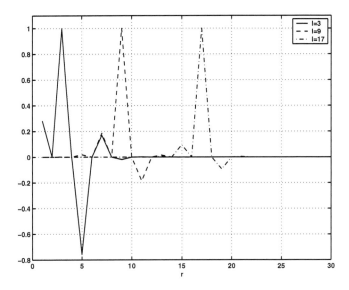

FIGURE 7.2. The coefficient ratio $d_r^{1l}(5)/d_l^{1l}(5)$ as a function of r (various l). The ratio decays rapidly when $r > l$.

as $l \to \infty$. Thus the expression (7.4) represents the prolate angle spheroidal function as a sum of a main term and an asymptotically small part (when $l \to \infty$). In this way we may separate the angle and wave dependence in the p.a.s.f. kernels.

The representation (7.4) together with the estimate (7.5) is the crucial step that makes the regularisation procedure work in spheroidal coordinates, reducing the problem of solving dual series equations with p.a.s.f. kernels to that of solving dual series equations with associated Legendre functions kernels; the latter task has been considered extensively in previous chapters (as well as in Volume I).

7.2 Acoustic Scattering by a Rigid Spheroidal Shell with a Circular Hole

Consider an infinitely thin prolate spheroidal shell, acoustically rigid and with one circular hole (see Figure 7.3); the aperture is located so that the structure is axisymmetric about the z-axis. For a suitable choice of ξ_0 and η_0 (hereafter fixed), the surface of the screen may be described in spheroidal coordinates (ξ, η, ϕ) by

$$\xi = \xi_0, \ \eta_0 \le \eta \le 1, \ 0 \le \phi < 2\pi;$$

the circular hole is defined by

$$\xi = \xi_0, \ -1 \le \eta < \eta_0, \ 0 \le \phi < 2\pi.$$

Throughout, it will be convenient to parametrise the angular coordinate via $\eta = \cos\theta$, where $0 \le \theta \le \pi$.

When illuminated by a plane acoustic wave propagating along the axis of symmetry (normal incidence on the aperture), the velocity potential of the incident field is given by

$$U^0 = \exp(ikz) = \exp(ic\eta\xi) = 2 \sum_{l=0}^{\infty} i^l \overline{S}_{0l}(c,1) \overline{S}_{0l}(c,\eta) R_{0l}^{(1)}(c,\xi) \qquad (7.6)$$

where $c = kd/2$, k is the wavenumber, d is the interfocal distance, $\overline{S}_{ml}(c,\eta)$ is the normalised prolate angle spheroidal function (with $m = 0$), and $R_{ml}^{(1)}(c,\xi)$ is the *prolate radial spheroidal function* (p.r.s.f.) of the first kind (with $m = 0$). A time harmonic dependence $e^{-i\omega t}$ has been assumed in (7.6) and is suppressed henceforth. Taking into account the normalisation of the prolate angle spheroidal functions we may rewrite the expression

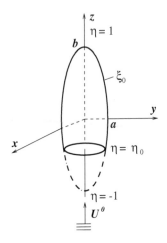

FIGURE 7.3. Spheroidal acoustic resonator.

(7.6) in the form

$$U^0 = 2 \sum_{l=0}^{\infty} i^l N_{0l}^{-2}(c) S_{0l}(c,1) S_{0l}(c,\eta) R_{0l}^{(1)}(c,\xi) \qquad (7.7)$$

where $S_{0l}(c,\eta)$ is the *prolate angle spheroidal function* (p.a.s.f.) and the norm $N_{0l}(c)$ was defined in Section 1.2.

The pertinent boundary value problem is formulated in a familiar way. The velocity potential U of the total diffracted field must satisfy the Helmholtz equation (1.144) and the mixed boundary conditions

$$U|_{\xi=\xi_0-0} = U|_{\xi=\xi_0+0}, \qquad \eta \in (-1, \eta_0) \qquad (7.8)$$

$$\frac{dU}{d\xi}|_{\xi=\xi_0-0} = \frac{dU}{d\xi}|_{\xi=\xi_0+0} = 0, \quad \eta \in (\eta_0, 1) \qquad (7.9)$$

are imposed on the aperture and the screen, respectively. The total field is considered as a superposition

$$U = U^0 + U^{sc}$$

of incident (U^0) and scattered (U^{sc}) fields; the scattered field behaves at infinity as an outgoing spherical wave

$$U^{sc} = \frac{e^{ikR}}{R} + O\left(\frac{1}{R^2}\right) \qquad (7.10)$$

as $\xi \to \infty$ or $R \to \infty$. Finally the scattered energy in any finite volume V of space must be finite:

$$W(V) = \iiint_V \left\{ k^2 |U^{sc}|^2 + |\text{grad } U^{sc}|^2 \right\} dV < \infty. \tag{7.11}$$

The application of this condition to volumes containing edges of the scatterer leads to the well-known *edge condition* uniquely defining the order of singularity of the scattered field at the edge of the screen. The uniqueness of the solution to our diffraction problem is guaranteed by this set of boundary, edge and radiation conditions [41], [93].

The physical meaning of the velocity potential U introduced here is well known: it is proportional to sound pressure, while its normal derivative $\frac{\partial U}{\partial n}$ is proportional to the normal component of ambient fluid velocity.

The Helmholtz equation (1.144) is separable in spheroidal coordinates (see Section 1.2) and we may expand the scattered field in terms of spheroidal wave harmonics. The radiation condition (7.10), and the continuity of the normal derivative of the velocity potential on the closed spheroidal surface $\xi = \xi_0$ forming the interface between interior and exterior regions, dictate that the solution has form

$$U^{sc}(\xi, \eta) =$$

$$\sum_{l=0}^{\infty} A_l S_{0l}(c, \eta) \begin{cases} R_{0l}^{(1)}(c, \xi), & \xi < \xi_0 \\ R_{0l}^{(3)}(c, \xi) R_{0l}^{(1)\prime}(c, \xi_0) / R_{0l}^{(3)\prime}(c, \xi_0), & \xi > \xi_0 \end{cases} \tag{7.12}$$

where the notation $R_{0l}^{(1)\prime}(c, \xi_0)$ means the derivative $\frac{d}{d\xi} R_{0l}^{(1)}(c, \xi)$ evaluated at $\xi = \xi_0$, and the unknown Fourier coefficients $\{A_l\}_{l=0}^{\infty}$ are to be determined.

Enforcement of the boundary conditions (7.8)–(7.9) on (7.12) produces the following dual series equations with p.a.s.f. kernels (of azimuthal index $m = 0$) for the unknown coefficients A_l,

$$\sum_{l=0}^{\infty} \frac{A_l}{R_{0l}^{(3)\prime}(c, \xi_0)} S_{0l}(c, \eta) = 0, \qquad \eta \in (-1, \eta_0) \tag{7.13}$$

$$\sum_{l=0}^{\infty} A_l R_{0l}^{(1)\prime}(c, \xi_0) S_{0l}(c, \eta) =$$

$$- 2 \sum_{l=0}^{\infty} i^l S_{0l}(c, 1) N_{0l}^{-2}(c) R_{0l}^{(1)\prime}(c, \xi_0) S_{0l}(c, \eta), \quad \eta \in (\eta_0, 1) \tag{7.14}$$

where the value of the Wronskian

$$W\left[R_{0l}^{(1)}\left(c,\xi_0\right),R_{0l}^{(3)}\left(c,\xi_0\right)\right] =$$

$$R_{0l}^{(1)\prime}\left(c,\xi_0\right)R_{0l}^{(3)}\left(c,\xi_0\right) - R_{0l}^{(1)}\left(c,\xi_0\right)R_{0l}^{(3)\prime}\left(c,\xi_0\right) = \frac{i}{c(\xi_0^2 - 1)} \quad (7.15)$$

has been employed.

Introducing the unknowns

$$x_l = \frac{A_l}{R_{0l}^{(3)\prime}\left(c,\xi_0\right)} \quad (7.16)$$

and the parameter

$$\varepsilon_l = 1 + i\frac{4c}{2l+1}(\xi_0^2 - 1)^{\frac{3}{2}} R_{0l}^{(1)\prime}\left(c,\xi_0\right) R_{0l}^{(3)\prime}\left(c,\xi_0\right) \quad (7.17)$$

that is asymptotically small ($\varepsilon_l = O\left(c^2/l^2\right)$ as $l \to \infty$), the system (7.13)–(7.14) becomes

$$\sum_{l=0}^{\infty} x_l S_{0l}\left(c,\eta\right) = 0, \quad \eta \in (-1,\eta_0) \quad (7.18)$$

$$\sum_{l=0}^{\infty} (2l+1) x_l (1-\varepsilon_l) S_{0l}\left(c,\eta\right) = \sum_{l=0}^{\infty} g_l S_{0l}\left(c,\eta\right), \eta \in (\eta_0,1) (7.19)$$

where

$$g_l = 8ci^{l+1}\left(\xi_0^2 - 1\right)^{\frac{3}{2}} S_{0l}\left(c,1\right) N_{0l}^{-2}(c) R_{0l}^{(1)\prime}\left(c,\xi_0\right). \quad (7.20)$$

We now employ the representation (7.4) to transform the dual series equations (7.18)–(7.19) with p.a.s.f. kernels to the following dual series equations with associated Legendre function kernels,

$$\sum_{l=0}^{\infty} y_l P_l\left(\eta\right) = -\sum_{l=0}^{\infty} y_l \sum_{\substack{r=0,1 \\ r\neq l}}^{\infty} {}' \frac{d_r^{0l}\left(c\right)}{d_l^{0l}\left(c\right)} P_r\left(\eta\right), \quad \eta \in (-1,\eta_0) \quad (7.21)$$

$$\sum_{l=0}^{\infty} (2l+1) y_l P_l\left(\eta\right) = \sum_{l=0}^{\infty} \left\{(2l+1)\varepsilon_l y_l + g_l d_l^{0l}(c)\right\} P_l\left(\eta\right) +$$

$$\sum_{l=0}^{\infty} \left\{(2l+1)\left(\varepsilon_l - 1\right) y_l + g_l d_l^{0l}(c)\right\} \sum_{\substack{r=0,1 \\ r\neq l}}^{\infty} {}' \frac{d_r^{0l}\left(c\right)}{d_l^{0l}\left(c\right)} P_r\left(\eta\right), \quad \eta \in (\eta_0,1),$$

$$(7.22)$$

with new unknowns

$$y_l = x_l d_l^{0l}(c).$$

(7.23)

It is possible to show that the condition (7.11) forces the unknown coefficient sequence $\{y_l\}_{l=0}^{\infty}$ to be square summable $(\{y_l\}_{l=0}^{\infty} \in l_2)$,

$$\sum_{l=0}^{\infty} |y_l|^2 < \infty.$$

(7.24)

To construct solutions to Equations (7.21)–(7.22) it is useful to compare
these dual series equations with equations (2.24)–(2.25) that arise from the
diffraction of an acoustic plane wave by an acoustically hard spherical cavity. From a formal point of view, both sets are equations of the same class,
though the right-hand side of (7.22) is rather more complex. Nonetheless,
this system may be regularised in the same way to obtain the analogue of
system (2.28) (for $m = 0$).

Integrate (7.22) employing the formula

$$\int P_n(z)\, dz = \frac{1}{2n+1}\{P_{n-1}(z) - P_{n+1}(z)\}$$

and then substitute the Mehler-Dirichlet representations for the Legendre
polynomials

$$P_{n-1}(\cos\theta) - P_{n+1}(\cos\theta) = \frac{2\sqrt{2}}{\pi}\int_0^\theta \frac{\sin\left(n+\frac{1}{2}\right)\alpha}{\sqrt{\cos\alpha - \cos\theta}}\sin\alpha\, d\alpha,$$

$$P_n(\cos\theta) = \frac{\sqrt{2}}{\pi}\int_\theta^\pi \frac{\sin\left(n+\frac{1}{2}\right)\alpha}{\sqrt{\cos\theta - \cos\alpha}}\, d\alpha,$$

into the series equations. Interchange the order of summation and integration to obtain two Abel integral equations that may be inverted in
the standard way; the validity of this operation is ensured by the uniform
convergence of the series equations that may be established from the condition (7.24) and the asymptotic behaviour of the Legendre polynomials P_l
as $l \to \infty$. This leads to the following equation,

$$\sum_{l=0}^{\infty} y_l \sin\left(l+\frac{1}{2}\right)\theta = \begin{cases} F_1(\theta), & \theta \in (\theta_0, 1) \\ F_2(\theta), & \theta \in (-1, \theta_0), \end{cases}$$

(7.25)

where $\eta = \cos\theta$, $\eta_0 = \cos\theta_0$ (so that θ_0 defines the angular size of the shell) and

$$F_1(\theta) = \sum_{l=0}^{\infty} \left\{ \varepsilon_l y_l + g_l d_l^{0l}(c) \frac{1}{(2l+1)} \right\} \sin\left(l + \frac{1}{2}\right)\theta +$$

$$\sum_{l=0}^{\infty} \left\{ (2l+1)(\varepsilon_l - 1) y_l + g_l d_l^{0l}(c) \right\} \sum_{\substack{r=0,1 \\ r \neq l}}^{\infty} {}' \frac{d_r^{0l}(c)}{d_l^{0l}(c)} \frac{\sin\left(r + \frac{1}{2}\right)\theta}{(2r+1)}, \quad (7.26)$$

$$F_2(\theta) = -\sum_{l=0}^{\infty} y_l \sum_{\substack{r=0,1 \\ r \neq l}}^{\infty} {}' \frac{d_r^{0l}(c)}{d_l^{0l}(c)} \sin\left(r + \frac{1}{2}\right)\theta. \quad (7.27)$$

To obtain a regularised i.s.l.a.e. for the unknown coefficients y_l, multiply both sides of the equation (7.25) by $\sin\left(s + \frac{1}{2}\right)\theta$ and integrate over $[0, \pi]$, yielding

$$y_s + \sum_{l=0}^{\infty} y_l \alpha_{sl}(c, \theta_0) = \beta_s(c, \theta_0), \quad (7.28)$$

where $s = 0, 1, 2, \ldots$, and

$$\beta_s(c, \theta_0) = \sum_{l=0}^{\infty} g_l d_l^{0l}(c) \sum_{r=0,1}^{\infty} {}' \frac{d_r^{0l}(c)}{d_l^{0l}(c)} \frac{R_{rs}(\theta_0)}{(2r+1)}, \quad (7.29)$$

$$\alpha_{sl}(c, \theta_0) = -\varepsilon_l R_{ls}(\theta_0) + (1 - \varepsilon_l)(2l+1) \sum_{\substack{r=0,1 \\ r \neq l}}^{\infty} {}' \frac{d_r^{0l}(c)}{d_l^{0l}(c)} \frac{R_{rs}(\theta_0)}{(2r+1)}$$

$$+ \sum_{\substack{r=0,1 \\ r \neq l}}^{\infty} {}' \frac{d_r^{0l}(c)}{d_l^{0l}(c)} [R_{rs}(\theta_0) - \delta_{rs}], \quad (7.30)$$

and

$$R_{rs}(\theta_0) = \hat{Q}_{r-1,s-1}^{\left(\frac{1}{2}, -\frac{1}{2}\right)}(\cos\theta_0) = \frac{1}{\pi} \left[\frac{\sin(r-s)\theta_0}{r-s} - \frac{\sin(r+s+1)\theta_0}{r+s+1} \right] \quad (7.31)$$

is the usual incomplete scalar product (the term $(\sin(r-s)\theta_0)/(r-s)$ is replaced by θ_0 if $r = s$).

The matrix operator of the system (7.28) is a compact perturbation of the identity operator in the functional space l_2. The proof of this statement, as well as the justification of splitting the operator in the original formulation of the problem into the singular and regular parts, relies on the asymptotic estimates of the p.a.s.f. (see [7.5]) and of the parameter ε_l defined by (7.17). Uniqueness of solutions to this systems, and of those encountered in susbsequent sections in this chapter, is established by adapting the uniqueness argument given in Section 2.1. As may be seen from (7.30)–(7.31), computation of the matrix elements in (7.28) simply requires the calculation of trigonometric functions and the coefficients $d_r^{0l}(c)$ derived from the expansion of the angular and radial spheroidal functions in series of Bessel or Legendre functions (see Section 1.2).

Equation (7.28) has many convenient features for numerical solution or further analytical investigation. It may be used to develop asymptotic expansions (with an error estimate) in the low-frequency range for arbitrary values of the parameter θ_0 (the angular size of the screen), including the limiting case of a "spheroidal" disk. In the resonant frequency range analytic approximations for the quasi-eigenvalues may be obtained for small apertures. These are used in the next section to construct a rigorous theory of the acoustical resonator (or Helmholtz resonator) of spheroidal form.

From a numerical point of view, second kind systems of the form (7.28) are readily amenable to reliable, efficient and accurate numerical solutions based on truncation methods. This will be the basis of a systematic analysis of the far-field radiation pattern and total cross-section of the acoustical spheroidal resonator.

7.3 Rigorous Theory of the Spheroidal Helmholtz Resonator.

The classical acoustic resonator coupled to free space by a small circular aperture was examined in Section 2.2; its resonant frequency f_0 was given by (2.61). In this section, we derive the complex resonant frequency f_0 and Q-factor for the vessel of spheroidal form that is coupled to free space by a circular hole.

Introducing the angular width of the aperture

$$\theta_1 = \pi - \theta_0, \tag{7.32}$$

so that the incomplete scalar product (7.31) has the form

$$R_{rs}(\theta_0) = R_{rs}(\pi - \theta_1) = \delta_{rs} - (-1)^{r-s} Q_{rs}(\theta_1) \tag{7.33}$$

where

$$Q_{rs}(\theta_1) = \frac{1}{\pi} \left[\frac{\sin(r-s)\theta_1}{r-s} + \frac{\sin(r+s+1)\theta_1}{r+s+1} \right] \tag{7.34}$$

(with the usual meaning when $r = s$), we reformulate the system (7.28) with coefficients expressed in terms of θ_1 as

$$y_s + \sum_{l=0}^{\infty} y_l \alpha'_{sl} (c, \theta_1) = \beta'_s (c, \theta_1) \tag{7.35}$$

where $s = 0, 1, \ldots, \alpha'_{sl} (c, \theta_1) = \alpha_{sl} (c, \pi - \theta_1)$ and $\beta'_s (c, \theta_1) = \beta_s (c, \pi - \theta_1)$. In this section we now replace α'_{sl} and β'_s by α_{sl} and β_s.

To derive the dispersion equation for the Helmholtz mode, consider the homogeneous system of equations derived from (7.35) by setting the right-hand side to zero. The quasi-eigenvalues of quasi-eigenoscillations are the complex frequencies at which nontrivial solutions of this homogeneous system exist; its matrix has the structure

$$\begin{bmatrix} 1 + \alpha_{00} & \alpha_{01} & \alpha_{02} & \ldots & \alpha_{0l} & \ldots \\ \alpha_{10} & 1 + \alpha_{11} & \alpha_{12} & \ldots & \alpha_{1l} & \ldots \\ \alpha_{20} & \alpha_{21} & 1 + \alpha_{22} & \ldots & \alpha_{2l} & \ldots \\ \ldots & \ldots & \ldots & \ldots & \ldots & \ldots \\ \alpha_{l0} & \alpha_{l1} & \alpha_{l2} & \ldots & 1 + \alpha_{ll} & \ldots \\ \ldots & \ldots & \ldots & \ldots & \ldots & \ldots \end{bmatrix} \tag{7.36}$$

and the associated determinant must vanish. Retaining terms correct to $O\left(c^3, \theta_1^6\right)$, the relevant determinant is

$$D = \prod_{s=0}^{\infty} [1 + \alpha_{ss} (c, \theta_1)] \left\{ 1 - \sum_{l=0, l\neq s}^{\infty} \frac{\alpha_{ls}(c, \theta_1)\alpha_{sl}(c, \theta_1)}{1 + \alpha_{ss} (c, \theta_1)} + O\left(c^3, \theta_1^6\right) \right\}. \tag{7.37}$$

From the infinite spectrum of quasi-eigenvalues we select only that mode defined by the dispersion equation obtained from the factor of index $s = 0$ in (7.37),

$$1 + \alpha_{00} (c, \theta_1) - \sum_{l=0}^{\infty} \alpha_{l0} (c, \theta_1) \alpha_{0l} (c, \theta_1) = 0. \tag{7.38}$$

When $\theta_1 \ll 1$, the equation (7.38) is solvable only with

$$c^2 \sim \theta_1. \tag{7.39}$$

Under this condition we expand (7.38) and discard the terms that are $O\left(c^6, \theta_1^3\right)$ to obtain the complex values of the parameters c, θ_1 of this Helmholtz mode. Suppressing a lengthy and awkward calculation, the result is

$$
\frac{2\theta_1}{\pi} - \frac{4}{3}\xi_0 \left(\xi_0^2 - 1\right)^{\frac{1}{2}} c^2 - i\frac{4}{9}\xi_0^2 \left(\xi_0^2 - 1\right)^{\frac{3}{2}} c^5
$$
$$
+ \xi_0 \left(\xi_0^2 - 1\right)^{\frac{1}{2}} \left[\frac{38}{75} \left(\xi_0^2 - 1\right) - \frac{14}{135} - \frac{2}{9}\left(\xi_0^2 - 1\right)^{\frac{1}{2}} \ln\left(\frac{\xi_0 + 1}{\xi_0 - 1}\right)\right] c^4
$$
$$
+ \frac{2}{3\pi}\left[4\xi_0 \left(\xi_0^2 - 1\right)^{\frac{1}{2}} - \frac{49}{75}\right] \theta_1 c^2 - \frac{4}{\pi^2}\left(\frac{\pi^2}{8} - 1\right)\theta_1^2 + O\left(c^6, \theta_1^3\right) = 0.
$$

$$(7.40)$$

An application of Newton's method gives the following result for the parameter $c_0 = k_0 d/2$ of the Helmholtz mode

$$
c_0 \approx \left[\frac{3\theta_1}{2\pi\xi_0 \left(\xi_0^2 - 1\right)^{\frac{1}{2}}}\right]^{\frac{1}{2}} - i\frac{3}{8}\left(\frac{\theta_1}{\pi}\right)^2. \tag{7.41}
$$

The real part of the expression (7.41) may be used to find the excitation frequency of the acoustical resonator, and its Q-factor equals

$$
Q = -\frac{\mathrm{Re}\,(c_0)}{2\,\mathrm{Im}\,(c_0)} = \sqrt{\frac{1}{3\xi_0 \left(\xi_0^2 - 1\right)^{\frac{1}{2}}}}\left(\frac{2\pi}{\theta_1}\right)^{\frac{3}{2}}. \tag{7.42}
$$

It is worth noting that with the identification

$$
V = \frac{4}{3}\pi\xi_0 \left(\xi_0^2 - 1\right) \left(\frac{d}{2}\right)^3, r = \frac{d}{2}\left(\xi_0^2 - 1\right)^{\frac{1}{2}}\theta_1, \tag{7.43}
$$

and

$$
c_0 = k_0 \frac{d}{2} = \frac{2\pi\,\mathrm{Re}(f_0)}{v_{sound}}\frac{d}{2},
$$

the real part of expression (7.41) obtained from our rigorous solution for the Helmholtz mode, equals the well-known acoustical expression (2.61).

7.4   Axial Electric Dipole Excitation of a Metallic Spheroidal Cavity.

With a time harmonic dependence $e^{-i\omega t}$ (that is henceforth assumed and suppressed), we consider Maxwell's equations in the symmetric form (1.58)–(1.59). For an axially symmetric situation (so that $\partial/\partial\phi \equiv 0$), Maxwell's

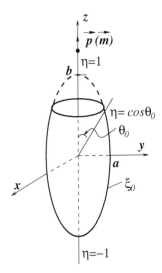

FIGURE 7.4. Prolate spheroidal shell (semi-axes a, b) with one circular aperture, excited by an axisymmetrically located electric dipole.

equations in prolate spheroidal coordinates separate into two independent systems of equations, given in Section 1.2. The system (1.94)–(1.96) describes a field of magnetic type (when only the components E_ϕ, H_ξ and H_η do not vanish) whereas the system (1.97)–(1.99) describes a field of electric type (when only the components H_ϕ, E_ξ and E_η do not vanish). An electric dipole located along the z-axis with moment aligned with that axis generates only a field of electric type governed by (1.97)–(1.99). The other nontrivial field components (E_ξ, E_η) are readily obtained from the component $H_\phi = H_\phi(\xi, \eta)$ that satisfies the second order partial differential equation (1.100).

Consider then an infinitely thin perfectly conducting spheroidal shell having one circular hole located so that the z-axis is normal to the aperture plane and the structure is axisymmetrical. It is excited by a vertical (axially aligned) electric dipole placed along the axis of rotational symmetry (the z-axis), at $\xi = \xi_1, \eta = 1$ (see Figure 7.4). The screen occupies the surface

$$\xi = \xi_0, \; -1 \le \eta < \eta_0, \; 0 \le \phi < 2\pi$$

and the aperture is given by

$$\xi = \xi_0, \; \eta_0 \le \eta \le 1, \; 0 \le \phi < 2\pi.$$

The magnetic field of the electric dipole is (see [50], [21])

$$H_\phi^{(0)}(\xi,\eta) = -\frac{pk^3}{4\pi\sqrt{\xi_1^2-1}} \times$$

$$\sum_{l=1}^{\infty} \frac{(-i)^{l+1}}{\chi_{1l}(c)\,N_{1l}(c)} S_{1l}(c,\eta) \begin{Bmatrix} R_{1l}^{(3)}(c,\xi_1)\,R_{1l}^{(1)}(c,\xi), & \xi < \xi_1 \\ R_{1l}^{(1)}(c,\xi_1)\,R_{1l}^{(3)}(c,\xi), & \xi > \xi_1 \end{Bmatrix} \quad (7.44)$$

where p is the dipole moment; $R_{1l}^{(1)}(c,\xi)$ and $R_{1l}^{(3)}(c,\xi)$ are the prolate radial spheroidal functions (p.r.s.f.) of first and third kind, respectively, $S_{1l}(c,\eta)$ is the prolate angular spheroidal function (p.a.s.f.), and the coefficients $N_{1l}(c)$ and $\chi_{1l}(c)$ are defined in Section 1.2.

The total field H_ϕ^{tot} is a superposition of the incident and the scattered field H_ϕ^{sc}:

$$H_\phi^{tot}(\xi,\eta) = H_\phi^{(0)}(\xi,\eta) + H_\phi^{sc}(\xi,\eta). \quad (7.45)$$

A rigorous formulation of the electromagnetic diffraction problem requires the total field to satisfy Maxwell's equations (1.58)–(1.59) subject to the following conditions: (i) the tangential components of the total electric field must be continuous on the spheroidal surface $\xi = \xi_0$,

$$E_\eta(\xi_0 - 0,\eta) = E_\eta(\xi_0 + 0,\eta), \qquad \eta \in (-1,1); \quad (7.46)$$

(ii) the *mixed* boundary conditions

$$\begin{aligned} E_\eta(\xi_0 - 0,\eta) &= E_\eta(\xi_0 + 0,\eta) = 0, & \eta \in (-1,\eta_0) \\ H_\phi(\xi_0 - 0,\eta) &= H_\phi(\xi_0 + 0,\eta), & \eta \in (\eta_0,1) \end{aligned} \quad (7.47)$$

are imposed on the tangential components of the total electric field on the screen and of the total magnetic field on the aperture, respectively; (iii) the scattered field satisfies radiation conditions, requiring that it behaves as an outgoing spherical wave at infinity,

$$H_\phi^{sc}(\xi,\eta) = \frac{e^{ic\xi}}{\xi} F^{sc}(\eta)\left(1 + O\left(\xi^{-1}\right)\right) \quad (7.48)$$

as $\xi \to \infty$, where $F^{sc}(\eta)$ is the radiation pattern of scattered field; and (iv) the scattered energy in any finite volume V of space must be finite:

$$W(V) = \frac{1}{2}\iiint_V \left\{\left|\overrightarrow{E^{sc}}\right|^2 + \left|\overrightarrow{H^{sc}}\right|^2\right\} dV < \infty. \quad (7.49)$$

This last condition is equivalent to the so-called *edge condition* prescribing the singularity of the scattered field at the edges of the screen. Uniqueness of the solution is guaranteed by this set of boundary, edge and radiation conditions [41], [93].

Let us represent the scattered field in terms of spheroidal harmonics. Enforcing the continuity condition (7.46) on the shell surface $\xi = \xi_0$, and the radiation condition (7.48) on the scattered field, yields the expansion

$$H_\phi^{(sc)}(\xi, \eta) = \frac{pk^3}{4\pi\sqrt{\xi_1^2 - 1}} \times$$

$$\sum_{l=1}^\infty A_l S_{1l}(c, \eta) \begin{Bmatrix} R_{1l}^{(1)}(c, \xi), & \xi < \xi_0 \\ R_{1l}^{(3)}(c, \xi) Z_l^{(1)}(c, \xi_0) / Z_l^{(3)}(c, \xi_0), & \xi > \xi_0 \end{Bmatrix} \quad (7.50)$$

where the unknown Fourier coefficients A_l are to be found, and

$$Z_l^{(i)}(c, \xi_0) = \frac{\partial}{\partial \xi}\left[(\xi^2 - 1)^{\frac{1}{2}} R_{1l}^{(i)}(c, \xi)\right]_{\xi = \xi_0}, \qquad i = 1 \text{ or } 3.$$

Imposing the mixed boundary conditions (7.48) on the expansion (7.50) produces the following dual series equations defined on subintervals corresponding to the screen surface and aperture,

$$\sum_{l=1}^\infty A_l Z_l^{(1)}(c, \xi_0) S_{1l}(c, \eta) = \sum_{l=1}^\infty B_l S_{1l}(c, \eta), \quad \eta \in (-1, \eta_0) \quad (7.51)$$

$$\sum_{l=1}^\infty \frac{A_l}{Z_l^{(3)}(c, \xi_0)} S_{1l}(c, \eta) = 0, \quad \eta \in (\eta_0, 1), \quad (7.52)$$

where the coefficients B_l determined by the incident field (7.44) equal

$$B_l = \frac{(-i)^{l+1}}{\chi_{1l}(c) N_{1l}(c)} \begin{Bmatrix} Z_l^{(1)}(c, \xi_0) R_{1l}^{(3)}(c, \xi_1), & \xi_1 > \xi_0 \\ Z_l^{(3)}(c, \xi_0) R_{1l}^{(1)}(c, \xi_1), & \xi_1 < \xi_0. \end{Bmatrix} \quad (7.53)$$

It is convenient to define new coefficients x_l by

$$A_l = \frac{2l + 1}{l(l+1)} Z_l^{(3)}(c, \xi_0) x_l \quad (7.54)$$

and to transform equations (7.51)–(7.52) to the equivalent system

$$\sum_{l=1}^\infty \frac{2l+1}{l(l+1)} x_l Z_l^{(1)}(c, \xi_0) Z_l^{(3)}(c, \xi_0) S_{1l}(c, \eta) = \sum_{l=1}^\infty B_l S_{1l}(c, \eta),$$

$$\eta \in (-1, \eta_0), \quad (7.55)$$

$$\sum_{l=1}^{\infty} \frac{2l+1}{l(l+1)} x_l S_{1l}(c,\eta) = 0, \qquad \eta \in (\eta_0, 1). \qquad (7.56)$$

This formulation of the canonical diffraction problem has therefore produced a set of dual series equations (7.55)–(7.56), containing the prolate angle spheroidal functions with the azimuthal index $m = 1$, to be solved for the coefficients x_l. We treat it in a similar manner to that used for the acoustic problems encountered in an earlier section.

Employing the asymptotic expansions established in [25] for the prolate radial spheroidal functions

$$R_{1l}^{(1)}(c,\xi) =$$

$$\left\{ \frac{1}{2} \left[\xi \left(\xi^2 - 1 \right)^{-\frac{1}{2}} + 1 \right] \right\}^{\frac{1}{2}} \cdot j_l \left\{ \frac{c}{2} \left[\xi + \left(\xi^2 - 1 \right)^{\frac{1}{2}} \right] \right\} \left\{ 1 + O\left(l^{-1} \right) \right\} \quad (7.57)$$

$$R_{1l}^{(3)}(c,\xi) =$$

$$\left\{ \frac{1}{2} \left[\xi \left(\xi^2 - 1 \right)^{-\frac{1}{2}} + 1 \right] \right\}^{\frac{1}{2}} \cdot h_l^{(1)} \left\{ \frac{c}{2} \left[\xi + \left(\xi^2 - 1 \right)^{\frac{1}{2}} \right] \right\} \left\{ 1 + O\left(l^{-1} \right) \right\}$$

$$(7.58)$$

as $l \to \infty$, where j_l and $h_l^{(1)}$ denote the spherical Bessel functions and spherical Hankel functions (of first kind), respectively, the following limit can be established:

$$\lim_{l \to \infty} \frac{2l+1}{l(l+1)} Z_l^{(1)}(c,\xi_0) Z_l^{(3)}(c,\xi_0) = \frac{i}{c \left(\xi_0^2 - 1 \right)^{\frac{1}{2}}}. \qquad (7.59)$$

This asymptotic result leads us to define the parameter

$$\varepsilon_l(c,\xi_0) = 1 + ic \left(\xi_0^2 - 1 \right)^{\frac{1}{2}} \frac{2l+1}{l(l+1)} Z_l^{(1)}(c,\xi_0) Z_l^{(3)}(c,\xi_0) \qquad (7.60)$$

that is asymptotically small: $\varepsilon_l = O\left(l^{-2} \right)$ as $l \to \infty$. Figure 7.5 illustrates the asymptotic behaviour of ε_l, for various values of ξ_0 and c. With this definition, equations (7.55)–(7.56) may be written

$$\sum_{l=1}^{\infty} x_l \left(1 - \varepsilon_l \right) S_{1l}(c,\eta) = -ic \left(\xi_0^2 - 1 \right)^{\frac{1}{2}} \sum_{l=1}^{\infty} B_l S_{1l}(c,\eta),$$

$$\eta \in (-1, \eta_0) \quad (7.61)$$

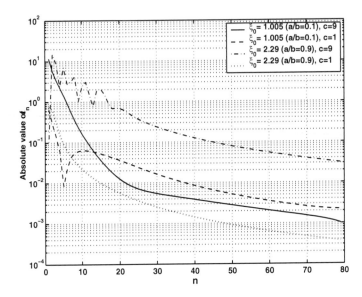

FIGURE 7.5. Behaviour of the asymptotically small parameter ϵ_n for various values of ξ_0 and c.

$$\sum_{l=1}^{\infty} \frac{2l+1}{l(l+1)} x_l S_{1l}(c,\eta) = 0, \qquad\qquad \eta \in (\eta_0, 1). \qquad (7.62)$$

The next step is to transform the dual series equations (7.61)–(7.62) to equations with associated Legendre function kernels that are amenable to solution by the Abel integral transform technique. The key is to insert the representation (7.4) in (7.61)–(7.62), and after setting $\eta = \cos\theta$, $\eta_0 = \cos\theta_0$ and slightly rearranging we obtain

$$\sum_{l=1}^{\infty} \frac{2l+1}{l(l+1)} y_l P_l^1(\cos\theta) = -\sum_{l=1}^{\infty} \frac{2l+1}{l(l+1)} y_l \sum_{\substack{r=0,1 \\ r\neq l-1}}^{\infty}{}' \frac{d_r^{1l}(c)}{d_{l-1}^{1l}(c)} P_{r+1}^1(\cos\theta),$$

$$\theta \in (0,\theta_0) \quad (7.63)$$

$$\sum_{l=1}^{\infty} y_l P_l^1 (\cos\theta) = \sum_{l=1}^{\infty} y_l \varepsilon_l \sum_{r=0,1}^{\infty} {}' \frac{d_r^{1l}(c)}{d_{l-1}^{1l}(c)} P_{r+1}^1 (\cos\theta) -$$

$$\sum_{l=1}^{\infty} y_l \sum_{\substack{r=0,1 \\ r\neq l-1}}^{\infty} {}' \frac{d_r^{1l}(c)}{d_{l-1}^{1l}(c)} P_{r+1}^1 (\cos\theta) + \sum_{l=1}^{\infty} g_l \sum_{r=0,1}^{\infty} {}' \frac{d_r^{1l}(c)}{d_{l-1}^{1l}(c)} P_{r+1}^1 (\cos\theta),$$

$$\theta \in (\theta_0, \pi) \quad (7.64)$$

where

$$y_l = x_l d_{l-1}^{1l}(c) \tag{7.65}$$

and $g_l = -ic\sqrt{\xi_0^2 - 1} d_{l-1}^{1l}(c) B_l.$

In operator notation, we may regard the left hand side of the equations (7.63)–(7.64) as representing the action of an operator L on the unknown Fourier coefficient sequence $y = \{y_l\}_{l=1}^{\infty}$ (i.e., Ly). Finding the inverse operator L^{-1} or equivalently, solving the equations, depends upon specifying the functional space in which y lies, i.e., specifying the behaviour of the coefficients y_l to be determined.

The finite energy condition (7.49) determines a suitable functional space for $y = \{y_l\}_{l=1}^{\infty}$. Select the volume of integration to be the interior of the *closed* spheroid $\xi = \xi_0$ and use the expressions (1.97)–(1.99) and (7.50) for the electromagnetic field components in the region $\xi \leq \xi_0$; the energy integral is

$$W = \frac{p^2 ck^3 \sqrt{\xi_0^2 - 1}}{8\pi (\xi_1^2 - 1)} \times$$

$$\sum_{l=1}^{\infty} \left\{ |A_l|^2 N_{1l}^2 (c) R_{1l}^{(1)} (c, \xi_0) Z_l^{(1)} (c, \xi_0) \right\} \left[1 + \frac{D_l (c, \xi_0)}{l^2} \right], \quad (7.66)$$

where $\lim_{l\to\infty} D_l (c, \xi_0) = M (c, \xi_0)$ is a bounded function for any values of c and ξ_0. Keeping in mind the rescaling (7.54) and (7.65), the finite energy condition (7.49) thus implies

$$\sum_{l=1}^{\infty} \frac{(2l+1)^2}{l^2 (l+1)^2} \frac{N_{1l}^2 (c)}{[d_{l-1}^{1l}(c)]^2} |y_l|^2 \left| Z_l^{(3)} (c, \xi_0) \right|^2 R_{1l}^{(1)} (c, \xi_0) Z_l^{(1)} (c, \xi_0) < \infty.$$

$$(7.67)$$

Using the limit (7.59), and the fact that $N_{1l}^2 (c) \backsim l$ as $l \to \infty$, we deduce

$$\sum_{l=1}^{\infty} |y_l|^2 < \infty, \tag{7.68}$$

that is, $\{y_l\}_{l=1}^{\infty} \in l_2$. This is the correct mathematical setting for the solution to the dual series equations (7.63)–(7.64), and ensures the mathematical validity of all operations for constructing the regularised solution, which is carried out in the standard manner similar to that employed in previous chapters.

Omitting its deduction, we state the final form of the regularised system:

$$y_s + \sum_{l=1}^{\infty} y_l \alpha_{sl}(c, \theta_0) = \beta_s(c, \theta_0),\qquad(7.69)$$

where $s = 1, 2, \ldots$,

$$\beta_s(c, \theta_0) = \sum_{l=1}^{\infty} g_l \sum_{r=0,1}^{\infty} {}' \frac{d_r^{1l}(c)}{d_{l-1}^{1l}(c)} [\delta_{r+1,s} - \Phi_{sr}(\theta_0)],\qquad(7.70)$$

$$\alpha_{sl}(c, \theta_0) = \frac{(2l+1)}{l(l+1)} \sum_{\substack{r=0,1 \\ r \neq l-1}}^{\infty} {}' \frac{d_r^{1l}(c)}{d_{l-1}^{1l}(c)} \frac{(r+1)(r+2)}{(2r+3)} \Phi_{sr}(\theta_0) -$$

$$\varepsilon_l \sum_{r=0,1}^{\infty} {}' \frac{d_r^{1l}(c)}{d_{l-1}^{1l}(c)} [\delta_{r+1,s} - \Phi_{sr}(\theta_0)] + \sum_{\substack{r=0,1 \\ r \neq l-1}}^{\infty} {}' \frac{d_r^{1l}(c)}{d_{l-1}^{1l}(c)} [\delta_{r+1,s} - \Phi_{sr}(\theta_0)],$$

$$(7.71)$$

and

$$\Phi_{sr}(\theta_0) = R_{r+1,s}(\theta_0) + \frac{R_{0s}(\theta_0)}{1 - R_{00}(\theta_0)} R_{r+1,0}(\theta_0)\qquad(7.72)$$

where $R_{rs}(\theta_0)$ is given by (7.31).

Using the asymptotic behaviour of the parameter ε_l and of the function $S_{1l}(c, \eta)$, and estimating the norm of the matrix operator $\{\alpha_{sl}\}_{s,l=1}^{\infty}$ it is easy to prove that the matrix operator of the system (7.69) is a compact perturbation of the identity operator in l_2. Thus this system of equations can be effectively solved by the truncation method, with a desired accuracy depending on the truncation number N_{tr}. Solutions of engineering accuracy are obtained if N_{tr} exceeds the maximal electrical size of the spheroid: $N_{tr} > kb = c\xi_0$, where b is the semi-major axis length.

The accuracy of this estimation is illustrated by the dependence of normalised error versus truncation number N_{tr} presented in Figure 7.6 (left). The error was computed in the maximum-norm sense,

$$e(N) = \frac{\max_{l \leq N} |y_l^{N+1} - y_l^N|}{\max_{l \leq N} |y_l^N|}\qquad(7.73)$$

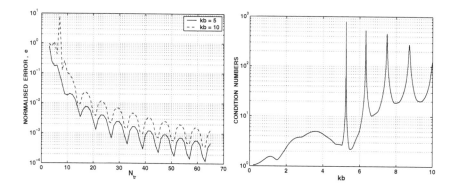

FIGURE 7.6. Dependence of normalised error $e(N_{tr})$ upon truncation number (left) and of condition number upon wavenumber (right); $a/b = 0.5$ ($\xi_0 = 1.155$), $\theta_0 = 30°$.

where $\left\{ y_l^N \right\}_{l=1}^N$ denotes the solution to (7.69), truncated to N equations. The Fredholm nature of this system guarantees that $e(N_{tr}) \rightarrow 0$ as $N_{tr} \rightarrow \infty$. The system is well conditioned, even near values of quasi-eigenfrequencies corresponding to internal cavity resonances as illustrated in Figure 7.6 (right). The sharp spikes in the plot correspond to internal cavity resonances that will be discussed below.

Let us use these numerical solutions to investigate radiation from an open spheroidal antenna, modelled as a spheroidal screen excited by a vertical electric dipole.

For this structure, the far field radiation pattern F is defined by

$$H_\phi = H_\phi^{(0)} + H_\phi^{(sc)} = \frac{e^{ic\xi}}{\xi} F(\eta) + O(\xi^{-2}), \tag{7.74}$$

as $\xi \rightarrow \infty$. Taking into account the asymptotic behaviour of the prolate radial spheroidal function $R_{1l}^{(3)}(c, \xi)$ (see Section 1.2), the magnetic field component has the form

$$H_\phi = -\frac{e^{ic\xi}}{\xi} \frac{pk^3}{4\pi c \sqrt{\xi_1^2 - 1}} \times$$

$$\sum_{l=1}^\infty \left[\frac{(-i)^{l+1} R_{1l}^{(1)}(c, \xi_1)}{\varkappa_{1l}(c) N_{1l}(c)} - \frac{1}{d_{l-1}^{1l}(c)} \frac{2l+1}{l(l+1)} y_l Z_l^{(1)}(c, \xi_0) \right] e^{-(l+1)\frac{\pi}{2}} S_{1l}(c, \eta)$$

$$+ O(\xi^{-2}), \tag{7.75}$$

and so the far field radiation pattern is

$$F\left(\eta\right) = \frac{pk^3}{4\pi c\sqrt{\xi_1^2 - 1}} \times$$

$$\sum_{l=1}^{\infty} \left[\frac{(-1)^{l+1} R_{1l}^{(1)}\left(c, \xi_1\right)}{\varkappa_{1l}(c) N_{1l}(c)} - \frac{(-i)^{l+1}}{d_{l-1}^{1l}(c)} \frac{2l+1}{l\left(l+1\right)} y_l Z_l^{(1)}\left(c, \xi_0\right) \right] S_{1l}(c, \eta).$$

$$(7.76)$$

The total power flux is, omitting inessential factors,

$$\overrightarrow{\Pi} = \frac{1}{2} \operatorname{Re} \iint_S \overrightarrow{E}_\eta \times \overrightarrow{H}_\phi^* dS. \tag{7.77}$$

In the far field zone $E_\eta = H_\phi + O\left(\xi^{-2}\right)$ and to this order, the power flux has only the radial component

$$\Pi = \Pi_\xi = \frac{1}{2} \iint_S |H_\phi|^2 dS = \frac{p^2 k^4}{32\pi^2 \left(\xi_1^2 - 1\right)} \times$$

$$\sum_{l=1}^{\infty} \left| \frac{(-1)^{l+1} R_{1l}^{(1)}\left(c, \xi_1\right)}{\varkappa_{1l}(c) N_{1l}(c)} - \frac{(-i)^{l+1}}{d_{l-1}^{1l}(c)} \frac{2l+1}{l\left(l+1\right)} y_l Z_l^{(1)}\left(c, \xi_0\right) \right|^2 N_{1l}^2(c). \quad (7.78)$$

The power flux Π_0 radiated by a dipole in free space corresponds to setting $y_l = 0$ in (7.78), giving

$$\Pi_0 = \frac{p^2 k^4}{32\pi^2 \left(\xi_1^2 - 1\right)} \sum_{l=1}^{\infty} \frac{\left[R_{1l}^{(1)}\left(c, \xi_1\right)\right]^2}{\left|\varkappa_{1l}(c)\right|^2}. \tag{7.79}$$

Radiation resistance R is defined as the ratio between the power radiated by the antenna to the power radiated by the dipole in free space,

$$R = \Pi/\Pi_0. \tag{7.80}$$

Our analysis of the open spheroidal antenna examines the computed radiation resistance $R(kb)$ (7.80) and far-field radiation pattern $|F\left(\eta\right)|$ (7.76) in the resonant frequency region $0 < kb \leq 10$. In all computations the position of a dipole is fixed as $\xi_1/\xi_0 = 1.1$ and the geometrical scale is defined by the major semi-axis of the spheroid taken to be $b = 1$ (so $d/2 = \xi_0^{-1}$ and $kb = c\xi_0$).

In Figure 7.7 the radiation resistance $R(kb)$ for a closed spheroid with $a/b = 0.2, \theta_0 = 0^0$ is plotted (as a dashed line). Resonant peaks correspond

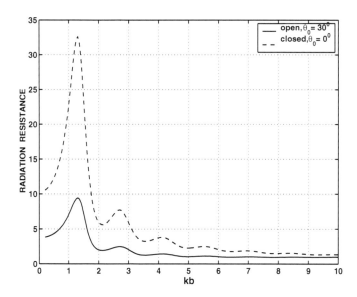

FIGURE 7.7. Radiation resistance for a closed spheroid (dashed line) and an open spheroid (solid line) with $\theta_0 = 30°$; aspect ratio $a/b = 0.2$.

to external quasi-eigenoscillations of electric type studied before [21]. Further examples of this phenomenon will be considered in Section 7.6. The radiation resistance $R(kb)$ for an open spheroidal shell of the same aspect ratio ($a/b = 0.2$), with a circular hole of angular size $\theta_0 = 30°$, is shown as a solid line. The introduction of the aperture does not significantly modify the behaviour of $R(kb)$. Maxima of $R(kb)$ are observed at approximately the same frequencies as for a closed shell. The lower average level of $R(kb)$ over the computed waveband is explained by the greater spacing of the feed from the antenna that causes a less efficient excitation. Note that no internal resonances are observed because the first resonance is excited when $ka \backsim 2.7$ (a is the minor axis). The latter conclusion is demonstrated by the radiation resistance plots in Figure 7.8 ($a/b = 0.5, \theta_0 = 30^0$). The occurrence of double extrema is obviously connected to the excitation of internal quasi-eigenoscillations of electric type in spheroidal cavity. A clear manifestation of the resonant effect due to cavity oscillations is observed for a thicker antenna (see Figure 7.9 where $a/b = 0.8, \theta_0 = 30^0$).

The presence of these additional resonances provides opportunities for modifying the radiation of spheroidal antenna by varying the frequency in a narrow band. An example is shown in Figure 7.10, where significant differences in the radiation patterns are observed in the vicinity of a double extremum: $kb = 3.3933$ (solid line) and $kb = 3.4$ (dashed line); for com-

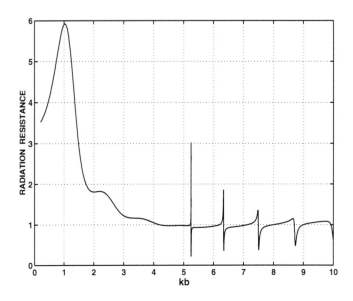

FIGURE 7.8. Radiation resistance for the open spheroid with $\theta_0 = 30°$; aspect ratio $a/b = 0.5$.

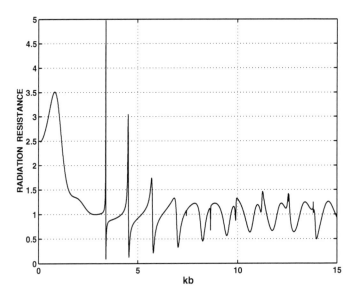

FIGURE 7.9. Radiation resistance for the open spheroid with $\theta_0 = 30°$; aspect ratio $a/b = 0.8$.

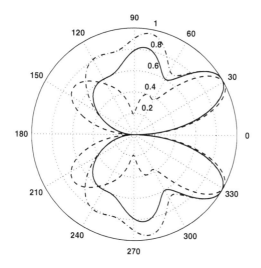

FIGURE 7.10. Radiation patterns for the open spheroid with aspect ratio $a/b = 0.8$, angle $\theta_0 = 30°$ and wavenumber kb equal to 3.3933 (solid), 3.4 (dashed) and 3.5 (dash-dot); the first two values of kb correspond to a double extremum.

parison the pattern at a nearby nonresonant frequency ($kb = 3.5$) is shown (dotted line).

7.5   Axial Magnetic Dipole Excitation of a Metallic Spheroidal Cavity.

In this section we consider the excitation of the spheroidal cavity with a single circular aperture by a magnetic dipole. The geometry of the problem is the same as in the previous section (see Figure 7.4), but the vertical electric dipole is replaced by a vertical magnetic dipole. The axially symmetric magnetic dipole field induces a total field of magnetic type: three of the components vanish and the remaining components (H_ξ, H_η, E_ϕ) are governed by the system of equations (1.94)–(1.96) deduced from Maxwell's equations. A second order partial differential equation similar to (1.100) determines the field component $E_\phi = E_\phi(\xi, \eta)$. The mixed boundary value problem for this component is formulated in the same way as was formulated for the magnetic field component H_ϕ when the cavity was excited by an electric dipole (see Section 7.4).

The total field E_ϕ^{tot} is a superposition of the incident $E_\phi^{(0)}$ and the scattered field E_ϕ^{sc}:

$$E_\phi^{tot}(\xi, \eta) = E_\phi^{(0)}(\xi, \eta) + E_\phi^{sc}(\xi, \eta). \tag{7.81}$$

A rigorous formulation of the diffraction problem requires the total field to satisfy Maxwell's equations (1.58)–(1.59) subject to the following conditions: (i) the tangential components of the total electric field must be continuous on the spheroidal surface $\xi = \xi_0$,

$$E_\phi(\xi_0 - 0, \eta) = E_\phi(\xi_0 + 0, \eta), \qquad \eta \in (-1, 1); \tag{7.82}$$

(ii) the *mixed* boundary conditions

$$\begin{aligned} E_\phi(\xi_0 - 0, \eta) &= E_\phi(\xi_0 + 0, \eta) = 0, & \eta &\in (-1, \eta_0) \\ H_\eta(\xi_0 - 0, \eta) &= H_\eta(\xi_0 + 0, \eta), & \eta &\in (\eta_0, 1) \end{aligned} \tag{7.83}$$

are imposed on the tangential components of the total electric field over the screen and on the total magnetic field over the aperture, respectively; (iii) the scattered field satisfies radiation conditions, requiring that it behaves as an outgoing spherical wave at infinity,

$$E_\phi^{sc}(\xi, \eta) = \frac{e^{ic\xi}}{\xi} F^{sc}(\eta)\left(1 + O\left(\xi^{-1}\right)\right) \tag{7.84}$$

as $\xi \to \infty$, where F^{sc} is the radiation pattern of scattered field; and (iv) the scattered energy defined by (7.49) must be finite in any finite volume V of space. This last condition is equivalent to the so-called *edge condition* prescribing the singularity of the scattered field at the edges of the screen. As usual, uniqueness of the solution is guaranteed by this set of boundary, edge and radiation conditions [41], [93].

Let us represent the incident and scattered field in terms of spheroidal harmonics. The field radiated by the magnetic dipole in free space is

$$E_\phi^{(0)}(\xi, \eta) = -\frac{mk^3}{\sqrt{\xi_1^2 - 1}} \times$$

$$\sum_{l=1}^{\infty} \frac{(-i)^{l+1}}{\chi_{1l}(c) N_{1l}(c)} S_{1l}(c, \eta) \left\{ \begin{array}{l} R_{1l}^{(3)}(c, \xi_1) R_{1l}^{(1)}(c, \xi), \ \xi < \xi_1 \\ R_{1l}^{(1)}(c, \xi_1) R_{1l}^{(3)}(c, \xi), \ \xi > \xi_1 \end{array} \right\} \tag{7.85}$$

where m is the dipole moment. Enforcing the continuity condition (7.83) on the shell surface $\xi = \xi_0$, and the radiation condition (7.84) on the scattered field, yields the expansion

$$E_\phi^{(sc)}(\xi,\eta) = \frac{mk^3}{4\pi\sqrt{\xi_1^2-1}} \times$$

$$\sum_{l=1}^{\infty} B_l S_{1l}(c,\eta) \left\{ \begin{array}{ll} R_{1l}^{(1)}(c,\xi), & \xi < \xi_0 \\ R_{1l}^{(3)}(c,\xi) R_{1l}^{(1)}(c,\xi_0)/R_{1l}^{(3)}(c,\xi_0), & \xi > \xi_0 \end{array} \right\} \quad (7.86)$$

where the unknown Fourier coefficients B_l are to be found.

Imposing the mixed boundary conditions (7.83) on the solution (7.86) produces the following dual series equations defined on subintervals corresponding to the screen surface and aperture,

$$\sum_{l=1}^{\infty} B_l R_{1l}^{(1)}(c,\xi_0) S_{1l}(c,\eta) = \sum_{l=1}^{\infty} \beta_l S_{1l}(c,\eta), \quad \eta \in (-1,\eta_0) \quad (7.87)$$

$$\sum_{l=1}^{\infty} \frac{B_l}{R_{1l}^{(3)}(c,\xi_0)} S_{1l}(c,\eta) = 0, \quad \eta \in (\eta_0,1) \quad (7.88)$$

where

$$\beta_l = \frac{(-i)^{l+1}}{\chi_{1l}(c)N_{1l}(c)} \left\{ \begin{array}{ll} R_{1l}^{(1)}(c,\xi_0) R_{1l}^{(3)}(c,\xi_1), & \xi_0 < \xi_1 \\ R_{1l}^{(3)}(c,\xi_0) R_{1l}^{(1)}(c,\xi_1), & \xi_0 > \xi_1. \end{array} \right. \quad (7.89)$$

Based on the representation of prolate radial spheroidal functions as a series of Bessel functions (see [7.57]–[7.58]), we deduce

$$R_{1l}^{(1)}(c,\xi_0) R_{1l}^{(3)}(c,\xi_0) = -\frac{i}{c(\xi_0^2-1)^{\frac{1}{2}}(2l+1)} + O(l^{-2}) \quad (7.90)$$

as $l \to \infty$, and are led to define the parameter

$$\mu_l(c,\xi_0) = 1 - ic(\xi_0^2-1)^{\frac{1}{2}}(2l+1) R_{1l}^{(1)}(c,\xi_0) R_{1l}^{(3)}(c,\xi_0) \quad (7.91)$$

that is asymptotically small: $\mu_l = O(l^{-2})$ as $l \to \infty$. After rescaling the unknowns via

$$y_l = \frac{B_l}{R_{1l}^{(3)}(c,\xi_0)} \quad (7.92)$$

and denoting

$$\gamma_l = ic(\xi_0^2-1)^{\frac{1}{2}} \beta_l \quad (7.93)$$

the equations (7.87)–(7.88) take the form

$$\sum_{l=1}^{\infty} y_l \left(1 - \mu_l\right) \frac{1}{2l+1} S_{1l}\left(c, \eta\right) = \sum_{l=1}^{\infty} \gamma_l S_{1l}\left(c, \eta\right), \quad \eta \in (-1, \eta_0) \qquad (7.94)$$

$$\sum_{l=1}^{\infty} y_l S_{1l}\left(c, \eta\right) = 0, \qquad \eta \in (\eta_0, 1). \qquad (7.95)$$

As in the previous section, the next step is to transform these functional equations to dual series equations with associated Legendre function kernels by inserting the identity (7.4), giving

$$\sum_{l=1}^{\infty} Y_l \frac{(1 - \mu_l)}{2l+1} \left\{ P_l^1\left(\cos\theta\right) + \sum_{\substack{r=0,1 \\ r\neq l-1}}^{\infty} {}'\frac{d_r^{1l}\left(c\right)}{d_{l-1}^{1l}\left(c\right)} P_{r+1}^1\left(\cos\theta\right) \right\} =$$

$$\sum_{l=1}^{\infty} \gamma_l \sum_{r=0,1}^{\infty} {}'d_r^{1l}\left(c\right) P_{r+1}^1\left(\cos\theta\right), \quad \theta \in (\theta_0, \pi) \qquad (7.96)$$

$$\sum_{l=1}^{\infty} Y_l \left\{ P_l^1\left(\cos\theta\right) + \sum_{\substack{r=0,1 \\ r\neq l-1}}^{\infty} {}'\frac{d_r^{1l}\left(c\right)}{d_{l-1}^{1l}\left(c\right)} P_{r+1}^1\left(\cos\theta\right) \right\} = 0, \quad \theta \in (0, \theta_0)$$

$$(7.97)$$

where

$$Y_l = d_{l-1}^{1l}(c) y_l. \qquad (7.98)$$

We now apply the Abel integral transform technique to reduce the pair of equations (7.96)–(7.97) to the following dual series equations with trigonometric function kernels and containing a constant C to be determined. This is similar to the transformation of equations (4.15)–(4.16) to equation (4.20).

$$\sum_{l=1}^{\infty} Y_l \left\{ \cos\left(l + \frac{1}{2}\right)\theta + \sum_{\substack{r=0,1 \\ r\neq l-1}}^{\infty} {}'\frac{d_r^{1l}\left(c\right)}{d_{l-1}^{1l}\left(c\right)} \cos\left(r + \frac{3}{2}\right)\theta \right\} = C \cos\frac{\theta}{2},$$

$$\theta \in (0, \theta_0) \quad (7.99)$$

$$\sum_{l=1}^{\infty} \frac{l(l+1)}{(2l+1)^2} Y_l (1-\mu_l) \left\{ \cos\left(l+\frac{1}{2}\right)\theta + \sum_{\substack{r=0,1 \\ r \neq l-1}}^{\infty} {}' \frac{d_r^{1l}(c)}{d_{l-1}^{1l}(c)} \cos\left(r+\frac{3}{2}\right)\theta \right\}$$

$$= \sum_{l=1}^{\infty} \gamma_l \sum_{r=0,1}^{\infty} {}' d_r^{1l}(c) \cos\left(r+\frac{3}{2}\right)\theta, \quad \theta \in (\theta_0, \pi). \quad (7.100)$$

There is a difference with the previous transform of equations (4.15)–(4.16) to the system (4.20). Here we integrate equation (7.97) (defined over $\theta \in (0, \theta_0)$) and then employ the representation of Mehler-Dirichlet type

$$P_n(\cos\theta) = \frac{\sqrt{2}}{\pi} \int_0^\theta \frac{\cos\left(n+\frac{1}{2}\right)\varphi}{\sqrt{\cos\varphi - \cos\theta}} d\varphi.$$

Considering Equations (7.99)–(7.100), it is convenient to redefine the asymptotically small parameter, introducing

$$M_l = \frac{4l(l+1)}{(2l+1)^2}\mu_l + \frac{1}{(2l+1)^2}, \quad (7.101)$$

and to redefine the independent variable

$$\vartheta = \pi - \theta, \quad (7.102)$$

with $\vartheta_1 = \pi - \theta_0$. Noting that $\cos\left(l+\frac{1}{2}\right)\theta = (-1)^l \sin\left(l+\frac{1}{2}\right)\vartheta$, the equations (7.99)–(7.100) may be rearranged in the form that allows us to equate a Fourier series with a function defined piecewise on two subintervals of $[0, \pi]$,

$$\sum_{l=1}^{\infty} (-1)^l Y_l \sin(l+\frac{1}{2})\vartheta = \begin{cases} F_1(\vartheta), & \vartheta \in (\vartheta_1, \pi) \\ F_2(\vartheta), & \vartheta \in (0, \vartheta_1) \end{cases} \quad (7.103)$$

where

$$F_1(\vartheta) = C \sin\frac{\vartheta}{2} - \sum_{l=1}^{\infty} Y_l \sum_{\substack{r=0,1 \\ r \neq l-1}}^{\infty} {}' \frac{d_r^{1l}(c)}{d_{l-1}^{1l}(c)} (-1)^{r+1} \sin\left(r+\frac{3}{2}\right)\vartheta$$

and

$$F_2(\vartheta) = 4\sum_{l=1}^{\infty} \gamma_l \sum_{r=0,1}^{\infty} {}' d_r^{1l}(c) (-1)^{r+1} \sin\left(r+\frac{3}{2}\right)\vartheta$$

$$+ \sum_{l=1}^{\infty} Y_l M_l \sum_{\substack{r=0,1 \\ r \neq l-1}}^{\infty} {}' \frac{d_r^{1l}(c)}{d_{l-1}^{1l}(c)} (-1)^{r+1} \sin\left(r+\frac{3}{2}\right)\vartheta. \quad (7.104)$$

Multiplying both sides of equation (7.103) by $\sin\left(s + \frac{1}{2}\right)\vartheta$ $(s = 1, 2, \ldots)$ and then integrating over the range $(0, \pi)$ produces the following i.s.l.a.e.,

$$Y_l + \sum_{l=1}^{\infty} Y_l \sum_{\substack{r=0,1 \\ r \neq l-1}}^{\infty} {}' \frac{d_r^{1l}(c)}{d_{l-1}^{1l}(c)} (-1)^{r+1} [\delta_{s,r+1} - \Phi_{s,r}(\vartheta_1)]$$

$$- \sum_{l=1}^{\infty} Y_l M_l \sum_{\substack{r=0,1 \\ r \neq l-1}}^{\infty} {}' \frac{d_r^{1l}(c)}{d_{l-1}^{1l}(c)} (-1)^{r+s+1} \Phi_{sr}(\vartheta_1)$$

$$= 4 \sum_{l=1}^{\infty} \gamma_l \sum_{r=0,1}^{\infty} {}' d_r^{1l}(c) (-1)^{r+s+1} \Phi_{sr}(\vartheta_1), \quad (7.105)$$

where

$$\Phi_{sr}(\theta_0) = R_{r+1,s}(\theta_0) + \frac{R_{0s}(\theta_0)}{1 - R_{00}(\theta_0)} R_{r+1,0}(\theta_0) \quad (7.106)$$

and $R_{rs}(\theta_0)$ is given by (7.31).

The system (7.105) has very similar features to the comparable i.s.l.a.e (7.69) derived for the electric dipole case, and equally satisfactory numerical methods are readily devised to solve this system for the unknown coefficients Y_l. The radiation resistance $R = R(c)$ may then be determined from

$$R = \Pi_0^{-1} \sum_{l=1}^{\infty} \left| \frac{(-1)^{l+1}}{\chi_{1l}(c) N_{1l}(c)} R_{1l}^{(1)}(c, \xi_1) - (-i)^{l+1} \frac{Y_l}{d_{l-1}^{1l}(c)} R_{1l}^{(1)}(c, \xi_0) \right| N_{1l}^2(c)$$
$$(7.107)$$

where

$$\Pi_0 = \sum_{l=1}^{\infty} \left| \frac{R_{1l}^{(1)}(c, \xi_1)}{\chi_{1l}(c)} \right|^2$$

is proportional to the power of the dipole radiated in free space. The far-field radiation pattern is

$$f(\eta) = \sum_{l=1}^{\infty} \left\{ \frac{(-1)^{l+1}}{\chi_{1l}(c) N_{1l}(c)} R_{1l}^{(1)}(c, \xi_1) - (-i)^{l+1} \frac{Y_l}{d_{l-1}^{1l}(c)} R_{1l}^{(1)}(c, \xi_0) \right\} S_{1l}(c, \eta).$$
$$(7.108)$$

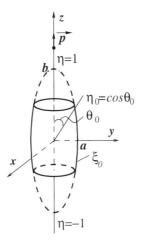

FIGURE 7.11. Spheroidal barrel excited by an axially located electric dipole.

7.6 Axial Electric Dipole Excitation of a Spheroidal Barrel.

In this section we consider a scatterer more complex than that in the previous sections, the spheroidal cavity with two symmetrically located circular apertures. In prolate spheroidal coordinates (ξ, η, ϕ) the infinitely thin spheroidal shell enclosing the cavity may be defined by

$$\xi = \xi_0, \quad -\eta_0 \le \eta \le \eta_0, \quad 0 \le \phi \le 2\pi$$

for suitable ξ_0 and η_0 and the circular apertures may be defined by

$$\xi = \xi_0, \quad \eta \in [0, \eta_0) \cup (-\eta_0, -1], \quad 0 \le \phi \le 2\pi.$$

The screen is assumed to be perfectly conducting and, as in Section 7.4, is excited by a vertical (axially aligned) electric dipole placed along the axis of rotational symmetry (the z-axis), at $\xi = \xi_1$ (see Figure 7.11). The formulation of this time-harmonic problem is rather similar to that described in Section 7.4, and leads to the following mixed boundary-value problem of diffraction theory.

Both total and scattered fields satisfy Maxwell's equations in the nondimensional form (1.58)–(1.59); the tangential components of total electric field must be continuous on the spheroidal surface $\xi = \xi_0$ (see [7.46]); and the scattered field satisfies the radiation condition (7.48) and the finite energy condition (7.49). The only difference arises in the specification of the subintervals on which the mixed boundary conditions for the total electric field (on the screen) and for the total magnetic field (over the apertures) are enforced:

$$E_\eta \left(\xi_0 - 0, \eta \right) = E_\eta \left(\xi_0 + 0, \eta \right) = 0, \quad \eta \in \left(-\eta_0, \eta_0 \right), \qquad (7.109)$$
$$H_\phi \left(\xi_0 - 0, \eta \right) = H_\phi \left(\xi_0 + 0, \eta \right), \quad \eta \in \left(-1, -\eta_0 \right) \cup \left(\eta_0, 1 \right). \, (7.110)$$

As usual, the total field is decomposed according to (7.45) as a superposition of the incident dipole field $H_\phi^{(0)} \left(\xi, \eta \right)$ (7.44), and the scattered field $H_\phi^{sc} \left(\xi, \eta \right)$ expressed in terms of spheroidal harmonics according to (7.50), with unknown Fourier coefficients $\{ A_l \}_{l=1}^\infty$ to be determined.

Enforcement of the mixed boundary conditions (7.109)–(7.110) leads to the following triple series equations defined on three subintervals corresponding to the screen and both apertures:

$$\sum_{l=1}^\infty \frac{A_l}{Z_l^{(3)} \left(c, \xi_0 \right)} S_{1l} \left(c, \eta \right) = 0, \quad \eta \in \left(-1, -\eta_0 \right) \cup \left(\eta_0, 1 \right) \qquad (7.111)$$

$$\sum_{l=1}^\infty A_l Z_l^{(1)} \left(c, \xi_0 \right) S_{1l} \left(c, \eta \right) = \sum_{l=1}^\infty \alpha_l S_{1l} \left(c, \eta \right), \quad \eta \in \left(-\eta_0, \eta_0 \right), \qquad (7.112)$$

where coefficients α_l are deduced from the expression (7.44) for the incident field,

$$\alpha_l = \frac{(-i)^{l+1}}{\chi_{1l} \left(c \right) N_{1l} \left(c \right)} \begin{cases} Z_l^{(1)} \left(c, \xi_0 \right) R_{1l}^{(3)} \left(c, \xi_1 \right), \; \xi_1 > \xi_0 \\ Z_l^{(3)} \left(c, \xi_0 \right) R_{1l}^{(1)} \left(c, \xi_1 \right), \; \xi_1 < \xi_0 \end{cases} \qquad (7.113)$$

(the dipole is located outside or inside the spheroidal surface $\xi = \xi_0$ according as $\xi_1 > \xi_0$ or $\xi_1 < \xi_0$).

The treatment of these equations closely follows the ideas of Section 7.4. The same asymptotically small parameter ε_l defined by (7.60) is employed. The unknowns are rescaled according to

$$A_l = \frac{2l+1}{l \left(l+1 \right)} Z_l^{(3)} \left(c, \xi_0 \right) B_l, \qquad (7.114)$$

and the equations (7.111)–(7.112) become

$$\sum_{l=1}^\infty \frac{2l+1}{l \left(l+1 \right)} B_l S_{1l} \left(c, \eta \right) = 0, \qquad \eta \in \left(-1, -\eta_0 \right) \cup \left(\eta_0, 1 \right), \quad (7.115)$$

$$\sum_{l=1}^\infty B_l \left(1 - \varepsilon_l \right) S_{1l} \left(c, \eta \right) = -ic \left(\xi_0^2 - 1 \right)^{\frac{1}{2}} \sum_{l=1}^\infty \alpha_l S_{1l} \left(c, \eta \right),$$

$$\eta \in \left(-\eta_0, \eta_0 \right). \quad (7.116)$$

These triple series equations fully determine the unique solution to the mixed boundary value problem formulated above, provided the unknown Fourier coefficients B_l are constrained by the finite energy condition (7.49). From (7.66) and (7.67) we deduce

$$\sum_{l=1}^{\infty} |B_l|^2 < \infty, \tag{7.117}$$

that is, $\{B_l\}_{l=1}^{\infty} \in l_2$. This provides the correct mathematical setting for the solution to the triple series equations (7.116), and ensures the validity of all operations in the construction of the solution.

The first step is to reduce the triple series equations (7.115)–(7.116) to two decoupled sets of dual series equations with Jacobi polynomial kernels. Invoking the p.a.s.f. symmetry [29]

$$S_{1l}(c, \eta) = (-1)^{l-1} S_{1l}(c, -\eta), \tag{7.118}$$

equations (7.115)–(7.116) may be rewritten as two decoupled pairs of equations for the odd and even index coefficients as

$$\sum_{l=0}^{\infty} \frac{l + \frac{3}{4}}{\left(l + \frac{1}{2}\right)(l+1)} B_{2l+1} S_{1,2l+1}(c, \eta) = 0, \quad \eta \in (-1, -\eta_0) \tag{7.119}$$

$$\sum_{l=0}^{\infty} B_{2l+1}(1 - \varepsilon_{2l+1}) S_{1,2l+1}(c, \eta) = \sum_{l=0}^{\infty} \beta_{2l+1} S_{1,2l+1}(c, \eta), \quad \eta \in (-\eta_0, 0)$$
$$\tag{7.120}$$

and

$$\sum_{l=1}^{\infty} \frac{l + \frac{1}{4}}{l\left(l + \frac{1}{2}\right)} B_{2l} S_{1,2l}(c, \eta) = 0, \quad \eta \in (-1, -\eta_0) \tag{7.121}$$

$$\sum_{l=1}^{\infty} B_{2l}(1 - \varepsilon_{2l}) S_{1,2l}(c, \eta) = \sum_{l=1}^{\infty} \beta_{2l} S_{1,2l}(c, \eta), \quad \eta \in (-\eta_0, 0), \tag{7.122}$$

where $\beta_l = -ic\left(\xi_0^2 - 1\right)^{\frac{1}{2}} \alpha_l$.

Insertion of the identity (7.4) transforms these pairs to dual equations with associated Legendre function kernels. The associated Legendre functions $P_l^{(1, \pm \frac{1}{2})}$ are related to Jacobi polynomials (see [90])

$$P_{2l+1}^1(\eta) = 2\left(l + \frac{1}{2}\right)\left(1 - \eta^2\right)^{\frac{1}{2}} P_l^{(1, -\frac{1}{2})}(2\eta^2 - 1), \tag{7.123}$$

$$P_{2l}^1(\eta) = 2\left(l + \frac{1}{2}\right)\eta\left(1 - \eta^2\right)^{\frac{1}{2}} P_{l-1}^{(1,\frac{1}{2})}\left(2\eta^2 - 1\right).\tag{7.124}$$

Thus introducing the new variable $u = 2\eta^2 - 1$ (with $u_0 = 2\eta_0^2 - 1$), and rescaling the unknowns via

$$x_{2l+1} = d_{2l}^{1,2l+1}(c)\, B_{2l+1}, \quad x_{2l} = d_{2l-1}^{1,2l}(c)\, B_{2l},\tag{7.125}$$

we may reduce each of the dual series equations (7.119)–(7.120), (7.121)–(7.122) to dual series equations with Jacobi polynomial kernels. The new variable u lies in the interval $(-1, 1)$. Note that both $\{x_{2l}\}_{l=1}^{\infty}$ and $\{x_{2l+1}\}_{l=0}^{\infty}$ lie in l_2. The odd index coefficients satisfy

$$\sum_{l=0}^{\infty}\left(l + \frac{1}{2}\right) x_{2l+1}\left(1 - \varepsilon_{2l+1}\right) \times$$

$$\left[P_l^{(1,-\frac{1}{2})}(u) + \sum_{\substack{r=0 \\ r \neq l}}^{\infty}{}' \frac{d_{2r}^{1,2l+1}(c)}{d_{2l}^{1,2l+1}(c)}\frac{r + \frac{1}{2}}{l + \frac{1}{2}} P_r^{(1,-\frac{1}{2})}(u)\right]$$

$$= \sum_{l=0}^{\infty}\beta_{2l+1}\sum_{r=0}^{\infty}{}'\left(r + \frac{1}{2}\right) d_{2r}^{1,2l+1}(c)\, P_r^{(1,-\frac{1}{2})}(u), \quad u \in (-1, u_0)\tag{7.126}$$

$$\sum_{l=0}^{\infty}\frac{\left(l + \frac{3}{4}\right)}{(l + 1)} x_{2l+1}\left[P_l^{(1,-\frac{1}{2})}(u) + \sum_{\substack{r=0 \\ r \neq l}}^{\infty}{}' \frac{d_{2r}^{1,2l+1}(c)}{d_{2l}^{1,2l+1}(c)}\frac{r + \frac{1}{2}}{l + \frac{1}{2}} P_r^{(1,-\frac{1}{2})}(u)\right] = 0,$$

$$u \in (u_0, 1)\tag{7.127}$$

and the even index coefficients satisfy

$$\sum_{l=1}^{\infty}\left(l + \frac{1}{2}\right) x_{2l}\left(1 - \varepsilon_{2l}\right)\left[P_{l-1}^{(1,\frac{1}{2})}(u) + \sum_{\substack{r=1 \\ r \neq l}}^{\infty}{}' \frac{d_{2r-1}^{1,2l}(c)}{d_{2l-1}^{1,2l}(c)}\frac{r + \frac{1}{2}}{l + \frac{1}{2}} P_{r-1}^{(1,\frac{1}{2})}(u)\right]$$

$$= \sum_{l=1}^{\infty}\beta_{2l}\sum_{r=1}^{\infty}{}'\left(r + \frac{1}{2}\right) d_{2r-1}^{1,2l}(c)\, P_{r-1}^{(1,\frac{1}{2})}(u), \quad u \in (-1, u_0)\tag{7.128}$$

$$\sum_{l=1}^{\infty}\frac{\left(l + \frac{1}{4}\right)}{l} x_{2l}\left[P_{l-1}^{(1,\frac{1}{2})}(u) + \sum_{\substack{r=1 \\ r \neq l}}^{\infty}{}' \frac{d_{2r-1}^{1,2l}(c)}{d_{2l-1}^{1,2l}(c)}\frac{r + \frac{1}{2}}{l + \frac{1}{2}} P_{r-1}^{(1,\frac{1}{2})}(u)\right] = 0,$$

$$u \in (u_0, 1).\tag{7.129}$$

Each set of dual series equations (7.126)–(7.127) and (7.128)–(7.129) is in standard form for a regularisation procedure that will convert it to a second-kind system. We consider each system separately.

As described in some detail in previous chapters, the core of the method is to transform each set to the form of a Fourier expansion in normalised Jacobi polynomials $\widehat{P}_l^{(\alpha,\beta)}$ that is equated to a function piecewise defined on two subintervals of $(-1,1)$.

7.6.1 The Series Equations with Odd Index Coefficients.

Integrate the equation (7.126) using the formula (1.174) of Volume I,

$$\int_{-1}^{t} (1+u)^{-\frac{1}{2}} P_l^{(1,-\frac{1}{2})}(u)\,du = \frac{(1+t)^{\frac{1}{2}}}{l+\frac{1}{2}} P_l^{(0,\frac{1}{2})}(t). \tag{7.130}$$

This equalises the convergence rate of both parts of the dual series equations (7.126)–(7.127). Then employ the following Abel type integral representations for Jacobi polynomials (see formulae [1.171] and [1.172] of Volume I)

$$P_l^{(0,\frac{1}{2})}(u) = (1+u)^{-\frac{1}{2}} \frac{\Gamma\left(l+\frac{3}{2}\right)}{\sqrt{\pi}\,\Gamma\left(l+1\right)} \int_{-1}^{u} \frac{P_l^{(\frac{1}{2},0)}(t)}{(u-t)^{\frac{1}{2}}}\,dt, \tag{7.131}$$

$$P_l^{(1,-\frac{1}{2})}(u) = (1-u)^{-1} \frac{\Gamma\left(l+2\right)}{\sqrt{\pi}\,\Gamma\left(l+\frac{3}{2}\right)} \int_{u}^{1} \frac{(1-t)^{\frac{1}{2}} P_l^{(\frac{1}{2},0)}(t)}{(t-u)^{\frac{1}{2}}}\,dt, \tag{7.132}$$

to convert the dual series equations to a pair of homogeneous Abel integral equations, one valid on the interval $(-1, u_0)$ and the other valid on $(u_0, 1)$.

After inversion of these Abel integral equations, and with some simple rescaling, we obtain the following equations. Let

$$p_l = 1 - \left(1+\frac{3}{4}\right)\left[\frac{\Gamma\left(l+1\right)}{\Gamma\left(l+\frac{3}{2}\right)}\right]^2 \tag{7.133}$$

denote the parameter that is asymptotically small: $p_l = O(l^{-2})$ as $l \to \infty$. The unknown coefficients are rescaled via

$$X_{2l+1} = \frac{\Gamma\left(l+\frac{3}{2}\right)}{\Gamma\left(l+1\right)} \left[h_l^{(\frac{1}{2},0)}\right]^{\frac{1}{2}} x_{2l+1}, \tag{7.134}$$

where $h_l^{(\frac{1}{2},0)}$ is the square of the norm of the Jacobi polynomials $P_l^{(\frac{1}{2},0)}$ (see the Appendix B.3, Volume I); it can easily be checked that $\{X_{2l+1}\}_{l=0}^{\infty} \in l_2$.

The transformed dual series employ the normalised Jacobi polynomials $\widehat{P}_l^{(\frac{1}{2},0)} = \left[h_l^{(\frac{1}{2},0)}\right]^{-\frac{1}{2}} P_l^{(\frac{1}{2},0)}$ and take the form

$$\sum_{l=0}^{\infty} X_{2l+1} \widehat{P}_l^{(\frac{1}{2},0)}(u) = \left\{ \begin{matrix} F_1(u), & u \in (-1, u_0) \\ F_2(u), & u \in (u_0, 1) \end{matrix} \right\} \tag{7.135}$$

where

$$F_1(u) = \sum_{l=0}^{\infty} X_{2l+1} \varepsilon_{2l+1} \widehat{P}_l^{(\frac{1}{2},0)}(u) - \sum_{l=0}^{\infty} X_{2l+1}(1 - \varepsilon_{2l+1}) \times$$

$$\sum_{\substack{r=0 \\ r \neq l}}^{\infty} {}' \frac{d_{2r}^{1,2l+1}(c)}{d_{2l}^{1,2l+1}(c)} \frac{\Gamma\left(r + \frac{3}{2}\right)}{\Gamma(r+1)} \frac{\Gamma(l+1)}{\Gamma\left(l + \frac{3}{2}\right)} \left[\frac{h_r^{(\frac{1}{2},0)}}{h_l^{(\frac{1}{2},0)}}\right]^{\frac{1}{2}} \widehat{P}_r^{(\frac{1}{2},0)}(u)$$

$$+ \sum_{l=0}^{\infty} \beta_{2l+1} \sum_{r=0}^{\infty} {}' d_{2r}^{1,2l+1}(c) \frac{\Gamma\left(r + \frac{3}{2}\right)}{\Gamma(r+1)} \left[h_r^{(\frac{1}{2},0)}\right]^{\frac{1}{2}} \widehat{P}_r^{(\frac{1}{2},0)}(u) \tag{7.136}$$

and

$$F_2(u) = \sum_{l=0}^{\infty} X_{2l+1} p_l \widehat{P}_l^{(\frac{1}{2},0)}(u) + \sum_{l=0}^{\infty} X_{2l+1}(1 - p_l) \times$$

$$\sum_{\substack{r=0 \\ r \neq l}}^{\infty} {}' \frac{d_{2r}^{1,2l+1}(c)}{d_{2l}^{1,2l+1}(c)} \frac{\Gamma(r+2)}{\Gamma\left(r + \frac{1}{2}\right)} \frac{\Gamma\left(l + \frac{1}{2}\right)}{\Gamma(l+2)} \left[\frac{h_r^{(\frac{1}{2},0)}}{h_l^{(\frac{1}{2},0)}}\right]^{\frac{1}{2}} \widehat{P}_r^{(\frac{1}{2},0)}(u). \tag{7.137}$$

The Jacobi polynomials $\widehat{P}_m^{(\frac{1}{2},0)}$ are orthonormal, so that multiplication of equation (7.135) by $(1 - u)^{\frac{1}{2}} \widehat{P}_m^{(\frac{1}{2},0)}(u)$ and integration over $(-1, 1)$ leads to the following Fredholm matrix equation of the second kind for the odd index unknown coefficients (X_{2l+1}),

$$(1 - \varepsilon_{2m+1}) X_{2m+1} + \sum_{l=0}^{\infty} X_{2l+1} H_{2l+1,2m+1}(c, \xi_0, u_0) = D_{2m+1}(c, \xi_1, u_0),$$

$$\tag{7.138}$$

where $m = 0, 1, 2, \ldots$,

$$D_{2m+1}(c, \xi_1, u_0) =$$

$$\sum_{l=0}^{\infty} \beta_{2l+1} \sum_{r=0}^{\infty} {}' d_{2r}^{1,2l+1}(c) \frac{\Gamma\left(r + \frac{3}{2}\right)}{\Gamma(r+1)} \left[h_r^{(\frac{1}{2},0)}\right]^{\frac{1}{2}} \left[\delta_{rm} - \widehat{Q}_{rm}^{(\frac{1}{2},0)}(u_0)\right] \tag{7.139}$$

and

$$H_{2l+1,2m+1}(c,\xi_0,u_0) = (\varepsilon_{2l+1} - p_l)\,\widehat{Q}_{lm}^{(\frac{1}{2},0)}(u_0) + (1 - \varepsilon_{2l+1}) \times$$

$$\sum_{\substack{r=0 \\ r\neq l}}^{\infty} {}'\frac{d_{2r}^{1,2l+1}(c)\,\Gamma\left(r+\frac{3}{2}\right)\,\Gamma(l+1)}{d_{2l}^{1,2l+1}(c)\,\Gamma(r+1)\,\Gamma\left(l+\frac{3}{2}\right)}\left[\frac{h_r^{(\frac{1}{2},0)}}{h_l^{(\frac{1}{2},0)}}\right]^{\frac{1}{2}}\left[\delta_{rm} - \widehat{Q}_{rm}^{(\frac{1}{2},0)}(u_0)\right]$$

$$+ (1 - p_l)\sum_{\substack{r=0 \\ r\neq l}}^{\infty} {}'\frac{d_{2r}^{1,2l+1}(c)\,\Gamma(r+2)\,\Gamma\left(l+\frac{1}{2}\right)}{d_{2l}^{1,2l+1}(c)\,\Gamma\left(r+\frac{1}{2}\right)\,\Gamma(l+2)}\left[\frac{h_r^{(\frac{1}{2},0)}}{h_l^{(\frac{1}{2},0)}}\right]^{\frac{1}{2}}\widehat{Q}_{rm}^{(\frac{1}{2},0)}(u_0).$$

$$(7.140)$$

The incomplete scalar product (see Appendix B.6, Volume I) has the usual meaning,

$$\widehat{Q}_{lm}^{(\frac{1}{2},0)}(u_0) = \int_{u_0}^{1}(1-x)^{\frac{1}{2}}\,\widehat{P}_l^{(\frac{1}{2},0)}(u)\,\widehat{P}_m^{(\frac{1}{2},0)}(u)\,du.$$

7.6.2   The Series Equations with Even Index Coefficients.

The solution of the system (7.128)–(7.129) for even index coefficients is constructed in the same way. To equalise the convergence rate, integrate the first (7.128) using the formula (see [1.174] of Volume I)

$$\int_{-1}^{t}(1+u)^{\frac{1}{2}}\,P_{l-1}^{(1,\frac{1}{2})}(u)\,du = \frac{(1+t)^{\frac{3}{2}}}{l+\frac{1}{2}}P_{l-1}^{(0,\frac{3}{2})}(t).\qquad(7.141)$$

The relevant integral representations of Abel type for the kernels in these equations are (see formulae [1.171] and [1.172] of Volume I)

$$P_{l-1}^{(1,\frac{1}{2})}(u) = (1-u)^{-1}\frac{\Gamma(l+1)}{\sqrt{\pi}\,\Gamma\left(l+\frac{3}{2}\right)}\int_{u}^{1}\frac{(1-x)^{\frac{1}{2}}\,P_{l-1}^{(\frac{1}{2},1)}(x)}{(x-u)^{\frac{1}{2}}}dx,\qquad(7.142)$$

$$\int_{-1}^{x}\frac{(1+u)^{\frac{1}{2}}\,P_{l-1}^{(1,\frac{1}{2})}(u)}{(x-u)^{\frac{1}{2}}}du = \sqrt{\pi}\,(1+x)\frac{\Gamma\left(l+\frac{1}{2}\right)}{\Gamma(l+1)}P_{l-1}^{(\frac{1}{2},0)}(x).\qquad(7.143)$$

The unknowns are rescaled according to

$$X_{2l} = \frac{\Gamma\left(l+\frac{3}{2}\right)}{\Gamma(l+1)}\left[h_{l-1}^{(\frac{1}{2},1)}\right]^{\frac{1}{2}}x_{2l},\qquad(7.144)$$

so that $\{X_{2l}\}_{l=0}^{\infty} \in l_2$. The asymptotically small parameter corresponding to p_l is

$$q_l = 1 - \left(l + \frac{1}{4}\right) \frac{i}{\left(l + \frac{1}{2}\right)} \left[\frac{\Gamma(l)}{\Gamma\left(l + \frac{3}{2}\right)}\right]^2; \qquad (7.145)$$

$q_l = O\left(l^{-2}\right)$ as $l \to \infty$. The regularised matrix equation takes the form

$$(1 - \varepsilon_{2m}) X_{2m} + \sum_{l=1}^{\infty} X_{2l} H_{2l,2m}\left(c, \xi_0, u_0\right) = D_{2m}\left(c, \xi_1, u_0\right), \qquad (7.146)$$

where $m = 1, 2, \ldots,$

$$D_{2m}\left(c, \xi_1, u_0\right) =$$

$$\sum_{l=1}^{\infty} \beta_{2l} \sum_{r=1}^{\infty} {}'d_{2r-1}^{1,2l}(c) \frac{\Gamma\left(r + \frac{3}{2}\right)}{\Gamma(r+1)} \left[h_{r-1}^{\left(\frac{1}{2},1\right)}\right]^{\frac{1}{2}} \left[\delta_{rm} - \widehat{Q}_{r-1,m-1}^{\left(\frac{1}{2},0\right)}(u_0)\right] \qquad (7.147)$$

and

$$H_{2l,2m}\left(c, \xi_0, u_0\right) = \left(\varepsilon_{2l} - q_l\right) \widehat{Q}_{l-1,m-1}^{\left(\frac{1}{2},1\right)}(u_0) + \left(1 - \varepsilon_{2l}\right) \times$$

$$\sum_{\substack{r=1 \\ r \neq l}}^{\infty} {}'\frac{d_{2r-1}^{1,2l}(c)}{d_{2l-1}^{1,2l}(c)} \frac{\Gamma\left(r + \frac{3}{2}\right)}{\Gamma(r+1)} \frac{\Gamma(l+1)}{\Gamma\left(l + \frac{3}{2}\right)} \left[\frac{h_{r-1}^{\left(\frac{1}{2},1\right)}}{h_{l-1}^{\left(\frac{1}{2},1\right)}}\right]^{\frac{1}{2}} \left[\delta_{rm} - \widehat{Q}_{r-1,m-1}^{\left(\frac{1}{2},1\right)}(u_0)\right]$$

$$+ \left(1 - q_l\right) \sum_{\substack{r=1 \\ r \neq l}}^{\infty} {}'\frac{d_{2r-1}^{1,2l}(c)}{d_{2l-1}^{1,2l}(c)} \frac{\Gamma(r+1)}{\Gamma\left(r - \frac{1}{2}\right)} \frac{\Gamma\left(l - \frac{1}{2}\right)}{\Gamma(l+1)} \left[\frac{h_{r-1}^{\left(\frac{1}{2},1\right)}}{h_{l-1}^{\left(\frac{1}{2},1\right)}}\right]^{\frac{1}{2}} \widehat{Q}_{r-1,m-1}^{\left(\frac{1}{2},1\right)}(u_0).$$

$$(7.148)$$

Equation (7.146) is a Fredholm matrix equation of the second kind, for the even index coefficients $\{X_{2l}\}_{l=1}^{\infty}$.

7.6.3 Numerical Results.

Let H^e, H^o denote the matrices $(H_{2l,2m}), (H_{2l+1,2m+1})$, respectively; let E^e, E^o denote the diagonal matrices with diagonal entries $(\varepsilon_{2l}), (\varepsilon_{2l+1})$, respectively. Then using the asymptotic estimates for ε_l, p_l, and q_l (see [7.60], [7.133] and [7.145], respectively), together with an estimate for the subdominant terms in the expansion (7.5) of the p.a.s.f., it is readily seen that the matrix operators

$$-E^e + H^e \text{ and } -E^o + H^o$$

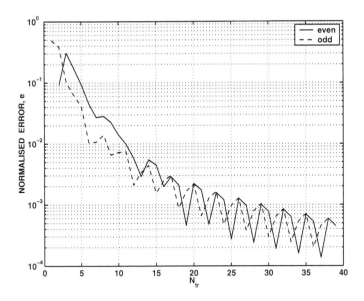

FIGURE 7.12. Normalised error $e(N_{tr})$ as a function of truncation number N_{tr}; $\xi_0 = 1.155$, $a/b = 0.5$, $c = 5$ and $\theta_0 = 30°$.

are square summable and represent compact operators on l_2. The sequences (D_{2m}) and (D_{2m+1}) also lie in l_2. Under such conditions the Fredholm theorems are valid, and it can be asserted that solutions to (7.138) and (7.146) both exist and are unique in l_2.

Due to the Fredholm nature of the matrix operator we can use a truncation method effectively for solving these regularised systems. Numerical solutions are always stable and converge to the exact solution as the truncation order $N_{tr} \to \infty$. The coefficients of the regularised system are accurately and rapidly computable. The desired accuracy of the truncated systems depends only on truncation order N_{tr}, and so it is limited only by the digital precision of the computer used. For a solution accuracy of 3 digits, the number of equations is taken to be $N_{tr} \geqslant c\xi_0 + 20$ independently of u_0 (the parameter determining the angular width of the holes). The accuracy of this estimation is illustrated by the plots of normalised error versus the truncation number, presented in Figure 7.12. The error was computed in the maximum-norm sense,

$$e(N) = \frac{\max_{l \leq N} \left| X_l^{N+1} - X_l^N \right|}{\max_{l \leq N} \left| X_l^N \right|} \tag{7.149}$$

where $\left\{ X_l^N \right\}_{l=1}^N$ denotes the solution to (7.138) and (7.146), truncated to N equations. The Fredholm nature of (7.138) and (7.146) guarantees that

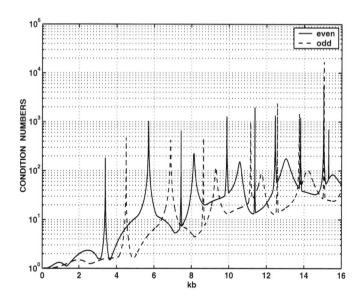

FIGURE 7.13. Condition numbers: both index systems ($a/b = 0.8$, $\theta_0 = 15°$).

$e(N) \to 0$ as $N \to \infty$. The systems are well conditioned, even near values of quasi-eigenfrequencies corresponding to internal cavity resonances as illustrated in Figure 7.13.

Note that the matrix elements and the right-hand sides of (7.138) and (7.146) do not require numerical integration; in particular, the incomplete scalar products $\hat{Q}_{lm}^{(\frac{1}{2},0)}(u_0)$ and $\hat{Q}_{lm}^{(\frac{1}{2},1)}(u_0)$ can be computed from recursion formulae (see Appendix B.6, Volume I). The analytical-numerical algorithm is reliable and efficient in a wide frequency range, from quasi-static to quasi-optics, and has no limitations on the geometrical parameters of the shell (either of angular measure of the holes, or of semi-axis ratio a/b).

Numerical solution of these systems may be used to investigate radiation characteristics of the spheroidal cavity when excited by an axial electric dipole field. The power radiated by the system is

$$P = \sum_{l=1}^{\infty} \left| \frac{(-i)^{l+1}}{\chi_{1l}(c) N_{1l}(c)} R_{1l}^{(1)}(c, \xi_1) - A_l \frac{Z_l^{(1)}(c, \xi_0)}{Z_l^{(3)}(c, \xi_0)} \right|^2 N_{1l}^2(c)$$

and the power radiated by the dipole in free space (simply set all A_l to zero) is

$$P_0 = \sum_{l=1}^{\infty} \left| \frac{R_{1l}^{(1)}(c, \xi_1)}{\chi_{1l}(c)} \right|^2. \tag{7.150}$$

The *radiation resistance* R then equals P/P_0. The far field radiation pattern is

$$
f(\eta) =
$$
$$
\sum_{l=1}^{\infty} (-i)^{l+1} S_{1l}(c,\eta) \left\{ \frac{(-i)^{l+1}}{X_{1l}(c) N_{1l}(c)} R_{1l}^{(1)}(c,\xi_1) - A_l \frac{Z_l^{(1)}(c,\xi_0)}{Z_l^{(3)}(c,\xi_0)} \right\}.
$$
$$
(7.151)
$$

These characteristics depend on the coefficients A_l, which are found after solving matrix equations (7.138), (7.146), taking into account the rescaling of (7.114), (7.125), (7.134) and (7.144).

In common with the single aperture cavity, the spheroidal cavity with two circular openings exhibits novel electromagnetic features that are absent in the closed spheroid: in addition to the external electric resonances of the spheroid, strong internal quasi-eigenoscillations of the cavity appear in the frequency response.

The spectrum of external electrical axisymmetric oscillations of an ideally conducting closed prolate spheroid comprises the roots $c = c_{res}(\xi_0)$ of the equation

$$
\frac{d}{d\xi} \left[\sqrt{\xi^2 - 1} R_{1l}^{(3)}(c,\xi) \right]_{\xi=\xi_0} = 0. \qquad (7.152)
$$

The Q-factor of the oscillation corresponding to a root c_{res} equals $Q = -2\,\mathrm{Re}\,(c_{res})\,/\,\mathrm{Im}\,(c_{res})$. These have been investigated in detail in [21] and [113].

In Figure 7.14 the radiation resistance $R = R(kb)$ as a function of electric size kb of the spheroid is plotted for various closed spheroids (correspond- ing to $u_0 = 1$ in [7.138] and [7.146]) of different aspect ratios; the dipole is located on the z-axis at $z_d = 1.1$. Note that for all numerical results presented in this chapter the major semi-axis b of the spheroid is fixed as $b = 1$. Resonant peaks have low Q-factor, and correspond exactly to external oscillations of electric type studied in [21] and [2].

In Figure 7.15 the radiation resistance $R(kb)$ of closed and open spheroi- dal shells (with holes of angular width $\theta_0 = 15°$ and aspect ratio $a/b = 0.8$) is displayed for the same dipole location ($z_d = 1.1$). The appearance of dou- blets, or double extrema, is obviously connected to the excitation of internal quasi-eigenoscillations of electric type in the spheroidal cavity. When the spheroid is fully closed, the roots $c = c(\xi_0)$ of the following equation define the spectrum of internal eigenoscillations:

$$
\frac{d}{d\xi} \left[\sqrt{\xi^2 - 1} R_{1l}^{(1)}(c,\xi) \right]_{\xi=\xi_0} = 0. \qquad (7.153)
$$

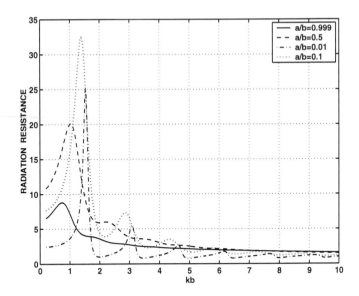

FIGURE 7.14. Radiation resistance for various closed spheroids.

As these roots have real values, the Q-factor of the internal eigenoscillations is infinite, i.e. steady state internal oscillations of the *closed* spheroid can be sustained. However open spheroidal shells always have a finite Q-value. The spectrum of quasi-eigenfrequencies of oscillations of the spheroidal cavity is defined by the complex roots of characteristic equations, obtained by setting the right hand sides of (7.138) and (7.146) to zero. As the aperture closes, the Q-factor increases, as illustrated in Figure 7.16 displaying $R(kb)$ against aperture size; the larger the aperture, the bigger the shift of resonant frequency from the corresponding closed cavity eigenfrequency.

Changing the dipole location highlights the resonant effect due to cavity oscillations. As may be observed in Figure 7.17, excitation at an internal location is much more efficient than at an external location.

The decoupling of the systems (7.138) and (7.146) for odd and even Fourier coefficients, respectively, indicates that odd and even oscillations may be independently excited in the cavity.

The presence of resonances creates opportunities to modify the radiation pattern of a spheroidal antenna in a narrow frequency band. Examples of radiation patterns corresponding to the parameters of a double extremum are presented in Figure 7.18. The shell partially screens radiation in certain spatial directions. By varying the shape of a spheroid, it is possible to maximise or minimise its radiation in a specific direction.

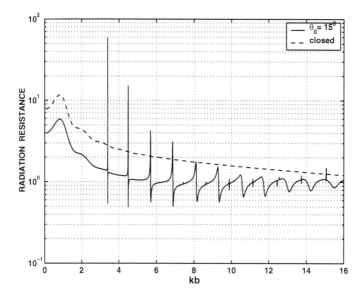

FIGURE 7.15. Radiation resistance of the open spheroid (solid line, $\theta_0 = 15°$) and the closed spheroid (dashed line) ; $a/b = 0.8$.

7.7   Impedance Loading of the Spheroidal Barrel.

Scattering by imperfectly conducting or absorbing cavities is an important subject in electromagnetics of relevance, for example, to electromagnetic signature reduction or to antenna design and performance. There is significant interest in modifying the signature of a metallic surface by coating it with a thin layer of material with appropriate dielectric and magnetic properties. On the other hand, various imperfectly conducting surfaces can serve as radiating antennas provided the feed is located on or near the surface.

In this section the canonical problem of diffraction from a spheroidal barrel with an impedance surface coating, when excited by an axially located electric dipole source (see Figure 7.11), is solved. The solution to the corresponding boundary value problem may be applied to demonstrate the effect of variously loading the cavity, including the possibility of differing internal and external surface impedances.

The shell is excited time harmonically by a vertical electric dipole of moment p, located on the z-axis at, say, $z = \frac{d}{2}\xi_1$ (d is the interfocal distance of the spheroid, see Figure 7.11). The formulation of the corresponding mixed boundary value problem is the same as for a perfectly conducting spheroidal barrel except that the boundary condition imposed on the

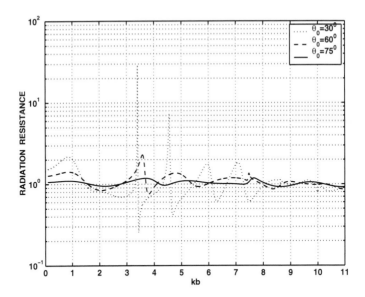

FIGURE 7.16. The dependence of radiation resistance upon aperture size $(a/b = 0.8, z_d = 0.9b)$.

perfectly conducting surface (7.109)–(7.110) is replaced by an impedance boundary condition. The interior and exterior of the shell are characterised by normalised surface impedances $Z^{(i)}$ and $Z^{(e)}$, respectively. The Leontovich boundary condition for the tangential field components $\mathbf{E}_{tan}, \mathbf{H}_{tan}$ on the face of an open impenetrable surface with outward normal $\mathbf{n}$ and normalised impedance Z has form

$$\mathbf{E}_{tan} - Z\mathbf{n} \times \mathbf{H}_{tan} = 0. \tag{7.154}$$

This implies that

$$E_\eta \left(\xi_0 + 0, \eta\right) + E_\eta \left(\xi_0 - 0, \eta\right) =$$
$$Z^{(i)} H_\phi \left(\xi_0 - 0, \eta\right) - Z^{(e)} H_\phi \left(\xi_0 + 0, \eta\right). \tag{7.155}$$

Imposition of the impedance boundary condition (7.155) on a series representation of the incident (7.44) and scattered (7.50) fields leads to the following *triple series equations* involving prolate angle spheroidal functions, with unknown coefficients A_l,

$$\sum_{l=1}^{\infty} \frac{A_l}{Z_l^{(3)}(c, \xi_0)} S_{1l}(c, \eta) = 0, \qquad \eta \in (-1, -\eta_0) \cup (\eta_0, 1) \tag{7.156}$$

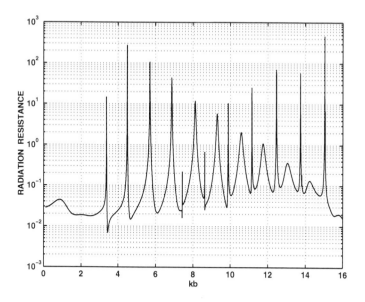

FIGURE 7.17. Radiation resistance for a dipole located at $z_d = 0.7b$ ($a/b = 0.8$, $\theta_0 = 15°$).

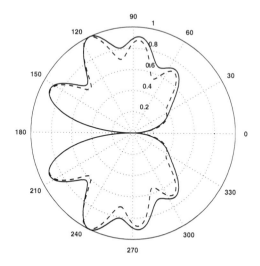

FIGURE 7.18. Radiation patterns for the open spheroid ($a/b = 0.8$ and $\theta_0 = 15°$) at wavenumbers $kb = 4.495$ (solid line) and $kb = 4.5$ (dashed line).

$$\sum_{l=1}^{\infty} A_l Z_l^{(1)}(c,\xi_0) S_{1l}(c,\eta) = \sum_{l=1}^{\infty} \alpha_l S_{1l}(c,\eta) + \sqrt{\xi_0^2 - \eta^2} \sum_{l=1}^{\infty} A_l \gamma_l S_{1l}(c,\eta),$$

$$\eta \in (-\eta_0, \eta_0) \quad (7.157)$$

where coefficients $\{\alpha_l\}_{l=1}^{\infty}$ are defined by (7.113) and

$$\gamma_l = \frac{ic}{2} \left\{ Z^{(i)} R_{1l}^{(1)}(c,\xi_0) - Z^{(e)} \frac{Z_l^{(1)}(c,\xi_0)}{Z_l^{(3)}(c,\xi_0)} R_{1l}^{(3)}(c,\xi_0) \right\}. \quad (7.158)$$

Introduction of the asymptotically small parameter ε_l given by (7.60), with the rescaling (7.114) of the unknowns, splits the equations (7.156)–(7.157) into singular and regular parts, in the form

$$\sum_{l=1}^{\infty} \frac{2l+1}{l(l+1)} B_l S_{1l}(c,\eta) = 0, \quad \eta \in (-1, -\eta_0) \cup (\eta_0, 1) \quad (7.159)$$

$$\sum_{l=1}^{\infty} B_l (1 - \varepsilon_l) S_{1l}(c,\eta) = -ic (\xi_0^2 - 1)^{\frac{1}{2}} \sum_{l=1}^{\infty} \alpha_l S_{1l}(c,\eta)$$

$$+ (\xi_0^2 - \eta^2)^{\frac{1}{2}} \sum_{l=1}^{\infty} B_l \Gamma_l S_{1l}(c,\eta), \quad \eta \in (-\eta_0, \eta_0) \quad (7.160)$$

where

$$\Gamma_l = \frac{c^2}{2} (\xi_0^2 - 1)^{\frac{1}{2}} \frac{2l+1}{l(l+1)} \times$$
$$\left\{ Z^{(i)} R_{1l}^{(1)}(c,\xi_0) Z_l^{(3)}(c,\xi_0) - Z^{(e)} R_{1l}^{(3)}(c,\xi_0) Z_l^{(1)}(c,\xi_0) \right\}. \quad (7.161)$$

So, covering the perfectly conducting surface by a thin impedance layer results in the appearance of an additional term in the right hand side of the second equation (7.160). The solution scheme for these equations is exactly the same as for the equations (7.116). However the asymmetric factor $(\xi_0^2 - \eta^2)^{\frac{1}{2}}$ introduces some complication in the application of the Abel integral transform technique, making it necessary to evaluate the integral

$$\int_{-1}^{u_0} du (1-u)^{\frac{1}{2}} P_m^{(\frac{1}{2},0)}(u) \int_{-1}^{u} \frac{(t_0 - x)^{\frac{1}{2}}}{(1+x)^{\frac{1}{2}} (u-x)^{\frac{1}{2}}} P_r^{(1,-\frac{1}{2})}(x) dx \quad (7.162)$$

where $t_0 = 2\xi_0^2 - 1$. This integral arises because when we integrate the equation corresponding to (7.126) we arrive at the analogue of (7.135) and

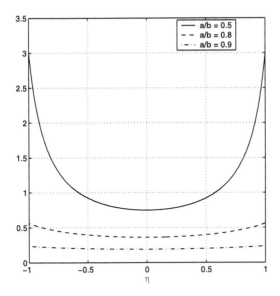

FIGURE 7.19. The function $f(\eta) = (\xi_0^2 - \eta^2)^{-1}$ for fixed ξ_0. Parameter values $(a/b, \xi_0)$ equal to $(0.5, 1.15)$, $(0.8, 1.67)$ and $(0.9, 2.29)$ are shown.

the final step of regularisation requires multiplication by $(1 - u)^{\frac{1}{2}} P_m^{(\frac{1}{2},0)}(u)$ followed by integration over $(-1, u_0)$. The integral (7.162) does not seem to have a closed form evaluation. However if we assume that surface impedance function has the following form,

$$Z^{(i,e)}(\xi_0, \eta) = \left(\frac{\xi_0^2 - \eta_0^2}{\xi_0^2 - \eta^2} \right) Z_0^{(i,e)} \qquad (7.163)$$

where $Z_0^{(i,e)}$ are constants, the modified form of the integral (7.162) can be easily evaluated analytically. The solution of the problem then reduces to the same equations as for a perfectly conducting screen (see [7.138] and [7.148]), except that the asymptotically small parameter ε_l (7.60) is replaced by

$$\hat{\varepsilon}_l = \varepsilon_l + \Gamma_l. \qquad (7.164)$$

The assumption (7.163) is a good approximation for *constant* surface impedance coatings on relatively thick spheroids when $\xi_0 \gg 1$; the dependence of $Z^{(i,e)}(\xi_0, \eta)$ on η is not significant and the impedance is distributed almost uniformly across the screen surface $Z^{(i,e)}(\xi_0, \eta) \sim Z_0^{(i,e)}$. For various fixed values of ξ_0 the impedance distribution $Z(\xi_0, \eta)/Z_0 = (\xi_0^2 - \eta^2)^{-1}$ is examined as a function of the surface screen position η in Figure 7.19.

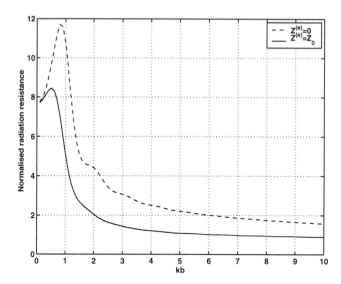

FIGURE 7.20. Normalised radiation resistance for the perfectly conducting $(Z_0^{(e)} = 0)$ and impedance loaded $(Z_0^{(e)} = Z_0)$ spheroid $(a/b = 0.8)$.

For elongated spheroids $(\xi_0 \sim 1)$ the formula (7.163) provides a highly nonuniform impedance, so if *constant* surface impedance coatings are to be investigated, the integral (7.162) be evaluated numerically. The final i.s.l.a.e. has a similar form to that obtained above, but the coefficients are considerably more complicated.

Let us illustrate the effect of variously loading the cavity with differing interior and exterior surface impedances with some numerical examples, using this simplifying assumption (7.163) on surface impedance. The truncation number is typically chosen to be $N_{tr} = kb + 12$ and gives results of acceptable engineering accuracy. The radiation resistance R may be computed as in the previous section (see Equation [7.150]).We consider a spheroid with aspect ratio of the semi-minor axis a to semi-major axis b of 0.8 and two circular apertures defined by $\theta_0 = 30°$ (so that $\eta_0 = \cos(\theta_0) = \sqrt{3}/2$).

As a preliminary, consider the *closed* spheroid. Figure 7.20 demonstrates the impact of loading the perfectly conducting surface (according to [7.163]) with $Z_0^{(e)} = Z_0 = 120\pi$. The dipole is located on the z-axis at $z = 1.1b$ (b is scaled to 1 and so $\xi_1 = 1.1\xi_0, \frac{d}{2} = \xi_0^{-1}$). The radiation resistance (normalised against that of the bare dipole) diminishes by a factor of about 2 for all but the smallest values of kb.

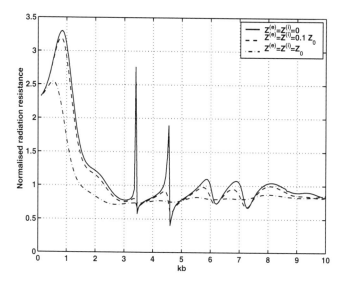

FIGURE 7.21. Radiation resistance of the shell, perfectly conducting or loaded $(a/b = 0.8, \theta_0 = 30°, z_d = 1.1b)$.

As explained in Section 7.6, the perfectly conducting cavity exhibits high Q-value resonances at values of kb corresponding approximately to the internal cavity oscillation frequencies of the *closed* spheroid. This is shown in Figure 7.21. The result of two loadings (equal interior and exterior impedances of the cavity) is also considered, equalling $0.1Z_0$ and Z_0. The main effect of the lighter loading is to reduce substantially the Q factor associated with each resonance; the heavier loading has the additional effect of producing a nearly constant response for $kb > 2$.

To illustrate the effect of differing interior and exterior loadings, the cavity was loaded only on one side with surface impedance Z_0. The results are shown in Figure 7.22. Loading only the interior in this way has a much stronger impact than if only an exterior load is applied. In the first case the high Q resonances are completely suppressed, and the response is similar to loading both interior and exterior with the same impedance of Z_0 (see Figure 7.20). In the second case there is some impact of the loading (compared to its absence), but the response is more nearly like that of the perfectly conducting shell.

Another illustration of differing interior and exterior loadings is given in Figure 7.23. The interior is unloaded, but the exterior has (normalised) impedance $Z_0^{(e)}/Z_0 = 0.3425 - i0.157$, corresponding to the choice of ma-

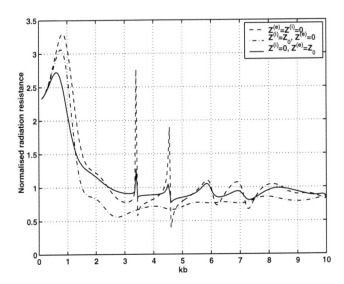

FIGURE 7.22. Radiation resistance of the shell, perfectly conducting or loaded ($a/b = 0.8$, $\theta_0 = 30°$, $z_d = 1.1b$).

terials and thickness of the coating described in [11]. This loading has a relatively small impact on the radiation resistance.

Perhaps most importantly, this provides a benchmark of guaranteed accuracy against which calculations, obtained by general purpose numerical codes for the simulation of scattering from coated cavities, can be validated.

7.8   Spheroid Embedded in a Spheroidal Cavity with Two Holes.

In this section we consider the spheroidal cavity enclosing another scattering object. Such a structure is intrinsically more complex than the scatterers studied in previous sections. This canonical problem exhibits several competing scattering mechanisms contributing to the overall field, arising from edge and aperture effects, the cavity region and the embedded scatterer. In the spirit of this chapter, we develop a rigorous solution of the corresponding mixed boundary-value problem for a certain second order partial differential equation derived from Maxwell's equations.

As an application we analyse the radiation characteristics of a spheroidal antenna, modelled as a perfectly conducting spheroid excited by a voltage

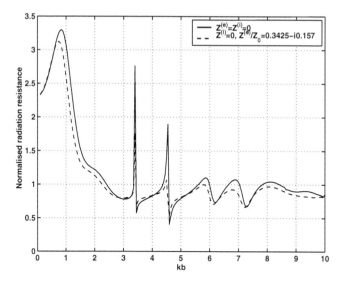

FIGURE 7.23. Radiation resistance of the shell, perfectly conducting or loaded ($a/b = 0.8$, $\theta_0 = 30°$, $z_d = 1.1b$).

applied across a narrow circumferential slot, in the presence of a metallic spheroidal screen having two circular holes (see Figure 7.24). The influence of the screen on radiation features of this class of spheroidal antennas can then be assessed. This antenna model is multiparametric: several features may be varied, including feed location, shape (spherical to practically cylindrical depending on aspect ratio), size and position of the outer screen. Radiation resistance and far-field radiation patterns are investigated, and attention is focused on resonance properties over the frequency band. As well as the intrinsic interest of this type of antenna, this rigorous solution provides a valuable benchmark for validation of more general purpose numerical codes.

The geometry of the problem is shown in Figure 7.24. In prolate spheroidal coordinates (ξ, η, ϕ), the enclosed spheroid is defined by $\xi = \xi_0$ with the feed slot located at $\eta = \eta_0$; the outer screen occupies that part of the spheroidal surface $\xi = \xi_1$ specified by the angular coordinate $\eta \in (-\eta_1, \eta_1)$.

The excitation field is assumed to be that due to an infinitesimally narrow slot across which is impressed a constant time harmonic voltage; it may be described [21] by

$$E_\eta = \Sigma \frac{1}{h_\eta} \delta (\eta - \eta_0) \tag{7.165}$$

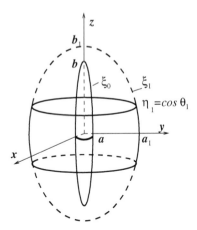

FIGURE 7.24. The spheroid partially enclosed in the larger spheroidal cavity.

where Σ is the electromotive force or voltage impressed on the slot, $h_\eta = (d/2)\left(\xi^2 - \eta^2\right)^{\frac{1}{2}}\left(1 - \eta^2\right)^{-\frac{1}{2}}$ is the appropriate Lamé coefficient in spheroidal coordinates. This axisymmetric excitation field is of electric type and, because the scatterer is axially symmetric, the scattered field is also axially symmetric and of electric type ($E_\xi \neq 0, H_\xi = 0$). It is governed by the system of equations (1.99) for the field components (E_ξ, E_η, H_ϕ) derived from Maxwell's equations (1.58)–(1.59). The field component H_ϕ satisfies second order partial differential equation (1.100) and the components E_ξ, E_η are obtained from H_ϕ. We obtain the component $H_\phi\left(\xi, \eta\right)$ as an expansion in spheroidal eigenfunctions.

Subdivide the region exterior to the closed metallic spheroid into two, regions 1 and 2 being defined by $\xi_0 \leq \xi \leq \xi_1$ and $\xi > \xi_1$, respectively. In the first region ($\xi_0 \leq \xi \leq \xi_1$) we seek the total field $H_\phi^{(1)}$ radiated by the system in the form

$$H_\phi^{(1)} = \sum_{l=1}^{\infty}\left\{A_l R_{1l}^{(1)}(c, \xi) + B_l R_{1l}^{(3)}(c, \xi)\right\} S_{1l}(c, \eta), \qquad (7.166)$$

where A_l and B_l are constants to be determined. In the second region ($\xi > \xi_1$) the radiation condition on the decay of the scattered field at infinity dictates that the solution form

$$H_\phi^{(2)} = \sum_{l=1}^{\infty} C_l R_{1l}^{(3)}(c, \xi) S_{1l}(c, \eta), \qquad (7.167)$$

where the constants are to be determined. On the outer screen $\xi = \xi_1$ the tangential electric component of the field is continuous

$$E_\eta^{(1)}(\xi_1 - 0, \eta) = E_\eta^{(2)}(\xi_1 + 0, \eta),\qquad (7.168)$$

for all $\eta \in (-1, 1)$. On the inner spheroid the tangential electric component of the field vanishes except at the feed gap,

$$E_\eta^{(1)}(\xi_0, \eta) = \Sigma \frac{1}{h_\eta} \delta(\eta - \eta_0),\qquad (7.169)$$

for all $\eta \in (-1, 1)$. The field must satisfy the following mixed boundary conditions derived from the continuity of the magnetic field component across the aperture and the vanishing of the tangential electric field on the outer screen,

$$H_\phi^{(1)}(\xi_1 - 0, \eta) = H_\phi^{(2)}(\xi_1 + 0, \eta), \qquad \eta \in (-1, -\eta_1) \cup (\eta_1, 1),\qquad (7.170)$$

$$E_\eta^{(1)}(\xi_1 - 0, \eta) = E_\eta^{(2)}(\xi_1 + 0, \eta) = 0, \qquad \eta \in (-\eta_1, \eta_1).\qquad (7.171)$$

Suitable functional equations to be solved follow from enforcing the conditions (7.168)–(7.171) on the expansions (7.166)–(7.167). The first step is to express the unknown coefficients A_l and B_l in terms of the unknown coefficients C_l. Using the relation (1.99),

$$E_\eta^{(1)} = \frac{1}{ic(\xi^2 - \eta^2)^{\frac{1}{2}}} \sum_{l=1}^{\infty} \left\{ A_l Z_l^{(1)}(c, \xi) + B_l Z_l^{(3)}(c, \xi) \right\} S_{1l}(c, \eta),\qquad (7.172)$$

and

$$E_\eta^{(2)} = \frac{1}{ic(\xi^2 - \eta^2)^{\frac{1}{2}}} \sum_{l=1}^{\infty} C_l Z_l^{(3)}(c, \xi) S_{1l}(c, \eta)\qquad (7.173)$$

where $Z_l^{(i)}(c, \xi) = \frac{\partial}{\partial \xi}\left[(\xi^2 - 1)^{\frac{1}{2}} R_{1l}^{(i)}(c, \xi) \right]$ $(i = 1, 3)$. Thus enforcing the boundary condition (7.169) on (7.172) produces the equation

$$\sum_{l=1}^{\infty} \left\{ A_l Z_l^{(1)}(c, \xi_0) + B_l Z_l^{(3)}(c, \xi_0) \right\} S_{1l}(c, \eta) =$$

$$ik\Sigma (1 - \eta^2)^{\frac{1}{2}} \delta(\eta - \eta_0), \qquad \eta \in (-1, 1).\qquad (7.174)$$

The completeness and orthogonality of the angle spheroidal functions (see Section 1.2) over $(-1, 1)$ implies that

$$A_l Z_l^{(1)}(c, \xi_0) + B_l Z_l^{(3)}(c, \xi_0) = \frac{ik\Sigma}{N_{1l}^2(c)} (1 - \eta_0^2)^{\frac{1}{2}} S_{1l}(c, \eta_0).\qquad (7.175)$$

From the conditions (7.168), (7.170) and (7.171) it follows that

$$A_l Z_l^{(1)}(c,\xi_1) + B_l Z_l^{(3)}(c,\xi_1) = C_l Z_l^{(3)}(c,\xi_1), \qquad (7.176)$$

and

$$\sum_{l=1}^{\infty} \left\{ A_l R_{1l}^{(1)}(c,\xi_1) + B_l R_{1l}^{(3)}(c,\xi_1) - C_l R_{1l}^{(3)}(c,\xi_1) \right\} S_{1l}(c,\eta) = 0,$$

$$\eta \in (-1,-\eta_1) \cup (\eta_1,1), \quad (7.177)$$

$$\sum_{l=1}^{\infty} \left\{ A_l Z_l^{(1)}(c,\xi_1) + B_l Z_l^{(3)}(c,\xi_1) \right\} S_{1l}(c,\eta) = \sum_{l=1}^{\infty} C_l Z_l^{(3)}(c,\xi_1) S_{1l}(c,\eta),$$

$$\eta \in (-\eta_1,\eta_1). \quad (7.178)$$

Thus, from (7.175) and (7.176) we deduce

$$A_l = \frac{Z_l^{(3)}(c,\xi_1)}{\Delta} \left\{ \frac{-ik\Sigma}{N_{1l}^2(c)} \left(1 - \eta_0^2\right)^{\frac{1}{2}} S_{1l}(c,\eta_0) - C_l Z_l^{(3)}(c,\xi_0) \right\} \qquad (7.179)$$

and

$$B_l = \frac{1}{\Delta} \left\{ C_l Z_l^{(3)}(c,\xi_1) Z_l^{(1)}(c,\xi_0) + \frac{ik\Sigma}{N_{1l}^2(c)} \left(1 - \eta_0^2\right)^{\frac{1}{2}} S_{1l}(c,\eta_0) Z_l^{(1)}(c,\xi_1) \right\}$$

$$(7.180)$$

where

$$\Delta = Z_l^{(3)}(c,\xi_1) Z_l^{(1)}(c,\xi_0) - Z_l^{(3)}(c,\xi_0) Z_l^{(1)}(c,\xi_1). \qquad (7.181)$$

Insertion of these expressions in equations (7.177)–(7.178) produces the following functional equations for the unknown coefficients C_l,

$$\sum_{l=1}^{\infty} C_l \frac{Z_l^{(3)}(c,\xi_0)}{\Delta} S_{1l}(c,\eta) = -ik\Sigma \left(1 - \eta_0^2\right)^{\frac{1}{2}} \sum_{l=1}^{\infty} \frac{S_{1l}(c,\eta_0)}{N_{1l}^2(c)\Delta} S_{1l}(c,\eta),$$

$$\eta \in (-1,-\eta_1) \cup (\eta_1,1) \quad (7.182)$$

$$\sum_{l=1}^{\infty} C_l Z_l^{(3)}(c,\xi_1) S_{1l}(c,\eta) = 0, \eta \in (-\eta_1,\eta_1). \qquad (7.183)$$

Rescaling the unknowns via

$$C_l = \frac{\Delta k\Sigma}{Z_l^{(3)}(c,\xi_0)} \frac{2l+1}{l(l+1)} b_l \qquad (7.184)$$

transforms the equations to the form

$$\sum_{l=1}^{\infty} \frac{2l+1}{l(l+1)} b_l S_{1l}(c,\eta) = \sum_{l=1}^{\infty} \gamma_l S_{1l}(c,\eta),$$

$$\eta \in (-1,-\eta_1) \cup (\eta_1,1) \quad (7.185)$$

$$\sum_{l=1}^{\infty} \frac{(2l+1) b_l}{l(l+1)} \left\{ Z_l^{(1)}(c,\xi_1) Z_l^{(3)}(c,\xi_1) - \frac{Z_l^{(1)}(c,\xi_0)}{Z_l^{(3)}(c,\xi_0)} \left[Z_l^{(3)}(c,\xi_1) \right]^2 \right\} \times$$

$$S_{1l}(c,\eta) = 0, \quad \eta \in (-\eta_1,\eta_1) \quad (7.186)$$

where

$$\gamma_l = i \left(1-\eta_0^2\right)^{\frac{1}{2}} \frac{S_{1l}(c,\eta_0)}{N_{1l}^2(c)\,\Delta}. \quad (7.187)$$

Thus, we have obtained triple series equations with spheroidal angle function kernels for the unknowns $\{b_l\}_{l=1}^{\infty}$. Once found, the magnetic component of scattered field may be calculated from the formulae (7.166) and (7.167), taking into account the relationships (7.179), (7.180), (7.184), and (7.187). The other field components (E_ξ, E_η) are then found from formulae (1.97)–(1.98).

We now aim to perform an analytical regularisation of the first kind series equations (7.185)–(7.186) to obtain a second kind infinite system of linear algebraic equations. We begin by introducing the parameter

$$\varepsilon_l = 1 + ic \left(\xi_1^2-1\right)^{\frac{1}{2}} \frac{2l+1}{l(l+1)} Z_l^{(1)}(c,\xi_1) Z_l^{(3)}(c,\xi_1)$$

$$- ic \left(\xi_1^2-1\right)^{\frac{1}{2}} \frac{2l+1}{l(l+1)} \frac{Z_l^{(1)}(c,\xi_0)}{Z_l^{(3)}(c,\xi_0)} \left[Z_l^{(3)}(c,\xi_1) \right]^2 \quad (7.188)$$

that is asymptotically small (as $l \to \infty$). It can be expressed as a sum of two terms,

$$\varepsilon_l^{(1)} = 1 + ic \left(\xi_1^2-1\right)^{\frac{1}{2}} \frac{2l+1}{l(l+1)} Z_l^{(1)}(c,\xi_1) Z_l^{(3)}(c,\xi_1),$$

and

$$\varepsilon_l^{(2)} = -ic \left(\xi_1^2-1\right)^{\frac{1}{2}} \frac{2l+1}{l(l+1)} \frac{Z_l^{(1)}(c,\xi_0)}{Z_l^{(3)}(c,\xi_0)} \left[Z_l^{(3)}(c,\xi_1) \right]^2.$$

The first term $(\varepsilon_l^{(1)})$ arose as an asymptotically small parameter in our analysis of diffraction from the empty spheroidal cavity (see Sections 7.4

or 7.6); the second term $\left(\varepsilon_l^{(2)}\right)$ reflects the interaction between the outer spheroidal shell and the inner closed spheroid. Their asymptotic behaviour may be deduced from the asymptotic behaviour of the radial spheroidal functions (see Section 1.2). Thus

$$\varepsilon_l^{(1)} = O\left(l^{-2}\right), \quad \varepsilon_l^{(2)} = O\left((\xi_0/\xi_1)^{2l+1}\right)$$

as $l \to \infty$ (note that $\xi_0 < \xi_1$).

The equations (7.185)–(7.186) are similar to those (see [7.116]) derived in Section 7.6 in the analysis of a spheroidal cavity excited by an electric dipole located on the z-axis. Here we consider diffraction from the same cavity but the excitation mechanism is provided by the inner closed spheroid with a voltage impressed across the slot. Thus it is natural that these problems lead to similar systems of equations, the only difference being the right-hand sides that depend on the excitation field. The regularisation process for the system of equations (7.185)–(7.186) is exactly the same as that for the equations (7.116); we follow the argument of Section 7.6 closely, indicating only the main stages.

Due to the symmetry of the p.a.s.f. about $\eta = 0$ the triple system may be reduced to two decoupled dual series equations; the odd index coefficients satisfy

$$\sum_{l=0}^{\infty} \frac{l + \frac{3}{4}}{\left(l + \frac{1}{2}\right)(l+1)} b_{2l+1} S_{1,2l+1}(c, \eta) = \sum_{l=0}^{\infty} \gamma_{2l+1} S_{1,2l+1}(c, \eta),$$

$$\eta \in (-1, -\eta_1) \quad (7.189)$$

$$\sum_{l=0}^{\infty} b_{2l+1} \left(1 - \varepsilon_{2l+1}\right) S_{1,2l+1}(c, \eta) = 0, \quad \eta \in (-\eta_1, 0), \quad (7.190)$$

and even index coefficients satisfy

$$\sum_{l=1}^{\infty} \frac{l + \frac{1}{4}}{l\left(l + \frac{1}{2}\right)} b_{2l} S_{1,2l}(c, \eta) = \sum_{l=1}^{\infty} \gamma_{2l} S_{1,2l}(c, \eta), \quad \eta \in (-1, -\eta_1) \quad (7.191)$$

$$\sum_{l=1}^{\infty} b_{2l} \left(1 - \varepsilon_{2l}\right) S_{1,2l}(c, \eta) = 0, \quad \eta \in (-\eta_1, 0). \quad (7.192)$$

These dual series are defined over half the full interval of the variable η, namely $(-1, 0)$. The next stage is to transform these equations with spheroidal angle function kernels to equivalent equations with associated Legendre function kernels, by inserting representation (7.4). Then convert each of the dual series equations to equations with Jacobi polynomials exactly as in Section 7.6. With the rescaling of the unknowns

$$x_{2l+1} = d_{2l}^{1,2l+1}(c) b_{2l+1}, \qquad x_{2l} = d_{2l-1}^{1,2l}(c) b_{2l} \qquad (7.193)$$

and setting

$$u = 2\eta^2 - 1, \qquad u_1 = 2\eta_1^2 - 1 \qquad (7.194)$$

we obtain the following dual series equations

$$\sum_{l=0}^{\infty} \left(l + \frac{1}{2}\right) x_{2l+1} \left(1 - \varepsilon_{2l+1}\right) \times$$

$$\left[P_l^{(1,-\frac{1}{2})}(u) + \sum_{\substack{r=0 \\ r \neq l}}^{\infty} {}' \frac{d_{2r}^{1,2l+1}(c)}{d_{2l}^{1,2l+1}(c)} \frac{r + \frac{1}{2}}{l + \frac{1}{2}} P_r^{(1,-\frac{1}{2})}(u) \right] = 0,$$

$$u \in (-1, u_1) \qquad (7.195)$$

$$\sum_{l=0}^{\infty} \frac{\left(l + \frac{3}{4}\right)}{(l+1)} x_{2l+1} \left[P_l^{(1,-\frac{1}{2})}(u) + \sum_{\substack{r=0 \\ r \neq l}}^{\infty} {}' \frac{d_{2r}^{1,2l+1}(c)}{d_{2l}^{1,2l+1}(c)} \frac{r + \frac{1}{2}}{l + \frac{1}{2}} P_r^{(1,-\frac{1}{2})}(u) \right]$$

$$= \sum_{l=0}^{\infty} \gamma_{2l+1} \sum_{r=0}^{\infty} {}' \left(r + \frac{1}{2}\right) d_{2r}^{1,2l+1}(c) P_r^{(1,-\frac{1}{2})}(u),$$

$$u \in (u_1, 1) \qquad (7.196)$$

for the odd index coefficients, and the system

$$\sum_{l=1}^{\infty} \left(l + \frac{1}{2}\right) x_{2l} \left(1 - \varepsilon_{2l}\right) \times$$

$$\left[P_{l-1}^{(1,\frac{1}{2})}(u) + \sum_{\substack{r=1 \\ r \neq l}}^{\infty} {}' \frac{d_{2r-1}^{1,2l}(c)}{d_{2l-1}^{1,2l}(c)} \frac{r + \frac{1}{2}}{l + \frac{1}{2}} P_{r-1}^{(1,\frac{1}{2})}(u) \right] = 0,$$

$$u \in (-1, u_1) \qquad (7.197)$$

$$\sum_{l=1}^{\infty} \frac{\left(l+\frac{1}{4}\right)}{l} x_{2l} \left[P_{l-1}^{\left(1,\frac{1}{2}\right)}(u) + \sum_{\substack{r=1 \\ r \neq l}}^{\infty} {}'\frac{d_{2r-1}^{1,2l}(c)}{d_{2l-1}^{1,2l}(c)} \frac{r+\frac{1}{2}}{l+\frac{1}{2}} P_{r-1}^{\left(1,\frac{1}{2}\right)}(u) \right]$$

$$= \sum_{l=1}^{\infty} \gamma_{2l} \sum_{r=1}^{\infty} {}' \left(r + \frac{1}{2} \right) d_{2r-1}^{1,2l}(c) P_{r-1}^{\left(1,\frac{1}{2}\right)}(u) ,$$

$$u \in (u_1, 1) \quad (7.198)$$

for the even index coefficients.

These transformed functional equations are dual series equations with Jacobi polynomial kernels, defined on $(-1, 1)$. We employ the Abel integral transform as described in Section 7.6 (see the deduction of the matrix equations [7.138] and [7.148] from the systems [7.127] and [7.129]). Note that in this section the angular extent of each aperture is η_1, whilst in Section 7.6 it equalled η_0. Omitting its deduction, the system (7.196) is transformed to the following Fredholm matrix equation (of second kind) for the odd index coefficients,

$$(1 - \varepsilon_{2m+1}) X_{2m+1} + \sum_{l=0}^{\infty} X_{2l+1} H_{2l+1, 2m+1} = D_{2m+1}, \quad (7.199)$$

where $m = 0, 1, 2, \ldots$,

$$X_{2l+1} = \frac{\Gamma\left(l + \frac{3}{2}\right)}{\Gamma(l+1)} \left\{ h_l^{\left(\frac{1}{2}, 0\right)} \right\}^{\frac{1}{2}} x_{2l+1},$$

the parameter

$$p_l = 1 - \left(l + \frac{3}{4} \right) \left[\frac{\Gamma(l+1)}{\Gamma\left(l+\frac{3}{2}\right)} \right]^2$$

is asymptotically small ($p_l = O\left(l^{-2}\right)$, as $l \to \infty$),

$$D_{2m+1} = \sum_{l=0}^{\infty} \gamma_{2l+1} \sum_{r=0}^{\infty} {}' d_{2r}^{1,2l+1}(c) \frac{\Gamma(r+2)}{\Gamma\left(r+\frac{1}{2}\right)} \left[h_r^{\left(\frac{1}{2}, 0\right)} \right]^{\frac{1}{2}} \widehat{Q}_{rm}^{\left(\frac{1}{2}, 0\right)}(u_1),$$

$$(7.200)$$

and

$$
H_{2l+1,2m+1} = \left(\varepsilon_{2l+1} - p_l\right) \widehat{Q}_{lm}^{\left(\frac{1}{2},0\right)}\left(u_1\right) + \left(1 - \varepsilon_{2l+1}\right) \times
$$

$$
\sum_{\substack{r=0 \\ r \neq l}}^{\infty} {}' \frac{d_{2r}^{1,2l+1}\left(c\right)}{d_{2l}^{1,2l+1}\left(c\right)} \frac{\Gamma\left(r + \frac{3}{2}\right) \Gamma\left(l + 1\right)}{\Gamma\left(r + 1\right) \Gamma\left(l + \frac{3}{2}\right)} \left[\frac{h_r^{\left(\frac{1}{2},0\right)}}{h_l^{\left(\frac{1}{2},0\right)}}\right]^{\frac{1}{2}} \left[\delta_{rm} - \widehat{Q}_{rm}^{\left(\frac{1}{2},0\right)}\left(u_1\right)\right]
$$

$$
+ \left(1 - p_l\right) \sum_{\substack{r=0 \\ r \neq l}}^{\infty} {}' \frac{d_{2r}^{1,2l+1}\left(c\right)}{d_{2l}^{1,2l+1}\left(c\right)} \frac{\Gamma\left(r + 2\right) \Gamma\left(l + \frac{1}{2}\right)}{\Gamma\left(r + \frac{1}{2}\right) \Gamma\left(l + 2\right)} \left[\frac{h_r^{\left(\frac{1}{2},0\right)}}{h_l^{\left(\frac{1}{2},0\right)}}\right]^{\frac{1}{2}} \widehat{Q}_{rm}^{\left(\frac{1}{2},0\right)}\left(u_1\right).
$$

$$(7.201)$$

The quantity

$$
\widehat{Q}_{lm}^{\left(\frac{1}{2},0\right)}\left(u_1\right) = \int_{u_1}^{1} \left(1 - x\right)^{\frac{1}{2}} \widehat{P}_l^{\left(\frac{1}{2},0\right)}\left(u\right) \widehat{P}_m^{\left(\frac{1}{2},0\right)}\left(u\right) du
$$

is the usual incomplete scalar product (see Appendix B.6, Volume I).
In a parallel fashion, the system (7.198) is transformed to the following Fredholm matrix equation of the second kind for the even index coefficients,

$$
\left(1 - \varepsilon_{2m}\right) X_{2m} + \sum_{l=1}^{\infty} X_{2l} H_{2l,2m} = D_{2m}, \qquad (7.202)
$$

where $m = 1, 2, \dots$,

$$
X_{2l} = \frac{\Gamma\left(l + \frac{3}{2}\right)}{\Gamma\left(l + 1\right)} \left\{h_{l-1}^{\left(\frac{1}{2},1\right)}\right\}^{\frac{1}{2}} x_{2l},
$$

the parameter

$$
q_l = 1 - \frac{l\left(l + \frac{1}{4}\right)}{l + \frac{1}{2}} \left[\frac{\Gamma\left(l\right)}{\Gamma\left(l + \frac{1}{2}\right)}\right]^2
$$

is asymptotically small ($q_l = O\left(l^{-2}\right)$, as $l \to \infty$),

$$
D_{2m} = \sum_{l=1}^{\infty} \gamma_{2l} \sum_{r=1}^{\infty} {}' d_{2r-1}^{1,2l}\left(c\right) \frac{\Gamma\left(r + 1\right)}{\Gamma\left(r - \frac{1}{2}\right)} \left[h_{r-1}^{\left(\frac{1}{2},1\right)}\right]^{\frac{1}{2}} \widehat{Q}_{r-1,m-1}^{\left(\frac{1}{2},1\right)}\left(u_1\right), \quad (7.203)
$$

and

$$
H_{2l,2m} = (\varepsilon_{2l} - q_l)\, \widehat{Q}^{(\frac{1}{2},1)}_{l-1,m-1}(u_1) + (1 - \varepsilon_{2l}) \times
$$

$$
\sum_{\substack{r=1 \\ r \neq l}}^{\infty} {}' \frac{d^{1,2l}_{2r-1}(c)}{d^{1,2l}_{2l-1}(c)} \frac{\Gamma\left(r + \frac{3}{2}\right)}{\Gamma\left(r + 1\right)} \frac{\Gamma\left(l + 1\right)}{\Gamma\left(l + \frac{3}{2}\right)} \left[\frac{h^{(\frac{1}{2},1)}_{r-1}}{h^{(\frac{1}{2},1)}_{l-1}}\right]^{\frac{1}{2}} \left[\delta_{rm} - \widehat{Q}^{(\frac{1}{2},1)}_{r-1,m-1}(u_1)\right]
$$

$$
+ (1 - q_l) \sum_{\substack{r=1 \\ r \neq l}}^{\infty} {}' \frac{d^{1,2l}_{2r-1}(c)}{d^{1,2l}_{2l-1}(c)} \frac{\Gamma\left(r + 1\right)}{\Gamma\left(r - \frac{1}{2}\right)} \frac{\Gamma\left(l - \frac{1}{2}\right)}{\Gamma\left(l + 1\right)} \left[\frac{h^{(\frac{1}{2},1)}_{r-1}}{h^{(\frac{1}{2},1)}_{l-1}}\right]^{\frac{1}{2}} \widehat{Q}^{(\frac{1}{2},1)}_{r-1,m-1}(u_1).
$$

$$
(7.204)
$$

The systems (7.199) and (7.202) are very suitable for numerical solution by a truncation method. The Fredholm nature of matrix operators in (7.199) and (7.202) ensures that the solution of the systems obtained by truncation to order N_{tr} always converges to the solution of infinite system as $N_{tr} \to \infty$. The choice $N_{tr} \geq kb + 20$ provides an accuracy of three digits in the solution. The matrices are well conditioned even for values of the frequency parameter c ($c = kd/2$) corresponding to quasi-eigenoscillations of the cavity. The matrix elements and the elements of the right-hand side do not require numerical integration or differentiation and are easily computed once the Fourier coefficients $d^{ll}_r(c)$ for the spheroidal functions have been found.

The solution of this boundary-value problem enables calculation of radiation features of the dipole spheroidal antenna (when modelled as a closed spheroid with a voltage impressed across the slot) in the presence of an outer open spheroidal screen. We examine radiation resistance R and far-field radiation pattern f in the resonant frequency range, when the electric size of the spheroidal antenna equals several wavelengths. As mentioned in the introduction to this chapter the algorithm used to compute the spheroidal functions restricted the frequency range to $c \leq 10$.

Radiation resistance R is defined as the ratio of power Π radiated by the system to power Π_o radiated by the dipole antenna in free space. Taking into account the asymptotic behaviour

$$
R^{(3)}_{1l}(c, \xi) = \frac{e^{ic\xi}}{c\xi}(-i)^{l+1} + O\left(\xi^{-2}\right),
\qquad (7.205)
$$

as $\xi \to \infty$,

$$
H_\phi = \frac{e^{ic\xi}}{c\xi} \sum_{l=1}^{\infty} (-i)^{l+1} C_l S_{1l}(c, \eta) + O\left(\xi^{-2}\right),
\qquad (7.206)
$$

as $\xi \to \infty$. Thus the power flux (see [7.77]) equals

$$\Pi = \frac{\pi}{k^2} \sum_{l=1}^{\infty} |C_l|^2 \left[N_{1l}(c)\right]^2 . \tag{7.207}$$

In free space the field of closed spheroid (excited by a voltage impressed on a narrow slot) has the spheroidal function expansion [21]

$$H_{\phi}^{(0)}(\xi, \eta) = \sum_{l=1}^{\infty} a_l S_{1l}(c, \eta) R_{1l}^{(3)}(c, \xi) ,$$

valid when $\xi \in (\xi_0, \infty)$, $\eta \in (-1, 1)$; from the condition (7.169)

$$E_{\eta} \big|_{\xi=\xi_0} = \frac{\Sigma}{h_{\eta}} \delta(\eta - \eta_0), \quad \eta \in (-1, 1)$$

it follows that

$$a_l = ik\Sigma \left(1 - \eta_0^2\right)^{\frac{1}{2}} \frac{S_{1l}(c, \eta_0)}{Z_l^{(3)}(c, \xi_0)}.$$

Thus

$$H_{\phi}^{(0)}(\xi, \eta) = ik\Sigma \left(1 - \eta_0^2\right)^{\frac{1}{2}} \sum_{l=1}^{\infty} \frac{S_{1l}(c, \eta_0)}{Z_l^{(3)}(c, \xi_0)} S_{1l}(c, \eta) R_{1l}^{(3)}(c, \xi) ,$$

for $\xi \in (\xi_0, \infty)$, $\eta \in (-1, 1)$; as $\xi \to \infty$

$$H_{\phi}^{(0)}(\xi, \eta) = \frac{e^{ic\xi}}{c\xi} ik\Sigma \left(1 - \eta_0^2\right)^{\frac{1}{2}} \sum_{l=1}^{\infty} \frac{S_{1l}(c, \eta_0)}{Z_l^{(3)}(c, \xi_0)} S_{1l}(c, \eta) + O\left(\xi^{-2}\right)$$

and the power flux of the bare spheroid in the far-field zone is

$$\Pi_0 = \pi \Sigma^2 \left(1 - \eta_0^2\right) \sum_{l=1}^{\infty} \frac{|S_{1l}(c, \eta_0)|^2 |N_{1l}(c)|^2}{\left|Z_l^{(3)}(c, \xi_0)\right|^2}.$$

Thus the expression for radiation resistance is

$$R = \frac{\Pi}{\Pi_0} = \frac{\pi \Sigma^2}{\Pi_0} \sum_{l=1}^{\infty} \left| \frac{2l+1}{l(l+1)} \frac{b_l \Delta_l}{Z_l^{(3)}(c, \xi_0)} \right|^2 |N_{1l}(c)|^2 . \tag{7.208}$$

The far-field radiation pattern, defined by (7.74), immediately follows from (7.206):

$$f(\eta) = \frac{1}{c} \sum_{l=1}^{\infty} (-i)^{l+1} C_l S_{1l}(c, \eta) . \tag{7.209}$$

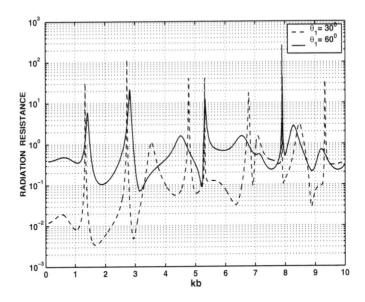

FIGURE 7.25. Radiation resistance of the spheroidal dipole antenna $(a/b = 0.1$ and $\theta_0 = 90°)$ in the presence of an outer spheroidal screen $(a_1/b_1 = 0.8)$ of varying aperture size (θ_1).

Note that the coefficients C_l (and b_l in [7.208]) are obtained by solving the matrix equations (7.199) and (7.202).

A highly elongated radiator behaves similarly to a wire antenna, possessing resonant features when its length $2b$ equals an integer number of half-wave lengths $\left(2b = \left(n + \frac{1}{2}\right)\lambda\right)$. The outer screen itself is an open resonant structure. The two spectra, arising from surface oscillations of the closed spheroid (oscillations of a dipole antenna), and from high Q-factor cavity resonances of the outer screen, interact. The shape and size of the screen affects the radiation properties of the dipole antenna in an essential way. As an example let us consider the radiation resistance $R = R(kb)$ of a thin dipole antenna with ratio of semi-axes $a/b = 0.1$ and voltage slot located at the centre ($\theta_0 = 90°$ or $\eta_0 = \cos\theta_0 = 0$), in the presence of a screen of fixed aspect ratio $a_1/b_1 = 0.8$ but variable aperture size $(\theta_1 = 30°, 60°; \eta_1 = \cos\theta_1)$. The results are displayed in Figure 7.25. Note that geometrical scale was fixed by setting $b = 1$. The resonances of the wire and of the internal cavity are visible, as well as a mixture of effects. The internal cavity resonances associated with the outer screen are excited when the electric size of the minor axis of the cavity ka_1 is about 2.7. An example of the far-field radiation pattern is presented in Figure 7.26.

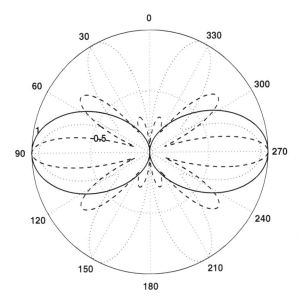

FIGURE 7.26. Radiation pattern of the spheroidal dipole antenna $(a/b = 0.1,$ $\theta_0 = 90°$) in the presence of an outer spheroidal screen $(a_1/b_1 = 0.8$) of varying aperture size: $\theta_1 = 89.9°$ (solid), $30°$ (dashed), and $60°$ (dotted); $kb = 2.74$.

We thus conclude our examination of this multiparametric problem. Several parameters can be varied, including voltage slot location, aspect ratio of the metallic spheroid, and the shape, size and position of the outer screen. The numerical solution of the regularised system allows a detailed analysis of the radiation features of this structure.

8

Wave Scattering Problems for Selected Structures.

In this chapter we examine wave-scattering problems for selected canonical structures with edges such as infinitely long strips, circular and elliptic discs, and hollow finite cylinders.

The classical problem of diffraction from thin strips is examined in Section 8.1; two polarisations of the incident electromagnetic wave (and their acoustic analogies) are considered. Axially slotted circular cylinders are considered in Section 8.2. The extension of regularisation ideas to two-dimensional scattering from axially slotted cylinders of arbitrary cross-section is studied in Section 8.3. This extension has much in common with the electrostatic potential problem for the same structures that was studied in Section 7.5 of Volume I. Diffraction from circular and elliptic discs is examined in Sections 8.4 and 8.5, respectively. The hollow finite cylinder (Section 8.6) completes the set of isolated scatterers for which regularised solutions are obtained.

Periodic structures are amenable to regularisation treatments. To illustrate the approach, two periodic arrays are studied in Section 8.7, the linear array of strips, and a linear array of hollow finite cylinders. Waveguide structures are similarly treatable, and a simple structure, the microstrip line in a waveguide of rectangular cross-section, is examined in Section 8.8 to illustrate this approach.

Whilst these problems are well known and have been examined by other methods, our purpose is to show that the diffraction problem for these canonical structures with edges is rigorously and uniquely solvable by the

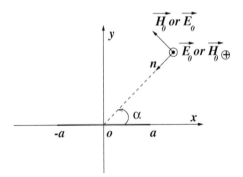

FIGURE 8.1. Infinitely long metallic strip excited by E- or H- polarised waves.

method of regularisation. Whilst the interest in this chapter is mainly methodological, the solutions of wave-scattering from elliptic discs and from thin-walled cylinders of arbitrary profile seem to be new.

8.1   Plane Wave Diffraction from Infinitely Long Strips.

There is an extensive literature on scattering by infinitely long strips; so we limit ourselves to a few references that illustrate different treatments of the problem. See for example [37], [27], [32], [125], [126], [56], [86] and [107].

Suppose an infinitely long thin metallic strip of width $2a$ is excited by an electromagnetic plane wave as shown in Figure 8.1. The plane wave propagates in the direction defined by the unit vector $\vec{n}$ that lies entirely in the xy plane. The strip occupies the region $|x| \leq a$, $y = 0$ and $-\infty < z < \infty$. We consider two polarisations, E-polarisation with the incident electric field $\vec{E}_0$ parallel to the z-axis (and the surface of the strip), and H-polarisation with the incident magnetic field $\vec{H}_0$ parallel to the z-axis. The field is incident at angle α so that $\vec{n} \cdot \vec{i}_x = \cos(\pi - \alpha)$. Under such circumstances the electromagnetic field does not depend upon the z coordinate and the problem is essentially two-dimensional. As described in Section 1.2, the E-polarised case corresponds to TM waves and is governed by the single component E_z with the remaining nonvanishing components determined by

$$H_x = \frac{1}{ik}\frac{\partial}{\partial y}E_z, \quad H_y = -\frac{1}{ik}\frac{\partial}{\partial x}E_z; \qquad (8.1)$$

the H-polarised case corresponds to TE waves and is governed by the single component H_z with the remaining nonvanishing components determined by

$$E_x = -\frac{1}{ik}\frac{\partial}{\partial y}H_z, \quad E_y = \frac{1}{ik}\frac{\partial}{\partial x}H_z. \tag{8.2}$$

Both components E_z and H_z are solutions Φ of the two-dimensional Helmholtz equation

$$\frac{\partial^2\Phi}{\partial x^2} + \frac{\partial^2\Phi}{\partial y^2} + k^2\Phi = 0. \tag{8.3}$$

The incident field is

$$E_z^0(x,y) = -H_z^0(x,y) = e^{-ik(x\cos\alpha + y\sin\alpha)}. \tag{8.4}$$

The total solution $E_z^t(x,y)$ or $H_z^t(x,y)$ is sought in the form

$$E_z^t(x,y) = E_z^0(x,y) + E_z^{sc}(x,y) \tag{8.5}$$

$$H_z^t(x,y) = H_z^0(x,y) + H_z^{sc}(x,y) \tag{8.6}$$

where $E_z^{sc}(x,y)$ and $H_z^{sc}(x,y)$ are scattered fields that obey the Sommerfeld radiation conditions. The boundary conditions for the E-polarised case are

$$E_z^t(x,+0) = E_z^t(x,-0) = 0, \quad |x| < a \tag{8.7}$$

$$\left.\frac{\partial E_z^t}{\partial y}\right|_{y=+0} = \left.\frac{\partial E_z^t}{\partial y}\right|_{y=-0}, \quad |x| > a \tag{8.8}$$

and those for the H-polarised case are

$$\left.\frac{\partial H_z^t}{\partial y}\right|_{y=+0} = \left.\frac{\partial H_z^t}{\partial y}\right|_{y=-0} = 0, \quad |x| < a \tag{8.9}$$

$$H_z^t(x,+0) = H_z^t(x,-0), \quad |x| > a. \tag{8.10}$$

It may be seen from these boundary conditions that the E-polarisation problem corresponds formally to the acoustic scattering problem for the soft strip, and the H-polarisation problem corresponds to the acoustic scattering problem for the hard strip.

As both E_z^{sc} and H_z^{sc} satisfy the Helmholtz equation let us seek their solutions in the form of single- and double-layer potentials ([1.292] and [1.298])

$$E_z^{sc} = -\int_{-a}^{a}\sigma_D(x')G_2(x,y;x',0)\,dx' \tag{8.11}$$

$$H_z^{sc} = -\int_{-a}^{a}\sigma_N(x')\frac{\partial}{\partial y'}G_2(x,y;x',0)\,dx' \cdot \operatorname{sgn}(y) \tag{8.12}$$

where σ_D and σ_N are the jump functions defined by

$$\sigma_D\left(x'\right) = \left.\frac{\partial E_z\left(x', y'\right)}{\partial y'}\right|_{y'=+0}^{y'=-0} = ik\left[H_x\left(x', -0\right) - H_x\left(x', +0\right)\right], \quad (8.13)$$

$$\sigma_N\left(x'\right) = \left.H_z\left(x', y'\right)\right|_{y'=-0}^{y'=+0}; \quad (8.14)$$

σ_D and σ_N are the z and x components, J_z and J_x, respectively, of the induced surface current.

The two-dimensional Green's function may be expressed in Cartesian coordinates by the formula (1.168)

$$G_2\left(x, y; x', y'\right) = \frac{i}{4}H_0^{(1)}\left(k\sqrt{\left(x - x'\right)^2 + \left(y - y'\right)^2}\right)$$

$$= \frac{1}{2\pi}\int_0^\infty \frac{\cos\nu\left(x - x'\right)}{\sqrt{\nu^2 - k^2}}e^{-\sqrt{\nu^2 - k^2}\left|y - y'\right|}d\nu \quad (8.15)$$

where $\text{Im}\left(\sqrt{\nu^2 - k^2}\right) \leq 0$.

First consider the E-polarised case. Enforce the boundary condition (8.7), setting in (8.11) $y = 0$, to obtain the Fredholm integral equation for σ_D,

$$\int_{-a}^a \sigma_D\left(x'\right) G_2\left(x, 0; x', 0\right) dx' = e^{ikx\cos\alpha}, \quad |x| < a \quad (8.16)$$

where

$$G_2\left(x, 0; x', 0\right) = \frac{i}{4}H_0^{(1)}\left(k\left|x - x'\right|\right) = \frac{1}{2\pi}\int_0^\infty \frac{\cos\nu\left(x - x'\right)}{\sqrt{\nu^2 - k^2}}d\nu. \quad (8.17)$$

We now use the mathematical technique developed in Section 1.5 to transform the integral equation (8.16) into a set of dual integral equations. First introduce the function

$$\psi\left(x'\right) = \left.\frac{\partial E_z\left(x', y'\right)}{dy'}\right|_{y'=-0} - \left.\frac{\partial E_z\left(x', y'\right)}{dy'}\right|_{y'=+0} = \begin{cases} \sigma_D\left(x'\right), & |x'| < a \\ 0, & |x'| > a. \end{cases} \quad (8.18)$$

Then represent its even and odd parts by Fourier cosine and sine transforms, respectively,

$$\psi\left(x'\right) = \int_0^\infty f\left(\nu\right)\cos\nu x' d\nu + \int_0^\infty g\left(\nu\right)\sin\nu x' d\nu. \quad (8.19)$$

Inserting the definition (8.18) into the integral equation (8.16) produces

$$\int_{-\infty}^{\infty} \psi\left(x'\right) G_2\left(x, 0; x', 0\right) dx' = e^{-ikx \cos \alpha}, \quad |x| < a. \tag{8.20}$$

Insert representations (8.19) and (8.17) into (8.20) and after obvious transformations obtain

$$\int_0^{\infty} \frac{f\left(\nu\right)}{\sqrt{\nu^2 - k^2}} \cos\left(\nu x\right) d\nu \; = \; \cos\left(kx \cos \alpha\right), \quad x \in (0, a) \tag{8.21}$$

$$\int_0^{\infty} \frac{g\left(\nu\right)}{\sqrt{\nu^2 - k^2}} \sin\left(\nu x\right) d\nu \; = \; -i \sin\left(kx \cos \alpha\right), \quad x \in (0, a). \tag{8.22}$$

Companion equations on the interval (a, ∞) follow directly from definition (8.18),

$$\int_0^{\infty} f\left(\nu\right) \cos\left(\nu x\right) d\nu \; = \; 0, \quad x \in (a, \infty) \tag{8.23}$$

$$\int_0^{\infty} g\left(\nu\right) \sin\left(\nu x\right) d\nu \; = \; 0, \quad x \in (a, \infty). \tag{8.24}$$

Thus the Fredholm integral equation of the first kind is equivalent to the following pair of dual integral equations for the functions f and g,

$$\int_0^{\infty} \frac{f\left(\nu\right)}{\sqrt{\nu^2 - k^2}} \cos\left(\nu x\right) d\nu \; = \; \cos\left(kx \cos \alpha\right), \quad x < a \tag{8.25}$$

$$\int_0^{\infty} f\left(\nu\right) \cos\left(\nu x\right) d\nu \; = \; 0, \quad x > a \tag{8.26}$$

and

$$\int_0^{\infty} \frac{g\left(\nu\right)}{\sqrt{\nu^2 - k^2}} \sin\left(\nu x\right) d\nu \; = \; -i \sin\left(kx \cos \alpha\right), \quad x < a \tag{8.27}$$

$$\int_0^{\infty} g\left(\nu\right) \sin\left(\nu x\right) d\nu \; = \; 0, \quad x > a. \tag{8.28}$$

In the case of H-polarisation, we enforce the boundary condition (8.9), use (8.4) and (8.12), and obtain an integro-differential equation for the unknown line current density σ_N,

$$\text{sgn}\left(y\right) \left\{ \frac{\partial}{\partial y} \int_{-a}^{a} \sigma_N\left(x'\right) \left[\frac{\partial}{\partial y'} G_2\left(x, y; x', y'\right) \right] \Bigg|_{y'=0} dx' \right\}_{y=0}$$
$$= ik \sin \alpha e^{-ikx \cos \alpha}, \quad |x| < a \tag{8.29}$$

where

$$\left.\frac{\partial}{\partial y'}G_2\left(x,y;x',y'\right)\right|_{y'=0} = \frac{1}{2\pi}\int_0^\infty \cos\nu\left(x-x'\right)e^{-\operatorname{sgn}(y)\sqrt{\nu^2-k^2}}d\nu. \quad (8.30)$$

Extend the integration limits in (8.29) to $(-\infty,\infty)$ after introducing the function

$$\Phi\left(x'\right) = H_z\left(x',+0\right) - H_z\left(x',-0\right) = \begin{cases} \sigma_N\left(x'\right), & |x'| < a \\ 0, & |x'| > a \end{cases} \quad (8.31)$$

with associated Fourier integral transform

$$\Phi\left(x'\right) = \int_0^\infty f\left(\mu\right)\cos\mu x'd\nu + \int_0^\infty g\left(\mu\right)\sin\mu x'd\nu. \quad (8.32)$$

Insert the representations (8.30) and (8.32) into the integral equation (8.29) and interchange the order of integration. Recognising that

$$\int_{-\infty}^\infty \left\{ \begin{matrix} \cos\mu x' \\ \sin\mu x' \end{matrix} \right\} \cos\nu\left(x-x'\right)dx' = 2\pi \left\{ \begin{matrix} \cos\nu x \\ \sin\nu x \end{matrix} \right\}\delta\left(\nu-\mu\right), \quad (8.33)$$

equation (8.29) is reduced to the integral equation

$$\operatorname{sgn}\left(y\right)\frac{\partial}{\partial y}\int_0^\infty \left(f\left(\nu\right)\cos\nu x + g\left(\nu\right)\sin\nu x\right)e^{-\operatorname{sgn}(y)\sqrt{\nu^2-k^2}y}d\nu\Big|_{y=0}$$
$$= ik\sin\alpha e^{-ikx\cos\alpha}, \quad |x| < a. \quad (8.34)$$

It can be shown that requirement of scattered energy boundedness imposes on f and g the conditions

$$f\left(\nu\right) = O\left(\nu^{-\frac{3}{2}}\right), \quad g\left(\nu\right) = O\left(\nu^{-\frac{3}{2}}\right) \quad (8.35)$$

as $\nu \to \infty$. It follows that the partial derivative with respect to y exists at each point $y \in (-\infty,\infty)$, including the point $y = 0$. Differentiation under the integral sign is permissible in (8.34) so that

$$\int_0^\infty \sqrt{\nu^2-k^2}f\left(\nu\right)\cos\nu x d\nu + \int_0^\infty \sqrt{\nu^2-k^2}g\left(\nu\right)\sin\nu x d\nu$$
$$= ik\sin\alpha e^{-ikx\cos\alpha}, \quad |x| < a. \quad (8.36)$$

Separation of the even and odd parts in (8.32) leads to a pair of dual integral equations for the functions f and g,

$$\int_0^\infty \sqrt{\nu^2-k^2}f\left(\nu\right)\cos\left(\nu x\right)d\nu = -ikx\cos\alpha\cos\left(kx\cos\alpha\right), \quad x < a \quad (8.37)$$

$$\int_0^\infty f(\nu) \cos(\nu x)\, d\nu = 0, \quad x > a \tag{8.38}$$

and

$$\int_0^\infty \sqrt{\nu^2 - k^2}\, g(\nu) \sin(\nu x)\, d\nu = -k \sin\alpha \sin(kx\cos\alpha), \quad x < a \tag{8.39}$$

$$\int_0^\infty g(\nu) \sin(\nu x)\, d\nu = 0, \quad x > a. \tag{8.40}$$

Regularise equations (8.37)–(8.38) and (8.39)–(8.40), making use of the Abel integral transform on the trigonometric kernels of these equations. At the first stage we obtain

$$\int_0^\infty \nu^{\frac{1}{2}} F(\nu) J_0(\nu x)\, d\nu =$$
$$\begin{cases} -ik\sin\alpha J_0(k\cos\alpha x) - \int_0^\infty \nu^{\frac{1}{2}} F(\nu)\, \varepsilon(\nu, k) J_0(\nu x)\, d\nu, & x < a \\ 0, & x > a \end{cases} \tag{8.41}$$

$$\int_0^\infty \nu^{\frac{1}{2}} G(\nu) J_1(\nu x)\, d\nu =$$
$$\begin{cases} -k\sin\alpha J_1(k\cos\alpha x) - \int_0^\infty \nu^{\frac{1}{2}} G(\nu)\, \varepsilon(\nu, k) J_1(\nu x)\, d\nu, & x < a \\ 0, & x > a \end{cases} \tag{8.42}$$

where

$$F(\nu) = \nu^{\frac{1}{2}} f(\nu), \quad G(\nu) = \nu^{\frac{1}{2}} g(\nu). \tag{8.43}$$

Notice that F and G lie in $L_2(0, \infty)$,

$$\int_0^\infty |F(\nu)|^2\, d\nu < \infty, \quad \int_0^\infty |G(\nu)|^2\, d\nu < \infty. \tag{8.44}$$

The parameter

$$\varepsilon(\nu, k) = \frac{\sqrt{\nu^2 - k^2}}{\nu} - 1 \tag{8.45}$$

is asymptotically small: $\varepsilon(\nu, k) = O(k^2/\nu^2)$ as $\nu \to \infty$.

An application of the Fourier-Bessel transform to Equations (8.41), (8.42) transforms them to the following pair of second kind Fredholm equations for the functions F_1 and G_1 that are related to the functions F and G as follows. Let $\varkappa = ka = \frac{2\pi}{\lambda}a$, $u = \mu a$, and $v = \nu a$, and set $F_1(u) \equiv F_1(\mu a) = a^{-\frac{1}{2}}F(\mu)$, and $G_1(u) \equiv G_1(\mu a) = a^{-\frac{1}{2}}G(\mu)$. Then F_1 satisfies

$$F_1(u) + \int_0^\infty F_1(v)K(v,u)\,dv = A(u), \qquad (8.46)$$

where the kernel is

$$K(v,u) = (v,u)^{\frac{1}{2}}\varepsilon(v,\varkappa)\frac{vJ_1(v)J_0(u) - uJ_0(v)J_1(u)}{v^2 - u^2} \qquad (8.47)$$

and

$$A(u) = -i\varkappa\sin\alpha\, u^{\frac{1}{2}}\frac{\varkappa\cos\alpha J_1(\varkappa\cos\alpha)J_0(u) - uJ_0(\varkappa\cos\alpha)J_1(u)}{\cos^2\alpha\cdot\varkappa^2 - u^2}; \qquad (8.48)$$

the transformed parameter has the form

$$\varepsilon(v,\varkappa) = v^{-1}\sqrt{v^2 - \varkappa^2} - 1. \qquad (8.49)$$

Also G_1 satisfies

$$G_1(u) - \int_0^\infty G_1(v)R(v,u)\,dv = B(u) \qquad (8.50)$$

where the kernel

$$R(v,u) = (uv)^{\frac{1}{2}}\varepsilon(v,\varkappa)\frac{vJ_0(v)J_1(u) - uJ_0(v)J_1(u)}{v^2 - u^2} \qquad (8.51)$$

and

$$B(u) = \varkappa\sin\alpha\cdot u^{\frac{1}{2}}\frac{\varkappa\cos\alpha J_0(\varkappa\cos\alpha)J_1(u) - uJ_0(u)J_1(\varkappa\cos\alpha)}{\varkappa^2\cos^2\alpha - u^2}. \qquad (8.52)$$

Making use of the asymptotic behaviour of the Bessel functions J_0 and J_1 one may justify that $F_1(u) = O(u^{-1})$ and $G_1(u) = O(u^{-1})$ as $u \to \infty$; so recalling the substitution (8.43), $f(\nu) = O\left(\nu^{-\frac{3}{2}}\right)$ and $g(\nu) = O\left(\nu^{-\frac{3}{2}}\right)$ as $\nu \to \infty$.

It is quite clear that as $\varkappa \to 0$ (or $a/\lambda \ll 1$), the kernels of equations (8.46), (8.50) are small and we can find the approximate analytical solutions of both equations; for arbitrary values of $\varkappa$ these equations may be solved numerically. A natural discretisation arises from the expansion of the functions F and G in Neumann series. Let us demonstrate this process in detail for equation (8.41), and simply state the final result for (8.42).

First, apply the Fourier-Bessel transform to equation (8.41) to obtain

$$\mu^{-\frac{1}{2}} F(\mu) = -ik \sin \alpha \int_0^a x J_0 (k \cos \alpha x) J_0 (\mu x) \, dx$$

$$- \int_0^a \nu^{\frac{1}{2}} F(\nu) \varepsilon (\nu, k) \int_0^a x J_0 (\nu x) J_0 (\mu x) \, dx \quad (8.53)$$

Let us seek a solution in the form of a Neumann series

$$F(\mu) = ika \sin \alpha \cdot \mu^{-\frac{1}{2}} \sum_{m=0}^{\infty} (4m + 2)^{\frac{1}{2}} x_m J_{2m+1} (\mu a), \quad (8.54)$$

where the coefficients x_m are determined by insertion into (8.53). This gives

$$\frac{1}{\mu} \sum_{m=0}^{\infty} (4m + 2)^{\frac{1}{2}} x_m J_{2m+1} (\mu a) = -\frac{1}{a} \int_0^a x J_0 (k \cos \alpha x) J_0 (\mu x) \, dx -$$

$$\sum_{m=0}^{\infty} (4m + 2)^{\frac{1}{2}} x_m \int_0^a \varepsilon (\nu, k) \left\{ \int_0^a x J_0 (\nu x) J_0 (\mu x) \, dx \right\} J_{2m+1} (\nu a) \, d\nu.$$

$$(8.55)$$

Making use of the integral formula [23], valid when $\nu > -1$,

$$\int_0^{\infty} t^{-1} J_{\nu+2n+1} (t) J_{\nu+2m+1} (t) \, dt = (4n + 2\nu + 2)^{-1} \delta_{nm}, \quad (8.56)$$

we multiply both sides of (8.55) by the factor $(4s + 2)^{\frac{1}{2}} J_{2s+1} (\mu a)$ and integrate over the semi-infinite interval $(0, \infty)$ and use the tabulated integrals

$$\int_0^{\infty} J_0 (\mu x) J_{2s+1} (\mu a) \, d\mu = \frac{1}{a} P_s \left(1 - 2 \frac{x^2}{a^2} \right), \quad (8.57)$$

$$\int_0^a x J_0 (\nu x) P_s \left(1 - 2 \frac{x^2}{a^2} \right) dx = a \frac{J_{2s+1} (\nu a)}{\nu}, \quad (8.58)$$

$$\int_0^a x J_0 (k \cos \alpha x) P_s \left(1 - 2 \frac{x^2}{a^2} \right) dx = a \frac{J_{2s+1} (ka \cos \alpha)}{k \cos \alpha}. \quad (8.59)$$

This yields the following i.s.l.a.e. of the second kind for the unknown coefficients x_n,

$$x_s + \sum_{m=0}^{\infty} \alpha_{ms}^{(1)} x_m = \beta_s^{(1)}, \quad (8.60)$$

where $s = 0, 1, 2, \ldots$, and

$$\alpha_{ms}^{(1)} = [(4m + 2)(4s + 2)]^{\frac{1}{2}} \int_0^\infty \frac{\varepsilon(v, \varkappa)}{v} J_{2m+1}(v) J_{2s+1}(v) \, dv, \quad (8.61)$$

$$\beta_s^{(1)} = -(4s + 2)^{\frac{1}{2}} J_{2s+1}(\varkappa \cos \alpha) / \varkappa \cos \alpha. \quad (8.62)$$

In a similar way the equations (8.39), (8.40) may be similarly transformed to a second kind system for the unknown coefficients y_n in the assumed Neumann series expansion

$$g(\nu) = \frac{\varkappa \sin \alpha}{\nu} \sum_{m=0}^\infty (4m + 4)^{\frac{1}{2}} y_m J_{2m+2}(\nu a). \quad (8.63)$$

The following i.s.l.a.e. is obtained,

$$y_s + \sum_{m=0}^\infty \alpha_{ms}^{(2)} y_m = \beta_s^{(2)}, \quad (8.64)$$

where $s = 0, 1, 2, \ldots$ and

$$\alpha_{ms}^{(2)} = [(4m + 4)(4s + 4)]^{\frac{1}{2}} \int_0^\infty \frac{\varepsilon(v, \varkappa)}{v} J_{2m+2}(v) J_{2s+2}(v) \, dv, \quad (8.65)$$

$$\beta_s^{(2)} = -(4s + 4)^{\frac{1}{2}} \frac{J_{2s+2}(\varkappa \cos \alpha)}{\varkappa \cos \alpha}. \quad (8.66)$$

It is interesting to note that the one-to-one correspondence between the second kind integral Fredholm equations (8.46) and (8.50) and the second kind matrix Fredholm equations (8.60) and (8.64) can be verified using the well-known relation

$$\sum_{k=1}^\infty (2k + \nu) J_{2k+\nu}(\omega) J_{2k+\nu}(z) =$$

$$\frac{\omega z}{2(\omega^2 - z^2)} [\omega J_\nu(\omega) J_{\nu-1}(z) - z J_{\nu-1}(\omega) J_\nu(z)] \quad (8.67)$$

and the representations (8.54) and (8.63).

Let us solve the equations (8.25)–(8.28) that describe the E-polarisation case. The equations (8.27)–(8.28) can be directly attacked with the Method of Regularisation. Rearrange them as follows,

$$\int_0^\infty G(\nu) \sin(\nu x) \, d\nu = -i \sin(kx \cos \alpha) - \int_0^\infty G(\nu) \mu(\nu, k) \sin(\nu x) \, d\nu,$$

$$x < a \quad (8.68)$$

$$\int_0^\infty \nu G(\nu) \sin(\nu x) \, d\nu = 0, \quad x > a \tag{8.69}$$

where

$$\mu(\nu, k) = \frac{\nu}{\sqrt{\nu^2 - k^2}} - 1 = -\frac{\varepsilon(\nu, k)}{1 + \varepsilon(\nu, k)} \tag{8.70}$$

is an asymptotically small parameter satisfying $\mu(\nu, k) = O(k^2/\nu^2)$ as $\nu \to \infty$.

A standard application of the Abel integral transform reduces (8.68), (8.69) to the functional equation

$$\int_0^\infty \nu G(\nu) J_0(\nu x) \, d\nu =$$
$$\begin{cases} -ik \cos \alpha \, J_0(k \cos \alpha x) - \int_0^\infty \nu G(\nu) \mu(\nu, k) J_0(\nu x) \, d\nu, & x < a \\ 0, & x > a. \end{cases} \tag{8.71}$$

In a similar way to that used before, a second kind matrix equation may be derived for the coefficients y_s of form,

$$y_s + \sum_{m=0}^\infty \alpha_{ms}^{(1)*} y_m = \beta_s^{(1)}, \tag{8.72}$$

where $s = 0, 1, 2, \ldots$, and the matrix element $\alpha_{ms}^{(1)*}$ differs from the element $\alpha_{ms}^{(1)}$ defined in (8.61) by the replacement of $\varepsilon(v, \varkappa)$ with $\mu(v, \varkappa)$, but $\beta_s^{(1)}$ coincides with that introduced in the formula (8.62).

The most interesting case arises when the equations (8.25)–(8.26) cannot be attacked directly by the Abel integral transform method (this situation was discussed in Section 2.6 of Volume I). An initial transform is required, to the form

$$\int_0^\infty \frac{g(\nu)}{\sqrt{\nu^2 - k^2}} \cos(\nu x) \, d\nu = -i\frac{\pi}{2} H_0^{(1)}(kx) f(0) + \cos(k \sin \alpha x),$$
$$x < a \tag{8.73}$$

$$\int_0^\infty g(\nu) \cos(\nu x) \, d\nu = 0, \quad x > a, \tag{8.74}$$

where

$$g(\nu) = f(\nu) - f(0) \tag{8.75}$$

The edge condition requires that the unknown spectral density function g must belong the functional space $L_2(-1)$; a direct corollary is

$$g(\nu) = O\left(\nu^{-\frac{1}{2}}\right) \tag{8.76}$$

as $\nu \to \infty$. Note that $g(0) = 0$.

The behaviour of the function g at $\nu = 0$ and $\nu \to \infty$ makes it clear that it has the Neumann series expansion

$$g(\nu) = \sum_{k=1}^{\infty} x_k J_{2k}(\nu a), \tag{8.77}$$

where the unknowns x_k are to be determined. It is evident that the equation (8.74) is satisfied automatically, and equation (8.73) may be transformed to the form

$$\frac{1}{2} \sum_{k=1}^{\infty} \frac{x_k}{k} T_{2k}\left(\sqrt{1-\rho^2}\right) = -i\frac{\pi}{2} H_0^{(1)}(\varkappa\rho) f(0) + \cos(\varkappa\cos\alpha\rho)$$

$$+ \sum_{k=1}^{\infty} x_k \int_0^{\infty} \frac{\mu(u,\varkappa)}{u} J_{2k}(u)\cos(u\rho)\, du, \quad 0 \le \rho < 1 \tag{8.78}$$

where $\rho = x/a$, $u = \nu a$, $\varkappa = ka$ and the parameter

$$\mu(u,\varkappa) = 1 - \frac{u}{\sqrt{u^2 - \varkappa^2}} = O\left(\varkappa^2 u^{-2}\right) \tag{8.79}$$

is asymptotically small as $u \to \infty$: $\mu(u,\varkappa) = O\left(\varkappa^2 u^{-2}\right)$.

Now multiply both sides of equation (8.78) by $\left(1-\rho^2\right)^{-\frac{1}{2}} T_{2n}\left(\sqrt{1-\rho^2}\right)$ and integrate over $(0,1)$, using the well-known orthogonal properties of the Chebyshev polynomials

$$\int_0^1 \frac{1}{\sqrt{1-\rho^2}} T_{2k}\left(\sqrt{1-\rho^2}\right) T_{2n}\left(\sqrt{1-\rho^2}\right) d\rho = \delta_{kn} \left\{ \begin{array}{l} \pi/4,\ n > 0 \\ \pi/2,\ n = 0 \end{array} \right\}. \tag{8.80}$$

Setting $n = 0$, we obtain an equation for the value $f(0)$ in terms of the coefficients x_k,

$$f(0) = -i\frac{2}{\pi} \frac{J_0(\varkappa\cos\alpha) + \sum_{k=1}^{\infty} x_k \int_0^{\infty} u^{-1}\mu(u,\varkappa) J_{2k}(u) J_0(u)\, du}{J_0(\varkappa/2) H_0^{(1)}(\varkappa/2)}. \tag{8.81}$$

When $n > 0$ we obtain the second kind i.s.l.a.e. for the rescaled coefficients $\{X_n\}_{n=1}^{\infty}$, that depend on $f(0)$,

$$X_n - \sum_{k=1}^{\infty} X_k \left\{ 4(kn)^{\frac{1}{2}} \int_0^{\infty} u^{-1} \mu(u, \varkappa) J_{2k}(u) J_{2n}(u)\, du \right\}$$

$$= 2\sqrt{n} J_{2n}(\varkappa \cos \alpha) - i\frac{\pi}{2} \left\{ 2\sqrt{n} J_n\left(\frac{\varkappa}{2}\right) H_n^{(1)}\left(\frac{\varkappa}{2}\right) \right\} f(0), \quad (8.82)$$

$n = 1, 2, \ldots$.

Substitute (8.81) into (8.82) and eliminate $f(0)$ to obtain

$$X_n - \sum_{k=1}^{\infty} X_k C_{kn} = d_n, \quad (8.83)$$

where $n = 1, 2, \ldots$ and

$$C_{kn} = 4\sqrt{kn} \int_0^{\infty} \frac{\mu(u, \varkappa)}{u} J_{2k} \left\{ J_{2n}(u) - J_0(u) \frac{J_n\left(\frac{\varkappa}{2}\right) H_n^{(1)}\left(\frac{\varkappa}{2}\right)}{J_0\left(\frac{\varkappa}{2}\right) H_0^{(1)}\left(\frac{\varkappa}{2}\right)} \right\},$$
$$(8.84)$$

$$d_n = 2\sqrt{n} \left\{ J_{2n}(\varkappa \cos \alpha) - J_0(\varkappa \cos \alpha) \frac{J_n\left(\frac{\varkappa}{2}\right) H_n^{(1)}\left(\frac{\varkappa}{2}\right)}{J_0\left(\frac{\varkappa}{2}\right) H_0^{(1)}\left(\frac{\varkappa}{2}\right)} \right\}. \quad (8.85)$$

Equation (8.83), taken with (8.81), completely defines the regularised system for an acoustically soft strip.

8.2   Axially Slotted Infinitely Long Circular Cylinders.

The first rigorous solution of the scattering of electromagnetic plane waves by an axially slotted infinitely long circular cylinder seems to be due to Koshparenok and Shestopalov [51] in 1971. Their solution employed the regularisation ideas that are extensively exploited in this book. However a notable difference is that Koshparenok and Shestopalov utilised the solution of the *Riemann-Hilbert problem* that is well known in the theory of analytical functions to effect the analytical inversion of a singular part of the relevant operator. In 1984, the same problem was studied in a similar way by Johnson and Ziolkowski [36]. In the limiting case of a narrow slot this problem has also been treated by Mautz and Harrington [59], [60]. Since the publication of [51] numerous other problems have been solved by

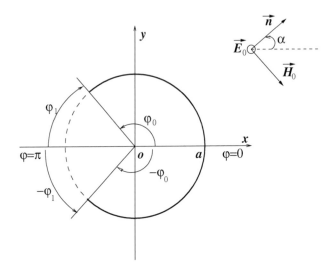

FIGURE 8.2. Excitation of an infinitely long slotted thin metallic cylinder by an E-polarised plane wave.

the same approach, including diffraction problems for infinitesimally thin circular cylinders with multiple slots and periodic arrays of slotted cylinders; a theory of waveguides based on this geometry was also examined in detail. The power of the method of regularisation is demonstrated in [109] where a more complex structure consisting of an infinitely long strip and a slotted cylinder is examined in the case of H-polarisation.

Whilst recognising the power of the Riemann-Hilbert approach, we take a fresh look at these problems, avoiding the apparatus of functions of complex variables. Our treatment is entirely based on real function methods and the regularisation ideas that are implemented by the Abel integral transform.

This section also provides an introduction to the next section, where we treat diffraction problems for slotted cylinders of an arbitrary cross-sectional profile. We thus consider the canonical problem of plane wave scattering by a perfectly conducting and infinitely long singly-slotted circular cylinder of fixed radius a. The slot is axially aligned and has constant cross-section. The structure is excited by an E- or H-polarised electromagnetic plane wave as shown in Figure 8.2. The slot semiwidth is described in polar cylindrical coordinates by an angle $\phi_1 = \pi - \phi_0$; the metallic part subtends an angle ϕ_0 at the axial line.

First consider the E-polarised case, $\vec{E}_0 = \vec{i}_z E_0$, with incident field prescribed by the z component

$$E_z^0 (\rho, \phi) = \exp\left[ik\rho \cos\left(\phi - \alpha\right)\right], \qquad (8.86)$$

where α is the angle of incidence measured from the plane $\phi = 0$; it has the cylindrical harmonic expansion

$$E_z^0 = \sum_{m=0}^{\infty} i^m \left(2 - \delta_m^0\right) J_m \left(k\rho\right) \cos m\alpha \cos m\phi$$

$$+ 2 \sum_{m=1}^{\infty} i^m J_m \left(k\rho\right) \sin m\alpha \sin m\phi. \quad (8.87)$$

The other nonvanishing components of the field are given by

$$H_\rho = \frac{1}{ik\rho} \frac{\partial E_z}{\partial \phi}, \quad H_\phi = -\frac{1}{ik} \frac{\partial E_z}{\partial \rho}. \quad (8.88)$$

As usual we decompose the total field as the sum

$$E_z^{tot} \left(\rho, \phi\right) = E_z^0 \left(\rho, \phi\right) + E_z^{sc} \left(\rho, \phi\right) \quad (8.89)$$

where the scattered field E_z^{sc} satisfies the Helmholtz equation and the Sommerfeld radiation conditions, and the total field is continuous on $\rho = a$. We may suppose that E_z^{sc} has the cylindrical harmonic expansion

$$E_z^{sc} = \sum_{m=0}^{\infty} \left(2 - \delta_m^0\right) x_m R_m \left(k\rho\right) \cos m\phi + 2 \sum_{m=1}^{\infty} y_m R_m \left(k\rho\right) \sin m\phi,$$

$$(8.90)$$

where the unknown coefficients x_m and y_m are to be determined and

$$R_m \left(k\rho\right) = \left\{ \begin{array}{ll} J_m \left(k\rho\right), & \rho < a \\ \frac{J_m(ka)}{H_m^{(1)}(ka)} H_m^{(1)} \left(k\rho\right), & \rho > a \end{array} \right\}. \quad (8.91)$$

The total field satisfies the mixed boundary conditions

$$E_z^{tot} \left(a + 0, \phi\right) = E_z^{tot} \left(a - 0, \phi\right) = 0, \quad \phi \in (-\phi_0, \phi_0) \quad (8.92)$$

$$H_z^{tot} \left(a + 0, \phi\right) = H_z^{tot} \left(a - 0, \phi\right), \quad \phi \in (\phi_0, \pi) \cup (-\phi_0, -\pi). \quad (8.93)$$

By virtue of definition (8.89), the condition (8.93) is equivalent to enforcing continuity of the scattered field on the aperture,

$$H_\phi^{sc} \left(a + 0, \phi\right) = H_\phi^{sc} \left(a - 0, \phi\right), \phi \in (\phi_0, \pi) \cup (-\phi_0, -\pi); \quad (8.94)$$

taking into account Formula (8.88), the condition is also equivalent to

$$\frac{\partial E_z^{sc}}{\partial \rho}\bigg|_{\rho=a+0} = \frac{\partial E_z^{sc}}{\partial \rho}\bigg|_{\rho=a-0}, \quad \phi \in (\phi_0, \pi) \cup (-\phi_0, -\pi). \quad (8.95)$$

The combination of (8.92) and (8.95) shows that the E-polarised problem is equivalent to the acoustical scattering problem for a *soft* open cylinder; both scattering problems are essentially Dirichlet mixed boundary value problems for the Helmholtz equation.

Enforcement of the boundary conditions (8.92) and (8.95) leads to dual series equations for the unknown coefficients x_n and y_n,

$$\frac{1}{2} J_0 (ka) x_0 + \sum_{m=1}^{\infty} x_m J_m (ka) \cos m\phi =$$

$$- J_0 (ka) - \sum_{m=1}^{\infty} i^m J_m (ka) \cos m\alpha \cos m\phi, \quad \phi \in (0, \phi_0) \quad (8.96)$$

$$\frac{x_0}{2 H_0^{(1)} (ka)} + \sum_{m=1}^{\infty} \frac{x_m}{H_m^{(1)} (ka)} \cos m\phi = 0, \quad \phi \in (\phi_0, \pi) \quad (8.97)$$

and

$$\sum_{m=1}^{\infty} y_m J_m (ka) \sin m\phi = - \sum_{m=1}^{\infty} i^m J_m (ka) \sin m\alpha \sin m\phi, \quad \phi \in (0, \phi_0)$$

$$(8.98)$$

$$\sum_{m=1}^{\infty} \frac{y_m}{H_m^{(1)} (ka)} \sin m\phi = 0, \quad \phi \in (\phi_0, \pi) . \quad (8.99)$$

We now extract the singular part of these equations by identifying a suitable asymptotically small parameter. Using the well-known asymptotics for Bessel functions [3]

$$J_m (z) = \frac{z^m}{2^m \Gamma (m + 1)} \left\{ 1 - \frac{z^2}{4 (m + 1)} + \frac{z^4}{32 (m + 1) (m + 2)} + O \left(\frac{z^6}{m^3} \right) \right\}$$

$$(8.100)$$

$$H_m^{(1)} (z) = - \frac{i}{\pi} \frac{2^m \Gamma (m)}{z^m} \left\{ 1 + \frac{z^2}{4 (m - 1)} + \frac{z^4}{32 (m - 1) (m - 2)} + O \left(\frac{z^6}{m^3} \right) \right\}$$

$$(8.101)$$

as $m \to \infty$, it follows that

$$i\pi m J_m (z) H_m^{(1)} (z) = 1 + \frac{z^2}{2 (m^2 - 1)} + O \left(\frac{z^4}{m^4} \right) \quad (8.102)$$

as $m \to \infty$. Define

$$\mu_m = 1 - i\pi m J_m (ka) H_m^{(1)} (ka) \tag{8.103}$$

so that $\mu_m = O\left((ka)^2 /m^2\right)$, as $m \to \infty$.

With the rescaling

$$X_m = \frac{x_m}{m H_m^{(1)} (ka)} \tag{8.104}$$

where $m \geq 1$, we may rearrange the dual series equations (8.96)–(8.97) as

$$\frac{1}{2} \pi J_0 (ka) (1 + x_0) + \sum_{m=1}^{\infty} X_m \{(1 - \mu_m) + q_m\} \cos m\phi = 0, \ \phi \in (0, \phi_0) \tag{8.105}$$

$$\frac{x_0}{2 H_0^{(1)} (ka)} + \sum_{m=1}^{\infty} m X_m \cos m\phi = 0, \phi \in (\phi_0, \pi) \tag{8.106}$$

where $q_m = i^{m+1} \pi J_m (ka) \cos m\alpha$ $(m \geq 1)$; with the rescaling

$$Y_m = \frac{y_m}{m H_m^{(1)} (ka)} \tag{8.107}$$

where $m \geq 1$ we may rearrange the dual series equations (8.98)–(8.99) as

$$\sum_{m=1}^{\infty} Y_m \{(1 - \mu_m) + p_m\} \sin m\phi = 0, \ \phi \in (0, \phi_0) \tag{8.108}$$

$$\sum_{m=1}^{\infty} m Y_m \sin m\phi = 0, \ \phi \in (\phi_0, \pi) \tag{8.109}$$

where $p_m = i^{m+1} \pi J_m (ka) \sin m\alpha$.

The coefficients x_m and y_m must satisfy the boundedness condition (1.287) on scattered energy. In the two-dimensional case the energy per length unit of the infinite cylinder is

$$W_l = \frac{1}{2} \int_0^a d\rho \, \rho \int_0^{2\pi} d\phi \left\{ |E_z^{sc}|^2 + |H_\rho^{sc}|^2 + |H_\phi^{sc}|^2 \right\}$$

$$= \pi a^2 |x_0|^2 \left\{ J_1^2 (ka) + J_0 (ka) J_1' (ka) \right\} +$$

$$2\pi a^2 \sum_{m=1}^{\infty} m^2 \left\{ |X_m|^2 + |Y_m|^2 \right\} \left| H_m^{(1)} (ka) \right|^2 \times$$

$$\left\{ \frac{1}{ka} J_m (ka) J_m' (ka) + J_m^2 (ka) - J_{m-1} (ka) J_{m+1} (ka) \right\}. \tag{8.110}$$

Making use of the asymptotic formulae (8.101) the boundedness condition for scattered energy implies

$$\sum_{m=1}^{\infty} m\,|X_m|^2 < \infty, \quad \sum_{m=1}^{\infty} m\,|Y_m|^2 < \infty, \qquad (8.111)$$

that is, the unknown coefficients $\{X_m\}_{m=1}^{\infty}$ and $\{Y_m\}_{m=1}^{\infty}$ belong to the functional class $l_2\,(1)$ (in the notation of Volume I).

The regularisation of this type of dual series equations is given in Section 2.2 of Volume I. A comparison of the pair (2.39)–(2.40) on page 66 of Volume I with the pair (8.105)–(8.106) shows that upon making the identification $x_n \mapsto X_n$, $x_0 \mapsto x_0$, $b \mapsto i\frac{\pi}{2}J_0\,(ka)$, $g_0 \mapsto -b = -i\frac{\pi}{2}J_0\,(ka)$, $q_n \mapsto \mu_n$, $g_0 \mapsto -q_n$, $a \mapsto \frac{1}{2}H_0^{(1)}\,(ka)$, $v_0 \mapsto \phi_0$, and setting $f_0 = f_n = r_n \equiv 0$, the regularised system is

$$x_0 = \frac{-i\pi J_0\,(ka)\,H_0^{(1)}\,(ka)}{\varkappa\,(ka,t_0)} +$$

$$(1+t_0)\frac{H_0^{(1)}\,(ka)}{\varkappa\,(ka,t_0)}\sum_{n=1}^{\infty} n^{-1}\,(Z_n\mu_n - G_n)\,\hat{P}_{n-1}^{(0,1)}\,(t_0), \quad (8.112)$$

$$Z_m -$$

$$\sum_{n=1}^{\infty}\{Z_n\mu_n - G_n\}\left\{\hat{Q}_{n-1,m-1}^{(1,0)}\,(t_0) + \frac{(1+t_0)^2}{\varkappa\,(ka,t_0)}\frac{\hat{P}_{n-1}^{(0,1)}\,(t_0)}{n}\frac{\hat{P}_{m-1}^{(0,1)}\,(t_0)}{m}\right\}$$

$$= -\frac{i\pi J_0\,(ka)}{\varkappa\,(ka,t_0)}\frac{1+t_0}{m}\hat{P}_{m-1}^{(0,1)}\,(t_0), \quad (8.113)$$

where $m = 1, 2, \ldots$,

$$\varkappa\,(ka,t_0) = i\pi J_0\,(ka)\,H_0^{(1)}\,(ka) - \ln\left(\frac{1-t_0}{2}\right)$$

and

$$\{Z_m, G_m\} = (2m)^{\frac{1}{2}}\{x_n, q_n\}. \qquad (8.114)$$

Analogously, a comparison of the pair (2.41)–(2.42) on page 66 of Volume I with the pair (8.108)–(8.109) shows that upon making the additional identification $y_n \mapsto Y_n$, $q_n \mapsto \mu_n$, $e_n \mapsto -p_n$, and setting $r_n = h_n \equiv 0$, the regularised system is

$$V_s + \sum_{n=1}^{\infty} V_n\mu_n\hat{Q}_{n-1,s-1}^{(0,1)}\,(t_0) = -\sum_{n=1}^{\infty} C_n\hat{Q}_{n-1,s-1}^{(0,1)}\,(t_0), \qquad (8.115)$$

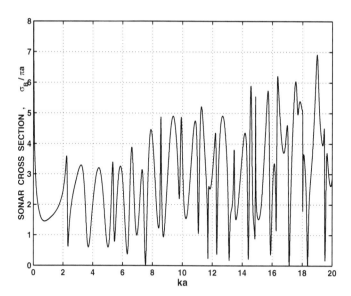

FIGURE 8.3. Radar cross-section of the axially slotted cylinder excited by a plane wave incident at $\alpha = 0°$.

where $s = 1, 2, \ldots$ and

$$\{V_s, G_s\} = \sqrt{\frac{s\pi}{2}} \{Y_n, p_n\}. \qquad (8.116)$$

As usual, $\hat{Q}_{sn}^{(\alpha,\beta)}(t)$ denotes the normalised incomplete scalar product defined in Appendix B.6 of Volume I.

Numerical calculations of the monostatic radar cross-sections of the open ($\phi_1 = 30°$) cylinder are given in Figures 8.3 and 8.4, where the angle α of incidence equals $0°$ and $45°$, respectively.

Let us now consider the case when the incident plane wave is H-polarised. If the role of the vectors $\vec{E}_0$ and $\vec{H}_0$ in the incident E-polarised plane wave (see Figure 8.2) is interchanged, the resultant plane wave correctly defines a H-polarised plane wave provided the propagation direction vector $\vec{n}$ is reversed. Hence the incident H-polarised plane wave has z component

$$H_z^0(\rho, \phi) = \exp\left[-ik\rho\cos(\phi - \alpha)\right]$$

$$= \sum_{m=0}^{\infty} (-i)^m (2 - \delta_m^0) J_m(k\rho)\cos m\alpha \cos m\phi$$

$$+ 2\sum_{m=1}^{\infty} (-i)^m J_m(k\rho)\sin m\alpha \sin m\phi. \quad (8.117)$$

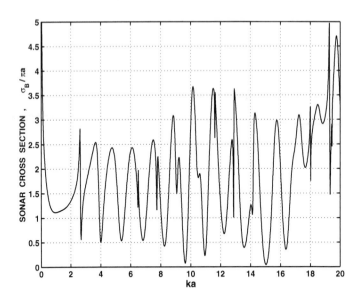

FIGURE 8.4. Radar cross-section of the axially slotted cylinder excited by a plane wave incident at $\alpha = 45°$.

As before, the total field component H_z^{tot} satisfies the Helmholtz equation and the other nonvanishing field components are completely determined by

$$E_\rho = -\frac{1}{ik\rho}\frac{\partial H_z^{tot}}{\partial \phi}, \quad E_\phi = \frac{1}{ik}\frac{\partial H_z^{tot}}{\partial \rho}. \tag{8.118}$$

The total field is decomposed as the sum

$$H_z^{tot} = H_z^0 + H_z^{sc}$$

where the scattered field satisfies the Sommerfeld radiation conditions and the total field component E_ϕ satisfies the continuity condition on the contour $\rho = a$,

$$E_\phi(a+0,\phi) = E_\phi(a-0,\phi), \quad \phi \in (0,2\pi). \tag{8.119}$$

Thus, we seek a scattered field solution in the form

$$H_z^{sc} = \sum_{m=0}^{\infty} \left(2 - \delta_m^0\right)(-i)^m x_m Q_m(k\rho)\cos m\phi$$

$$+ 2\sum_{m=0}^{\infty}(-i)^m y_m Q_m(k\rho)\sin m\phi \tag{8.120}$$

where the coefficients x_m and y_m are to be determined, and

$$Q_m(k\rho) = \begin{cases} J_m(k\rho), & \rho < a \\ \frac{J'_m(ka)}{H'_m(ka)} H_m^{(1)}(k\rho), & \rho > a. \end{cases} \tag{8.121}$$

The mixed boundary conditions are

$$E_\phi^{tot}(a+0,\phi) = E_\phi^{tot}(a-0,\phi) = 0, \phi \in (-\phi_0, \phi_0), \tag{8.122}$$

$$H_z^{tot}(a+0,\phi) = H_z^{tot}(a-0,\phi), \phi \in (\phi_0, \pi) \cup (-\pi, -\phi_0). \tag{8.123}$$

The relationship (8.118) shows that (8.122) can be replaced by the equivalent form

$$\frac{\partial H_z}{\partial \rho}\bigg|_{\rho=a+0} = \frac{\partial H_z}{\partial \rho}\bigg|_{\rho=a-0} = 0, \phi \in (-\phi_0, \phi_0). \tag{8.124}$$

Thus the H-polarised case corresponds to the acoustical scattering problem for the hard (or rigid) slotted circular cylinder.

Enforcement of the boundary conditions (8.122)–(8.123) leads to the dual series equations

$$-\frac{x_0}{2H_1^{(1)}(ka)} + \sum_{m=1}^{\infty} i^m \frac{x_m}{H_m^{(1)'}(ka)} \cos m\upsilon = 0, \upsilon \in (0, \phi_1), \tag{8.125}$$

$$-\frac{1}{2}J_1(ka)(1+x_0) + \sum_{m=1}^{\infty} i^m x_m J'_m(ka) \cos m\upsilon$$

$$= -\sum_{m=1}^{\infty} i^m J'_m(ka) \cos m\alpha \cos m\upsilon, \quad \upsilon \in (\phi_1, \pi) \tag{8.126}$$

and

$$\sum_{m=1}^{\infty} i^m \frac{x_m}{H_m^{(1)'}(ka)} \sin m\upsilon = 0, \upsilon \in (0, \phi_1), \tag{8.127}$$

$$\sum_{m=1}^{\infty} i^m y_m J'_m(ka) \sin m\upsilon = -\sum_{m=1}^{\infty} i^m J'_m(ka) \sin m\alpha \sin m\upsilon, \quad \upsilon \in (\phi_1, \pi) \tag{8.128}$$

where $\upsilon = \pi - \phi$ and $\phi_1 = \pi - \phi_0$. The introduction of the variable υ is necessary to conform to the scheme developed in Section 2.2 of Volume I.

Further manipulation of these dual series equations is of a familiar character.

Introduce the rescaled unknown coefficients

$$\{X_m, Y_m\} = \frac{i^m}{H_m^{(1)\prime}(ka)} \{x_m, y_m\}. \tag{8.129}$$

From the asymptotic expansions (8.101),

$$J_m'(ka) H_m^{(1)\prime}(ka) = \frac{i}{\pi} \frac{m}{(ka)^2} \left[1 + O\left(\frac{(ka)}{m^2}\right)\right] \tag{8.130}$$

as $m \to \infty$, so that the parameter

$$\varepsilon_m = 1 + i\pi \frac{(ka)^2}{m} J_m'(ka) H_m^{(1)\prime}(ka) \tag{8.131}$$

is asymptotically small as $m \to \infty$: $\varepsilon_m = O\left((ka)/m^2\right)$.

Thus equations (8.125)–(8.126) and (8.127)–(8.128) may be transformed to

$$-\frac{x_0}{2H_1^{(1)}(ka)} + \sum_{m=1}^{\infty} X_m \cos mv = 0, \ v \in (0, \phi_1) \tag{8.132}$$

$$\frac{i}{2}\pi (ka)^2 J_1(ka)(1 + x_0) + \sum_{m=1}^{\infty} m\{X_m(1 - \varepsilon_m) - p_m^c\}\cos mv = 0,$$
$$v \in (\phi_1, \pi) \quad (8.133)$$

and

$$\sum_{m=1}^{\infty} Y_m \sin mv = 0, \ v \in (0, \phi_1) \tag{8.134}$$

$$\sum_{m=1}^{\infty} m\{Y_m(1 - \varepsilon_m) - p_m^s\}\sin mv = 0, \ v \in (\phi_1, \pi), \tag{8.135}$$

where

$$\left\{\begin{array}{c} p_m^c \\ p_m^s \end{array}\right\} = \frac{i}{m}\pi (ka)^2 J_m'(ka)\left\{\begin{array}{c} \cos m\alpha \\ \sin m\alpha \end{array}\right\}. \tag{8.136}$$

In the same way as before, one may establish that the finite energy condition requires that both $\{X_m\}_{m=1}^{\infty}$ and $\{Y_m\}_{m=1}^{\infty}$ lie in $l_2(1)$.

The regularised second kind infinite systems of linear algebraic equations may be obtained from the results of Chapter 2 of Volume I. The equations (2.39)–(2.40) are identified with equations (8.132)–(8.133); making the replacements $b \mapsto -\frac{1}{2} H_1^{(1)}(ka)$, $x_n \mapsto X_n$, $a \mapsto -f_0 = \frac{i}{2}\pi(ka)^2 J_1(ka)$, $r_n \mapsto \varepsilon_n$, $f_n \mapsto p_n^c$, and setting $g_0 = g_n = q_n \equiv 0$ leads to the regularised system

$$x_0 = \frac{1 - \varkappa(ka, t_1)}{\varkappa(ka, t_1)} + \frac{H_1^{(1)}(ka)}{\varkappa(ka, t_1)} \sum_{n=1}^{\infty} \{Z_n \varepsilon_n + G_n\} \Phi_n(t_1), \qquad (8.137)$$

$$(1 - \varepsilon_m) Z_m + \sum_{n=1}^{\infty} Z_n \varepsilon_n S_{nm}(ka, t_1) =$$

$$\frac{\xi(ka)}{\varkappa(ka, t_1) H_1^{(1)}(ka)} \Phi_m(t_1) + G_m - \sum_{n=1}^{\infty} G_m S_{nm}(ka, t_1), \quad (8.138)$$

where $m = 1, 2, \ldots$, $t_1 = \cos \phi_1$, and

$$\{Z_n, G_n\} = \sqrt{2n} \{X_n, p_n^c\}, \qquad (8.139)$$

$$
\begin{aligned}
S_{nm}(ka, t_1) &= \hat{Q}_{n-1,m-1}^{(1,0)}(t_1) - \frac{\xi(ka)}{\varkappa(ka, t_1)} \Phi_n(t_1) \Phi_m(t_1), \\
\xi(ka) &= i\pi (ka)^2 J_1(ka) H_1^{(1)}(ka), \\
\varkappa(ka, t_1) &= 1 + i\pi (ka)^2 J_1(ka) H_1^{(1)}(ka) \ln\left[\frac{1 - t_1}{2}\right], \\
\Phi_s(t_1) &= (1 + t_1) \frac{\hat{P}_{s-1}^{(0,1)}(t_1)}{s} \quad (s \geq 1).
\end{aligned}
\qquad (8.140)
$$

We may compare equations (2.41)–(2.42) on page 61 of Volume I with the pair (8.134)–(8.135); making the identifications $y_n \mapsto Y_n$, $r_n \mapsto \varepsilon_n$, $h_n \mapsto p_n^s$, and setting $q_n = e_n \equiv 0$ leads us to the regularised system

$$(1 - \varepsilon_m) V_m + \sum_{n=1}^{\infty} V_m \varepsilon_m \hat{Q}_{n-1,m-1}^{(0,1)}(t_1) = R_m - \sum_{n=1}^{\infty} R_n \hat{Q}_{n-1,m-1}^{(0,1)}(t_1)$$

$$(8.141)$$

where $m = 1, 2, \ldots$, and

$$\{V_m, R_m\} = \sqrt{\frac{m\pi}{2}} \{y_m, p_m^s\}.$$

In this polarisation it is remarkable that when $\phi_1 \ll 1$, $ka < 1$ the open cylinder acts as a Helmholtz resonator. An approximate characteristic equation for the Helmholtz mode is easily derived by inspection of equations (8.137)–(8.138); the desired equation is

$$\varkappa\left(ka, t_1\right) = 0, \tag{8.142}$$

or in analytical form,

$$1 + i\pi\left(ka\right)^2 J_1\left(ka\right) H_1^{(1)}\left(ka\right) \ln\left[\frac{1 - t_1}{2}\right] = 0. \tag{8.143}$$

Under condition $ka \ll 1$, $\phi_1 \ll 1$ (and hence $1 - t_1 \ll 1$) we can replace (8.143) by its approximation

$$1 + 2\left(ka\right)^2 \ln\left(\frac{\phi_1}{2}\right) = 0, \tag{8.144}$$

so that a simple approximation for the relative wavenumber of the Helmholtz mode is

$$\left(ka\right)_R = \left(-2\ln\left(\frac{\phi_1}{2}\right)\right)^{-\frac{1}{2}}. \tag{8.145}$$

Refinements to expression (8.145) can be obtained using the technique developed in Chapter 2.

8.3   Axially Slotted Cylinders of Arbitrary Profile.

The studies of the previous section suggest that the regularisation technique based on the Abel integral transform might be extended to determine diffraction from a larger class of scatterers. In this section we demonstrate its applicability to cylindrical screens of arbitrarily shaped profile. The method and its mathematical justification were developed by Yu.A. Tuchkin [80], [94] for various scattering problems for infinitely long cylinders with arbitrarily shaped cross-section. We use his approach to construct the solution algorithm to the particular diffraction problem of scattering of a plane TM polarised electromagnetic wave by a thin metallic open cavity. We illustrate the method by computing the radar cross section for a particular cavity. Our purpose is to demonstrate that the methods developed for canonical scatterers work in a wider context.

The algorithm has many similarities to that developed for the corresponding calculation of the electrostatic potential problem of an arbitrarily

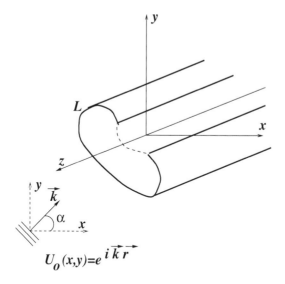

$$U_0(x,y){=}e^{i\vec{k}\vec{r}}$$

FIGURE 8.5. Axially slotted cylinder with cross-section of arbitrary form.

shaped conductor that was discussed in Section 7.5 of Volume I, but with some adjustments for the wave problem. We shall describe the solution scheme in some details, even somewhat repeating the results of Volume I.

We consider two dimensional TM scattering from an infinitely long cylindrical surface with cross-section that is independent of one axial direction, fixed to be the z-axis. The surface or screen is assumed to be infinitely thin, perfectly conducting, and is open, i.e., has an aperture. Its cross-sectional profile in the xy plane may be arbitrarily shaped (see Figure 8.5), though sufficiently smooth for the Fourier series representation given below in (8.158) to converge.

The screen is illuminated by the z-independent plane electromagnetic wave

$$E_z^{inc}\,\vec{i_z} = U_0(x,y)\,\vec{i_z} = e^{i\vec{k}\cdot\vec{r}}\,\vec{i_z}$$

where $\vec{r}$ is the position vector $\vec{r} = x\,\vec{i_x} + y\,\vec{i_y}$, $\vec{k}$ is the wave vector with magnitude k related to wavelength λ by $\left|\vec{k}\right| = k = 2\pi/\lambda$, and a harmonic time dependence factor $e^{-i\omega t}$, $(\omega = ck)$ is assumed and suppressed throughout.

The electromagnetic field is of transverse magnetic type (E-polarisation, the only non-zero components are E_z, H_x, H_y); the electrical component of the electromagnetic field is oriented along the z-axis. We wish to find the

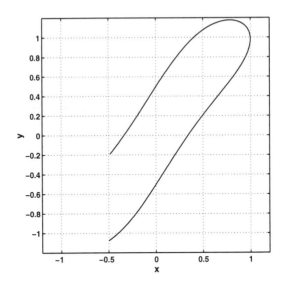

FIGURE 8.6. Cross-section of the axially slotted cylinder.

total electromagnetic field

$$E_z \overrightarrow{i_z} = U(x, y) \overrightarrow{i_z}$$

or, equivalently, to find the scattered field

$$E_z^s \overrightarrow{i_z} = U_s(x, y) \overrightarrow{i_z} = (U(x, y) - U_0(x, y)) \overrightarrow{i_z}$$

resulting from the scattering of the incident wave by the screen. This physical problem, thus defined, is described by the boundary-value problem formulated below in terms of the Helmholtz equation, with Dirihlet boundary conditions, for the longitudinal component of the electric field $E_z = U$ (x, y). The two other non-zero field components are found from the relations

$$Z_0 H_x = \frac{1}{ik} \frac{\partial E_z}{\partial y}, \, Z_0 H_y = -\frac{1}{ik} \frac{\partial E_z}{\partial x}$$

where Z_0 denotes the impedance of free space.

Let L denote the two-dimensional cross-section of the arbitrarily shaped open screen (enclosing the cavity) in the xy plane (see Figure 8.6). Let $p = p(x, y)$ denote a point in the plane R^2. The incident field is described by the function $U_0(p)$. The scattered field $U_s(p)$ to be found must satisfy the Helmholtz equation (1.144) at each $p \in R^2$ subject to (i) the Dirichlet boundary conditions enforced at each point p on L,

$$U_s(p)|_{p=p-0} = U_s(p)|_{p=p+0} = -U_0(p), \tag{8.146}$$

(ii) the radiation conditions (1.282)–(1.283) ensuring that the scattered field must be like an outgoing cylindrical wave at infinity, and (iii) the edge condition on the endpoints p_1, p_2 of the screen requiring that $\operatorname{grad} U_s(p)$ is of the form

$$\operatorname{grad} U_s(p) = \frac{h(p)}{\sqrt{|p - p_1||p - p_2|}} \tag{8.147}$$

where h is a smooth (vector) function which is bounded on the whole plane. The mathematical theory of diffraction shows that this set of the conditions ensures that a unique solution corresponding to the physical situation exists.

A well-known form of solution to the problem formulated is the single layer potential representation of the scattered field in terms of the line current J_z,

$$E_z^s(q) = ikZ_0 \int_L G_2(k\,|p - q|) J_z(p) dl_p, \qquad q \in R^2.$$

Here dl_p is the differential of arc length at the point $p \in L$, q is a point at which the scattered field is considered, $G_2(kR) = -\frac{1}{4}iH_0^{(1)}(kR)$ is the Hankel function of first kind and order zero that is the free space Green's function of the Helmholtz equation, and $R = |p - q|$ is the distance between the point p on the screen and the observation point q. It is convenient to set

$$Z(p) = ikZ_0 J_z(p), \tag{8.148}$$

so that the scattered field representation may be written in terms of the unknown normalised line current density $Z(p)$ as

$$U_s(q) = \int_L G_2(k\,|p - q|) Z(p) dl_p, \qquad q \in R^2. \tag{8.149}$$

$Z(p)$ may be interpreted as the jump in the normal derivative of the potential integral (8.149) from the interior to the exterior of the screen at the point p

$$Z(p) = \frac{\partial U_s^{(+)}(p)}{\partial n} - \frac{\partial U_s^{(-)}(p)}{\partial n}. \tag{8.150}$$

Applying the boundary condition (8.146) to Equation (8.149) yields the functional equation

$$\int_L G_2(k\,|p - q|) Z(p) dl_p = -U_0(q), \qquad q \in L \tag{8.151}$$

for the unknown current density $Z(p)$. Once this is found, the scattered field at any point q can be found from (8.149).

Our reformulation of the integral equation begins by regarding the open screen L as part of a larger *closed* structure S, that is parametrised by the functions $x(\theta), y(\theta)$ where $\theta \in [-\pi, \pi]$; the parametrising functions are periodic so that $x(-\pi) = x(\pi), y(-\pi) = y(\pi)$. The screen L is parametrised by the subinterval $[-\theta_0, \theta_0]$,

$$L = \{(x(\theta), y(\theta)), \theta \in [-\theta_0, \theta_0]\},$$

whilst the aperture is created by the removal from S of the segment

$$L' = \{(x(\theta), y(\theta)), \theta \in [-\pi, -\theta_0] \cup [\theta_0, \pi]\}.$$

Although $S = L \cup L'$, the choice of L' is not unique, and may be chosen as conveniently as possible for the problem at hand; however, it must be chosen so that the surface S is smooth, particularly at the joining points of L and L'.

With this parametrisation, the differential of arc length is

$$l(\tau) = \sqrt{(x'(\tau))^2 + (y'(\tau))^2},$$

and the functional equation (8.151) takes the form

$$\int_{-\theta_0}^{\theta_0} G_2(kR(\theta, \tau))z_0(\tau)d\tau = u(\theta), \quad \theta \in [-\theta_0, \theta_0] \tag{8.152}$$

where $R(\theta, \tau) = \sqrt{[x(\theta) - x(\tau)]^2 + [y(\theta) - y(\tau)]^2}$ is the distance between two points of the cavity parametrised by θ and τ, $z_0(\tau) = Z(x(\tau), y(\tau)) l(\tau)$, and $u(\theta) = -U_0(x(\theta), y(\theta))$.

Without loss of generality we may introduce the new unknown function $z(\tau)$ defined by

$$z(\tau) = \begin{cases} z_0(\tau), & \tau \in [-\theta_0, \theta_0], \\ 0, & \tau \in [-\pi, -\theta_0] \cup [\theta_0, \pi]. \end{cases} \tag{8.153}$$

and transform the functional equation (8.152) for this new unknown as an integral over the full interval $[-\pi, \pi]$ of the angular coordinate θ,

$$\int_{-\pi}^{\pi} G_2(kR(\theta, \tau))z(\tau)d\tau = u(\theta), \quad \theta \in [-\theta_0, \theta_0] \tag{8.154}$$

Equation (8.154) together with the requirement that z vanishes outside the interval $[-\theta_0, \theta_0]$ is completely equivalent to equation (8.152).

The first stage of the solution scheme is to obtain the integral equation in the equivalent form of a dual series equation with exponential functions $e^{in\theta}$. Split the kernel of the integral equation (8.154) into singular and regular parts,

$$H_0^{(1)}(kR(\theta, \tau)) = \frac{2i}{\pi} \ln(2 \left|\sin \frac{\theta - \tau}{2}\right|) + H(\theta, \tau). \qquad (8.155)$$

It will be supposed that the surface S is such that $H(\theta, \tau)$ is smooth and continuously differentiable with respect to θ and τ; this allows its expansion in a double Fourier series,

$$H(\theta, \tau) = \sum_{p=-\infty}^{\infty} \sum_{n=-\infty}^{\infty} h_{np} e^{i(n\theta + p\tau)}, \qquad \theta, \tau \in [-\pi, \pi], \qquad (8.156)$$

where

$$\sum_{p=-\infty}^{\infty} \sum_{n=-\infty}^{\infty} (1 + |p|^2)(1 + |n|^2) |h_{np}|^2 < \infty.$$

Here the coefficients h_{np} are given by

$$h_{np} = \frac{1}{4\pi^2} \int_{-\pi}^{\pi} \int_{-\pi}^{\pi} H(\theta, \tau) e^{-i(n\theta + p\tau)} d\theta d\tau \qquad (8.157)$$

Represent the incident function $u(\theta)$ and the solution $z(\tau)$ in the form of the Fourier series,

$$-2u(\theta) \sum_{n=-\infty}^{\infty} = \sum_{n=-\infty}^{\infty} g_n e^{in\theta}, \qquad \theta \in [-\pi, \pi] \qquad (8.158)$$

$$z(\tau) = \sum_{n=-\infty}^{\infty} \varsigma_n e^{in\tau}, \qquad \tau \in [-\pi, \pi] \qquad (8.159)$$

For the singular part of the Green's function we may use the expansion

$$\ln(2 \left|\sin \frac{\theta - \tau}{2}\right|) = -\frac{1}{2} \sum_{n=-\infty}^{\infty} {}' \frac{1}{|n|} e^{in(\theta - \tau)}, \qquad \theta, \tau \in [-\pi, \pi] \quad (8.160)$$

where the prime (') over the summation sign means that $n \neq 0$.

Inserting (8.155)-(8.160) into equation (8.154) and recalling that $z(\tau)$ vanishes outside the interval $[-\theta_0, \theta_0]$, we obtain a dual series equation with exponential functions:

$$\sum_{n=-\infty}^{\infty}{}' |n|^{-1} \varsigma_n e^{in\theta} - 2 \sum_{n=-\infty}^{\infty} e^{in\theta} \sum_{p=-\infty}^{\infty} h_{n,-p} \varsigma_p = \sum_{n=-\infty}^{\infty} g_n e^{in\theta},$$
$$\theta \in [-\theta_0, \theta_0], \quad (8.161)$$

$$\sum_{n=-\infty}^{\infty} \varsigma_n e^{in\theta} = 0, \qquad \theta \in [-\pi, -\theta_0] \cup [\theta_0, \pi]. \quad (8.162)$$

Thus the integral equation (8.154) is converted to an equivalent dual series equation defined on two subintervals $[-\pi, \pi]$ with the unknowns $\{\varsigma_n\}_{n=-\infty}^{\infty}$ to be found.

To transform the system (8.161)–(8.162) with exponential kernels to a system with trigonometric functions ($\cos(n\theta)$ and $\sin(n\theta)$), introduce the new unknowns

$$x_n = (\zeta_n + \zeta_{-n})/|n|, \qquad y_n = (\zeta_n - \zeta_{-n})/|n|, \quad (8.163)$$

where $n = 1, 2, \ldots$. Set

$$g_n^+ = g_n + g_{-n}, \qquad g_n^- = g_n - g_{-n}, \quad (n = 1, 2, \ldots), \quad (8.164)$$

and define the matrices from the coefficients $\{h_{np}\}_{n,p=-\infty}^{\infty}$ (8.157) by

$$\begin{aligned}
k_{np}^{(++)} &= \left[(h_{n,p} + h_{n,-p}) + (h_{-n,p} + h_{-n,-p})\right]/(2 + 2\delta_{n0}), & n, p \geq 0; \\
k_{np}^{(+-)} &= \left[(h_{n,p} - h_{n,-p}) + (h_{-n,p} - h_{-n,-p})\right]/(2 + 2\delta_{n0}), & n \geq 0, p \geq 1; \\
k_{np}^{(-+)} &= \left[(h_{n,p} + h_{n,-p}) - (h_{-n,p} + h_{-n,-p})\right]/2, & n \geq 1, p \geq 0; \\
k_{np}^{(--)} &= \left[(h_{n,p} - h_{n,-p}) - (h_{-n,p} - h_{-n,-p})\right]/2, & n, p \geq 1.
\end{aligned}$$
$$(8.165)$$

We reduce the system of equations (8.161)–(8.162) to two coupled systems of dual series equations with trigonometric function kernels:

$$\begin{aligned}
\sum_{n=1}^{\infty} x_n \cos n\theta &= a_0 + \sum_{n=1}^{\infty} a_n \cos n\theta, & \theta \in [0, \theta_0], \\
\sum_{n=1}^{\infty} n x_n \cos n\theta &= -\zeta_0, & \theta \in [\theta_0, \pi], \quad (8.166)
\end{aligned}$$

and

$$\begin{aligned}
\sum_{n=1}^{\infty} y_n \sin n\theta &= \sum_{n=1}^{\infty} c_n \sin n\theta, & \theta \in [0, \theta_0], \\
\sum_{n=1}^{\infty} n y_n \sin n\theta &= 0, & \theta \in [\theta_0, \pi], \quad (8.167)
\end{aligned}$$

where

$$
a_0 = g_0 + 2k_{00}^{(++)}\zeta_0 + 2\sum_{p=1}^{\infty} p(k_{0p}^{(++)}x_p - k_{0p}^{(+-)}y_p),
$$

$$
a_n = g_n^+ + 2k_{n0}^{(++)}\zeta_0 + 2\sum_{p=1}^{\infty} p(k_{np}^{(++)}x_p - k_{np}^{(+-)}y_p),
$$

$$
c_n = g_n^- + 2k_{n0}^{(-+)}\zeta_0 - 2\sum_{p=1}^{\infty} p(k_{np}^{(--)}y_p - k_{np}^{(-+)}x_p). \qquad (8.168)
$$

These series equations (8.166)–(8.167) are in a standard form, and the regularisation procedure described in Section 2.2 of Volume I may be applied to obtain two infinite systems of linear algebraic equations for the following re-scaled unknowns,

$$
X_n = x_n\sqrt{2n}, \ Y_n = y_n\sqrt{2n}, \ X_0 = 2\zeta_0. \qquad (8.169)
$$

Setting $t_0 = \cos\theta_0$, the systems are

$$
Y_m + \sum_{p=1}^{\infty}\sqrt{2p}\sum_{n=1}^{\infty}\sqrt{2n}\left[Y_p k_{np}^{(--)} - X_p k_{np}^{(-+)}\right]\hat{Q}_{n-1,m-1}^{(0,1)}(t_0)
$$

$$
= \sum_{n=1}^{\infty}\sqrt{2n}(X_0 k_{n0}^{(-+)} + g_n^-)\hat{Q}_{n-1,m-1}^{(0,1)}(t_0), \qquad (8.170)
$$

and

$$
X_m - \sum_{p=1}^{\infty}\sqrt{2p}\sum_{n=1}^{\infty}\sqrt{2n}\left[X_p k_{np}^{(++)} - Y_p k_{np}^{(+-)}\right]\hat{Q}_{n-1,m-1}^{(1,0)}(t_0)
$$

$$
= \sum_{n=1}^{\infty}\sqrt{2n}(X_0 k_{n0}^{(++)} + g_n^+)\hat{Q}_{n-1,m-1}^{(1,0)}(t_0) + X_0(1+t_0)\frac{1}{m}\hat{P}_{m-1}^{(0,1)}(t_0),
$$

$$
\qquad (8.171)
$$

where $m = 1, 2, \ldots$; an additional equation, which has to be solved together with the equations (8.170)–(8.171) is

$$\sum_{p=1}^{\infty} \sqrt{2p}(X_p k_{0p}^{(++)} - Y_p k_{0p}^{(+-)}) +$$

$$\frac{(1+t_0)}{2} \sum_{p=1}^{\infty} \sqrt{2p} \sum_{n=1}^{\infty} \sqrt{\frac{2}{n}} \left[X_p k_{np}^{(++)} - Y_p k_{np}^{(+-)} \right] \hat{P}_{n-1}^{(0,1)}(t_0)$$

$$= -\frac{(1+t_0)}{2} \sum_{n=1}^{\infty} \sqrt{\frac{2}{n}} (X_0 k_{n0}^{(++)} + g_n^+) \hat{P}_{n-1}^{(0,1)}(t_0)$$

$$- g_0 - X_0 \left[k_{00}^{(++)} + \frac{1}{2} \ln(\frac{(1-t_0)}{2}) \right]. \quad (8.172)$$

Here

$$\hat{Q}_{n,m}^{(0,1)}(t_0) = \int_{t_0}^{1} (1+t) \hat{P}_n^{(0,1)}(t) \hat{P}_m^{(0,1)}(t) dt \quad (8.173)$$

is the usual normalised incomplete scalar product.

The equations (8.170)–(8.171) and (8.172) form a Fredholm equation of *second* kind. The matrix operator $H = (H_{mn})$ of this equation is compact in the functional space of square summable sequences l_2 and the unknowns X_n, Y_n are asymptotically $O(n^{-1})$ as $n \to \infty$. We can use a truncation method effectively for solving the final matrix equation. As the truncation number N_{tr} and the finite number N_{tr} of equations to be solved increases, its solution rapidly converges to the exact solution of infinite system. The associated matrix is well conditioned and the algorithm is numerically stable. The matrix elements are in closed form; they are simple functions of the Fourier coefficients h_{np} that may be computed practically and efficiently using the fast Fourier transform.

As an illustration, let us calculate the scattering by a thin metallic open cavity approximately modelling an aircraft engine duct specified with two parameters a and q by

$$x = a \cos \theta, \quad y = a \left[\arctan(\frac{3}{2} \cos \theta) + q \sin \theta \right].$$

The parameter a is fixed, $a = 1$ and q lies in the range $0 < q < 1$. The cavity surface (or metal) corresponds to the parameter range $\theta \in [-\theta_0, \theta_0]$, whereas the aperture is described by the range $\theta \in [-\pi, -\theta_0] \cup [\theta_0, \pi]$. Note that the point $\theta_0 = 0$ always lies on the metal. The parameter q determines the width of the duct; its electrical length, fully closed, is approximately $2\sqrt{2}ka$. Once the truncation order N_{tr} exceeds a threshold approximately

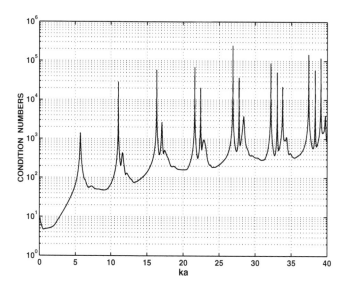

FIGURE 8.7. Condition numbers for cavity with $\theta_0 = 120°$, $q = 0.5$.

equal to the electrical length of the scatterer, the solution of truncated system of equations rapidly converges to the exact solution of infinite system.

The matrix equations (8.170)–(8.171) and (8.172) are well-conditioned even near those frequencies corresponding to quasi-eigenvalues of the cavity. An example of the dependence of condition number on wavenumber is shown in Figure 8.7. The sharp spikes on the plot correspond to quasi-eigenvalues of the cavity.

The solution obtained to (8.170)–(8.171) and (8.172) allows the calculation of line current density and radar cross-section (RCS). Recalling the definitions (8.169) and (8.163), we may find the function $z(\tau)$ in equation (8.153) from the expansion (8.159)

$$z(\tau) = \varsigma_0 + \sum_{n=1}^{\infty} \varsigma_{-n} e^{-in\tau} + \sum_{n=1}^{\infty} \varsigma_n e^{in\tau}, \qquad \tau \in [-\pi, \pi] \qquad (8.174)$$

and calculate a line current density $Z(\theta)$ that is normalized by the wave number k,

$$\frac{1}{k} Z(\theta) = \frac{1}{kl(\theta)} z(\theta), \qquad \theta \in [-\pi, \pi] \qquad (8.175)$$

In considering the representation (8.163) and (8.169) of the coefficients $\{\varsigma_n\}_{n=-\infty}^{\infty}$ derived from the set $\{X_n, Y_n\}_{n=-\infty}^{\infty}$ it is easy to estimate that $\varsigma_n = O(n^{-\frac{1}{2}})$ as $n \to \infty$ and the general term in the series (8.174) has the same estimate. The series (8.174) is slowly convergent; its convergence may be accelerated using the technique described in the Section 4.2 of Volume I.

The radar cross-section σ_B is defined by

$$\sigma_B = \lim_{|p| \to \infty} 2\pi |p| |U_s(p)|^2 \tag{8.176}$$

and, when normalized by the factor πa, is found by the formula

$$\frac{\sigma_B}{\pi a} = \frac{\pi a^2}{ka} \left| \sum_{m=-\infty}^{\infty} e^{im\tau} \sum_{n=-\infty}^{\infty} \varsigma_{-n} \, \beta_{nm} \right|^2. \tag{8.177}$$

Here the coefficients ς_n are the same as in (8.174), a is characteristic linear size of the screen (such as the radius of a circular screen etc.) and the coefficients β_n are defined via the Fourier expansion of the incident plane wave,

$$e^{-ika(x\cos\tau + y\sin\tau)} = \sum_{n=-\infty}^{\infty} \sum_{m=-\infty}^{\infty} \beta_{nm} e^{i(n\theta + m\tau)} \tag{8.178}$$

where $\tau = \pi + \alpha$, and α is the angle of incidence of the plane wave to the x-axis.

In contradistinction to the current density (8.174) the calculation of the scattered field in far-field zone (and the associated RCS (8.177)) presents few difficulties. The general series term β_{nm} in (8.178) is exponentially decreasing when $|n|$ or $|m| > kL$, where kL is electrical size of the screen. So, to obtain accurate results for the RCS calculation in equation (8.177), it is enough to take the number of terms in both set of unknowns $\{X_n\}_{n=0}^{\frac{N}{2}-1}$, $\{Y_n\}_{n=1}^{\frac{N}{2}-1}$ to exceed kL; thus, allowing for a small margin, a sensible choice of truncation number is $N_{tr} > 2kL + 20$.

The RCS was computed for the open duct with the wave directly incident on the aperture ($\alpha = 0$, Figure 8.8), over the range $1 < ka < 40$. The spikes in the condition number plot (Figure 8.7) clearly define quasi-eigenvalues of the cavity. The condition numbers at these frequencies lie in the range 10^4 to 10^5, thus establishing that even at these nearly resonant frequencies, the matrix is well conditioned and the solution is stable.

8.4   Diffraction from Circular Discs.

The subject of diffraction from circular discs has a long history. It has been intensively studied by numerous authors and by many different methods:

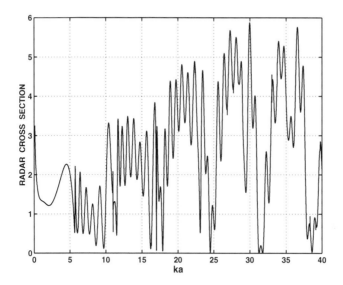

FIGURE 8.8. Radar cross-section of the cavity with $\theta_0 = 120°$, $q = 0.5$.

see, for example, [18], [15], [39], [40], [124], [85], [127], [110], [108] and [47]. Most papers deal with specific frequency range, mostly the Rayleigh scattering region ($\lambda \gg a$) or the high-frequency region ($\lambda \ll a$), where a is the radius of the disc. Such approaches are understandable from the physical point of view, and in both regimes the dominant scattering mechanisms are fairly clear and predictable.

In the resonance region the disc structure exhibits some noteworthy features such as low-amplitude oscillations in its frequency response for total or radar cross-sections ($\sigma(ka)$ or $\sigma_B(ka)$). It is convenient to consider methods that cover all frequency bands. The regularisation methods developed in this book are well suited to this purpose, providing an algorithm that is valid in any frequency region, from low frequencies to quasi-optics ($\lambda \ll a$). Of course, as frequency increases, the order of the matrix (derived by truncation of the relevant regularised i.s.l.a.e.) also increases, but at a reasonable rate (about $O(ka)$) rather than the higher rates typical of general purpose scattering algorithms.

We begin with the acoustic scenario. Suppose a soft or rigid disc of radius a lies in the plane $z = 0$ with centre located at the origin (see Figure 8.9). The incident plane wave propagates in the direction $\overrightarrow{n}$ given by $\overrightarrow{n} \cdot \overrightarrow{i}_z = \cos\alpha$, and its velocity potential is

$$U_0 = \exp\{ik[\rho\sin\alpha\cos(\varphi - \varphi_0) + z\cos\alpha]\} \qquad (8.179)$$

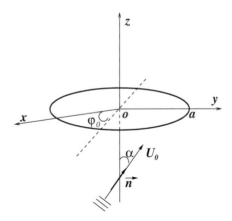

FIGURE 8.9. Soft or rigid circular disc, excited by a plane wave.

where, as usual, the time-harmonic dependence $\exp\left(-i\omega t\right)$ is suppressed. We consider both soft and rigid (i.e., hard) discs, and denote the scattered fields by $U_{sc}^{(S)}$ and $U_{sc}^{(R)}$, respectively; as usual the scattered fields are solutions of the Helmholtz equation obeying the Sommerfeld radiation conditions.

We express the solution as the sum

$$U^t = U_0 + U_{sc}^{(S,R)} \tag{8.180}$$

and seek solutions for the scattered field in the form

$$U_{sc}^{(S)} = \sum_{m=0}^{\infty} \left(2 - \delta_m^0\right) \cos m\left(\varphi - \varphi_0\right) \int_0^{\infty} g_m^{(S)}(\nu) J_m(\nu\rho)\, e^{-\sqrt{\nu^2 - k^2}|z|}\, d\nu, \tag{8.181}$$

$$U_{sc}^{(R)} =$$
$$\operatorname{sgn}\left(z\right) \sum_{m=0}^{\infty} \left(2 - \delta_m^0\right) \cos m\left(\varphi - \varphi_0\right) \int_0^{\infty} g_m^{(R)}(\nu) J_m(\nu\rho)\, e^{-\sqrt{\nu^2 - k^2}|z|}\, d\nu, \tag{8.182}$$

where $\operatorname{Im}\left(\sqrt{\nu^2 - k^2}\right) \le 0$, and the coefficients $g_m^{(S)}(\nu), g_m^{(R)}(\nu)$ are to be determined.

Boundary conditions are formulated in the plane $z = 0$. In the soft case we require

$$U^t\left(\rho, \varphi, +0\right) = U^t\left(\rho, \varphi, -0\right) = 0, \rho < a \tag{8.183}$$

$$\left.\frac{\partial U_{sc}^{(S)}}{\partial z}\right|_{z=+0} = \left.\frac{\partial U_{sc}^{(S)}}{\partial z}\right|_{z=-0}, \quad \rho > a, \qquad (8.184)$$

for all $\varphi \in (0, 2\pi)$; in the rigid case we require

$$\left.\frac{\partial U^t}{\partial z}\right|_{z=+0} = \left.\frac{\partial U^t}{\partial z}\right|_{z=-0} = 0, \quad \rho < a, \qquad (8.185)$$

$$U_{sc}^{(R)}(\rho, \varphi, +0) = U_{sc}^{(R)}(\rho, \varphi, -0), \quad \rho > a, \qquad (8.186)$$

for all $\varphi \in (0, 2\pi)$. Enforcement of the appropriate boundary conditions leads to the dual integral equations

$$\int_0^\infty g_m^{(S)}(\nu) J_m(\nu\rho) d\nu = -J_m(k\rho \sin \alpha), \quad \rho < a \qquad (8.187)$$

$$\int_0^\infty \sqrt{\nu^2 - k^2} g_m^{(S)}(\nu) J_m(\nu\rho) d\nu = 0, \quad \rho > a \qquad (8.188)$$

in the soft case, and

$$\int_0^\infty \sqrt{\nu^2 - k^2} g_m^{(R)}(\nu) J_m(\nu\rho) d\nu = ik \cos \alpha J_m(k\rho \sin \alpha), \quad \rho < a \qquad (8.189)$$

$$\int_0^\infty g_m^{(R)}(\nu) J_m(\nu\rho) d\nu = 0, \quad \rho > a \qquad (8.190)$$

in the rigid case; the equations (8.187)–(8.190) hold for all $m = 0, 1, 2, \ldots$.
The appropriate asymptotically small parameters are (see also Section 8.1)

$$\varepsilon(\nu, k) = \frac{\sqrt{\nu^2 - k^2}}{\nu} - 1, \quad \mu(\nu, k) = \frac{\varepsilon(\nu, k)}{1 + \varepsilon(\nu, k)} = 1 - \frac{\nu}{\sqrt{\nu^2 - k^2}}. \qquad (8.191)$$

Application of the Abel integral transform method transforms the dual integral equations (8.187)–(8.188) to the form

$$\int_0^\infty \nu^{\frac{1}{2}} G_m^{(S)}(\nu) J_{m-\frac{1}{2}}(\nu\rho) d\nu =$$
$$\begin{cases} -(k \sin \alpha)^{\frac{1}{2}} J_{m-\frac{1}{2}}(k \sin \alpha \rho) + \\ \int_0^\infty \nu^{\frac{1}{2}} G_m^{(S)}(\nu) \mu(\nu, k) J_{m-\frac{1}{2}}(\nu\rho) d\nu, & \rho < a \\ 0, & \rho > a \end{cases} \qquad (8.192)$$

where

$$G_m^{(S)}(\nu) = \{1 + \varepsilon(\nu, k)\}\, g_m^{(S)}(\nu) = \frac{\sqrt{\nu^2 - k^2}}{\nu}\, g_m^{(S)}(\nu). \qquad (8.193)$$

Likewise, the dual integral equations (8.189)–(8.190) are transformed to the form

$$\int_0^\infty \nu^{\frac{1}{2}} g_m^{(R)}(\nu) J_{m+\frac{1}{2}}(\nu\rho) d\nu =$$

$$\begin{cases} i\cot\alpha\,(k\sin\alpha)^{\frac{1}{2}} J_{m+\frac{1}{2}}(k\sin\alpha\rho) - \\ \int_0^\infty \nu^{\frac{1}{2}} g_m^{(R)}(\nu)\varepsilon(\nu, k)\, J_{m+\frac{1}{2}}(\nu\rho) d\nu, & \rho < a \\ 0, & \rho > a \end{cases} \qquad (8.194)$$

We now convert (8.192)–(8.193) to a second kind i.s.l.a.e. in a familiar way that relies on the well-known discontinuous integrals of the Weber-Schafheitlin type

$$\int_0^\infty J_{m-\frac{1}{2}}(\nu\rho) J_{2n+m+\frac{1}{2}}(\nu a) d\nu = \begin{cases} \frac{\rho^{m-\frac{1}{2}}}{a^{m+\frac{1}{2}}} P_n^{(m-\frac{1}{2},0)}(1 - 2\rho^2/a^2), & \rho < a \\ 0, & \rho > a \end{cases}$$

$$\qquad (8.195)$$

and

$$\int_0^\infty J_{m+\frac{1}{2}}(\nu\rho) J_{2n+m+\frac{3}{2}}(\nu a) d\nu = \begin{cases} \frac{a^{m+\frac{1}{2}}}{\rho^{m+\frac{3}{2}}} P_n^{(m+\frac{1}{2},0)}(1 - 2\rho^2/a^2), & \rho < a \\ 0, & \rho > a. \end{cases}$$

$$\qquad (8.196)$$

Expand the unknown functions $G_m^{(S)}(\nu)$ or $g_m^{(R)}(\nu)$ in the Neumann series

$$G_m^{(S)}(\nu) = a(k\sin\alpha)^{\frac{1}{2}}\nu^{-\frac{1}{2}} \sum_{p=0}^\infty (4p + 2m + 1)^{\frac{1}{2}}\, x_p^m\, J_{2p+m+\frac{1}{2}}(\nu a) \qquad (8.197)$$

$$g_m^{(R)}(\nu) = ia\cot\alpha(k\sin\alpha)^{\frac{1}{2}}\nu^{-\frac{1}{2}} \sum_{p=0}^\infty (4p + 2m + 3)^{\frac{1}{2}}\, y_p^m\, J_{2p+m+\frac{3}{2}}(\nu a)$$

$$\qquad (8.198)$$

where the unknown coefficients x_p^m and y_p^m are to be determined.

Substitute these expansions in (8.192)–(8.194) and simplify using the Weber-Schafheitlin integrals (8.195)–(8.196). For fixed m, the polynomials

$P_n^{(m\pm\frac{1}{2},0)}(1-2z^2)$ are orthogonal with respect to a certain weight function on $(0,1)$:

$$\int_0^1 z^{2m+1+(\mp1)} P_n^{(m\mp\frac{1}{2},0)}(1-2z^2) P_s^{(m\mp\frac{1}{2},0)}(1-2z^2) dz =$$

$$\frac{\delta_{ns}}{2n+2m+1+(1\mp1)}. \quad (8.199)$$

Using this property we may deduce the second kind i.s.l.a.e. satisfied by the unknown coefficients x_n^m or y_n^m. For fixed m, the systems are

$$x_n^m - \sum_{p=0}^{\infty} x_p^m \alpha_{pn}^m = \beta_n^m \quad (8.200)$$

$$y_n^m - \sum_{p=0}^{\infty} y_p^m \gamma_{pn}^m = \xi_n^m \quad (8.201)$$

where $n = 0, 1, \ldots$ and

$$\begin{aligned}
\alpha_{pn}^m &= [(4p+2m+1)(4n+2m+1)]^{\frac{1}{2}} A_{pn}^m, \\
\beta_n^m &= -(4n+2m+1)^{\frac{1}{2}} \frac{J_{2n+m+\frac{1}{2}}(\varkappa\sin\alpha)}{\varkappa\sin\alpha}, \\
\gamma_{pn}^m &= [(4p+2m+3)(4n+2m+3)]^{\frac{1}{2}} B_{pn}^m, \\
\xi_n^m &= (4n+2m+3)^{\frac{1}{2}} \frac{J_{2n+m+\frac{3}{2}}(\varkappa\sin\alpha)}{\varkappa\sin\alpha}, \quad (8.202)
\end{aligned}$$

with $\varkappa = ka = 2\pi a/\lambda$ denoting relative wave number and

$$A_{pn}^m = \int_0^\infty \frac{\mu(\xi,\varkappa)}{\xi} J_{2p+m+\frac{1}{2}}(\xi) J_{2n+m+\frac{1}{2}}(\xi) d\xi \quad (8.203)$$

$$B_{pn}^m = \int_0^\infty \frac{\varepsilon(\xi,\varkappa)}{\xi} J_{2p+m+\frac{3}{2}}(\xi) J_{2n+m+\frac{3}{2}}(\xi) d\xi; \quad (8.204)$$

recall that $\varepsilon(\xi,\varkappa) = \xi^{-1}\sqrt{\xi^2-\varkappa^2}-1$ and $\mu(\xi,\varkappa) = \varepsilon(\xi,\varkappa)/(1+\varepsilon(\xi,\varkappa))$. Notice that the integrands in equations (8.203) and (8.204) are $O(\xi^{-4})$ as $\xi \to \infty$.

Let us obtain formulae for the jump functions

$$j^{(S)} = \left.\frac{\partial U^{(S)}}{\partial z}\right|_{z=+0} - \left.\frac{\partial U^{(S)}}{\partial z}\right|_{z=-0} \quad (8.205)$$

$$j^{(R)} = U^{(R)}(\rho, \varphi, +0) - U^{(R)}(\rho, \varphi, -0); \qquad (8.206)$$

the functions $j^{(S)}$ and $j^{(R)}$ may be termed *acoustic surface current densities*. In terms of the functions $G_m^{(S)}(\nu)$ or $g_m^{(R)}(\nu)$ the surface current densities take the form

$$j^{(S)} = -2 \sum_{m=0}^{\infty} (2 - \delta_m^0) \cos m(\varphi - \varphi_0) S_m^{(S)}(\rho), \qquad (8.207)$$

$$j^{(R)} = 2 \sum_{m=0}^{\infty} (2 - \delta_m^0) \cos m(\varphi - \varphi_0) S_m^{(R)}(\rho), \qquad (8.208)$$

where

$$S_m^{(S)}(\rho) = \int_0^{\infty} \nu G_m^{(S)}(\nu) J_m(\nu\rho) d\nu, \qquad (8.209)$$

$$S_m^{(R)}(\rho) = \int_0^{\infty} g_m^{(R)}(\nu) J_m(\nu\rho) d\nu. \qquad (8.210)$$

Upon inserting the expansions (8.197)–(8.198) we obtain

$$S_m^{(S)}(\rho) = a(k \sin \alpha)^{\frac{1}{2}} \sum_{p=0}^{\infty} (4p + 2m + 1)^{\frac{1}{2}} x_p^m W_{2p+m+\frac{1}{2},m}^{\frac{1}{2}}(a, \rho), \quad (8.211)$$

$$S_m^{(R)}(\rho) = ia \cot \alpha (k \sin \alpha)^{\frac{1}{2}} \sum_{p=0}^{\infty} (4p + 2m + 3)^{\frac{1}{2}} y_p^m W_{2p+m+\frac{3}{2},m}^{-\frac{1}{2}}(a, \rho),$$
$$(8.212)$$

where $W_{2p+m+\frac{1}{2},m}^{\frac{1}{2}}(a, \rho)$ and $W_{2p+m+\frac{3}{2},m}^{-\frac{1}{2}}(a, \rho)$ are the following readily evaluated integrals of Weber-Schafheitlin type (with $\delta = 0, 1$),

$$W_{2p+m+\frac{1}{2}+\delta,m}^{\frac{1}{2}-\delta}(a, \rho) = \int_0^{\infty} \nu^{\frac{1}{2}-\delta} J_{2p+m+\frac{1}{2}+\delta}(\nu a) J_m(\nu\rho) d\nu.$$

These integrals vanish when $\rho > a$, and when $\rho < a$ have value

$$(-1)^m 2^{-m+\frac{1}{2}-\delta} a^{-\frac{1}{2}} \frac{\Gamma(p+1)}{\Gamma(p+m+\frac{1}{2}+\delta)} (a^2 - \rho^2)^{\frac{(\delta-1)}{2}} P_{2p+m+\delta}^m(\sqrt{1 - \rho^2/a^2}).$$
$$(8.213)$$

Thus the acoustic surface current densities vanish when $\rho > a$, and when $\rho < a$ have the final forms

$$S_m^{(S)}(\rho) = (-1)^m 2^{-m+\frac{1}{2}} \frac{(ka\sin\alpha)^{\frac{1}{2}}}{\sqrt{a^2 - \rho^2}} \times$$

$$\sum_{p=0}^{\infty} (4p + 2m + 1)^{\frac{1}{2}} \frac{\Gamma(p+1)}{\Gamma(p + m + \frac{1}{2})} x_p^m P_{2p+m}^m(\sqrt{1 - \rho^2/a^2}), \quad (8.214)$$

$$S_m^{(R)}(\rho) = i\cot\alpha(ka\sin\alpha)^{\frac{1}{2}}(-1)^m 2^{-m-\frac{1}{2}} \times$$

$$\sum_{p=0}^{\infty} (4p + 2m + 3)^{\frac{1}{2}} \frac{\Gamma(p+1)}{\Gamma(p + m + \frac{3}{2})} y_p^m P_{2p+m+1}^m(\sqrt{1 - \rho^2/a^2}) \quad (8.215)$$

in the soft and rigid disc cases, respectively.

The representations are very useful for two reasons. First the behaviour of the surface current density near the edge is clearly indicated. The second is that the boundary conditions (8.184) and (8.186) are clearly satisfied.

We now turn to electromagnetic scattering from circular discs. The simplest problem concerns electromagnetic scattering by an ideally conducting circular disc with axially symmetric excitation. As shown in Chapter 1, Maxwell's equations separate into two independent groups, TM type ($E_z \neq 0, H_z = 0$) and TE type ($H_z \neq 0, E_z = 0$): see Equations (1.86)–(1.87). Fields of TM type and TE type are produced by the electric and magnetic vertical dipole, respectively and it suffices to determine the azimuthal components H_φ and E_φ, respectively.

We therefore find the total electromagnetic field resulting from the interaction of the vertical dipole with a perfectly conducting circular disc. The configuration is shown in Figure 8.10. The disc has radius a and lies in the plane $z = 0$. The electric vertical dipole (VED) with moment $\overrightarrow{p}$ or magnetic vertical dipole (VMD) with moment $\overrightarrow{m}$ are located on the z-axis at the point $z = d$. It follows from the results of Section 1.4 that the incident fields for both polarisations are

$$\left\{\begin{matrix} H_\varphi^0 \\ E_\varphi^0 \end{matrix}\right\} = \frac{1}{4\pi}\left\{\begin{matrix} -ikp \\ ikm \end{matrix}\right\}\int_0^\infty \frac{e^{-\sqrt{\nu^2 - k^2}|z-d|}}{\sqrt{\nu^2 - k^2}}\nu^2 J_1(\nu\rho)d\nu. \quad (8.216)$$

As always, we decompose the total field into the sums

$$H_\varphi^{tot} = H_\varphi^0 + H_\varphi^{sc} \quad (8.217)$$

$$E_\varphi^{tot} = E_\varphi^0 + E_\varphi^{sc} \quad (8.218)$$

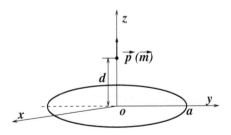

FIGURE 8.10. Metallic disc excited by an electric or magnetic dipole.

in the TM and TE cases, respectively, where the oft-mentioned conditions (of continuity, Sommerfeld radiation conditions, and boundedness of scattered energy) dictate the following forms for the scattered field

$$H_\varphi^{sc} = \frac{p}{4\pi} \operatorname{sgn}(z) \int_0^\infty A(\nu) J_1(\nu\rho) e^{-\sqrt{\nu^2-k^2}|z|} d\nu \qquad (8.219)$$

and

$$E_\varphi^{sc} = \frac{m}{4\pi} \int_0^\infty B(\nu) J_1(\nu\rho) e^{-\sqrt{\nu^2-k^2}|z|} d\nu; \qquad (8.220)$$

the Sommerfeld radiation conditions require that $\operatorname{Im}\left(\sqrt{\nu^2-k^2}\right) \leq 0$, and $A(\nu), B(\nu)$ are unknown spectral density functions to be determined.

The boundary conditions for TM waves are

$$\left.\frac{\partial H_\varphi^{tot}}{\partial z}\right|_{z=+0} = \left.\frac{\partial H_\varphi^{tot}}{\partial z}\right|_{z=-0} = 0, \ \rho < a \qquad (8.221)$$

$$H_\varphi^{sc}(\rho,+0) = H_\varphi^{sc}(\rho,-0), \ \rho > a \qquad (8.222)$$

and those for TE waves are

$$E_\varphi^{tot}(\rho,+0) = E_\varphi^{tot}(\rho,-0) = 0, \ \rho < a \qquad (8.223)$$

$$\left.\frac{\partial E_\varphi^{sc}}{\partial z}\right|_{z=+0} = \left.\frac{\partial E_\varphi^{sc}}{\partial z}\right|_{z=-0}, \ \rho > a. \qquad (8.224)$$

Enforcement of the boundary conditions leads to the dual integral equations

$$\int_0^\infty \sqrt{\nu^2 - k^2} A(\nu) J_1(\nu\rho) d\nu = -ik \int_0^\infty e^{-\sqrt{\nu^2 - k^2} d} \nu^2 J_1(\nu\rho) d\nu, \ \rho < a$$

$$(8.225)$$

$$\int_0^\infty A(\nu) J_1(\nu\rho) d\nu = 0, \ \rho > a \qquad (8.226)$$

in the TM case, and to

$$\int_0^\infty B(\nu) J_1(\nu\rho) d\nu = -ik \int_0^\infty \frac{e^{-\sqrt{\nu^2 - k^2} d}}{\sqrt{\nu^2 - k^2}} \nu^2 J_1(\nu\rho) d\nu, \ \rho < a \qquad (8.227)$$

$$\int_0^\infty \sqrt{\nu^2 - k^2} B(\nu) J_1(\nu\rho) d\nu = 0, \rho > a \qquad (8.228)$$

in the TE case.

We now make use of the previously developed mathematical tools to convert each set of dual series equations to their regularised form. In the TE case, we obtain

$$\int_0^\infty \nu^{\frac{1}{2}} A(\nu) J_{\frac{3}{2}}(\nu\rho) d\nu =$$
$$\begin{cases} -\int_0^\infty \nu^{\frac{1}{2}} \alpha(\nu) J_{\frac{3}{2}}(\nu\rho) d\nu - \int_0^\infty \nu^{\frac{1}{2}} A(\nu) \varepsilon(\nu, k) J_{\frac{3}{2}}(\nu\rho) d\nu, \rho < a \\ 0, \qquad \rho > a \end{cases} \qquad (8.229)$$

and in the TM case we obtain

$$\int_0^\infty \nu^{\frac{1}{2}} B_1(\nu) J_{\frac{1}{2}}(\nu\rho) d\nu =$$
$$\begin{cases} -\int_0^\infty \nu^{\frac{1}{2}} \beta(\nu) J_{\frac{1}{2}}(\nu\rho) d\nu + \int_0^\infty \nu^{\frac{1}{2}} B_1(\nu) \mu(\nu, k) J_{\frac{1}{2}}(\nu\rho) d\nu, \rho < a \\ 0, \qquad \rho > a \end{cases}$$
$$(8.230)$$

where $B_1(\nu) = \nu^{-1}\sqrt{\nu^2 - k^2} B(\nu)$, the parameters $\varepsilon(\nu, k)$ and $\mu(\nu, k)$ are given by the formulae (8.191), and

$$\alpha(\nu) = ik\nu e^{-\sqrt{\nu^2 - k^2} d}, \ \beta(\nu) = ik\nu^2 \frac{e^{-\sqrt{\nu^2 - k^2} d}}{\sqrt{\nu^2 - k^2}}. \qquad (8.231)$$

As both $A(\nu)$ and $B(\nu)$ belong to the class $L_2(0,\infty)$, we may expand them in the Neumann series

$$A(\nu) = a^{-\frac{5}{2}}\nu^{-\frac{1}{2}}\sum_{n=0}^{\infty}(4n+5)^{\frac{1}{2}}\,x_n J_{2n+\frac{5}{2}}(\nu a), \qquad (8.232)$$

$$B_1(\nu) = a^{-\frac{5}{2}}\nu^{-\frac{1}{2}}\sum_{n=0}^{\infty}(4n+3)^{\frac{1}{2}}\,y_n J_{2n+\frac{3}{2}}(\nu a), \qquad (8.233)$$

where the coefficients x_n, y_n are to be determined. A standard argument of the type used many times previously leads to the following second kind i.s.l.a.e. in the TM case

$$x_m + \sum_{n=0}^{\infty}x_n a_{nm} = b_m, \qquad (8.234)$$

where $m = 0, 1, 2, \ldots$,

$$a_{nm} = [(4n+5)(4m+5)]^{\frac{1}{2}}\int_0^{\infty}\frac{\varepsilon(\xi,\varkappa)}{\xi}J_{2n+\frac{5}{2}}(\xi)J_{2m+\frac{5}{2}}(\xi)d\xi \qquad (8.235)$$

$$b_m = -i\varkappa(4m+5)^{\frac{1}{2}}\int_0^{\infty}\xi^{\frac{1}{2}}e^{-\sqrt{\xi^2-\varkappa^2}q}J_{2m+\frac{5}{2}}(\xi)d\xi \qquad (8.236)$$

and $\varkappa = ka$, $q = d/a$; in the TE case we obtain the i.s.l.a.e.

$$y_m - \sum_{n=0}^{\infty}y_n c_{nm} = f_m, \qquad (8.237)$$

where $m = 0, 1, 2, \ldots$, and

$$c_{nm} = [(4n+3)(4m+3)]^{\frac{1}{2}}\int_0^{\infty}\frac{\mu(\xi,\varkappa)}{\xi}J_{2n+\frac{3}{2}}(\xi)J_{2m+\frac{3}{2}}(\xi)d\xi, \qquad (8.238)$$

$$f_m = -i\varkappa(4m+3)^{\frac{1}{2}}\int_0^{\infty}\xi^{\frac{3}{2}}\frac{e^{-\sqrt{\xi^2-\varkappa^2}q}}{\sqrt{\xi^2-\varkappa^2}}J_{2m+\frac{3}{2}}(\xi)d\xi. \qquad (8.239)$$

8.5   Diffraction from Elliptic Plates.

Compared to the literature on circular discs, the diffraction from elliptic discs has been rather less well studied. There appear to be only a handful of papers dealing with this subject; for example see [53] and [9].

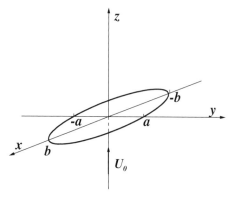

FIGURE 8.11. Diffraction of a plane wave by a soft or rigid elliptic disc.

In Section 8.2 of Volume I we studied some potential problems for elliptic plates. The principle of regularisation relies on analytic inversion of the
static (or singular) part of the appropriate operator to solve dynamic (or
wave-scattering) problems. In other words solutions of Laplace's equation
can be exploited to treat wave-scattering problems described by the Helmholtz equation or Maxwell's equations. Diffraction from elliptic plates is
no exception and we will exploit useful features discovered in the potential
theory setting to treat the wave-scattering problem.

To make the ideas clearer, we consider the rather simple situations of
diffraction of a normally incident acoustic plane wave from soft or rigid
elliptic plate.

Let the elliptic plate with minor and major semi-axes a and b, respectively, lie in the plane $z = 0$ with centre located at the origin (see Figure
8.11); let $q = a/b \, (\leq 1)$. It is excited by the normally incident plane wave
$(\overrightarrow{n} = \overrightarrow{i}_z)$ with velocity potential

$$U_0 = \exp(ikz). \tag{8.240}$$

Let us seek the total field solution in Cartesian coordinates; it has the
obvious symmetry

$$U^{tot}(x, y, z) = U^{tot}(-x, y, z) = U^{tot}(x, -y, z).$$

Decompose as usual the total field as the sum

$$U^{tot} = U_0 + U^{sc}, \tag{8.241}$$

where the scattered field has the form

$$U^{sc}(x, y, z) = \int_0^\infty d\nu \cos \nu x \int_0^\infty d\mu f(\nu, \mu) \cos(\mu y) e^{-\sqrt{\nu^2 + \mu^2 - k^2}|z|}$$

$$\tag{8.242}$$

in the soft case, and the form

$$U^{sc}(x, y, z) = sign(z) \int_0^\infty d\nu \cos \nu x \int_0^\infty d\mu g(\nu, \mu) \cos(\mu y) e^{-\sqrt{\nu^2 + \mu^2 - k^2}|z|}$$

(8.243)

in the rigid case. The unknown spectral density functions f and g are to be determined; the Sommerfeld radiation conditions require the choice $\text{Im}(\sqrt{\nu^2 + \mu^2 - k^2}) \le 0$.

First, consider the soft case. The boundary conditions are

$$U^{tot}(x, y, +0) = U^{tot}(x, y, -0) = 0, \quad \sqrt{x^2/b^2 + y^2/a^2} < 1,$$

(8.244)

$$\left.\frac{\partial U^{sc}}{\partial z}\right|_{z=+0} = \left.\frac{\partial U^{sc}}{\partial z}\right|_{z=-0}, \quad \sqrt{x^2/b^2 + y^2/a^2} > 1.$$

(8.245)

Enforcement of the boundary conditions leads to the two-dimensional dual integral equations for f,

$$\int_0^\infty d\nu \cos \nu x \int_0^\infty d\mu f(\nu, \mu) \cos(\mu y) = -1, \quad \sqrt{x^2/b^2 + y^2/a^2} < 1 \quad (8.246)$$

$$\int_0^\infty d\nu \cos \nu x \int_0^\infty d\mu \sqrt{\nu^2 + \mu^2 - k^2} f(\nu, \mu) \cos(\mu y) = 0,$$

$$\sqrt{x^2/b^2 + y^2/a^2} > 1. \quad (8.247)$$

It is convenient to introduce the parametrisation $x = b\rho \cos \varphi$, $y = a\rho \sin \varphi$ so that the equations (8.246)–(8.247) become

$$\int_0^\infty d\nu \cos(\nu b\rho \cos \varphi) \int_0^\infty d\mu f(\nu, \mu) \cos(\mu a\rho \sin \varphi) = -1, \quad \rho < 1, \quad (8.248)$$

$$\int_0^\infty d\nu \cos(\nu b\rho \cos \varphi) \int_0^\infty d\mu \sqrt{\nu^2 + \mu^2 - k^2} f(\nu, \mu) \cos(\mu a\rho \sin \varphi) = 0,$$

$$\rho > 1 \quad (8.249)$$

for all $\varphi \in (0, 2\pi)$. Guided by the results of section 8.2 of Volume I we expand

$$\cos(\nu b\rho \cos \varphi) \cos(\mu a\rho \sin \varphi) =$$

$$\sum_{m=0}^\infty (2 - \delta_m^0) J_{2m}(\sqrt{\nu^2 + q^2\mu^2} b\rho) T_{2m}\left(\frac{q\mu}{\sqrt{\nu^2 + q^2\mu^2}}\right) T_{2m}(\cos \varphi).$$

(8.250)

The completeness and orthogonality of the functions $T_{2m}(\cos\varphi)$ on $\left(0, \frac{\pi}{2}\right)$ implies that Equations (8.248)–(8.249) are equivalent to

$$\int_0^\infty d\nu \int_0^\infty d\mu\, f(\nu, \mu) J_0\left(\sqrt{\nu^2 + q^2\mu^2}\, b\rho\right) = -1, \quad \rho < 1 \qquad (8.251)$$

$$\int_0^\infty d\nu \int_0^\infty d\mu \sqrt{\nu^2 + \mu^2 - k^2}\, f(\nu, \mu) J_0\left(\sqrt{\nu^2 + q^2\mu^2}\, b\rho\right) = 0, \quad \rho > 1.$$
$$(8.252)$$

Let us transform these equations by introducing the new spectral variable

$$\tau = \sqrt{\nu^2 + q^2\mu^2} \qquad (8.253)$$

and making use of Dirichlet's formula (see [1.135] of Volume I), to

$$\int_0^\infty d\tau \cdot \tau J_0\left(\tau b\rho\right) \int_0^\tau d\nu \frac{F(\nu, \tau)}{\sqrt{\tau^2 - \nu^2}} = -q, \quad \rho < 1 \qquad (8.254)$$

$$\int_0^\infty d\tau \cdot \tau J_0\left(\tau b\rho\right) \int_0^\tau d\nu F(\nu, \tau) \sqrt{\frac{\tau^2 - (1 - q^2)\nu^2 - q^2 k^2}{\tau^2 - \nu^2}} = 0, \quad \rho > 1$$
$$(8.255)$$

where $F(\nu, \tau) = f(\nu, \sqrt{\tau^2 - \nu^2}/q)$.

Taking into account the edge condition and anticipating a form that automatically satisfies (8.255), we expand F in a Neumann series

$$F(\nu, \tau) = \frac{(2b)^{\frac{1}{2}}}{\tau^{\frac{1}{2}}} \left[\tau^2 - (1 - q^2)\nu^2 - q^2 k^2\right]^{-\frac{1}{2}} \sum_{k=0}^\infty x_k J_{2k+\frac{1}{2}}(\tau b) \qquad (8.256)$$

where the coefficients x_k are to be determined. Substitution of this form into equation (8.255) leads to an equation containing integrals of Weber-Schafheitlin type

$$\frac{2}{\pi} \sum_{k=0}^\infty x_k \int_0^\infty \tau^{\frac{1}{2}} J_0(\tau b\rho) J_{2k+\frac{1}{2}}(\tau b) d\tau = 0, \, \rho > 1 \qquad (8.257)$$

because, in fact, equation (8.257) is automatically satisfied term by term, as

$$\int_0^\infty \tau^{\frac{1}{2}} J_0(\tau b\rho) J_{2k+\frac{1}{2}}(\tau b) d\tau = \frac{2^{\frac{1}{2}} \Gamma(n+1)}{b^{\frac{3}{2}} \Gamma(n+\frac{1}{2})} \frac{P_{2k}\left(\sqrt{1-\rho^2}\right)}{\sqrt{1-\rho^2}} H(1-\rho).$$
$$(8.258)$$

The substitution of the representation (8.256) into (8.254) leads to

$$\sum_{k=0}^{\infty} x_k \int_0^{\infty} \tau^{\frac{1}{2}} J_0(\tau b \rho) J_{2k+\frac{1}{2}}(\tau b) R(\tau; q, k) d\tau = -q, \quad \rho < 1 \qquad (8.259)$$

where

$$R(\tau; q, k) = \int_0^{\tau} \frac{d\nu}{\sqrt{(\tau^2 - \nu^2)[\tau^2 - (1 - q^2)\nu^2 - q^2 k^2]}}. \qquad (8.260)$$

At the static extreme $(k \to 0)$ we have

$$R(\tau; q, 0) = \int_0^{\tau} \frac{d\nu}{\sqrt{(\tau^2 - \nu^2)[\tau^2 - (1 - q^2)\nu^2]}} = \frac{K\left(\sqrt{1 - q^2}\right)}{\tau} \qquad (8.261)$$

where K denotes the complete elliptic integral of first kind. In accordance with the concept of regularisation let us split (8.259). Let

$$Q(\tau; q, k) = 1 - \frac{\tau}{K\left(\sqrt{1 - q^2}\right)} R(\tau; q, k) \qquad (8.262)$$

so that $Q(\tau; q, k)$ is an asymptotically small parameter obeying $Q(\tau; q, k) = O\left(k^2/\tau^2\right)$ as $\tau \to \infty$. Carrying out the obvious transformations (using [8.258]) we rewrite (8.259) as

$$\sum_{k=0}^{\infty} \frac{\Gamma(k + \frac{1}{2})}{\Gamma(k + 1)} x_k P_{2k}\left(\sqrt{1 - \rho^2}\right) = -\frac{q}{K\left(\sqrt{1 - q^2}\right)} +$$

$$(2b)^{\frac{1}{2}} \sum_{k=0}^{\infty} x_k \int_0^{\infty} \tau^{-\frac{1}{2}} Q(\tau; q, k) J_0(\tau b \rho) J_{2k+\frac{1}{2}}(\tau b) d\tau, \quad \rho < 1. \quad (8.263)$$

Using the orthogonality of the functions $P_{2k}\left(\sqrt{1 - \rho^2}\right)$ with respect to the weight function $\rho\left(1 - \rho^2\right)^{-\frac{1}{2}}$ on $(0, 1)$, we may reduce equation (8.262) to a second kind i.s.l.a.e. for the rescaled unknown coefficients $X_n = (4n + 1)^{-\frac{1}{2}} x_n$,

$$X_n - \sum_{k=0}^{\infty} X_k S_{kn} = W_n, \qquad (8.264)$$

where $n = 0, 1, 2, \ldots$, and

$$S_{kn} = [(4k + 1)(4n + 1)]^{\frac{1}{2}} \int_0^{\infty} \tau^{-1} Q(\tau; q, k) J_{2n+\frac{1}{2}}(\tau b) J_{2k+\frac{1}{2}}(\tau b) d\tau,$$
$$(8.265)$$

$$W_n = -\frac{q}{K\left(\sqrt{1-q^2}\right)} \frac{\Gamma(n+1)}{\Gamma(n+\frac{1}{2})} \frac{\delta_{0n}}{\sqrt{4n+1}}. \tag{8.266}$$

An accurate evaluation of the function $Q(\tau; q, k)$ contained in the integrand of (8.262) is required; in fact it is simply expressible in terms of elliptic integrals of the first kind. When the disc is circular ($q = 1$),

$$R(\tau; 1, k) = \frac{\pi}{2}\left(\tau^2 - k^2\right)^{-\frac{1}{2}}$$

and $Q(\tau; 1, k)$ degenerates to a form that is characteristic for scattering from a soft circular disc,

$$Q(\tau; 1, k) = 1 - \frac{\tau}{\sqrt{\tau^2 - k^2}} = \frac{\varepsilon(\tau, k)}{1 + \varepsilon(\tau, k)} = \mu(\tau, k) \tag{8.267}$$

where $\varepsilon(\tau, k)$ and $\mu(\tau, k)$ are defined by (8.251). Recall that the square root must be chosen so that $\operatorname{Im} Q(\tau; 1, k) \leq 0$, so that

$$Q(\tau; 1, k) = \begin{cases} 1 - i\tau/\sqrt{k^2 - \tau^2}, & \tau < k \\ 1 - \tau/\sqrt{\tau^2 - k^2}, & \tau > k. \end{cases} \tag{8.268}$$

The form of the parameter $Q(\tau; 1, k)$ reflects the fact that when $\tau < k$ the spectral contribution is due to propagating waves, whereas when $\tau > k$ the spectral contribution comprises evanescent waves that are increasingly exponentially damped as $\tau \to \infty$.

Explicit evaluation of $R(\tau; q, k)$ depends upon whether $\tau < qk$, $qk < \tau < k$ or $\tau > k$. When $\tau < qk$,

$$R(\tau; q, k) = \frac{i}{\sqrt{1-q^2}} \int_0^\tau \left[(\tau^2 - \nu^2)\left(\frac{q^2 k^2 - \tau^2}{1-q^2} + \nu^2\right)\right]^{-\frac{1}{2}} d\nu$$

$$= \frac{i}{q\sqrt{k^2 - \tau^2}} K\left(\frac{\sqrt{1-q^2}\,\tau}{q\sqrt{k^2 - \tau^2}}\right). \tag{8.269}$$

When $\tau \in (qk, k)$, let $c(\tau) = \sqrt{\frac{\tau^2 - q^2 k^2}{1 - q^2}}$ and then

$$\sqrt{1-q^2}\,R(\tau; q, k) =$$

$$\int_0^{c(\tau)} \frac{d\nu}{\sqrt{(\tau^2 - \nu^2)[c^2(\tau) - \nu^2]}} + i \int_{c(\tau)}^\tau \frac{d\nu}{\sqrt{(\tau^2 - \nu^2)[\nu^2 - c^2(\tau)]}}$$

$$= K\left(\frac{\sqrt{\tau^2 - q^2 k^2}}{\tau\sqrt{1-q^2}}\right) + iK\left(\frac{q\sqrt{k^2 - \tau^2}}{\tau\sqrt{1-q^2}}\right). \tag{8.270}$$

Finally when $\tau > k$

$$R(\tau; q, k) = \frac{1}{\sqrt{\tau^2 - q^2 k^2}} K\left(\frac{\tau\sqrt{1-q^2}}{\sqrt{\tau^2 - q^2 k^2}}\right). \tag{8.271}$$

Thus, setting $\xi = \tau b$, $\varkappa_a = ka$ and $\varkappa_b = kb$, the function $\tau R(\tau; q, k)$ is representable as

$$\tau R(\tau; q, k) =$$

$$\begin{cases} \dfrac{i\xi}{q\sqrt{\varkappa_b^2 - \xi^2}} K\left(\dfrac{\sqrt{1-q^2}}{q}\dfrac{\xi}{\sqrt{\varkappa_b^2 - \xi^2}}\right), \xi < \varkappa_a \\[3mm] \dfrac{1}{\sqrt{1-q^2}}\left\{K\left(\dfrac{\sqrt{\xi^2 - \varkappa_a^2}}{\xi\sqrt{1-q^2}}\right) + iK\left(\dfrac{q}{\sqrt{1-q^2}}\dfrac{\sqrt{\varkappa_b^2 - \xi^2}}{\xi}\right)\right\}, \varkappa_a < \xi < \varkappa_b \\[3mm] \dfrac{\xi}{\sqrt{\xi^2 - \varkappa_a^2}} K\left(\sqrt{1-q^2}\dfrac{\xi}{\sqrt{\xi^2 - \varkappa_a^2}}\right), \xi > \varkappa_b \end{cases} \tag{8.272}$$

We now consider the rigid case. The boundary conditions are

$$\left.\frac{\partial U^{tot}}{\partial z}\right|_{z=+0} = \left.\frac{\partial U^{tot}}{\partial z}\right|_{z=-0} = 0, \quad \sqrt{x^2/b^2 + y^2/a^2} < 1 \tag{8.273}$$

$$U^{sc}(x, y, +0) = U^{sc}(x, y, -0), \quad \sqrt{x^2/b^2 + y^2/a^2} > 1 \tag{8.274}$$

where U^{sc} is sought in the form (8.243). In a similar way to the soft case, the rigid disc problem generates the following dual integral equations to be solved

$$\int_0^\infty d\tau\, \tau J_0(\tau b\rho) \int_0^\tau d\nu G(\nu, \tau)\sqrt{\frac{\tau^2 - (1-q^2)\nu^2 - q^2 k^2}{\tau^2 - \nu^2}} = ikq^2, \quad \rho < 1 \tag{8.275}$$

and

$$\int_0^\infty d\tau\, \tau J_0(\tau b\rho) \int_0^\tau d\nu \frac{G(\nu, \tau)}{\sqrt{\tau^2 - \nu^2}} = 0, \quad \rho > 1, \tag{8.276}$$

where $G(\nu, \tau) = g(\nu, \sqrt{\tau^2 - \nu^2}/q)$.

The transformation of (8.275)–(8.276) to a second kind i.s.l.a.e. is quite similar to that for the system occurring with the soft elliptic plate. The relevant expansion of G in a Neumann series is

$$G(\nu, \tau) = \frac{b^{\frac{1}{2}}}{(2\pi)^{\frac{1}{2}}\tau^{\frac{3}{2}}} \sum_{k=0}^\infty y_k J_{2k+\frac{3}{2}}(\tau b) \tag{8.277}$$

where the coefficients y_k are to be determined. This representation auto-matically satisfies the equation (8.276) and transforms equation (8.275) into the equivalent form

$$\pi^{-\frac{1}{2}}b^{-1}\left(1-\rho^2\right)^{-\frac{1}{2}}\sum_{k=0}^{\infty}\frac{\Gamma(k+\frac{3}{2})}{\Gamma(k+1)}y_k\,P_{2k+1}\left(\sqrt{1-\rho^2}\right) = ik\frac{q^2}{E\left(\sqrt{1-q^2}\right)}$$

$$+\frac{b^{\frac{1}{2}}}{(2\pi)^{\frac{1}{2}}}\sum_{k=0}^{\infty}y_k\int_0^{\infty}d\tau\,\tau^{\frac{1}{2}}J_{2k+\frac{3}{2}}\left(\tau b\right)Q(\tau;q,k)J_0(\tau b\rho),\ \rho<1$$

where

$$Q(\tau;q,k) = 1 - \frac{1}{\tau E\left(\sqrt{1-q^2}\right)}\int_0^{\tau}\sqrt{\frac{\tau^2-(1-q^2)\,\nu^2-q^2k^2}{\tau^2-\nu^2}}\,d\nu. \quad (8.278)$$

$Q(\tau;q,k)$ is an asymptotically small parameter that obeys $Q(\tau;q,k) = O(k^2/\tau^2)$ as $\tau\to\infty$.

As before we use the orthogonality of the functions $P_{2k+1}\left(\sqrt{1-\rho^2}\right)$ with respect to the weight function $\rho\left(1-\rho^2\right)^{-1}$ on $(0,1)$, i.e.,

$$\int_0^1\frac{\rho}{\sqrt{1-\rho^2}}P_{2k+1}\left(\sqrt{1-\rho^2}\right)P_{2n+1}\left(\sqrt{1-\rho^2}\right)d\rho$$

$$=\int_0^1\frac{\rho}{\sqrt{1-\rho^2}}P_{2k+1}\left(x\right)P_{2n+1}\left(x\right)dx = \frac{\delta_{kn}}{4n+3},$$

and also use the tabulated values

$$\int_0^1\rho J_0(\tau b\rho)P_{2n+1}\left(\sqrt{1-\rho^2}\right)d\rho = \frac{2^{\frac{1}{2}}}{(\tau b)^{\frac{3}{2}}}\frac{\Gamma(n+\frac{3}{2})}{\Gamma(n+1)}J_{2n+\frac{3}{2}}(\tau b),$$

$$\int_0^1\rho P_{2n+1}\left(\sqrt{1-\rho^2}\right)d\rho = \frac{\delta_{0n}}{4n+3}. \quad (8.279)$$

A standard argument produces the second kind i.s.l.a.e. for the rescaled coefficients $Y_n = (4n+3)^{-\frac{1}{2}}y_n$,

$$Y_n - \sum_{k=0}^{\infty}Y_k S_{kn}^{(R)} = W_n^{(R)}, \quad (8.280)$$

where $n = 0,1,2,\dots$, and

$$S_{kn}^{(R)} = [(4n+3)\,(4k+3)]^{\frac{1}{2}}\int_0^{\infty}\tau^{-1}Q(\tau;q,k)J_{2k+\frac{3}{2}}\left(\tau b\right)J_{2n+\frac{3}{2}}\left(\tau b\right)d\tau,$$

$$(8.281)$$

$$W_n^{(R)} = \frac{2i}{\sqrt{3}} \frac{q^2 \varkappa_b}{E\left(\sqrt{1-q^2}\right)} \delta_{0n}. \tag{8.282}$$

The integral

$$R^{(R)}\left(\tau; q, k\right) \overset{def}{=} \frac{1}{\tau} \int_0^\tau \sqrt{\frac{\tau^2 - \left(1-q^2\right)\nu^2 - q^2 k^2}{\tau^2 - \nu^2}}\, d\nu \tag{8.283}$$

contained in (8.278) may be evaluated in terms of complete elliptic integrals of the first and second kind, depending upon the relationship between τ and the wavenumbers qk and k. When $\tau < qk$,

$$R^{(R)}\left(\tau; q, k\right) = -i\sqrt{1-q^2} \int_0^1 \sqrt{\frac{b^2 + x^2}{1 - x^2}}\, dx$$

$$= -iq\frac{\sqrt{k^2 - \tau^2}}{\tau} E\left(\frac{\sqrt{1-q^2}\tau}{q\sqrt{k^2-\tau^2}}\right) \tag{8.284}$$

where $b^2 = \left(q^2 k^2 - \tau^2\right) / \left(1 - q^2\right)\tau^2$. When $qk < \tau < k$, set $a^2 = -b^2$, and then

$$R\left(\tau; q, k\right) = \sqrt{1-q^2} \int_0^a \sqrt{\frac{a^2 - x^2}{1-x^2}}\, dx - i\sqrt{1-q^2} \int_a^1 \sqrt{\frac{x^2 - a^2}{1-x^2}}\, dx$$

$$= \sqrt{1-q^2}\left\{ E\left(\frac{\sqrt{\tau^2 - q^2 k^2}}{\sqrt{1-q^2}\tau}\right) - \frac{q^2\left(k^2 - \tau^2\right)}{\left(1-q^2\right)\tau^2} K\left(\frac{\sqrt{\tau^2 - q^2 k^2}}{\sqrt{1-q^2}\tau}\right) \right\}$$

$$- i\sqrt{1-q^2}\left\{ E\left(\frac{q\sqrt{k^2 - \tau^2}}{\sqrt{1-q^2}\tau}\right) - \frac{\tau^2 - q^2 k^2}{\left(1-q^2\right)\tau^2} K\left(\frac{q\sqrt{k^2 - \tau^2}}{\sqrt{1-q^2}\tau}\right) \right\}. \tag{8.285}$$

When $\tau > k$,

$$R^{(R)}\left(\tau; q, k\right) = \sqrt{1-q^2} \int_0^1 \sqrt{\frac{a^2 - x^2}{1-x^2}}\, dx$$

$$= \frac{\sqrt{\tau^2 - q^2 k^2}}{\tau} E\left(\frac{\sqrt{1-q^2}\tau}{\sqrt{\tau^2 - q^2 k^2}}\right). \tag{8.286}$$

It should be noted that function $R\left(\tau; q, k\right)$ is continuous for $\tau \in (0, \infty)$. When $qk < \tau < k$ it is possible to represent $R\left(\tau; q, k\right)$ more compactly, using the relation [30]

$$\frac{dK(k)}{dk} = \frac{1}{k}\left[\frac{1}{(k')^2} E(k) - K(k)\right] \tag{8.287}$$

where, as usual, $k' = \sqrt{1 - k^2}$ denotes the complementary modulus; thus

$$R(\tau; q, k) = \sqrt{1 - q^2} \left\{ k_1 \left(k_1'\right)^2 \frac{dK(k_1)}{dk_1} - ik_2 \left(k_2'\right)^2 \frac{dK(k_2)}{dk_2} \right\} \quad (8.288)$$

where

$$k_1 = \frac{\sqrt{\tau^2 - q^2 k^2}}{\sqrt{1 - q^2}\tau}, \quad k_2 = \frac{\sqrt{\tau^2 - q^2 k^2}}{\sqrt{1 - q^2}\tau}.$$

It can be shown that the *jump function* is given by

$$\left.\frac{\partial U^{sc}}{\partial z}\right|_{z=+0} - \left.\frac{\partial U^{sc}}{\partial z}\right|_{z=-0} =$$

$$-2 \int_0^\infty d\nu \cos(\nu x) \int_0^\infty d\mu f(\nu, \mu) \sqrt{\nu^2 + \mu^2 - k^2} \cos(\mu y) =$$

$$-\frac{2\pi}{q^2 b} \frac{H(1-\rho)}{\sqrt{1-\rho^2}} \sum_{k=0}^\infty (4k+1)^{\frac{1}{2}} \frac{\Gamma(k+1)}{\Gamma\left(k+\frac{1}{2}\right)} X_k P_{2k}\left(\sqrt{1-\rho^2}\right), \quad (8.289)$$

in the soft case, and by

$$U^{sc}(x, y, +0) - U^{sc}(x, y, -0) =$$

$$2 \int_0^\infty d\nu \cos(\nu x) \int_0^\infty d\mu g(\nu, \mu) \cos(\mu y) =$$

$$-\frac{\sqrt{\pi}}{2q} H(1-\rho) \sum_{k=0}^\infty (4k+3)^{\frac{1}{2}} \frac{\Gamma(k+1)}{\Gamma\left(k+\frac{3}{2}\right)} Y_k P_{2k}\left(\sqrt{1-\rho^2}\right) \quad (8.290)$$

in the hard case. We recall that the boundary of the elliptic disc is given by $\rho = 1$. Also we note that the Heaviside function is

$$H(1-\rho) = \begin{cases} 1, \rho < 1 \\ 0, \rho > 1 \end{cases}.$$

We conclude with some illustrative calculations for the rigid elliptic plate. The distributions of the normalised field $|U^{tot}/U_0|$ along the major axis (in normalised units x/λ) on the illuminated and shadowed sides of the elliptic plate are given in Figures 8.12 and 8.13, respectively. Three values of the aspect ratio q are examined. When $q = 1$, the plate is a circular disc and the results are in excellent agreement with those of [13] (p. 547).

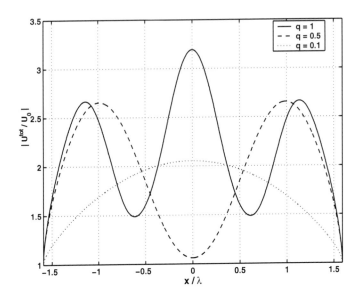

FIGURE 8.12. Normalised field distribution on the illuminated side of the rigid elliptic plate, for various aspect ratios (normal incidence).

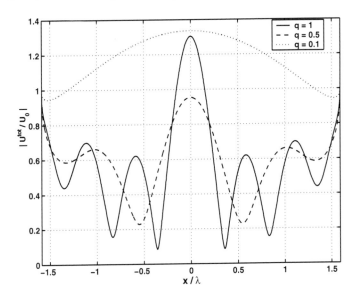

FIGURE 8.13. Normalised field distribution on the shadowed side of the rigid elliptic plate, for various aspect ratios (normal incidence).

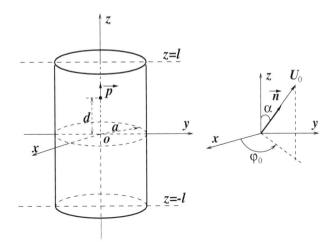

FIGURE 8.14. Plane wave scattering by a hollow finite cylinder.

8.6 Wave Scattering Problems for Hollow Finite Cylinders.

From a topological point of view, a screen shaped as a hollow cylinder of finite length is rather more complicated than a disc. The disc is parameterised by a single parameter (its radius), but the hollow cylinder is specified by its radius a and length $2l$; from both diffraction and computational points of view the magnitude of the ratio a/l is extremely important. Some useful references on this subject are to be found in [123], [45], [70] and [55].

We continue to exploit polar cylindrical coordinates in solving some wave-scattering problems for structures with a cavity and edges.

Consider the hollow finite cylinder with length $2l$ and of circular cross-section with radius a. It is convenient to locate it symmetrically with respect to the z-axis, so that it is described by

$$\rho = a, \ \varphi \in (0, 2\pi), \ z \in (-l, l)$$

as shown in Figure 8.14. An acoustic plane wave is incident at angle α with velocity potential

$$U_0 \equiv U_0(\rho, \varphi, z) = e^{ik\{\rho \sin \alpha \cos(\varphi - \varphi_0) + z \cos \alpha\}}. \qquad (8.291)$$

Without loss of a generality we may set $\varphi_0 = 0$, and then

$$U_0 = e^{ikz \cos \alpha} \sum_{m=0}^{\infty} \left(2 - \delta_m^0\right) i^m J_m(k\rho \sin \alpha) \cos m\varphi. \qquad (8.292)$$

The total solution is decomposed in a familiar way as

$$U^{tot} = U_0 + U^{sc}, \tag{8.293}$$

where the scattered field U^{sc} satisfies the Helmholtz equation. Now suppose that the surface of the finite hollow cylinder is acoustically soft; the scattered velocity potential is continuous across the surface $\rho = a$ for all z and $\varphi \in (0, 2\pi)$,

$$U^{sc}(a+0, \varphi, z) = U^{sc}(a-0, \varphi, z).$$

Taking into account the Sommerfeld radiation condition, we seek the solution in the form

$$U^{sc} = \sum_{m=0}^{\infty} \left(2 - \delta_m^0\right) \cos m\varphi \int_0^{\infty} [A_m(\nu)\cos\nu z + B_m(\nu)\sin\nu z] \times$$
$$\left\{ \begin{array}{ll} J_m\left(\sqrt{k^2 - \nu^2}\rho\right), & \rho < a \\ \frac{J_m\left(\sqrt{k^2-\nu^2}a\right)}{H_m^{(1)}\left(\sqrt{k^2-\nu^2}a\right)} H_m^{(1)}\left(\sqrt{k^2-\nu^2}a\right), & \rho > a \end{array} \right\} d\nu \tag{8.294}$$

where $\mathrm{Im}\left(\sqrt{k^2 - \nu^2}\right) \geq 0$. Note when $k < \nu$,

$$J_m\left(\sqrt{k^2 - \nu^2}\rho\right) = J_m\left(i\sqrt{\nu^2 - k^2}\rho\right) = i^m I_m\left(\sqrt{\nu^2 - k^2}\rho\right)$$

$$H_m^{(1)}\left(\sqrt{k^2 - \nu^2}\rho\right) = H_m^{(1)}\left(i\sqrt{\nu^2 - k^2}\rho\right)$$
$$= \frac{2}{\pi}(-i)^{m+1} K_m\left(\sqrt{\nu^2 - k^2}\rho\right) \tag{8.295}$$

where I_m and K_m are modified Bessel functions of first and third kind, respectively.

For the acoustically soft hollow finite cylinder the boundary conditions take the form

$$U^{tot}(a+0, \varphi, z) = U^{tot}(a-0, \varphi, z) = 0, \quad |z| < l \tag{8.296}$$

$$\left.\frac{\partial U^{sc}}{\partial \rho}\right|_{\rho=a+0} = \left.\frac{\partial U^{sc}}{\partial \rho}\right|_{\rho=a-0}, \quad |z| > l \tag{8.297}$$

for all $\varphi \in (0, 2\pi)$. After imposing the boundary conditions we obtain the following pair of dual integral equations for the unknown spectral density functions $A_m(\nu)$ and $B_m(\nu)$,

$$\int_0^{\infty} A_m(\nu) J_m\left(\sqrt{k^2 - \nu^2}a\right)\cos\nu z d\nu =$$
$$-i^m J_m(ka\sin\alpha)\cos(k\cos\alpha z), \quad z < l \tag{8.298}$$

$$\int_0^\infty \frac{A_m\left(\nu\right)}{H_m^{(1)}\left(\sqrt{k^2-\nu^2}a\right)}\cos\nu z d\nu = 0, \quad z > l \tag{8.299}$$

and

$$\int_0^\infty B_m\left(\nu\right)J_m\left(\sqrt{k^2-\nu^2}a\right)\sin\nu z d\nu =$$
$$i^{m-1}J_m\left(ka\sin\alpha\right)\sin\left(k\cos\alpha z\right), \quad z < l \tag{8.300}$$

$$\int_0^\infty \frac{B_m\left(\nu\right)}{H_m^{(1)}\left(\sqrt{k^2-\nu^2}a\right)}\sin\nu z d\nu = 0, \quad z > l. \tag{8.301}$$

For an acoustically rigid (hard) hollow finite cylinder we require continuity of the normal derivative of velocity potential

$$\left.\frac{\partial U}{\partial\rho}\right|_{\rho=a+0} = \left.\frac{\partial U}{\partial\rho}\right|_{\rho=a-0}, \quad \varphi\in\left(0,2\pi\right), |z| < \infty \tag{8.302}$$

and the appropriate form of solution is

$$U^{sc} = \sum_{m=0}^\infty \left(2-\delta_m^0\right)\cos m\varphi\int_0^\infty \left[A_m\left(\nu\right)\cos\nu z + B_m\left(\nu\right)\sin\nu z\right]\times$$
$$\left\{\begin{array}{ll} J_m\left(\sqrt{k^2-\nu^2}\rho\right), & \rho < a \\ \frac{J_m'\left(\sqrt{k^2-\nu^2}a\right)}{H_m^{(1)'}\left(\sqrt{k^2-\nu^2}a\right)}H_m^{(1)}\left(\sqrt{k^2-\nu^2}\rho\right), & \rho > a \end{array}\right\}d\nu. \tag{8.303}$$

The boundary conditions to be enforced across the cylindrical surface $\rho = a$, $\varphi\in\left(0,2\pi\right)$ are

$$\left.\frac{\partial U^{tot}}{\partial\rho}\right|_{\rho=a+0} = \left.\frac{\partial U^{tot}}{\partial\rho}\right|_{\rho=a-0}, \quad |z| < l \tag{8.304}$$

$$U^{sc}\left(a+0,\varphi,z\right) = U^{sc}\left(a-0,\varphi,z\right), \quad |z| > l \tag{8.305}$$

for all $\varphi\in\left(0,2\pi\right)$, and the corresponding pair of dual integral equations for the unknown spectral density functions $A_m\left(\nu\right)$ and $B_m\left(\nu\right)$ to be determined are

$$\int_0^\infty \sqrt{k^2-\nu^2}A_m\left(\nu\right)J_m'\left(\sqrt{k^2-\nu^2}a\right)\cos\nu z d\nu =$$
$$-i^m k\sin\alpha J_m'\left(ka\sin\alpha\right)\cos\left(k\cos\alpha z\right), \quad z < l \tag{8.306}$$

$$\int_0^\infty \frac{A_m(\nu)}{\sqrt{k^2 - \nu^2} H_m^{(1)\prime}\left(\sqrt{k^2 - \nu^2}a\right)} \cos \nu z \, d\nu = 0, \qquad z > l \qquad (8.307)$$

and

$$\int_0^\infty \sqrt{k^2 - \nu^2}\, B_m(\nu)\, J_m'\left(\sqrt{k^2 - \nu^2}a\right) \sin \nu z \, d\nu =$$
$$- i^{m=1} k \sin \alpha J_m'\left(ka \sin \alpha\right) \sin\left(k \cos \alpha z\right), \quad z < l \quad (8.308)$$

$$\int_0^\infty \frac{B_m(\nu)}{\sqrt{k^2 - \nu^2} H_m^{(1)\prime}\left(\sqrt{k^2 - \nu^2}a\right)} \sin \nu z \, d\nu = 0, \qquad z > l. \qquad (8.309)$$

The dual integral equations are of the same type as those solved in Section 8.1; the difference lies in the choice of asymptotically small parameters determined by the products $I_m(z) K_m(z)$ and $I_m'(z) K_m'(z)$. When $\nu > k$,

$$J_m\left(\sqrt{k^2 - \nu^2}a\right) H_m^{(1)}\left(\sqrt{k^2 - \nu^2}a\right) =$$
$$- i\frac{2}{\pi} I_m\left(\sqrt{\nu^2 - k^2}a\right) K_m\left(\sqrt{\nu^2 - k^2}a\right), \quad (8.310)$$

and using the well-known asymptotic formulae (see, for example, equations [B.164]–[B.165] of Volume I) we find that

$$i\pi\nu a J_m\left(\sqrt{k^2 - \nu^2}a\right) H_m^{(1)}\left(\sqrt{k^2 - \nu^2}a\right) = 1 + O\left(\frac{(ka)^2 + m^4}{(\nu a)^2}\right) \qquad (8.311)$$

as $\nu a \to \infty$. For the soft case, the asymptotic parameter is thus chosen to be

$$\varepsilon(\nu, k) = 1 - i\pi\nu a J_m\left(\sqrt{k^2 - \nu^2}a\right) H_m^{(1)}\left(\sqrt{k^2 - \nu^2}a\right). \qquad (8.312)$$

Also when $\nu > k$,

$$J_m'\left(\sqrt{k^2 - \nu^2}a\right) H_m^{(1)\prime}\left(\sqrt{k^2 - \nu^2}a\right) =$$
$$i\frac{2}{\pi} I_m'\left(\sqrt{\nu^2 - k^2}a\right) K_m'\left(\sqrt{\nu^2 - k^2}a\right) \qquad (8.313)$$

and

$$i\pi\nu a J_m'\left(\sqrt{k^2 - \nu^2}a\right) H_m^{(1)\prime}\left(\sqrt{k^2 - \nu^2}a\right) = 1 + O\left(\frac{(ka)^2 + m^4}{(\nu a)^2}\right) \qquad (8.314)$$

as $\nu a \to \infty$. The appropriate asymptotically small parameter for the rigid case is chosen to be

$$\mu(\nu, k) = 1 - i\pi\nu a J'_m\left(\sqrt{k^2 - \nu^2}a\right) H_m^{(1)'}\left(\sqrt{k^2 - \nu^2}a\right). \qquad (8.315)$$

We now solve equations (8.300)–(8.301) after writing them in the form

$$\int_0^\infty b_m(\nu)\left[1 - \varepsilon(\nu, k)\right]\sin\nu z d\nu = i^m \pi J_m(ka\sin\alpha)\sin(kz\cos\alpha), \ z < l$$
$$(8.316)$$

$$\int_0^\infty \nu b_m(\nu)\sin\nu z d\nu = 0, \ z > l \qquad (8.317)$$

where

$$B_m(\nu) = \nu a H_m^{(1)}\left(\sqrt{k^2 - \nu^2}a\right) b_m(\nu). \qquad (8.318)$$

A standard argument transforms equations (8.316)–(8.317) to the form

$$\int_0^\infty \nu b_m(\nu) J_0(\nu z) d\nu = \begin{cases} i^m \pi k \cos\alpha J_m(ka\sin\alpha) J_0(k\cos\alpha z) + \\ \int_0^\infty \nu b_m(\nu)\varepsilon(\nu, k) J_0(\nu z) d\nu, & z < l \\ 0, & z > l. \end{cases}$$
$$(8.319)$$

Making the substitution

$$b_m(\nu) = \nu^{-1}\sum_{s=0}^\infty (4s + 2)^{\frac{1}{2}} y_s^m J_{2s+1}(\nu l) \qquad (8.320)$$

leads to the second kind i.s.l.a.e. for the unknown coefficients y_n^m, for each fixed $m = 0, 1, 2, \ldots$,

$$y_n^m - \sum_{s=0}^\infty y_s^m \gamma_{sn} = \varkappa_n^m \qquad (8.321)$$

where $n = 0, 1, 2, \ldots$ and

$$\gamma_{sn} = \left[(4s + 2)(4n + 2)\right]^{\frac{1}{2}} \int_0^\infty \nu^{-1}\varepsilon(\nu, k) J_{2s+1}(\nu l) J_{2n+1}(\nu l) d\nu, \qquad (8.322)$$

$$\varkappa_n^m = i^m (4n + 2)^{\frac{1}{2}} \pi J_m(ka\sin\alpha) J_{2n+1}(kl\cos\alpha). \qquad (8.323)$$

The solution of the corresponding equations (8.298)–(8.299) for the soft case is quite analogous to the solution for the soft strip (see [8.25]–[8.26]). First, introduce the rescaled spectral density function

$$f_m(\nu) = A_m(\nu) / H_m^{(1)}\left(\sqrt{k^2 - \nu^2}a\right) \tag{8.324}$$

and set

$$g_m(\nu) = f_m(\nu) - f_m(0). \tag{8.325}$$

Then Equations (8.298)–(8.299) take the form

$$\int_0^\infty \nu^{-1} g_m(\nu) \cos \nu z d\nu =$$
$$- i\pi a f_m(0) \int_0^\infty J_m\left(\sqrt{k^2 - \nu^2}a\right) H_m^{(1)}\left(\sqrt{k^2 - \nu^2}a\right) \cos \nu z d\nu$$
$$- i^{m+1} \pi a J_m(ka \sin \alpha) \cos(k \cos \alpha z)$$
$$+ \int_0^\infty \nu^{-1} g_m(\nu) \varepsilon(\nu, k) \cos \nu z d\nu, \qquad z < l, \tag{8.326}$$

$$\int_0^\infty g_m(\nu) \cos \nu z d\nu = 0, \qquad z > l. \tag{8.327}$$

Now use the representation (8.77) and, suppressing details of its deduction, the regularised system is (for each fixed $m = 0, 1, 2, \ldots$)

$$X_n^m - \sum_{k=1}^\infty X_k^m \alpha_{kn} = -i\pi W_n^m f(0) + \beta_n^m, \tag{8.328}$$

where $n = 1, 2, \ldots$, and

$$- \sum_{k=1}^\infty X_k^m \alpha_{k0} = -i\pi W_0^m f(0) + \beta_0^m; \tag{8.329}$$

in these equations we set $p = a/l$, $\varkappa = kl$ and

$$g_m(\nu) = \sum_{k=1}^\infty 2\sqrt{k} X_k^m J_{2k}(\nu l), \tag{8.330}$$

$$\alpha_{kn} = 4(kn)^{\frac{1}{2}} \int_0^\infty u^{-1} \varepsilon(u, \varkappa) J_{2k}(u) J_{2n}(u) du,$$
$$\beta_n^m = -i^{m+1} \pi p l J_m(\varkappa p \sin \alpha) J_{2n}(\varkappa \cos \alpha),$$
$$W_n^m = p \int_0^\infty J_m\left(\sqrt{\varkappa^2 - v^2}p\right) H_m^{(1)}\left(\sqrt{\varkappa^2 - v^2}p\right) J_{2n}(v) dv,$$

and

$$\varepsilon\left(u,\varkappa\right) = 1 - i\pi p J_m\left(\sqrt{\varkappa^2 - u^2}p\right) H_m^{(1)}\left(\sqrt{\varkappa^2 - u^2}p\right). \qquad (8.331)$$

We now turn to the solution of the equations (8.306)–(8.309) describing the rigid case. The noteworthy feature of the regularisation process in this case is a two-stage extraction of the singular part. At the first stage we rescale the unknown spectral densities according to

$$\{a_m\left(\nu\right), b_m\left(\nu\right)\} = \frac{\{A_m\left(\nu\right), B_m\left(\nu\right)\}}{\sqrt{k^2 - \nu^2}H_m^{(1)\prime}\left(\sqrt{k^2 - \nu^2}a\right)} \qquad (8.332)$$

and transform the original equations to

$$\int_0^\infty \sqrt{\nu^2 - k^2}a_m\left(\nu\right)\left[1 - \varepsilon_1\left(\nu, k\right)\right]\cos\nu z d\nu =$$

$$i^{m+1}\pi ka\sin\alpha J_m'\left(ka\sin\alpha\right)\cos\left(k\cos\alpha z\right), \quad z < l \quad (8.333)$$

$$\int_0^\infty a_m\left(\nu\right)\cos\nu z d\nu = 0, \ z > l \qquad (8.334)$$

and

$$\int_0^\infty \sqrt{\nu^2 - k^2}b_m\left(\nu\right)\left[1 - \varepsilon_1\left(\nu, k\right)\right]\sin\nu z d\nu =$$

$$- i^m\pi ka\sin\alpha J_m'\left(ka\sin\alpha\right)\sin\left(k\cos\alpha z\right), \quad z < l \quad (8.335)$$

$$\int_0^\infty b_m\left(\nu\right)\sin\nu z d\nu = 0, \ \ z > l. \qquad (8.336)$$

Next extract from both equations (8.334) and (8.336) the previously encountered asymptotically small parameter

$$\varepsilon_2\left(\nu, k\right) = \frac{\sqrt{\nu^2 - k^2}}{\nu} - 1 = O\left(\nu^2/k^2\right), \text{ as } \nu \to \infty \qquad (8.337)$$

and transform (8.334)–(8.336) to

$$\int_0^\infty \nu a_m\left(\nu\right)\cos\nu z d\nu = i^{m+1}\pi ka\sin\alpha J_m'\left(ka\sin\alpha\right)\cos\left(k\cos\alpha z\right) +$$

$$\int_0^\infty \nu a_m\left(\nu\right)\sigma\left(\nu, k\right)\cos\nu z d\nu, \quad z < l \quad (8.338)$$

$$\int_0^\infty a_m(\nu) \cos \nu z \, d\nu = 0, z > l \tag{8.339}$$

and

$$\int_0^\infty \nu b_m(\nu) \sin \nu z \, d\nu = -i^m \pi k a \sin \alpha J_m'(ka \sin \alpha) \sin(k \cos \alpha z)$$

$$+ \int_0^\infty \nu b_m(\nu) \sigma(\nu, k) \sin \nu z \, d\nu, \quad z < l \tag{8.340}$$

$$\int_0^\infty b_m(\nu) \sin \nu z \, d\nu = 0, \quad z > l, \tag{8.341}$$

where

$$\sigma(\nu, k) = \varepsilon_1(\nu, k) - \varepsilon_2(\nu, k)[1 - \varepsilon_1(\nu, k)]. \tag{8.342}$$

A standard application of the Abel integral transform converts (8.338)–(8.340) to the pre-regularised forms

$$\int_0^\infty \nu a_m(\nu) J_0(\nu z) \, d\nu = \begin{cases} i^{m+1} \pi k a \sin \alpha J_m'(ka \sin \alpha) J_0(k \cos \alpha z) \\ + \int_0^\infty \nu a_m(\nu) \sigma(\nu, k) J_0(\nu z) \, d\nu, & z < l \\ 0, & z > l \end{cases} \tag{8.343}$$

and

$$\int_0^\infty \nu b_m(\nu) J_1(\nu z) \, d\nu = \begin{cases} -i^m \pi k a \sin \alpha J_m'(ka \sin \alpha) J_1(k \cos \alpha z) \\ + \int_0^\infty \nu b_m(\nu) \sigma(\nu, k) J_1(\nu z) \, d\nu, & z < l \\ 0, & z > l. \end{cases} \tag{8.344}$$

In the familiar way, we seek solutions in the form of a Neumann series expansion

$$a_m(\nu) = \frac{1}{k\nu} \sum_{s=0}^\infty (4s+2)^{\frac{1}{2}} x_s^m J_{2s+1}(\nu l), \tag{8.345}$$

$$b_m(\nu) = \frac{1}{k\nu} \sum_{s=0}^\infty (4s+4)^{\frac{1}{2}} y_s^m J_{2s+2}(\nu l), \tag{8.346}$$

where the coefficients x_s^m, y_s^m are determined as the solution of the second kind i.s.l.a.e. (for each fixed $m = 0, 1, 2, \ldots$),

$$x_n^m - \sum_{s=0}^\infty x_s^m \gamma_{sm}^m = \beta_n^m, \tag{8.347}$$

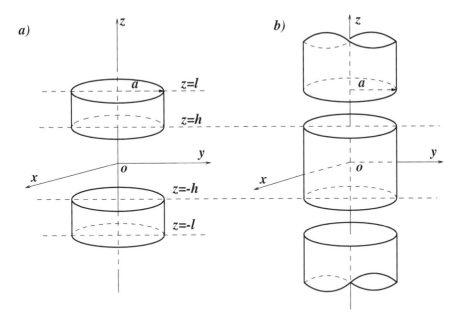

FIGURE 8.15. Plane wave scattering by a) two equal hollow finite cylinders, and b) two equal cuts in an infinitely long cylindrical tube.

$$y_n^m - \sum_{s=0}^{\infty} y_s^m \, \xi_{sm}^m = \zeta_n^m, \qquad (8.348)$$

where $n = 0, 1, 2, \ldots$ and

$$\gamma_{sn}^m = [(4s + 2)(4n + 2)]^{\frac{1}{2}} \int_0^{\infty} \nu^{-1} \sigma(\nu, k) J_{2s+1}(\nu l) J_{2n+1}(\nu l) \, dl$$

$$\beta_n^m = i^{m+1} \pi k a \tan \alpha J_m'(ka \sin \alpha) J_{2n+1}(kl \cos \alpha)(4n + 2)^{\frac{1}{2}}$$

$$\xi_{sm}^m = 4[(s + 1)(n + 1)]^{\frac{1}{2}} \int_0^{\infty} \nu^{-1} \sigma(\nu, k) J_{2s+2}(\nu l) J_{2n+2}(\nu l) \, d\nu$$

$$\zeta_n^m = -2(n + 1)^{\frac{1}{2}} i^m \pi k a \tan \alpha J_m'(ka \sin \alpha) J_{2n+2}(kl \cos \alpha) \, d\nu.$$

This concludes our examination of the acoustic scattering problem.

Let us now, briefly, describe how to solve the analogous wave-scattering problems for more complex structures sho wn in Figure 8.15. The orientation of the incident plane wave is the same as shown in Figure 8.14. From the variety of possible situations we consider simple plane wave-scattering by two equal and acoustically soft finite hollow cylinders. Related structures may be treated in a similar w ay.

After imposing the corresponding boundary conditions the selected problem produces a pair of triple integral equations for the unknown spectral density functions $g_m(\nu)$ and $b_m(\nu)$ (see equations [8.316]–[8.317] and [8.326]–[8.327]),

$$\int_0^\infty g_m(\nu)\cos(\nu z)\,d\nu = -\pi f_m(0)\,\delta(z), \quad 0 \le z < h \qquad (8.349)$$

$$\int_0^\infty \nu^{-1} g_m(\nu)\cos(\nu z)\,d\nu =$$
$$-i\pi a f_m(0)\int_0^\infty J_m\left(\sqrt{k^2-\nu^2}\,a\right) H_m^{(1)}\left(\sqrt{k^2-\nu^2}\,a\right)\cos(\nu z)\,d\nu$$
$$-i^{m+1}\pi a J_m(ka\sin\alpha\cos(k\cos\alpha z))$$
$$+\int_0^\infty \nu^{-1} g_m(\nu)\,\varepsilon(\nu,k)\cos(\nu z)\,d\nu, \quad h < z < l \quad (8.350)$$

$$\int_0^\infty g_m(\nu)\cos(\nu z)\,d\nu = 0, \; l < z < \infty \qquad (8.351)$$

and

$$\int_0^\infty \nu b_m(\nu)\sin(\nu z)\,d\nu = 0, \; 0 \le z < h \qquad (8.352)$$

$$\int_0^\infty g_m(\nu)\sin(\nu z)\,d\nu = i^m \pi J_m(ka\sin\alpha)\sin(kz\cos\alpha)$$
$$+\int_0^\infty b_m(\nu)\,\varepsilon(\nu,k)\sin(\nu z)\,d\nu, \quad h < z < l \quad (8.353)$$

$$\int_0^\infty \nu b_m(\nu)\sin(\nu z)\,d\nu = 0, \; l < z < \infty. \qquad (8.354)$$

We now assume a Neumann series form of solution for $g_m(\nu)$ defined by (8.330), and for $b_m(\nu)$ defined by (8.320). Equations (8.351) and (8.354) are then automatically satisfied. The triple integral equations are reduced to the dual series equations. If we set

$$x_k^m = \sqrt{k}\,\hat{x}_k^m, \qquad (8.355)$$

and $\rho = z/l$ (so that $\rho_0 = h/l < 1$), the rescaled coefficients satisfy

$$\sum_{k=1}^{\infty} k \widehat{x}_k^m T_{2k} \left(\sqrt{1 - \rho^2} \right) = -\frac{\pi}{2} l \sqrt{1 - \rho^2} f_m (0) \delta (\rho) , \quad \rho \in (0, \rho_0) \quad (8.356)$$

$$\sum_{k=1}^{\infty} \widehat{x}_k^m T_{2k} \left(\sqrt{1 - \rho^2} \right) =$$

$$- i\pi a f_m (0) V_m (\rho) + F_m (\rho) + 2 \sum_{k=1}^{\infty} k \widehat{x}_k^m \Phi_k^m (\rho) , \quad \rho \in (\rho_0, 1) \quad (8.357)$$

where

$$V_m (\rho) = \int_0^{\infty} J_m \left(\sqrt{k^2 - \nu^2} a \right) H_m^{(1)} \left(\sqrt{k^2 - \nu^2} a \right) \cos (\nu l \rho) \, d\nu$$

$$F_m (\rho) = -i^{m+1} \pi a J_m (ka \sin \alpha) \cos (kl \cos \alpha \rho)$$

$$\Phi_k^m (\rho) = \int_0^{\infty} \nu^{-1} \varepsilon (\nu, k) J_{2k+1} (\nu l) \cos (\nu l \rho) \, d\nu. \quad (8.358)$$

In a similar way the rescaled variables given by

$$y_k^m = \left(k + \frac{1}{2} \right)^{\frac{1}{2}} \widehat{y}_k^m$$

satisfy

$$\sum_{k=0}^{\infty} \left(k + \frac{1}{2} \right) \widehat{y}_k^m U_{2k} \left(\sqrt{1 - \rho^2} \right) = 0, \quad \rho \in (0, \rho_0) \quad (8.359)$$

$$\sum_{k=0}^{\infty} \widehat{y}_k^m U_{2k} \left(\sqrt{1 - \rho^2} \right) = W_m (\rho) + 2 \sum_{k=0}^{\infty} \left(k + \frac{1}{2} \right) \widehat{y}_k^m R_k^m (\rho) , \quad \rho \in (\rho_0, 1)$$

$$(8.360)$$

where

$$W_m (\rho) = i^m \pi J_m (ka \sin \alpha) \frac{\sin (kl \cos \alpha \rho)}{\rho}$$

$$R_k^m (\rho) = \rho^{-1} \int_0^{\infty} \nu^{-1} \varepsilon (\nu, k) J_{2k+1} (\nu l) \sin (\nu l \rho) \, d\nu. \quad (8.361)$$

It is now obvious that the trigonometric substitution,

$$\cos \frac{1}{2} \varphi = \sqrt{1 - \rho^2}, \quad (8.362)$$

transforms this pair to dual series equations with the trigonometric ker-
nels $\cos(k\varphi)$ and $\sin\left(\left(k+\frac{1}{2}\right)\varphi\right)$. Setting $\cos\frac{1}{2}\varphi_0 = \sqrt{1 - \rho_0^2}$, equations
(8.359)–(8.360) become

$$\sum_{k=0}^{\infty} \left(k + \frac{1}{2}\right) \widehat{y}_k^m \sin\left(k + \frac{1}{2}\right)\varphi = 0, \quad \varphi \in (0, \varphi_0)$$

$$\sum_{k=0}^{\infty} \widehat{y}_k^m \sin\left(k + \frac{1}{2}\right)\varphi =$$

$$W_m\left(\sin\frac{\varphi}{2}\right)\sin\frac{\varphi}{2} + 2\sin\frac{\varphi}{2}\sum_{k=0}^{\infty}\left(k + \frac{1}{2}\right)\widehat{y}_k^m R_k^m\left(\sin\frac{\varphi}{2}\right), \varphi \in (\varphi_0, \pi)$$

$$(8.363)$$

and equations (8.356)–(8.357) become

$$\sum_{k=1}^{\infty} k\widehat{x}_k^m \cos k\varphi = -\frac{\pi}{2}l\cos\frac{\varphi}{2}f_m(0)\delta(\varphi), \quad \varphi \in (0, \varphi_0)$$

$$\sum_{k=1}^{\infty} \widehat{x}_k^m \cos k\varphi = -i\pi a f_m(0) V_m\left(\sin\frac{\varphi}{2}\right) + F_m\left(\sin\frac{\varphi}{2}\right)$$

$$+ 2\sum_{k=1}^{\infty} k\widehat{x}_k^m \Phi_k^m\left(\sin\frac{\varphi}{2}\right), \quad \varphi \in (\varphi_0, \pi). \quad (8.364)$$

The regularisation of these equations is a routine application of the procedure developed in Chapter 2 of Volume I that we omit.

We conclude this section by the consideration of an electromagnetic problem, the excitation of a hollow finite cylinder by a vertical electric dipole.
It should be noted that excitation by a magnetic vertical dipole may be
treated similarly.

The problem geometry is shown in Figure 8.16. Based on the results,
obtained in Chapter 1, the electromagnetic field radiated by a vertical
electric dipole (VED) is described by the z-component of an electric Hertz
vector

$$\overrightarrow{\Pi}_0 = \overrightarrow{i}_z \Pi_0 p \qquad (8.365)$$

where $p = |\overrightarrow{p}|$ is the dipole moment of the VED, and Π_0 is the three-
dimensional Green's function

$$\Pi_0 = \frac{1}{4\pi}\frac{e^{ikR}}{R} = \frac{i}{4\pi}\int_0^\infty H_0^{(1)}\left(\sqrt{k^2 - \nu^2}\rho\right)\cos[\nu(z - d)]d\nu. \qquad (8.366)$$

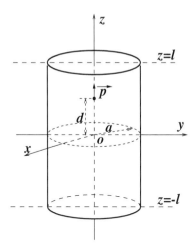

FIGURE 8.16. Excitation of a hollow finite cylinder by a vertical electric dipole.

where $R = \sqrt{\rho^2 + (z - d)^2}$. Henceforth we set $p = 1$. The field components of the dipole are

$$E_\rho^0 = \frac{1}{\rho}\frac{\partial^2 \Pi_0}{\partial \rho \partial z}, \quad H_\varphi^0 = ik\frac{\partial \Pi_0}{\partial \rho}, \quad E_z^0 = \left(\frac{\partial^2}{\partial^2 z} + k^2\right)\Pi_0. \quad (8.367)$$

As usual, we seek a solution in the form

$$\Pi_z^{tot} = \Pi_z^0 + \Pi_z^{sc}$$

where the scattered potential Π_z^{sc} satisfies the Helmholtz equation (and the components of the scattered field are defined by the analogue of [8.367]). The continuity condition for the tangential electric field component E_z^{sc} across $\rho = a$ is

$$E_z^{sc}(a + 0, z) = E_z^{sc}(a - 0, z), |z| < \infty \quad (8.368)$$

and bearing in mind the Sommerfeld radiation conditions the solution form may be chosen as

$$\Pi_z^{sc} = \frac{p}{4\pi}\int_0^\infty \left\{ \begin{array}{l} H_0^{(1)}\left(\sqrt{k^2 - \nu^2}a\right) J_0\left(\sqrt{k^2 - \nu^2}\rho\right), \rho < a \\ J_0\left(\sqrt{k^2 - \nu^2}a\right) H_0^{(1)}\left(\sqrt{k^2 - \nu^2}\rho\right), \rho > a \end{array} \right\} \times$$
$$[A(\nu)\cos\nu z + B(\nu)\sin\nu z]\, d\nu. \quad (8.369)$$

After imposing the boundary conditions

$$E_z^{tot}(a + 0, z) = E_z^{tot}(a - 0, z) = 0, \quad |z| < l \quad (8.370)$$

$$H_\varphi^{sc}\left(a+0,z\right)=H_\varphi^{sc}\left(a-0,z\right),\ \left|z\right|>l \tag{8.371}$$

we obtain the pair of dual integral equations

$$\int_0^\infty \left(k^2-\nu^2\right)A\left(\nu\right)J_0\left(\sqrt{k^2-\nu^2}a\right)H_0^{(1)}\left(\sqrt{k^2-\nu^2}a\right)\cos\left(\nu z\right)d\nu$$
$$=-i\int_0^\infty \left(k^2-\nu^2\right)H_0^{(1)}\left(\sqrt{k^2-\nu^2}a\right)\cos\left(\nu d\right)\cos\left(\nu z\right)d\nu,\ z\in\left(0,l\right) \tag{8.372}$$

$$\int_0^\infty A\left(\nu\right)\cos\left(\nu z\right)d\nu=0,\ z\in\left(l,\infty\right), \tag{8.373}$$

and

$$\int_0^\infty \left(k^2-\nu^2\right)B\left(\nu\right)J_0\left(\sqrt{k^2-\nu^2}a\right)H_0^{(1)}\left(\sqrt{k^2-\nu^2}a\right)\sin\left(\nu z\right)d\nu$$
$$=-i\int_0^\infty \left(k^2-\nu^2\right)H_0^{(1)}\left(\sqrt{k^2-\nu^2}a\right)\sin\left(\nu d\right)\sin\left(\nu z\right)d\nu,\ z\in\left(0,l\right) \tag{8.374}$$

$$\int_0^\infty B\left(\nu\right)\sin\left(\nu z\right)d\nu=0,\ z\in\left(l,\infty\right). \tag{8.375}$$

Using previous results we may transform these dual integral equations to the following,

$$\int_0^\infty \nu \begin{Bmatrix} A\left(\nu\right) \\ B\left(\nu\right) \end{Bmatrix} \begin{Bmatrix} \cos\left(\nu z\right) \\ \sin\left(\nu z\right) \end{Bmatrix} d\nu =$$
$$\pi a\int_0^\infty \left(\nu^2-k^2\right)H_0^{(1)}\left(\sqrt{k^2-\nu^2}a\right)\begin{Bmatrix} \cos\left(\nu d\right)\cos\left(\nu z\right) \\ \sin\left(\nu d\right)\sin\left(\nu z\right) \end{Bmatrix}d\nu$$
$$+\int_0^\infty \nu \begin{Bmatrix} A\left(\nu\right) \\ B\left(\nu\right) \end{Bmatrix}\sigma\left(\nu,k\right)\begin{Bmatrix} \cos\left(\nu z\right) \\ \sin\left(\nu z\right) \end{Bmatrix}d\nu,\ z\in\left(0,l\right) \tag{8.376}$$

$$\int_0^\infty \begin{Bmatrix} A\left(\nu\right) \\ B\left(\nu\right) \end{Bmatrix} \begin{Bmatrix} \cos\left(\nu z\right) \\ \sin\left(\nu z\right) \end{Bmatrix} d\nu=0,\ z\in\left(l,\infty\right) \tag{8.377}$$

where $\sigma\left(\nu,k\right)$ is defined in (8.342).

A standard argument transforms the dual integral equations (8.376)–(8.377) to the second kind i.s.l.a.e.

$$x_n-\sum_{s=0}^\infty \alpha_{sn}^{(1)}x_s=\beta_n^{(1)}, \tag{8.378}$$

$$y_n - \sum_{s=0}^{\infty} \alpha_{sn}^{(2)} y_s = \beta_n^{(2)}, \tag{8.379}$$

where $n = 0, 1, 2, \ldots$, and

$$\alpha_{sn}^{(1)} = [(4s+2)(4n+2)]^{\frac{1}{2}} \int_0^{\infty} \nu^{-1} \sigma(\nu, k) J_{2s+1}(\nu l) J_{2n+1}(\nu l) \, d\nu$$

$$\alpha_{sn}^{(2)} = 4[(s+2)(n+2)]^{\frac{1}{2}} \int_0^{\infty} \nu^{-1} \sigma(\nu, k) J_{2s+2}(\nu l) J_{2n+2}(\nu l) \, d\nu$$

$$\beta_n^{(1)} = \pi a \sqrt{4n+2} \int_0^{\infty} \frac{\nu^2 - k^2}{\nu} H_0^{(1)}\left(\sqrt{k^2 - \nu^2} a\right) J_{2n+1}(\nu l) \cos(\nu d) \, d\nu$$

$$\beta_n^{(2)} = 2\pi a \sqrt{n+1} \int_0^{\infty} \frac{\nu^2 - k^2}{\nu} H_0^{(1)}\left(\sqrt{k^2 - \nu^2} a\right) J_{2n+1}(\nu l) \sin(\nu d) \, d\nu.$$

Of course there is a great variety of scattering problems for structures composed of hollow finite cylinders. Here we just described the major points of a rigorous approach to the analysis of such structures.

8.7 Wave Scattering Problems for Periodic Structures.

In this section we demonstrate the applicability of regularisation methods for the analysis of scattering by an infinite array of identical canonical scatterers, so-called *periodic* structures. Periodic structures have been the subject of many studies in the literature (see for example [17]), and have many important applications such as frequency selective surfaces, antenna array design, and so on.

Of course this theme is too vast to be studied thoroughly within one single section, so we concentrate on two instructive examples that illustrate the applicability of regularisation treatments. The first is the periodic structure consisting of infinitely long parallel strips of the same width (see Figure 8.17); the second is a periodic linear array of identical hollow finite cylinders (see Figure 8.18). The first example is a typical two-dimensional problem, whereas the second is three-dimensional.

8.7.1 Periodic Linear Array of Strips.

We consider plane wave-scattering for a periodic linear array of either acoustically soft or rigid strips; this corresponds to the electromagnetic scattering of E- or H-polarized plane waves.

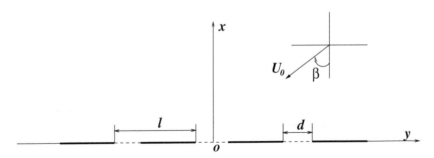

FIGURE 8.17. Scattering of an incident plane wave by a periodic array of infinitely long strips.

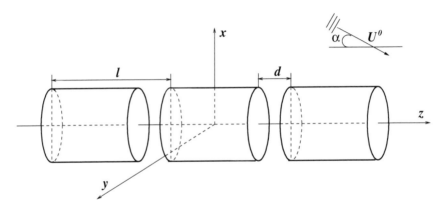

FIGURE 8.18. Scattering of an incident plane wave by a periodic array of identical hollow finite cylinders.

The geometry of the problem is shown in Figure 8.17, where l denotes the period of structure, and d is the gap between two neighbouring strips; the strip width is $l - d$. If the plane wave is incident at angle β, its velocity potential is

$$U^0 (x, y) = e^{-ik(x \cos \beta + y \sin \beta)}. \tag{8.380}$$

As the structure is periodic with period l then the scattered velocity potential $U^{sc} (x, y)$ is invariant under translation in the y-coordinate by distances that are multiples of the periodic length l. Thus, we seek a continuous function $U (x, y)$, that satisfies everywhere except on the strips the homogeneous Helmholtz equation

$$\Delta U + k^2 U = 0, \tag{8.381}$$

and the periodicity condition

$$U(x, y) = U(x, y \pm nl) \tag{8.382}$$

where n is an arbitrary integer. Also $U(x, y)$ satisfies the boundary conditions for acoustically soft strips,

$$U^{tot}(+0, y) = U^{tot}(-0, y) = 0, \quad \frac{d}{2} < |y| < \frac{l}{2} \tag{8.383}$$

$$\left. \frac{\partial U^{tot}}{\partial x} \right|_{x=+0} = \left. \frac{\partial U^{tot}}{\partial x} \right|_{x=-0}, \quad |y| < \frac{d}{2}. \tag{8.384}$$

The boundary conditions also apply at translates of the function $U(x, y)$ according to (8.382).

Taking into account the Sommerfeld radiation condition, we seek the total potential in the form

$$U^{tot}(x, y) = \begin{cases} U^0(x, y) + \sum_{n=-\infty}^{\infty} a_n e^{i\sqrt{k^2 - \gamma_n^2} x} e^{i\gamma_n y}, & x > 0 \\ \sum_{n=-\infty}^{\infty} b_n e^{-i\sqrt{k^2 - \gamma_n^2} x} e^{i\gamma_n y}, & x < 0 \end{cases} \tag{8.385}$$

where $\gamma_n = 2\pi n/l$ and $\mathrm{Im}\left(\sqrt{k^2 - \gamma_n^2}\right) \geq 0$.

The continuity condition for $U^{tot}(x, y)$ at $x = 0$ shows that

$$(-1)^n \frac{\sin\left(\frac{1}{2}kl \sin \beta\right)}{\frac{1}{2}kl \sin \beta + n\pi} + a_n = b_n; \tag{8.386}$$

in the case of normal incidence ($\beta = 0$) we have

$$\begin{aligned} 1 + a_0 &= b_0, \\ a_n &= b_n, \; n \neq 0. \end{aligned} \tag{8.387}$$

After imposing the boundary conditions (8.383)–(8.384) we obtain the dual series equations

$$\sum_{n \neq 0}^{\infty} \left(k^2 - \gamma_n^2\right)^{\frac{1}{2}} a_n e^{i\gamma_n y} = -ka_0, \quad |y| < \frac{d}{2} \tag{8.388}$$

$$\sum_{n \neq 0}^{\infty} (a_n - \alpha_n) e^{i\gamma_n y} = -a_0 - \alpha_0, \quad \frac{d}{2} < |y| < \frac{l}{2} \tag{8.389}$$

where

$$\alpha_n = (-1)^n \frac{\sin\left(\frac{1}{2}kl \sin \beta\right)}{\frac{1}{2}kl \sin \beta + n\pi}. \tag{8.390}$$

Equation (8.388) can be easily transformed to the form

$$\sum_{n\neq0}^{\infty} |n|\, a_n \sqrt{1 - \varkappa^2/n^2\pi^2}\, e^{i\gamma_n y} = \frac{i}{\pi}\varkappa a_0, \quad |y| < \frac{d}{2} \tag{8.391}$$

where $\varkappa = \frac{1}{2}kl$. Then, identifying the asymptotically small parameter

$$\varepsilon_n = 1 - \sqrt{1 - \varkappa^2/n^2\pi^2} \tag{8.392}$$

that obeys $\varepsilon_n = O\left(\varkappa^2 n^{-2}\right)$ as $n \to \infty$, we may split equation (8.391) into static and dynamic parts,

$$\sum_{n\neq0}^{\infty} |n|\, a_n\,(1 - \varepsilon_n)\, e^{i\gamma_n y} = i\varkappa a_0, \quad |y| < \frac{d}{2}. \tag{8.393}$$

Simple algebraic manipulation separates the pair (8.393)–(8.389) into two independent systems of dual series equations with trigonometric kernels,

$$\sum_{n=1}^{\infty} n a_n^+\,(1 - \varepsilon_n)\cos n\phi = \frac{1}{2\pi}i\varkappa a_0, \quad \phi \in (0, \phi_0) \tag{8.394}$$

$$\sum_{n=1}^{\infty} a_n^+ \cos n\phi = -\frac{1}{2}(a_0 + \alpha_0) - \sum_{n=1}^{\infty} \alpha_n^+ \cos n\phi, \quad \phi \in (\phi_0, \pi) \tag{8.395}$$

and

$$\sum_{n=1}^{\infty} n a_n^-\,(1 - \varepsilon_n)\sin n\phi = 0, \quad \phi \in (0, \phi_0) \tag{8.396}$$

$$\sum_{n=1}^{\infty} a_n^- \sin n\phi = -\sum_{n=1}^{\infty} \alpha_n^- \sin n\phi, \quad \phi \in (\phi_0, \pi) \tag{8.397}$$

where $\phi = 2\pi y/l$ and $\phi_0 = \pi d/l$, and

$$a_n^+ = \frac{1}{2}(a_n + a_{-n}),$$

$$a_n^- = \frac{1}{2}(a_n - a_{-n}), \tag{8.398}$$

$$\alpha_n^+ = \frac{1}{2}(\alpha_n + \alpha_{-n}) = (-1)^n\,\frac{\varkappa\sin\beta\sin(\varkappa\sin\beta)}{\varkappa^2\sin^2\beta - n^2\pi^2},$$

$$\alpha_n^- = \frac{1}{2}(\alpha_n - \alpha_{-n}) = (-1)^{n+1}\,\frac{\sin(\varkappa\sin\beta)}{\varkappa^2\sin^2\beta - n^2\pi^2}. \tag{8.399}$$

The substitutions $v = \pi - \phi$ and $v_0 = \pi - \phi_0$ reduce equations (8.394)–(8.397) to the standard form (see [2.39]–[2.42] in Chapter 2 of Volume I),

$$\frac{1}{2}a_0 + \frac{1}{2}\alpha_0 + \sum_{n=1}^{\infty} (-1)^n \left(a_n^+ + \alpha_n^+\right) \cos nv = 0, \; v \in (0, v_0) \quad (8.400)$$

$$-\frac{1}{2}i\varkappa Q_0 + \sum_{n=1}^{\infty} (-1)^n na_n^+ (1 - \varepsilon_n) \cos nv = 0, \; v \in (v_0, \pi) \quad (8.401)$$

and

$$\sum_{n=1}^{\infty} (-1)^n \left(a_n^- + \alpha_n^-\right) \sin nv = 0, \; v \in (0, v_0) \quad (8.402)$$

$$\sum_{n=1}^{\infty} (-1)^n na_n^- (1 - \varepsilon_n) \sin nv = 0, \; v \in (v_0, \pi). \quad (8.403)$$

In formulae (2.39)–(2.40) of Volume I we make the replacements $b \mapsto \frac{1}{2}$, $x_0 \mapsto a_0$, $g_0 \mapsto -\frac{1}{2}\alpha_0$, $x_n \mapsto (-1)^n a_n^+$, $q_n \mapsto 0$, $g_n \mapsto -(-1)^n \alpha_n^+$, $a \mapsto -\frac{1}{2}i\varkappa$ $f_0 \mapsto 0$, $r_n \mapsto \varepsilon_n$, and $f_n \mapsto 0$ and the final regularised form for (8.400)–(8.401) is given by (2.62) and (2.60) of Volume I, whereas the regularised form for (8.402)–(8.403) is given by (2.46) of Volume I.

We now investigate the companion scattering problem for rigid strips. The problem statement is the same as for soft strips, except that the boundary conditions are replaced by

$$\left.\frac{\partial U^{tot}}{\partial x}\right|_{x=+0} = \left.\frac{\partial U^{tot}}{\partial x}\right|_{x=-0} = 0, \; \frac{d}{2} < |y| < \frac{l}{2}, \quad (8.404)$$

$$U^{tot}(+0, y) = U^{tot}(-0, y), \; 0 < |y| < \frac{d}{2}. \quad (8.405)$$

As before we may seek a solution in the form (8.385). Due to the continuity conditions on the normal derivatives of the total velocity potential across $x = 0$,

$$\left.\frac{\partial U^{tot}}{\partial x}\right|_{x=+0} = \left.\frac{\partial U^{tot}}{\partial x}\right|_{x=-0}, \; |y| < \frac{l}{2} \quad (8.406)$$

the relation between the coefficients a_n and b_n is

$$-\varkappa \cos \beta \frac{\sin (\varkappa \sin \beta + n\pi)}{\sqrt{\varkappa^2 - n^2\pi^2} (\varkappa \sin \beta + n\pi)} + a_n = b_n. \quad (8.407)$$

Enforcement of the boundary conditions (8.404)–(8.405) leads to the dual series equations

$$\sum_{n=-\infty}^{\infty} (a_n + f_n) e^{i\gamma_n y} = 0, \quad |y| < \frac{d}{2} \tag{8.408}$$

$$\sum_{n=-\infty}^{\infty} a_n \sqrt{\varkappa^2 - n^2\pi^2} e^{i\gamma_n y} = \sum_{n=-\infty}^{\infty} g_n e^{i\gamma_n y}, \quad \frac{d}{2} < |y| < \frac{l}{2} \tag{8.409}$$

where

$$f_n = \frac{1}{2} \frac{\sin(\varkappa \sin\beta + n\pi)}{\varkappa \sin\beta + n\pi} \left[1 - \frac{\varkappa \cos\beta}{\sqrt{\varkappa^2 - n^2\pi^2}} \right], \tag{8.410}$$

$$g_n = \varkappa \cos\beta \frac{\sin(\varkappa \sin\beta + n\pi)}{\varkappa \sin\beta + n\pi}; \tag{8.411}$$

in the case of normal incidence $(\beta = 0)$, $f_n = 0$, $g_0 = \varkappa$, and $g_n = 0$ when $n \geq 1$.

It may be shown that Equations (8.408)–(8.409) are equivalent to a pair of dual series equations

$$\frac{1}{2}(a_0 + f_0) + \sum_{n=1}^{\infty} (a_n^+ + f_n^+) \cos n\phi = 0, \quad \phi \in (0, \phi_0) \tag{8.412}$$

$$\frac{-i}{2\pi}(\varkappa a_0 - g_0) + \sum_{n=1}^{\infty} na_n^+ (1 - \varepsilon_n) \cos n\phi =$$

$$- \frac{i}{\pi} \sum_{n=1}^{\infty} g_n^+ \cos n\phi, \quad \phi \in (\phi_0, \pi) \tag{8.413}$$

and

$$\sum_{n=1}^{\infty} (a_n^- + f_n^-) \sin n\phi = 0, \quad \phi \in (0, \phi_0)$$

$$\sum_{n=1}^{\infty} na_n^- (1 - \varepsilon_n) \sin n\phi = -\frac{i}{\pi} \sum_{n=1}^{\infty} g_n^- \sin n\phi, \quad \phi \in (\phi_0, \pi) \tag{8.414}$$

where, as before, $a_n^+ = \frac{1}{2}(a_n + a_{-n})\,;\, a_n^- = \frac{1}{2}(a_n - a_{-n})$. Furthermore,

$$
f_n^+ = \frac{1}{2}(f_n + f_{-n}) = \frac{1}{2}\left[1 - \frac{\varkappa\cos\beta}{\sqrt{\varkappa^2 - n^2\pi^2}}\right]\frac{(-1)^n\,\varkappa\sin\beta\sin(\varkappa\sin\beta)}{\varkappa^2\sin^2\beta - n^2\pi^2}
$$

$$
g_n^+ = \frac{1}{2}(g_n + g_{-n}) = \frac{1}{2}(-1)^n\,\varkappa^2\sin^2\beta\frac{\sin(\varkappa\sin\beta)}{\varkappa^2\sin^2\beta - n^2\pi^2}
$$

$$
f_n^- = \frac{1}{2}(f_n - f_{-n}) = (-1)^{n+1}\left[1 - \frac{\varkappa\cos\beta}{\sqrt{\varkappa^2 - n^2\pi^2}}\right]\frac{n\pi\sin(\varkappa\sin\beta)}{\varkappa^2\sin^2\beta - n^2\pi^2}
$$

$$
g_n^- = \frac{1}{2}(g_n - g_{-n}) = (-1)^{n+1}\,n\pi\frac{\varkappa\cos\beta\sin(\varkappa\sin\beta)}{\varkappa^2\sin^2\beta - n^2\pi^2}. \qquad (8.415)
$$

Again, the equations (8.413)–(8.414) may be identified with the standard equations (2.39)–(2.40) and (2.41)–(2.42) of Volume I, respectively, and the regularised solution is immediately deduced from that given in Section 2.2 of Volume I.

In both soft and hard cases it is possible to find approximate solutions under the conditions $\phi_0 \ll 1$ (narrow slots) or $\pi - \phi_0 \ll 1$ (narrow strips), including the additional restriction $\varkappa\sin\phi_0 \ll 1$ or $\varkappa\sin\phi_1 \ll 1$, where $\phi_1 = \pi - \phi_0$. Also analytical approximate formulae can be obtained for the Rayleigh scattering regime ($\varkappa \ll 1$) for an arbitrary ratio d/l. These results are ready deductions from the second kind i.s.l.a.e..

Moreover the simple structure of ε_n (see definition [8.392]) allows further analytical treatment of the dual series equations to accelerate the convergence rate of the resultant second kind i.s.l.a.e. using the same ideas as explained on pages 165–166 of Volume I.

8.7.2 Periodic Linear Array of Hollow Finite Cylinders.

In contrast with the previous problem, the periodic structure to be considered now consists of a linear array of three-dimensional elements, hollow finite cylinders. It is a fully three-dimensional problem in the sense that the electromagnetic scattering problem is not reducible to some equivalent acoustic problem, or vice versa; both wave-scattering problems must be examined separately. For the sake of simplicity we consider the acoustic plane wave diffraction problem for the linear array of acoustically soft hollow finite cylinders forming a periodic structure. The problem geometry is shown in Figure 8.18. The orientation of the incident plane wave is the same as that for the analogous scattering problem for a single element (see Figure 8.14).

Thus, we suppose that an infinite number of hollow finite cylinders form a periodic structure with a period l. The length of each cylinder is $l - d$, where d is the gap between neighbouring cylinders. The acoustical plane wave is described by the velocity potential

$$U^0 = e^{ikz \cos \alpha} e^{ik\rho \sin \alpha \cos(\phi - \phi_0)} \tag{8.416}$$

where, due to the problem symmetry, we set $\phi_0 = 0$, without loss of generality.

As usual, we decompose the total solution as a sum

$$U^{tot} = U^0 + U^{sc}, \tag{8.417}$$

where the scattered potential U^{sc} satisfies the Helmholtz equation and the Sommerfeld radiation condition; the solution is periodic in the z-coordinate, i.e.,

$$U^{tot}(\rho, \phi, z) = U^{tot}(\rho, \phi, z \pm nl) \tag{8.418}$$

for all integers n. Thus we seek the scattered potential in the form

$$U^{sc}(\rho, \phi, z) = \sum_{m=0}^{\infty} \left(2 - \delta_m^0\right) \cos m\phi \times$$

$$\sum_{n=-\infty}^{\infty} a_n^m \left\{ \begin{array}{ll} I_m\left(\sqrt{\gamma_n^2 - k^2}\rho\right), & \rho < a \\ \dfrac{I_m\left(\sqrt{\gamma_n^2 - k^2}a\right)}{K_m\left(\sqrt{\gamma_n^2 - k^2}a\right)} K_m\left(\sqrt{\gamma_n^2 - k^2}\rho\right), & \rho > a \end{array} \right\} e^{i\gamma_u z} \tag{8.419}$$

where $\gamma_n = 2\pi n/l$ and a_n^m are the unknown Fourier coefficients to be determined.

The incident potential may be expressed as the product

$$U^0(\rho, \phi, z) = U^0(\rho, \phi) U^0(z) \tag{8.420}$$

where

$$U^0(\rho, \phi) = e^{ik\rho \sin \alpha \cos \alpha} = \sum_{m=0}^{\infty} \left(2 - \delta_m^0\right) J_m(k\rho \sin \alpha) \cos m\phi \tag{8.421}$$

and

$$U^0(z) = e^{ikz \cos \alpha} = \sum_{m=-\infty}^{\infty} \frac{\sin(\varkappa \cos \alpha - n\pi)}{\varkappa \cos \alpha - n\pi} e^{i\gamma_n z}, \tag{8.422}$$

where $\varkappa = kl/2$. Due to the periodicity it is sufficient to impose the boundary conditions on the desired solution only within the single spatial period $|z| < \frac{l}{2}$,

$$U^t(a + 0, \phi, z) = U^t(a - 0, \phi, z) = 0, \quad |z| < \frac{l - d}{2} \tag{8.423}$$

$$\left.\frac{\partial U^t}{\partial \rho}\right|_{\rho = a+0} = \left.\frac{\partial U^t}{\partial \rho}\right|_{\rho = a-0}, \quad \frac{l - d}{2} < |z| < \frac{l}{2} \tag{8.424}$$

for all $\phi \in (0, 2\pi)$.

Imposition of the boundary conditions leads to the dual series equations

$$(-i)^m J_m(ka) a_0^m - J_m(ka \sin \alpha) \beta_0 + \sum_{n \neq 0} a_n^m I_m \left(\sqrt{\gamma_n^2 - k^2} a \right) e^{i\gamma_n z}$$

$$= J_m(ka \sin \alpha) \sum_{n \neq 0}^{\infty} \beta_n e^{i\gamma_n z}, \quad |z| < \frac{l}{2} - \frac{d}{2} \quad (8.425)$$

$$\frac{2}{\pi} (-i)^{m+1} \frac{a_0^m}{H_m^{(1)}(ka)} + \sum_{n \neq 0}^{\infty} \frac{a_n^m}{K_m \left(\sqrt{\gamma_n^2 - k^2} a \right)} e^{i\gamma_n z} = 0, \quad \frac{l-d}{2} < |z| < \frac{l}{2}$$

$$(8.426)$$

where

$$\beta_n = \frac{\sin(\varkappa \cos \alpha - n\pi)}{\varkappa \cos \alpha - n\pi}. \quad (8.427)$$

Introducing the parameter

$$\varepsilon_n = 1 - 4\pi \frac{a}{l} |n| I_m \left(\sqrt{\gamma_n^2 - k^2} a \right) K_m \left(\sqrt{\gamma_n^2 - k^2} a \right)$$

that is asymptotically small ($\varepsilon_n = O(n^{-2})$ as $n \to \infty$), we may split (8.425)–(8.426) into singular and regular parts,

$$\sum_{n \neq 0} A_n^m e^{i\gamma_n z} = \beta_{0m} - (-i)^m J_m(ka) a_0^m + \sum_{n \neq 0}^{\infty} (\beta_{nm} + \varepsilon_n A_n^m) e^{i\gamma_n z},$$

$$|z| < \frac{l-d}{2} \quad (8.428)$$

$$\sum_{n \neq 0} |n| A_n^m e^{i\gamma_n z} = -\frac{2}{\pi} (-i)^{m+1} \frac{a_0^m}{H_m^{(1)}(ka)}, \quad \frac{l-d}{2} < |z| < \frac{l}{2}, \quad (8.429)$$

where

$$a_n^m = |n| K_m \left(\sqrt{\gamma_n^2 - k^2} a \right) A_n^m 4\pi \frac{a}{l},$$

$$\beta_{nm} = J_m(ka \sin \alpha) \beta_n. \quad (8.430)$$

It can be readily shown that equations (8.428)–(8.429) are equivalent to the following pair of dual series equations containing trigonometric kernels

$$\sum_{n=1}^{\infty} x_n^m \cos n\phi = \frac{1}{2}\beta_{0m} - \frac{1}{2}(-1)^m J_m(ka) a_0^m +$$

$$+ \sum_{n=1}^{\infty} \left(x_n^m \varepsilon_n + \beta_{nm}^+\right) \cos n\phi, \ \phi \in (0, \phi_0) \quad (8.431)$$

$$\sum_{n=1}^{\infty} n x_n^m \cos n\phi = -\frac{1}{\pi}(-i)^{m+1} \frac{a_0^m}{H_m^{(1)}(ka)}, \ \phi \in (\phi_0, \pi) \quad (8.432)$$

and

$$\sum_{n=1}^{\infty} y_n^m \sin n\phi \ = \ \sum_{n=1}^{\infty} \left(y_n^m \varepsilon_n + \beta_{nm}^-\right) \sin n\phi, \phi \in (0, \phi_0) \quad (8.433)$$

$$\sum_{n=1}^{\infty} n y_n^m \sin n\phi \ = \ 0, \phi \in (\phi_0, \pi) \quad (8.434)$$

where $\phi = 2\pi z/l$, $\phi_0 = \pi(1 - d/l)$ and

$$\{x_n^m, y_n^m\} \ = \ \frac{1}{2}\left(A_n^m \pm A_{-n}^m\right),$$

$$\left(\beta_{nm}^+, \beta_{nm}^-\right) \ = \ \frac{1}{2}\left(\beta_{nm} \pm \beta_{-n,m}\right). \quad (8.435)$$

The desired regularisation is again provided by the solution of the canonical equations (2.39)–(2.40) and (2.41)–(2.42) of Volume I in the form of a second kind i.s.l.a.e.

8.8   Shielded Microstrip Lines.

In this section we consider shielded microstrip lines and examine a particular case selected from the variety of related problems for these structures. Our aim is to demonstrate that the method of regularisation can equally be used for waveguide problems as well as wave-scattering problems, and may be used to develop a full-wave analysis of such structures.

The particular geometry is shown in Figure 8.19. A thin conducting strip of a width w is placed over lossless dielectric material of height h and relative dielectric constant ε. The structure is shielded by perfectly conducting walls. The region above the microstrip line ($|x| < \frac{1}{2}L_x$, $h <$

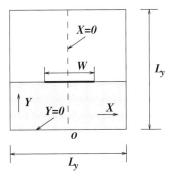

FIGURE 8.19. The waveguide cross-section (shielded microstrip line).

$y < L_y$) is the vacuum region. The problem is to find accurate values of the propagation constants that characterise wave transmission along the waveguide (in the z-direction). This aim can be realised by employing a full modal expansion, and imposing the boundary conditions; regularisation methods lead to a fast converging and stable matrix equation that can be solved in some cases analytically and in the remainder may be solved numerically with guaranteed accuracy.

Waveguides are discussed by Jones [38] and Collin [17]. The electromagnetic fields in such a structure are described completely by the z-components of the electric and magnetic Hertz vectors

$$\vec{\Pi} = \vec{i}_z \Pi_z, \ \vec{\Pi}^{(m)} = \vec{i}_z \Pi_z^{(m)}. \tag{8.436}$$

In the guided wave scenario we seek electromagnetic waves that propagate along the positive z-axis, so that the functions Π_z and $\Pi_z^{(m)}$ are representable in the form

$$
\begin{aligned}
\Pi_z (x, y, z) &= \Pi (x, y) \, e^{i\beta z}, \\
\Pi_z^{(m)} (x, y, z) &= \vec{\Pi}^{(m)} (x, y) \, e^{i\beta z},
\end{aligned}
\tag{8.437}
$$

where β is a *propagation constant*.

Using the symmetric form of Maxwell's equations, the electromagnetic field components are defined by the formulae

$$E_x = i\beta \frac{\partial \Pi}{\partial x} + ik \frac{\partial \Pi^{(m)}}{\partial y}$$

$$E_y = i\beta \frac{\partial \Pi}{\partial y} - ik \frac{\partial \Pi^{(m)}}{\partial x}$$

$$E_z = g^2 \Pi$$

$$H_x = -ik \frac{\partial \Pi}{\partial y} + i\beta \frac{\partial \Pi^{(m)}}{\partial x}$$

$$H_y = ik \frac{\partial \Pi}{\partial x} + i\beta \frac{\partial \Pi^{(m)}}{\partial y}$$

$$H_z = g^2 \Pi^{(m)} \tag{8.438}$$

where the z-subscript has been suppressed and $g^2 = k^2 - \beta^2$. The wavenumber k refers to the vacuum or the dielectric medium as appropriate. Furthermore, in (8.438) we have suppressed the time harmonic dependence $e^{-i\omega t}$ and the common factor $e^{i(\beta z - \omega t)}$.

The traditional approach to the analysis of microstrip lines describes the field in terms of the so-called scalar potentials ψ and Φ

$$E_x = \frac{\partial \psi}{\partial x} + \frac{k}{\beta} \frac{\partial \Phi}{\partial y}; H_x = -\frac{k}{\beta} \frac{\partial \psi}{\partial y} + \frac{\partial \Phi}{\partial x}$$

$$E_y = \frac{\partial \psi}{\partial y} - \frac{k}{\beta} \frac{\partial \Phi}{\partial x}; H_y = \frac{k}{\beta} \frac{\partial \psi}{\partial x} + \frac{\partial \Phi}{\partial y}$$

$$E_z = -i\frac{g^2}{\beta} \psi; H_z = -i\frac{g^2}{\beta} \Phi \tag{8.439}$$

where

$$\psi = i\beta \Pi, \quad \Phi = i\beta \Pi^{(m)}. \tag{8.440}$$

The space inside the waveguide is divided into two regions, (1) $|x| < \frac{1}{2}L_x$; $0 \leq y < h$ and (2) $|x| < \frac{1}{2}L_x$; $h < y < L_y$. Satisfaction of the Helmholtz equation, for each region, and of the boundary conditions, on the walls of the guide leads us to seek solutions for the scalar potentials $\psi^{(i)}$ and $\Phi^{(i)}$ ($i = 1, 2$) in the form

$$\psi^{(1)}(x, y) = \sum_{n=1}^{\infty} A_n \sin(\gamma_n x) \sinh\left(\alpha_n^{(1)} y\right), \tag{8.441}$$

$$\Phi^{(1)}(x, y) = B_0 \cosh\left(\alpha_0^{(1)} y\right) + \sum_{n=1}^{\infty} B_n \cos(\gamma_n x) \cosh\left(\alpha_n^{(1)} y\right), \tag{8.442}$$

$$\psi^{(2)}\left(x, y\right) = \sum_{n=1}^{\infty} C_n \sin\left(\gamma_n x\right) \sinh\left[\alpha_n^{(2)}\left(L_y - y\right)\right], \qquad (8.443)$$

$$\Phi^{(2)}\left(x, y\right) = D_0 \cosh\left[\alpha_0^{(2)}\left(L_y - y\right)\right] +$$
$$\sum_{n=1}^{\infty} D_n \cos\left(\gamma_n x\right) \cosh\left(\alpha_n^{(1)}\left(L_y - y\right)\right), \qquad (8.444)$$

where A_n, B_n, C_n, D_n are unknown coefficients to be determined; furthermore we use the notations

$$\gamma_n = \frac{2n\pi}{L_x}; \alpha_n^{(1)} = \sqrt{\gamma_n^2 + \beta^2 - k_1^2}; \alpha_n^{(2)} = \sqrt{\gamma_n^2 + \beta^2 - k_0^2} \qquad (8.445)$$

where $k_0 = \omega/c$ and $k_1 = \sqrt{\varepsilon} k_0$.

The continuity conditions for tangential components of the electric field at the interface $|x| < L_x/2$, $y = h$ are

$$E_x^{(1)}\left(x, h\right) = E_x^{(2)}\left(x, h\right), \qquad (8.446)$$
$$E_z^{(1)}\left(x, h\right) = E_z^{(2)}\left(x, h\right), \qquad (8.447)$$

implying

$$\frac{\partial}{\partial x}\left(\psi^{(1)} - \psi^{(2)}\right) + \frac{k_0}{\beta}\frac{\partial}{\partial y}\left(\Phi^{(1)} - \Phi^{(2)}\right) = 0, \qquad (8.448)$$
$$\left(k_1^2 - \beta^2\right)\psi^{(1)} = \left(k_0^2 - \beta^2\right)\psi^{(2)} \qquad (8.449)$$

at the interface. It follows instantly from (8.449) that

$$C_n = \frac{k_1^2 - \beta^2}{k_0^2 - \beta^2}\frac{\sinh\left(\alpha_n^{(1)}h\right)}{\sinh\left[\alpha_n^{(2)}\left(L_y - h\right)\right]}A_n. \qquad (8.450)$$

An additional relation follows from (8.448)

$$D_n = \frac{\beta\gamma_n}{k_0\alpha_n^{(2)}}\frac{k_1^2 - k_0^2}{k_0^2 - \beta^2}\frac{\sinh\left(\alpha_n^{(1)}h\right)}{\sinh\left[\alpha_n^{(2)}\left(L_y - h\right)\right]}A_n$$
$$- \frac{k_1}{k_0}\frac{\alpha_n^{(1)}}{\alpha_n^{(2)}}\frac{\sinh\left(\alpha_n^{(1)}h\right)}{\sinh\left[\alpha_n^{(2)}\left(L_y - h\right)\right]}B_n. \qquad (8.451)$$

The last relation reflects the coupling of TE and TM waves in the microstrip line. The coupling vanishes when the interior of the waveguide is a vacuum (or is completely filled by homogeneous dielectric material). In this limiting case we have

$$C_n = \frac{\sinh\left(\alpha_n^{(1)}h\right)}{\sinh\left[\alpha_n^{(1)}(L_y - h)\right]} A_n,$$

$$D_n = -\frac{\sinh\left(\alpha_n^{(1)}h\right)}{\sinh\left[\alpha_n^{(1)}(L_y - h)\right]} B_n. \tag{8.452}$$

Also, the continuity conditions, (8.448)–(8.449), may be re-formulated in decoupled form,

$$\psi^{(1)}(x,h) = \psi^{(2)}(x,h), \quad |x| < \frac{L_x}{2}$$

$$\left.\frac{\partial}{\partial y}\Phi^{(1)}(x,y)\right|_{y=h} = \left.\frac{\partial}{\partial y}\Phi^{(2)}(x,y)\right|_{y=h}, \quad |x| < \frac{L_x}{2}. \tag{8.453}$$

The boundary conditions in the limiting case above are

$$\psi^{(1)}(x,h) = \psi^{(2)}(x,h) = 0, \quad |x| < \frac{w}{2} \tag{8.454}$$

$$\left.\frac{\partial}{\partial y}\psi^{(1)}(x,y)\right|_{y=h} = \left.\frac{\partial}{\partial y}\psi^{(2)}(x,y)\right|_{y=h}, \quad \frac{w}{2} < |x| < \frac{L_x}{2} \tag{8.455}$$

$$\left.\frac{\partial\Phi^{(1)}(x,y)}{\partial y}\right|_{y=h} = \left.\frac{\partial}{\partial y}\Phi^{(2)}(x,y)\right|_{y=h} = 0, \quad |x| < \frac{w}{2} \tag{8.456}$$

$$\Phi^{(1)}(x,h) = \Phi^{(2)}(x,h), \quad \frac{w}{2} < |x| < \frac{L_x}{2}. \tag{8.457}$$

After imposing the boundary conditions we obtain the following decoupled dual series equations for the coefficients A_n and B_n

$$\sum_{n=1}^{\infty} A_n \sinh\left(\alpha_n^{(1)}h\right) \sin(\gamma_n x) = 0, \quad |x| < \frac{w}{2} \tag{8.458}$$

$$\sum_{n=1}^{\infty} A_n \alpha_n^{(1)} \frac{\sinh\left(\alpha_n^{(1)}L_y\right)}{\sinh\left[\alpha_n^{(1)}(L_y - h)\right]} \sin(\gamma_n x) = 0, \quad \frac{w}{2} < |x| < \frac{L_x}{2} \tag{8.459}$$

and

$$\alpha_0^{(1)} \sinh\left(\alpha_0^{(1)} h\right) B_0 + \sum_{n=1}^{\infty} B_n \alpha_n^{(1)} \sinh\left(\alpha_n^{(1)} h\right) \cos\left(\gamma_n x\right) = 0, \quad |x| < \frac{w}{2}$$

(8.460)

$$\frac{\sinh\left(\alpha_0^{(1)} L_y\right)}{\sinh\left[\alpha_0^{(1)}\left(L_y - h\right)\right]} B_0 + \sum_{n=1}^{\infty} B_n \frac{\sinh\left(\alpha_n^{(1)} L_y\right)}{\sinh\left[\alpha_n^{(1)}\left(L_y - h\right)\right]} \sin\left(\gamma_n x\right) = 0,$$

$$\frac{w}{2} < |x| < \frac{L_x}{2}. \quad (8.461)$$

We commence the regularization process by the rescaling

$$a_n = \frac{\alpha_n^{(1)}}{n} \frac{\sinh\left(\alpha_n^{(1)} L_y\right)}{\sinh\left[\alpha_n^{(1)}\left(L_y - h\right)\right]} A_n,$$

(8.462)

$$b_n = \frac{\sinh\left(\alpha_n^{(1)} L_y\right)}{\sinh\left[\alpha_0^{(1)}\left(L_y - h\right)\right]} B_n.$$

(8.463)

Furthermore, set $\phi = 2\pi x / L_x$ and $\phi_0 = \pi w / L_x$. Then we introduce the asymptotically small parameters

$$\varepsilon_n = 1 - \frac{4\pi}{L_x} \frac{n \sinh\left[\alpha_n^{(1)}\left(L_y - h\right)\right] \sinh\left(\alpha_n^{(1)} h\right)}{\alpha_n^{(1)} \sinh\left(\alpha_n^{(1)} h\right)},$$

(8.464)

$$\mu_n = 1 - \frac{L_x}{\pi} \frac{\alpha_n^{(1)} \sinh\left[\alpha_n^{(1)}\left(L_y - h\right)\right] \sinh\left(\alpha_n^{(1)} h\right)}{n \sinh\left(\alpha_n^{(1)} L_y\right)};$$

(8.465)

$\varepsilon_n, \mu_n = O\left(n^{-2}\right)$ as $n \to \infty$.

Taken together, we may reduce equations (8.458)–(8.459) and (8.460)–(8.461) to the desired form

$$\sum_{n=1}^{\infty} a_n \left(1 - \varepsilon_n\right) \sin n\phi = 0, \quad \phi \in \left(0, \phi_0\right),$$

(8.466)

$$\sum_{n=1}^{\infty} n a_n \sin n\phi = 0, \quad \phi \in \left(\phi_0, \pi\right)$$

(8.467)

and

$$\frac{L_x}{\pi} \alpha_0^{(1)} \sinh\left(\alpha_0^{(1)} h\right) B_0 + \sum_{n=1}^{\infty} n b_n \left(1 - \mu_n\right) \cos n\phi = 0, \ \phi \in (0, \phi_0),$$

(8.468)

$$\frac{\sinh\left(\alpha_0^{(1)} L_y\right)}{\sinh\left[\alpha_0^{(1)} (L_y - h)\right]} B_0 + \sum_{n=1}^{\infty} b_n \cos n\phi = 0, \ \phi \in (\phi_0, \pi)$$

(8.469)

The regularised system to be derived from these dual series equations can be immediately deduced by reference to the standard equations (2.39)–(2.40) and (2.41)–(2.42) analysed in Volume I. With the identification

$$v_0 \mapsto \phi_0; y_n \mapsto a_n; q_n \mapsto \varepsilon_n; l_n \mapsto 0; r_n \mapsto 0; h_n \mapsto 0$$

(8.470)

the desired solution is given by the system (2.46) of Volume I.

To make a proper identification of the equations (2.39)–(2.40) of Volume I with (8.468)–(8.469) we set $\phi = \pi - v$; $\phi_0 = \pi - v_0$ and make the identification

$$x_n \ \mapsto \ (-1)^n b_n; x_0 \mapsto B_0; b \mapsto \frac{\sinh\left(\alpha_0^{(1)} L_y\right)}{\sinh\left[\alpha_0^{(1)} (L_y - h)\right]};$$

$$g_0 \ = \ q_n = g_n \mapsto 0; a \mapsto \frac{L_x}{\pi} \alpha_0^{(1)} \sin\left(\alpha_0^{(1)} h\right);$$

$$f_0 \ = \ f_n \mapsto 0; r_n \mapsto \mu_n.$$

(8.471)

In this case the desired solution is given by the formulae (2.60) and (2.62) of Volume I.

Thus in both cases we have obtained the well-conditioned homogeneous i.s.l.a.e of the second kind. The propagation constants are the values of β for which there are nontrivial solutions of these homogeneous equations. Practically, this requires the finding of the roots of the determinantal equation formed from the corresponding matrix of the truncated i.s.l.a.e., of form

$$\det\{A\} = 0$$

(8.472)

for an appropriate matrix A. It most cases these equations are solvable numerically. In a very few cases it is possible to treat them analytically at some extreme values of the parameters.

References

[1] Vinogradov, S.S., Smith, P.D. and Vinogradova, E.D., *Canonical Problems in Scattering and Potential Theory, Part I: Canonical Structures in Potential Theory*, Chapman & Hall/CRC, Boca Raton, FL (2001).

[2] Abramov, A.A., Vainshtein, L.A., Dyshko, A.L. and Konyukhova, N.B., "Numerical investigations of the free electrical axisymmetrical oscillations of an ideally conducting prolate spheroid," *USSR Comput. Maths. Math. Phys. (English Transl.)*, **29**(2), 140-153 (1989).

[3] Abramowitz, M. and Stegun, I.A., *Handbook of Mathematical Functions*, Dover, New York, NY (1965).

[4] ApRhys, T.L., "The design of radially symmetric lenses," *IEEE Trans. Antennas Propagat.*, AP-**18**(4), (1970).

[5] Asano, S. and Yamamoto, G., "Light scattering by a spheroidal particle," *Appl. Opt.*, **14**(1), 29-49 (1975).

[6] Bekefi, G. and Farnell, G.W., "A homogeneous dielectric sphere as a microwave lens," *Can. J. Phys.*, **34**, 790-803 (1956).

[7] Belander, P.A. and Couture, M., "Boundary diffraction of an inhomogeneous wave," *J. Opt. Soc. Am.*, **73**, 446-450 (1983).

[8] Bérenger, J.P. "A perfectly matched layer for the absorption of electromagnetic waves," *J. Computat. Phys.* **114**, 185-200 (1994).

[9] Bjorkberg, J. and Kristensson, G., "Electromagnetic scattering by a perfectly conducting elliptic disc," *Can. J. Phys.*, **65**, 723-734 (1987).

[10] Blaschak, J.G. and Kriegsmann, G.A. "A comparative study of absorbing boundary conditions," *J. Comp. Phys.* **77**(1), 109-139 (1988).

[11] Bleszynski, E., Bleszynski, M. and Jaroszewicz, T., "Surface integral equations for electromagnetic scattering for impenetrable and penetrable sheets," *Antennas & Propagation Mag.*, **35**(6), 14-25 (1993).

[12] Born, M. and Wolf, E., *Principles of Optics,* 3rd edition, Pergamon Press, Oxford UK (1965).

[13] Bowman, J.J., Senior, T.B.A. and Uslenghi, P.L.E., *Electromagnetic and Acoustic Scattering by Simple Shapes,* Hemisphere Publishing Corp., revised printing, New York, NY (1987).

[14] Byrd, P.F. and Friedman, M.D., *Handbook of Elliptic Integrals for Engineers and Scientists,* 2nd edition (revised), Springer-Verlag, Berlin, Heidelberg, New York NY (1971).

[15] Chako, N., "Diffraction of electromagnetic waves by circular apertures and discs. Integral representation method," *Acta Phys. Polon.,* **24**(5), 621-627 (1963).

[16] Clemmow, P.C., *The Plane Wave Spectrum Representation of Electromagnetic Fields,* Oxford University Press, Oxford, UK (1966).

[17] Collin, R.E., *Field Theory of Guided Waves,* Second edition, IEEE Press, New York, NY (1991).

[18] Collins, W.D., "On the solution of some axisymmetric boundary value problems by means of integral equations. V. Some scalar diffraction problems for circular discs," *Quart. J. Mech. Appl. Math.,* **14**(1), 101-117 (1961).

[19] Collins, W.D., "Some scalar diffraction problems for a spherical cap," *Arch. Ration. Mech. Analysis,* **10**(3), 249-266 (1962).

[20] Drabowitch, S., Papiernik, A., Griffiths, H., Encinas, J. and Smith, B.L., *Modern Antennas,* Chapman & Hall, London, UK (1998).

[21] Dyshko, A.L. and Konyukhova, N.B., "Numerical investigations of forced electrical axisymmetrical oscillations of a perfectly conducting prolate spheroid," *USSR Comput. Maths. Math. Phys. (English Transl.)*, **35**(5), 753-771 (1995).

[22] Engquist, B. and Majda, A., "Absorbing boundary conditions for the numerical simulation of waves," *Math. Comp.* **31**, 629-651 (1977).

[23] Erdelyi, A., Magnus, W., Oberhittinger, F. and Tricomi, F.G., *Higher Transcendental Functions*, vols. 1-3, Bateman Manuscript Project, McGraw-Hill, New York NY (1953).

[24] Erdelyi, A., Magnus, W., Oberhittinger, F. and Tricomi, F.G., *Tables of Integral Transforms*, Vols. 1 and 2, Bateman Manuscript Project, McGraw-Hill, New York, NY (1954).

[25] Farafonov, V.G., "Diffraction of a plane electromagnetic-wave at a dielectric spheroid," *Differential Equations (English Transl.)*, **19**(10), 1319-1329 (1983).

[26] Farafonov, V.G., "Electromagnetic-wave scattering on a perfectly conducting spheroid," *Radiotechnika I Electronika (in Russian)*, **29**(10), 1857-1863 (1984).

[27] Faulkner, T.R., "Diffraction of an electromagnetic plane wave by a metallic strip," *J. Inst. Math. Applic.*, **1**(2), 149-163 (1965).

[28] Felsen, L.B. and Marcuvitz , N. *Radiation and Scattering of Waves*, Prentice-Hall, London, UK (1973).

[29] Flammer, C., *Spheroidal Wave Functions*, Stanford University Press, Stanford, CA (1957).

[30] Gradshteyn, I.S. and Rhyzik, I.M., *Tables of Integrals, Series and Products*, 5th edition (edited by A. Jeffrey), Academic Press, San Diego, CA (1994).

[31] Harrington, R.F., *Field Computation by Moment Methods*, Macmillan, New York, NY (1968).

[32] Heins, A.E., "On an integral equation in diffraction theory," *Composito Math.*, **18**(1-2), 49-54 (1966).

[33] Hyge, G. and Spencer, R.C., "Studies of the focal region of a spherical reflector: polarisation effects," *IEEE Trans. Antennas Propagat.*, AP-**16**(4), 399-404 (1968).

[34] Jain, D.L. and Kanwal, R.P., "Acoustic diffraction by a rigid annular spherical cap," *Trans. ASME*, **E39**(1), 139-147 (1971).

[35] James, G.L., *The Geometrical Theory of Diffraction*, Institution of Electrical Engineers, London UK (1976).

[36] Johnson, W.A. and Ziolkowski, R.W., "The scattering of H-polarised plane waves from an axially slotted infinite cylinder: a dual series approach," *Radio Sci.*, **19**(1), 275-291 (1984).

[37] Jones, D.S. and Noble, B., "The low-frequency scattering by a perfectly conducting strip," *Proc. Camb. Phil. Soc.*, **57**(2), 364-366 (1961).

[38] Jones, D.S., *The Theory of Electromagnetism*, Pergamom Press, Oxford, UK (1964).

[39] Jones, D.S., "Diffraction at high frequencies by a circular disc," *Proc. Camb. Phil. Soc.*, **65**(1), 223-245 (1965).

[40] Jones, D.S., "Diffraction of a high-frequency plane electromagnetic wave by a perfectly conducting circular disc," *Proc. Camb. Phil. Soc.*, **61**(1), 241-270 (1965).

[41] Jones, D.S., *Acoustic and Electromagnetic Waves*, Oxford University Press, Oxford, UK (1986).

[42] Jones, D.S., *Methods in Electromagnetic Wave Propagation*, Oxford University Press, Oxford, UK (1994).

[43] Jones, R.K. and Shumpert, T.H., "Surface currents and RCS of a spherical shell with a circular aperture," *IEEE Trans. Antennas Propagat.*, **AP-28**(1), 128-132 (1980).

[44] Kantorovich, L.V. and Akilov, G.P., *Functional analysis in normed spaces*, Pergamom Press, Oxford, UK (1974).

[45] Kao, C.C., "Electromagnetic scattering from a finite tubular cylinder: numerical solutions," *Radio Sci.*, **5**(3), 617-624 (1970).

[46] Keller, J.B., "Geometric theory of diffraction," *J. Opt. Soc. Am.*, 1962, **52**, 116-130.

[47] Khizhnyak, A.N. and Vinogradov, S.S., "Scalar plane wave diffraction by finite number of equidistant coaxial circular discs," *Microwave and Optical Technology Letters*, **17**(5), 328-332 (1998).

[48] Kildal, P.S., "Studies of elements patterns and excitations of the line feeds of the spherical reflector antenna in Arecibo," *IEEE Trans. Antennas Propagat.*, AP-**34**(2), 197-207 (1986).

[49] King, R.W.P. and Wu, T.T., *The Scattering and Diffraction of Waves*, Harvard University Press, Cambridge, MA (1959).

[50] Komarov, I.V., Ponomarev, L.I. and Slavyanov, S.Yu., *Spheroidal and Coulomb's Spheroidal Functions*, Nauka, Moscow, Russia (1976).

[51] Koshparenok, V.N. and Shestopalov, V.P., "Diffraction of a plane electromagnetic wave by a circular cylinder with a longitudinal slot," *USSR Comput. Maths. Math. Phys. (English Transl.)*, **11**(3), 222-243 (1971).

[52] Kraus, J.D., *Antennas*, Second edition, McGraw-Hill, New York, NY (1988).

[53] Kristensson, G., "Acoustic scattering by a soft elliptic disc," *J. Sound and Vibration*, **103**(4), 487-498.

[54] Kumar, A. and Hristov, H.D., *Microwave Cavity Antennas*, Artech House (1989).

[55] Lapta, S.I. and Sologub, V.G., "Scattering of a dipole field by a short segment of circular waveguide," *Izv. Vuz. Radiofiz. (in Russian)*, **16**(10), 1588-1598 (1973).

[56] Lebedev, N.N and Skalskata, I.P., "Applying dual integral equations to the electromagnetic diffraction problem for thin conducting strip," *Zh. Tekh. Fiz. (in Russian)*, **42**(4), 681-690 (1972).

[57] Li, L.W., Leong, M.S., Yeo, T.S., Kooi, P.S. and Tan, K.Y., "Computations of spheroidal harmonics with complex arguments: a review with an algorithm," *Physical Review E*, **58**(5), 6792-6806 (1998).

[58] Luneberg, R.K., *The Mathematical Theory of Optics*, Brown University Press, Providence, RI (1944).

[59] Mautz, J.R. and Harrington, R.F., "Electromagnetic penetration into a conducting circular cylinder through a narrow slot, TM case," *J. Electromagn. Waves Appl.*, **2**(3/4), 269-293 (1988).

[60] Mautz, J.R. and Harrington, R.F., "Electromagnetic penetration into a conducting circular cylinder through a narrow slot, TE case," *J. Electromagn. Waves Appl.*, **3**(4), 307-336 (1989).

[61] Mieras, H., "Radiation pattern computation of a spherical lens using Mie series," *IEEE Trans. Antennas Propagat.*, AP-**30**(6), 1221-1224 (1982).

[62] Miles, J.W., "Potential and Rayleigh–scattering theory for a spherical cap," *Quart. Appl. Math.*, **29**(1), 109-123 (1971).

[63] Miles, J.W., "Scattering by a spherical cap," *J. Acoust. Soc. Am.*, **50**(3), 892-903 (1971).

[64] Miller, E.K., Medgyesi-Mitschang, L. and Newman, E.H. (Editors), *Computational Electromagnetics: Frequency-Domain Method of Moments*, IEEE Press, New York, NY (1992).

[65] Morse, P.M. and Feshbach, H., *Methods of Theoretical Physics*, Part II, McGraw-Hill, New York, NY (1953).

[66] Nag, S. and Sinha, B.P., "Electromagnetic plane wave scattering by a system of two uniformly lossy dielectric prolate spheroids in arbitrary orientation," *IEEE Trans. Antennas* Propagat., AP-**43**(3), 322-327 (1995).

[67] Nicolsky, V.V., *Electromagnetics and Propagation of Radio Waves (in Russian)*, Nauka, Moscow, Russia (1973).

[68] Norris, A.N., "Complex point-source representation of real point sources and the Gaussian beam summation method," *J. Opt. Soc. Am.*, **3**, 2005-2010 (1986).

[69] Norris, A.N. and Hansen, T.B., "Exact complex source representation of time–harmonic radiation," *Wave Motion*, **25**, 127-141 (1997).

[70] Pereira, C.S. and Tai, C.T., "Cylindrical ground-clutter shield," *IEEE Trans. Antennas Propagat.*, AP-**24**(2), 208-216 (1976).

[71] Peterson, A.F., Ray, S.L. and Mittra, R., *Computational Methods for Electromagnetics*, IEEE Press, New York, NY (1998).

[72] Petropoulos, P.G., "Fourth-order accurate staggered finite difference schemes for the time-dependent Maxwell equations," in Smith, P.D. and Jarvis, R.J., (Editors), *Ordinary and Partial Differential Equations V*, Pitman Research Notes in Mathematics **370**, 85-104, Addison Wesley Longman, Harlow, UK (1997).

[73] Radin, A.M. and Shestopalov, V.P., "Diffraction of waves by sphere with holes," *Doklady Akademii Nauk SSSR (in Russian)*, **212**(4), 838-841 (1973).

[74] Ramo, S., Whinnery, J.R. and Van Duzer, T., *Fields and Waves in Communication Electronics*, Third edition, John Wiley & Sons, Inc., New York, NY (1994).

[75] Lord Rayleigh, "Theory of the Helmholtz resonator," *Proc. Roy. Soc. Lond.*, Ser. A, **92**, 265-275 (1915).

[76] Rudge, A.W., Milne, K., Olver, A.D. and Knight, P., *The handbook of antenna design*, vol. 1, Peter Peregrinus Ltd, London, UK (1982).

[77] Ruck, G.T., Barrick, D.E., Stuart, W.D. and Krichbaum, C.K., *Radar-Cross Section Handbook*, Vols. I and II, Plenum Press, New York NY (1970).

[78] Sakurai, H., Hashidate, T., Ohki, M., et al., "Electromagnetic scattering by the Luneberg lens with reflecting cap," *IEEE Trans. Electromag. Compat.*, EMC-**40**(2), 94-96 (1998).

[79] Shestopalov, V.P., *Series Equations in Modern Diffraction Theory*, Naukova dumka, Kiev, Ukraine (1983).

[80] Shestopalov, V.P., Tuchkin, Yu.A., Poedinchuk, A.E. and Sirenko, Yu.K., *New Methods of Solving Direct and Inverse Diffraction Problems (in Russian)*, Osnova, Kharkiv, Ukraine (1997).

[81] Smith, P.D., "Time domain integral equation techniques in electromagnetic scattering," in Serbest, A.H. and Cloude, S.R. (Editors), *Direct and Inverse Electromagnetic Scattering*, Pitman Research Notes in Mathematics **361**, 171-188, Addison Wesley Longman, Harlow, UK (1996).

[82] Smith, P.D. and Vinogradova E.D., "Radiation from an open spheroidal antenna with an impedance surface coating," *Proc. Int. Conf. Electromagnetics in Advanced Applications (ICEAA 99)*, 293-296, Torino, Italy, September 1999.

[83] Smith, P.D. and Vinogradova, E.D., "Mixed boundary value problems of diffraction theory: the spheroidal cavity," *Proc. Int. Conf. Mathematical and Numerical Aspects of Wave Propagation (WAVES'2000)*, Santiago de Compostela, Spain, July 2000, 246-252.

[84] Sneddon, I.N., *Mixed Boundary Value Problems in Potential Theory*, North-Holland Publishing Company, Amsterdam, Holland (1966).

[85] Sologub, V.G., "High frequency asymptotics of a solution of the diffraction problem for a circular disc," *USSR Comput. Maths. and Math. Phys. (English Transl.)*, **12**(2), 388-412 (1972).

[86] Spahn, R.J., "The diffraction of a plane wave by an infinite slit," *Quart. Appl. Math.*, **40**(1), 105-110 (1982).

[87] Spencer, R.C. and Hyge, G., "Studies of the focal region of a spherical reflector: geometric optics," *IEEE Trans. Antennas Propagat.*, AP-**16**(3), 317-324 (1968).

[88] Stratton, J.A., *Electromagnetic Theory*, McGraw-Hill, New York, NY (1941).

[89] Suedan, G.A. and Jull, E.V., "Scalar beam diffraction by a wide circular aperture," *J. Opt. Soc. Am.*, **5**(10), 1629-1634 (1988).

[90] Szegö, G., *Orthogonal polynomials*, American Mathematical Society, Colloquium publications, Vol. **23**, Providence, RI (1939).

[91] Tai, C.T., "Electromagnetic theory of spherical Luneberg lens," *Appl. Sci. Res.*, Ser B, **7**, 113-130 (1958).

[92] Thomas, D.P., "Diffraction by a spherical cap," *Proc. Camb. Phil. Soc.*, **59**(1), 197-209 (1963).

[93] Tikhonov, A.N. and Samarskii, A.A., *Equations of Mathematical Physics*, Pergamon Press, Oxford UK (1963).

[94] Tuchkin, Yu.A., "Wave scattering by open cylindrical screens of arbitrary profile with Dirichlet boundary conditions," *Sov. Physics Doklady (English Transl.)* **30**(12), 1027–1029 (1985).

[95] Uslenghi, P.L.E. and Zich, R.E., "Radiation and scattering from isorefractive bodies of revolution," *IEEE Trans. Antennas Propagat.*, AP-**46**(11), 1606-1611 (1998).

[96] Vaid, B.K. and Jain, D.L., "Acoustic diffraction by two concentric coaxial soft spherical caps," *J. Eng. Math.*, **8**(2), 81-88 (1974).

[97] Van de Hulst, H.C., *Light Scattering by Small Particles*, John Wiley and Sons, New York, NY (1957).

[98] Vinogradov, S.S., Radin, A.M. and Shestopalov, V.P., "Diffraction of the field of a vertical dipole by a spherical segment," *Doklady Akademii Nauk Ukrainskoi SSR (in Russian)*, Ser. A, N8, 2741-2745 (1976).

[99] Vinogradov, S.S. and Shestopalov, V.P., "Solution of a vectorial scattering problem for a sphere with a hole," *Doklady Akademii Nauk SSSR (in Russian)*, **237**(1), 60-63 (1977).

[100] Vinogradov, S.S., Tuchkin, Yu.A. and Shestopalov, V.P., "An effective solution of paired summation equations with kernels in the form of Legendre associated functions," *Sov. Physics Doklady (English Transl.)*, **23**(9), 650-651 (1978).

[101] Vinogradov, S.S., Tuchkin, Yu.A. and Shestopalov, V.P., "An effective solution of dual series equations involving associated Legendre functions," *Doklady Akademii Nauk SSSR (in Russian)*, **242**(1), 80-83 (1978).

[102] Vinogradov, S.S., "Soft spherical cap irradiated by a plane sound wave," *USSR J. Math. Physics and Comp. Math. (English Transl.)*, **5** (1978).

[103] Vinogradov, S.S., Tuchkin, Yu.A. and Shestopalov, V.P., "Summator equations with kernels in the form of Jacobi polynomials," *Sov. Phys. Doklady (English Transl.)*, **25**(7), 231-232 (1980).

[104] Vinogradov, S.S., Tuchkin, Yu.A. and Shestopalov, V.P., "Investigation of the dual series equations involving Jacobi polynomials," *Doklady Akademii Nauk SSSR (in Russian)*, **253**(2), 318-321 (1980).

[105] Vinogradov, S.S., *Wave Scattering by Unclosed Spherical Shells (in Russian)*, Kharkov University Press, Dissertation for Degree of Candidate of Science in the Physical and Mathematical Sciences, Kharkhov State University Press (1980).

[106] Vinogradov, S.S., Tuchkin, Yu.A. and Shestopalov, V.P., "On the theory of the scattering of waves by nonclosed screens of spherical shape," *Sov. Physics Doklady (English Transl.)*, **26**(2), 169-171 (1981).

[107] Vinogradov, S.S., Tuchkin, Yu.A. and Shestopalov, V.P., "On the Abel transformation in diffraction problem for a thin strip," *Doklady Akademii Nauk SSSR (in Russian)*, **267**(2), 330-334 (1982).

[108] Vinogradov, S.S., "Plane wave diffraction by the structure consisting of an absolutely rigid disc and a spherical mirror," *Doklady Akademii Nauk Ukrainskoi SSR (in Russian)*, Ser. A, N7, 55-59 (1983).

[109] Vinogradov, S.S. and Shestopalov, V.P., "On one class of integro-series equations of diffraction theory," *Sov. Physics Doklady (English Transl.)*, **28**(6), 449-450 (1983).

[110] Vinogradov, S.S., "On one method of solving the diffraction problem for a thin disc," *Doklady Akademii Nauk Ukrainskoi SSR (in Russian)*, Ser. A, N6, 37-40 (1983).

[111] Vinogradov, S.S., "Reflectivity of a spherical shield," *Radiophys. Quantum Electron. (English Transl.)*, **26**(1), 78-88 (1983).

[112] Vinogradov, S.S. and Sulima, A.V., "Investigation of the absorption cross-section for a partially screened dielectric sphere," *Radiophys. Quantum Electron. (English transl.)*, **26**(10), 1276-1281 (1983).

[113] Vinogradov, S.S. and Lutsenko, E.D., "Calculation of electrostatic fields of spheroidal shells with two circular holes," *Elektrichestvo (in Russian)* **2**, 52-56, (1988).

[114] Vinogradov, S.S. and Sulima, A.V., "Investigation of a Poynting vector flux inside a partially shielded dielectric sphere," *Radiophys. Quantum Electron. (English Transl.)*, **32**(2) (1989).

[115] Vinogradov, S.S., Vinogradova, E.D., Nosich, A.I., et al., "Analytical regularisation based analysis of a spherical reflector symmetrically illuminated by an acoustic beam," *J. Acoust. Soc. Am.*, **107**(6), 2999-3005 (2000).

[116] Vinogradova, E.D., "Plane wave diffraction on an ideally rigid thin spheroidal screen," *Doklady Akademii Nauk Ukrainskoi SSR (in Russian)*, Ser. A, N5, 36-40 (1991).

[117] Vinogradova, E.D., "On the theory of resonant spheroidal antenna," *Proc. Electromagnetic Theory Symposium*, 501-502, St.Petersburg, Russia, May 1995.

[118] Vinogradova, E.D., "Analysis of radiation of a surface spheroidal antenna," *IEEE Antennas & Propagation Society Symposium*, Baltimore, July 1996.

[119] Vinogradova, E.D. and Smith, P.D., "Impact of a hollow spheroidal cylinder on electric dipole radiation features," *Proc. Int. Conf. Mathematical Methods in EM Theory (MMET*98)*, 760-762, Kharkov, Ukraine, June 1998.

[120] Vinogradova, E.D. and Smith, P.D., "Radiation of the Shielded Spheroidal Antenna," *IEEE Antennas & Propagation Society Symposium*, 296, Atlanta, GA, June 1998.

[121] Volakis, J.M., Senior, T.B.A., Legault, S.R., Özdemir, T. and Casciato, M., "Artificial absorbers for truncating finite mesh elements," in Serbest, A.H. and Cloude, S.R. (Editors), *Direct and Inverse Electromagnetic Scattering*, Pitman Research Notes in Mathematics **361**, 15-24, Addison Wesley Longman, Harlow, UK (1996).

[122] Wait, J.R., "Electromagnetic radiation from spheroidal structures," in Collin, R.E. and Zucker, F.J., *Antenna theory*, vol. 2, 523-529, McGraw-Hill, New York, NY (1969).

[123] Williams, W.E., "Diffraction by a cylinder of finite length," *Proc. Camb. Phil. Soc.*, **52**, 322-335 (1956).

[124] Williams, W.E., "High frequency diffraction by a circular disc," *Proc. Camb. Phil. Soc.*, **71**(2), 423-430 (1972).

[125] Wolfe, P., "The diffraction of waves by slits and strips," *SIAM J. Appl. Math.*, **19**(1), 20-32 (1970).

[126] Wolfe, P., "Diffraction of a plane wave by a strip: exact and asymptotic solutions," *SIAM J. Appl. Math.*, **23**(1), 118-132 (1972).

[127] Wolfe, P., "Diffraction of a plane wave on a circular disc," *J. Math. Anal. Appl.*, **67**(1), 35-57 (1979).

[128] Ziolkowski, R.W. and Johnson, W.A., "Electromagnetic scattering of an arbitrary plane wave from a spherical shell with a circular aperture," *J. Math. Phys.*, **28**(6), 1293-1324 (1987).

[129] Ziolkowski, R.W., Marsland, D.P., Libelo, L.F. and Pisane, G.E., "Scattering from an open spherical shell having a circular aperture and enclosing a concentric dielectric sphere," *IEEE Trans. Antennas Propagat.*, **AP-36**, 985-999 (1988).

Index

Printed and bound by CPI Group (UK) Ltd, Croydon, CR0 4YY

28/10/2024

01780251-0001